Plant Microbe Interaction 2017

Plant Microbe Interaction 2017

Special Issue Editors

Jan Schirawski
Michael H. Perlin

MDPI • Basel • Beijing • Wuhan • Barcelona • Belgrade

MDPI

Special Issue Editors
Jan Schirawski
RWTH Aachen University
Germany

Michael H. Perlin
University of Louisville
USA

Editorial Office
MDPI
St. Alban-Anlage 66
Basel, Switzerland

This is a reprint of articles from the Special Issue published online in the open access journal *International Journal of Molecular Sciences* (ISSN 1422-0067) from 2017 to 2018 (available at: https://www.mdpi.com/journal/ijms/special_issues/plant_microbe_interaction_2017)

For citation purposes, cite each article independently as indicated on the article page online and as indicated below:

LastName, A.A.; LastName, B.B.; LastName, C.C. Article Title. *Journal Name* **Year**, *Article Number*, Page Range.

ISBN 978-3-03897-328-7 (Pbk)
ISBN 978-3-03897-329-4 (PDF)

Cover image courtesy of Michael H. Perlin.

Contents

About the Special Issue Editors

Jan Schirawski is Professor of Microbial Genetics at the RWTH Aachen University in Germany. Starting in 1988 as a chemistry student at the Heinrich-Heine University Düsseldorf, Germany, he opened a new career path when he ventured abroad (University of North Carolina at Chapel Hill, USA) for his Diploma thesis. Returning to Germany, he joined the group of Prof. Dr. Gottfried Unden at the Johannes Gutenberg University to work on bacterial metabolism and was awarded his PhD in 1997. After a post-doctoral period in France (Institut Jacques Monod) where he worked on RNA plant viruses, and a post-doctoral period in Ireland (National Food Biotechnology Centre) where he worked on bacterial defense strategies against bacteriophages, he joined the group of Prof. Dr. Regine Kahmann at the Max Planck Institute for Terrestrial Microbiology in Marburg, Germany, in 2001 to work on plant–fungal interactions. He developed his own research group studying infection mechanisms of smut fungi and was employed as a Professor for Molecular Biology of Plant-Microbe Interactions at the Georg August University in Göttingen, Germany in 2009. He changed to Aachen University in 2012, where he is still actively studying the infection specificity mechanisms of his favorite pet, the smut fungus *Sporisorium reilianum*, on its host plants maize and sorghum.

Michael H. Perlin is a Professor of Biology at the University of Louisville, USA. He began his college studies at the University of Chicago in Biological Sciences and received his A.B. degree in 1978. He went on to obtain both MS and PhD training in Microbiology from the University of Chicago, where he graduated in 1983. Throughout this period, he worked with Dr. Stephen Lerner, examining the evolution of bacterial antibiotic resistance, primarily to the aminoglycoside group of antimicrobial compounds. He continued to do a short post-doctoral fellowship in Infectious Diseases at the University of Chicago and to teach at Loyola University, before accepting a position as an Assistant Professor at the University of Louisville in 1984. Since receiving tenure and promotion, he has also had Visiting Professorships at both Cornell University in Ithaca, New York, and at the Universidad de Salamanca in Spain. While at Louisville, his program has developed into a focus on host-pathogen interactions, with particular emphasis on smut fungi, including the *Microbotryum violaceum* complex and *Ustilago maydis*, the pathogen of maize.

Preface to "Plant Microbe Interaction 2017"

Plants have inhabited this earth for more than 500 million years. During that time, a diverse array of different plants have developed, each adapted to its particular habitat. A particular habitat is not only defined by its physical or chemical conditions like illumination, humidity, temperature, availability of nutrients, or toxic compounds, but also by its microbial diversity. Not all plants are only passively colonized by microbes. It turns out that plants have elaborated mechanisms to defend themselves against parasitic or detrimental microbes, and also to support the growth of beneficial microbes. As such, plants are involved in shaping the environment to which they are exposed.

Plants are also the most significant modifiers of the recent earth's atmosphere. The ability for photosynthesis, the fixation of carbon dioxide as biomass, the use of light as an energy source, and the generation of oxygen are among the most important qualities for making and maintaining this earth as a habitable environment. Therefore, plants are the most important contributors to the development of balanced ecosystems that support the large diversity of organisms present on earth today.

Humans have learned to use and exploit plants for their own benefit. Plants are used as food, as feed, as building materials, as sources of medicines, as air conditioners, as fertilizers, as carbon dioxide binders, as embellishment, as heating materials, as detoxifiers, and much more. Recently, human activity has severely imbalanced the atmosphere of the earth, resulting in a very rapid climate change that does not leave enough time for the evolution of robust adaptations. Therefore, we need to quickly understand how plants are interacting with their environment, and which effects which environmental factors have on plant development. A better understanding of the complex interactions plants have with their environment will be needed to develop strategies for supporting plant growth under these increasingly adverse conditions.

This book shall contribute to the necessary understanding of how plants are affected by their environment that is also shaped by microbes. The fifteen articles of the book highlight the newest research results in the area of plant–microbe interaction. The articles cover different aspects of plant–microbe interaction, and highlight studies for unraveling the mechanisms of plant disease, for understanding the beneficial functions of microbes and for elucidating complete microbiomes in an effort to correlate microbial diversity with plant health. We hope that the reader will find the narrative of these fifteen articles both informative and inspiring, that it will open the reader's horizon on novel aspects, and lead to the creation of new and innovative ideas that will eventually help in preserving our earth as a balanced and comfortable place to be.

Jan Schirawski, Michael H. Perlin
Special Issue Editors

International Journal of
Molecular Sciences

MDPI

Editorial

Plant–Microbe Interaction 2017—The Good, the Bad and the Diverse

Jan Schirawski [1,*] and Michael H. Perlin [2]

[1]   Microbial Genetics, Institute of Applied Microbiology, RWTH Aachen University, Worringerweg 1,
    52074 Aachen, Germany
[2]   Department of Biology, Program on Disease Resistance, University of Louisville, Louisville, KY 40292, USA;
    michael.perlin@louisville.edu
*   Correspondence: jan.schirawski@rwth-aachen.de; Tel.: +49-241-802-6616

Received: 18 April 2018; Accepted: 2 May 2018; Published: 5 May 2018

Abstract: Of the many ways that plants interact with microbes, three aspects are highlighted in this issue: interactions where the plant benefits from the microbes, interactions where the plant suffers, and interactions where the plant serves as habitat for microbial communities. In this editorial, the fourteen articles published in the Special Issue Plant–Microbe Interaction 2017 are summarized and discussed as part of the global picture of the current understanding of plant-microbe interactions.

Keywords: plant–microbe interactions; microbiome; transcriptome; effectors; comparative methods; *Streptomyces*; plant growth-promoting bacteria; phytoremediation; rhodopsins

1. Introduction

Often not visible to the naked eye, interactions between plants and microorganisms occur in many different ways and on many different levels. Virtually all organs of the plant interact with microorganisms at a certain stage of their life, and this interaction is not necessarily negative for the plant. Indeed, there are plenty of interactions where the plant benefits either through direct or through indirect effects of the associated microbes. In these interactions, plants serve as sheltered habitats for the microorganisms that may colonize apoplastic spaces, plant surface areas or areas adjacent to the plant surface, e.g., the rhizosoil, the soil in the vicinity of roots. In addition to a sheltered habitat and a future source of nutrients that are liberated upon plant death, many plants release compounds that attract and feed the associated microbes. The associated microbes may in turn secrete compounds that favor plant growth, they may make the plant more resistant to abiotic or biotic stress, or they may defend the plant against more malignant microbes.

With the development of techniques to identify and quantify the microbial diversity associated with plants, we begin to grasp the immensity of the interactions to which plants are exposed. Complete microbiomes can be evaluated that are associated with different parts of the plant, and microbes can be found wherever they are looked for. Since functional effects of these multidimensional interactions are difficult to disentangle, most researchers stick to easier tractable and less complex interactions that can be experimentally tackled. This holds also true for researchers studying plant–microbe interactions that are clearly negative for the plant and result in the development of plant disease symptoms. Research on these negative interactions focuses at identifying microbial and plant factors necessary for establishment of the disease and elucidating their molecular function.

The sum of all these research endeavors will lead to an increased, detailed and more and more complex understanding of the multidimensional interactions that plants keep with microbes. In this issue, three aspects of plant–microbe interaction are highlighted: the good, the bad and the diverse. In twelve original research articles and two reviews, we will learn about different interactions where the plant profits from the interaction with the microbes, where it suffers, and where it serves as a habitat for a microbial community. Below, we will summarize the most interesting highlights.

2. The Good: Soil Microbes Positively Affecting Plant Growth

Of the microorganisms colonizing the rhizosoil, *Streptomyces* species are special. They grow filamentously and can colonize not only soil but also roots and aerial parts of the plants; they are active producers of antibiotics and can save the plant from attack by more dangerous bacteria; and they produce volatile organic compounds that give rise to the typical fragrance of fresh forest soil. As such, they qualify as biocontrol agents in several cropping systems, and strains serving as antagonists of various plant pathogens can be identified. The versatile *Streptomyces* species also have plant growth-promoting abilities and can be used as biofertilizers. Because of their ability to form spores and survive adverse conditions in the soil, they are also more competitive than other microbes. In addition, they produce various lytic enzymes that can break down insoluble organic polymers and generate nutrients that can be used by plants. These fascinating aspects of *Streptomyces* species are nicely summarized in the review by Vurukonda et al. [1].

Plant growth-promoting bacteria can also be used in phytoremediation of metal-contaminated soils. Montalbán et al. show that exposing *Helianthus tuberosus*, a high biomass crop used for bio-ethanol production, to particular plant growth-promoting bacteria that were isolated from plants growing on a metal-contaminated soil increased the ability of the plant to sustain elevated concentrations of cadmium and zinc [2]. The bacteria were shown to grow endophytically in the root and resulted in a significantly increased cadmium uptake into the plant. In presence of the bacteria, the plant showed a decrease of metal-induced stress and an improved growth. Thus, these plant growth-promoting bacteria can help both in phytoremediation and in sustainable biomass production [2].

Plant growth-promoting bacteria can induce drought and salt tolerance. Two articles study the effect of plant growth-promoting bacteria on perennial ryegrass (*Lolium perenne*), an important cool-season perennial grass species for pasture, forage and turf with high yield and good turf quality such as a dense root system, superior tillering, and regeneration ability. Unfortunately, this popular grass species is not very tolerant to drought or to high salinity. Su et al. show that the beneficial soil bacterium *Bacillus amyloliquefaciens* GB03 together with a water-retaining agent consisting of super absorbent hydrogels used for soil erosion control can significantly improve the drought resistance of perennial ryegrass; this was true even compared to application of single components that already significantly improve drought resistance of the plant relative to control [3]. He et al. used a novel bacterium isolated from a C4 perennial succulent xerohalophyte shrub with excellent drought and salt tolerance to significantly increase both the growth and salt tolerance of perennial ryegrass [4]. In addition, they sequenced the bacterial genome and identified several genes putatively involved in plant growth-promoting traits and abiotic stress tolerance [4].

Zhang et al. studied the positive role of the arbuscular mycorrhizal fungus *Rhizophagus irregularis* CD1 on plant growth promotion and the *Verticillium* wilt resistance of cotton [5]. They determined the symbiotic efficiency of 17 cotton varieties to *R. irregularis*. The best one, Lumian 1, was used for a two-year field trial. Presence of the mycorrhizal fungus significantly increased plant growth and plant disease resistance against *Verticillium dahliae* wilt. While the negative effect on *V. dahliae* colonization could be due to mycorrhiza-induced resistance, the authors show that growth of *R. irregularis* may directly inhibit growth of *V. dahliae* by releasing as yet unknown volatiles [5].

Thus, microorganisms can be used to positively change growth capacities of plants and to make them more resistant against biotic and abiotic stresses like draught and salt, stresses that will likely occur much more often with progressing climate change. We will need to increase our knowledge about how these systems function in order to tackle the challenges of the future and ensure plant fitness under increasingly adverse conditions.

3. The Bad: Elucidating Mechanistic Strategies of Plant Pathogens

Of the plant-pathogenic microorganisms, fungi are an enormous threat to plant health. While a lot of plant-pathogenic fungi are highly host-specific, host switching events are often at the base of emerging fungal diseases. Therefore, the elucidation of host-specificity factors that allow fungal proliferation and disease formation on particular host plants is one of the hot research topics in plant pathology. In the review by Borah et al., comparative methods for the molecular determination of host-specificity factors are discussed [6]. It turns out that the elucidation of host-specificity factors requires several successive steps, for each of which several comparative methods exist. In most cases, comparison of molecular characteristics of different host-specific strains or species resulted at best in a list of target genes potentially involved in host specific virulence that await verification and functional validation. The authors indicate that intelligent combination of classical genetics, genomics, and transcriptomics covering both the pathogen and the host may lead to host-specificity factor identification and a mechanistic understanding of host specificity [6].

Biotrophic plant-pathogenic fungi live in close intimacy with the plant because they feed on living plant tissue and have to subvert the defense systems of the plant. One of their strategies for survival in the hostile plant tissue environment is the secretion of effector proteins that interact with plant proteins to the advantage of the pathogen. In their contribution, Kuppireddy et al. have analyzed the genome of *Microbotryum lychnidis-dioicae*, a biotrophic fungus causing anther smut on a common weed, *Silene latifolia*, to identify putative effector proteins [7]. Out of 50 identified putative effectors, they showed for four that they are indeed secreted proteins. Interaction analysis revealed a plant protein with homology to a protein involved in pollen germination. Considering that *M. lychnis-dioicae* forms spores exclusively in anthers, the places of pollen generation, this interaction may lead to tantalizing insights into the interaction of the fungus with its host tissue [7]. Gao et al. also identified effectors but in a different pathosystem [8]. They generated and analyzed the transcriptome of *Fusarium proliferatum*, the causal agent of a destructive tomato disease during which dark brown necrotic spots appear on leaves and stems that grow and cause stems to soften and wilt, often leading to death of the entire tomato plant. In the absence of a published genome sequence, they resorted to de-novo assembly of the sequenced transcriptome and analyzed gene expression to identify 184 putative effector candidates, most displaying elevated expression during plant colonization [8]. In a related study, Wang et al. analyzed the transcriptome of Kiwifruit in response to infection by the bacterial canker pathogen *Pseudomonas syringae* pv. *actinidiae* (Psa) [9]. Gene expression analysis of the infected kiwifruit plant revealed upregulation of several genes. These included key genes for defense compound (terpene) biosynthesis and the generation of secondary metabolites, genes involved in plant immunity (pathogen-associated molecular pattern-induced immunity and effector-triggered immunity), as well as a change in expression of metabolic processes that may all have a role in suppressing spread of Psa [9].

Adam et al. analyzed a completely different aspect in the interaction of fungal pathogens with plants [10]. They found that phytopathogenic and phyto-associated ascomycetes contain rhodopsin-encoding genes. The rice plant pathogen *Fusarium fujikuroi* contains two different rhodopsins, CarO and OpsA. While CarO was previously shown to be a light-driven proton pump, here the authors show that CarO is positively regulated by presence of indole-3-acetic acid and of sodium acetate. Intriguingly, they showed that deletion of the CarO-encoding gene from the genome of *F. fujikuroi* resulted in a hypervirulent strain with more severe bakanae symptoms than the reference strain, indicating that CarO has a role in attenuating the disease potential of the fungus [10]. Thus, although our knowledge on how plant pathogens infect host plants and on how the plants react to pathogen attack steadily increases, much remains still unknown. As the last example shows, research often reveals unexpected results that stimulate further research and necessitate an adjustment of the current plant–pathogen interaction models.

4. The Diverse: Microbiomes of Seeds and Roots

Plants are covered by microbes: some of them cause disease, some have a positive influence on plant growth, and some microbes may just be there with an as-yet undiscovered role in microbial ecology. Roots are surrounded by a thick layer of associated microbes in the rhizosoil, and not even seeds are sterile. Microbes associated with seeds can have a profound influence on plant development, since they are present upon seed germination and can affect plant ecology, health and productivity. The seed microbiome comprises both endophytic microbes as well as microbes present on the seed surface. Chen et al. investigated whether there was a core microbiome associated with the seeds of a medicinal plant, *Salvia miltiorrhiza*, used for traditional treatment of coronary and cerebrovascular diseases [11]. The plant is also known to contain active secondary metabolites, such as salvianolic acid and tanshinone, a diterpenoid quinone. They collected seeds from different geographic cultivation areas and determined the total seed-associated microbiomes. They compared the seed microbiomes of the different locations and also between those of *S. miltiorrhiza* and other commonly cultivated crop plants. The authors found a clear overlap of microbial taxa associated with seeds of *S. miltiorrhiza*. In contrast, the overlap in microbiomes of seeds of different plants was limited to a few microbial species. Interestingly, the authors found that in the core bacterial microbiome, genes for secondary metabolism were overrepresented, including genes encoding prenyltransferases, terpenoid backbone biosynthesis enzymes, as well as enzymes for degradation of limonene, pinene and geraniol. This suggests a possible contribution of the microbiome to the secondary metabolite profile of medicinal plants [11].

In a parallel study, Sánchez-López et al. investigated whether seed-associated microbes could be transmitted vertically through several generations [12]. They investigated seeds of *Crotalaria pumila*, a pioneer plant in metal-contaminated soils. The most prominent community member of the seed-associated microbiome of *C. pumila* over three generations was a *Methylobacterium* sp. Cp3 [12]. The authors could show that root inoculation of flowering plants with strain Cp3 led to the occurrence of Cp3 in the seeds. Using tagged strains, the authors followed the bacteria colonizing the root cortical cells and the xylem vessels in the stem of *C. pumila* under metal stress. They present evidence consistent with a positive role of strain Cp3 for seed germination and seedling development [12]. This shows that the seed microbiome may contribute significantly to the general fitness of the plant and may even be involved in adaptation of the plant to adverse living conditions like metal-contaminated soils.

That roots are associated with microbes is common knowledge. What is less well-known is that the microbes in the rhizospheres of different plants affect each other to the advantage of the plants. Li et al. compared the root-associated microbiomes of maize and peanut either grown in monoculture or in intercropping [13]. It turns out that intercropping resulted in a higher microbial diversity with a higher accumulation of beneficial bacteria in the soil that led to increased levels of soil-available nutrients and an increase in plant biomass [13]. While this study showed the advantage of intercropping, in many areas, plant monocultures are cultivated successively on the same field. Plantations of tea (*Camella sinensis*) can be grown for over 30 years on the same field. Compared to new tea fields planted only two years ago, the older tea fields endure poor growth, chlorosis, wilting, and ratooning problems. Arafat et al. have compared the rhizosoils of young and old tea monoculture soils by measuring the bacterial diversity, the physicochemical properties of the soil and the content of plant exudates (metabolites leaking from the plants into the soil via the roots) [14]. While the physicochemical properties of the soils were nearly identical, the authors noticed an enhancement of catechin-containing compounds and a lowering of the pH of the soils with continued tea monoculture, which affected microbial distribution patterns. The authors suspect that plant exudates influence the bacterial community of the associated soil, which might lead to the described problems in yield reduction [14].

5. Conclusions

In this issue of Plant–Microbe Interactions 2017, two interesting reviews and twelve research articles highlight three aspects of plant–microbe interaction. Studying the beneficial interactions can enable us to increase plant fitness without the application of plant protection chemicals. These discoveries have a direct influence on agricultural practices, which justifies research to mechanistically understand how the plant growth-promoting microorganisms exert their beneficial effects. Of at least equal importance is understanding how plant pathogens can cause disease. Whereas effector proteins are suspected to be responsible for manipulating the plant's defense systems and the plant's metabolism to the advantage of the pathogen, the multitude of newly discovered effectors leaves a big gap in the understanding of how the particular effector proteins function. In addition, other genes of the pathogen might also affect their plant infection capacities, putatively opening the door to the development of novel plant protection strategies. Finally, the diversity of microbial communities is shown here to be not only responsible for ecosystem stability but have multiple positive effects on plant growth, disease resistance or tolerance towards abiotic stresses. Further research in this field is needed to finally understand the network interactions within microbial communities living on, in or near plants and influencing plant fitness on several levels. Therefore, in spite of new insights, research on plant–microbe interaction will continue to provide unexpected discoveries that help in understanding the microbial interaction network of plants. This in turn will empower us to optimize plant cultivation and provide food for an ever-growing population.

Acknowledgments: The authors acknowledge funding from the German Science Foundation (DFG; J.S.), the National Science Foundation (NSF; M.H.P.) and the National Institutes of Health (NIH; sub-award #OGMB131493C1 to M.H.P. from grant P20GM103436 (Nigel Cooper, PI)).

Conflicts of Interest: The authors declare no conflict of interest.

References

1. Vurukonda, S.S.K.P.; Giovanardi, D.; Stefani, E. Plant growth promoting and biocontrol activity of *Streptomyces* spp. as endophytes. *Int. J. Mol. Sci.* **2018**, *19*, 952. [CrossRef] [PubMed]
2. Montalbán, B.; Thijs, S.; Lobo, M.C.; Weyens, N.; Ameloot, M.; Vangronsveld, J.; Pérez-Sanz, A. Cultivar and metal-specific effects of endophytic bacteria in *Helianthus tuberosus* exposed to Cd and Zn. *Int. J. Mol. Sci.* **2017**, *18*, 2026. [CrossRef] [PubMed]
3. Su, A.-Y.; Niu, S.-Q.; Liu, Y.-Z.; He, A.-L.; Zhao, Q.; Paré, P.; Li, M.-F.; Han, Q.-Q.; Ali Khan, S.; Zhang, J.-L. Synergistic effects of *Bacillus amyloliquefaciens* (GB03) and water retaining agent on drought tolerance of perennial ryegrass. *Int. J. Mol. Sci.* **2017**, *18*, 2651. [CrossRef] [PubMed]
4. He, A.-L.; Niu, S.-Q.; Zhao, Q.; Li, Y.-S.; Gou, J.-Y.; Gao, H.-J.; Suo, S.-Z.; Zhang, J.-L. Induced salt tolerance of perennial ryegrass by a novel bacterium strain from the rhizosphere of a desert shrub *Haloxylon ammodendron*. *Int. J. Mol. Sci.* **2018**, *19*, 469. [CrossRef] [PubMed]
5. Zhang, Q.; Gao, X.; Ren, Y.; Ding, X.; Qiu, J.; Li, N.; Zeng, F.; Chu, Z. Improvement of *Verticillium* wilt resistance by applying arbuscular mycorrhizal fungi to a cotton variety with high symbiotic efficiency under field conditions. *Int. J. Mol. Sci.* **2018**, *19*, 241. [CrossRef] [PubMed]
6. Borah, N.; Albarouki, E.; Schirawski, J. Comparative methods for molecular determination of host-specificity factors in plant-pathogenic fungi. *Int. J. Mol. Sci.* **2018**, *19*, 863. [CrossRef] [PubMed]
7. Kuppireddy, V.; Uversky, V.; Toh, S.; Tsai, M.-C.; Beckerson, W.; Cahill, C.; Carman, B.; Perlin, M. Identification and initial characterization of the effectors of an anther smut fungus and potential host target proteins. *Int. J. Mol. Sci.* **2017**, *18*, 2489. [CrossRef] [PubMed]
8. Gao, M.; Yao, S.; Liu, Y.; Yu, H.; Xu, P.; Sun, W.; Pu, Z.; Hou, H.; Bao, Y. Transcriptome analysis of tomato leaf spot pathogen *Fusarium proliferatum*: De novo assembly, expression profiling, and identification of candidate effectors. *Int. J. Mol. Sci.* **2018**, *19*, 31. [CrossRef] [PubMed]
9. Wang, T.; Wang, G.; Jia, Z.-H.; Pan, D.-L.; Zhang, J.-Y.; Guo, Z.-R. Transcriptome analysis of kiwifruit in response to *Pseudomonas syringae* pv. *actinidiae* infection. *Int. J. Mol. Sci.* **2018**, *19*, 373. [CrossRef] [PubMed]

Int. J. Mol. Sci. **2018**, *19*, 1374

10. Adam, A.; Deimel, S.; Pardo-Medina, J.; García-Martínez, J.; Konte, T.; Limón, M.; Avalos, J.; Terpitz, U. Protein activity of the *Fusarium fujikuroi* rhodopsins CarO and OpsA and their relation to fungus–plant interaction. *Int. J. Mol. Sci.* **2018**, *19*, 215. [CrossRef] [PubMed]

11. Chen, H.; Wu, H.; Yan, B.; Zhao, H.; Liu, F.; Zhang, H.; Sheng, Q.; Miao, F.; Liang, Z. Core microbiome of medicinal plant *Salvia miltiorrhiza* seed: A rich reservoir of beneficial microbes for secondary metabolism? *Int. J. Mol. Sci.* **2018**, *19*, 672. [CrossRef] [PubMed]

12. Sánchez-López, A.; Pintelon, I.; Stevens, V.; Imperato, V.; Timmermans, J.-P.; González-Chávez, C.; Carrillo-González, R.; Van Hamme, J.; Vangronsveld, J.; Thijs, S. Seed endophyte microbiome of *Crotalaria pumila* unpeeled: Identification of plant-beneficial methylobacteria. *Int. J. Mol. Sci.* **2018**, *19*, 291. [CrossRef] [PubMed]

13. Li, Q.; Chen, J.; Wu, L.; Luo, X.; Li, N.; Arafat, Y.; Lin, S.; Lin, W. Belowground interactions impact the soil bacterial community, soil fertility, and crop yield in maize/peanut intercropping systems. *Int. J. Mol. Sci.* **2018**, *19*, 622. [CrossRef] [PubMed]

14. Arafat, Y.; Wei, X.; Jiang, Y.; Chen, T.; Saqib, H.; Lin, S.; Lin, W. Spatial distribution patterns of root-associated bacterial communities mediated by root exudates in different aged ratooning tea monoculture systems. *Int. J. Mol. Sci.* **2017**, *18*, 1727. [CrossRef] [PubMed]

International Journal of
Molecular Sciences

MDPI

Review

Plant Growth Promoting and Biocontrol Activity of *Streptomyces* spp. as Endophytes

Sai Shiva Krishna Prasad Vurukonda *, Davide Giovanardi and Emilio Stefani *

Department of Life Sciences, University of Modena and Reggio Emilia, via Amendola 2,
42122 Reggio Emilia, Italy; davide.giovanardi@unimore.it
* Correspondence: saishivakrishnaprasad.vurukonda@unimore.it (S.S.K.P.V.); emilio.stefani@unimore.it (E.S.);
Tel.: +39-052-252-2062 (S.S.K.P.V.); +39-052-252-2013 (E.S.)

Received: 18 February 2018; Accepted: 16 March 2018; Published: 22 March 2018

Abstract: There has been many recent studies on the use of microbial antagonists to control diseases incited by soilborne and airborne plant pathogenic bacteria and fungi, in an attempt to replace existing methods of chemical control and avoid extensive use of fungicides, which often lead to resistance in plant pathogens. In agriculture, plant growth-promoting and biocontrol microorganisms have emerged as safe alternatives to chemical pesticides. *Streptomyces* spp. and their metabolites may have great potential as excellent agents for controlling various fungal and bacterial phytopathogens. Streptomycetes belong to the rhizosoil microbial communities and are efficient colonizers of plant tissues, from roots to the aerial parts. They are active producers of antibiotics and volatile organic compounds, both in soil and *in planta*, and this feature is helpful for identifying active antagonists of plant pathogens and can be used in several cropping systems as biocontrol agents. Additionally, their ability to promote plant growth has been demonstrated in a number of crops, thus inspiring the wide application of streptomycetes as biofertilizers to increase plant productivity. The present review highlights *Streptomyces* spp.-mediated functional traits, such as enhancement of plant growth and biocontrol of phytopathogens.

Keywords: actinobacteria; streptomycetes; plant growth promoting rhizobacteria; microbe–microbe interactions; microbial biocontrol agents

1. Introduction

Plants are extensively colonized by a range of beneficial microorganisms and acquire a variety of plant–microbe interactions. Some of these interactions are beneficial, whereas some are detrimental to the plant. The microorganisms grow on plants as a resource of nutrients or habitat niche. In one such symbiotic interaction, the roots of many plants are infected by specific fungi (mycorrhizal association), rhizobia, and actinobacteria (particularly streptomycetes) that help the plant to acquire nutrients from the soil [1,2].

Currently, microbial endophytic communities are the focus of several studies aimed at unraveling and clarifying their role as plant growth promoters and their involvement in plant health. Several different bacterial species have been identified colonizing plant tissues and vessels, from the root system up to the stem, leaves, and other plant organs. Most of them are described as producers of metabolites positively interfering with plant life, for example, by enhancing nutrient acquisition or by stimulating plant defense mechanisms towards pathogens [3]. Rhizobacteria and mycorrhizal fungi are among the microorganisms that have proved to be of highest efficacy in promoting plant growth and, therefore, crop productivity. Rhizosphere bacteria are able to enhance nutrient uptake from the rhizosoil by the plants that they colonize. For this reason, they might be considered efficient biofertilizers. In most cases, such growth-promoting rhizosphere bacteria belong to the following species: *Alcaligenes*, *Arthrobacter*, *Azospirillum*, *Azotobacter*, *Bacillus*, *Burkholderia*,

Enterobacter, *Klebsiella*, *Pseudomonas*, and *Serratia* [4,5]. *Streptomyces* spp. also belongs to the rhizospheric microbial communities, but only very recently their ability to act as plant growth promoters has been emphasized [6]. Rhizobacteria are also frequently found endophytically in roots and other plant parts, showing their ability to colonize their hosts. In such cases, their plant-stimulating activity does not cease, but can continue in the colonized plant tissues [7,8]. Mycorrhizal associations (ecto- and endomycorrhizae) are also pivotally important in ensuring plant growth and biomass through increasing nutrient and water uptake and enhancing plant resistance to abiotic and biotic stresses [9].

Mycorrhiza and rhizobia are, therefore, natural miniature fertilizer factories, an economical and safe source of plant nutrients compared to synthetic chemical fertilizers, which substantially contribute to environmental pollution. These microorganisms can increase agricultural production and improve soil fertility and, therefore, have great potential as a supplementary, renewable, and environmentally friendly source of plant nutrients.

Streptomyces spp. include many saprophytes, some of them becoming beneficial plant endosymbionts, but also include a few plant pathogens. The filamentous and sporulating nature of *Streptomyces* allows them to survive during unfavorable environmental conditions. Therefore, they appear to compete more efficiently against many other microorganisms present in the rhizosoil. Streptomycetes produce various lytic enzymes during their metabolic processes. Such enzymes are able to degrade insoluble organic polymers, such as chitin and cellulose, breaking them to substituent sugars for binding and uptake by multiple ABC transporters [10–13].

Plant growth promotion and productivity stimulated by microbial endophytic communities are often associated with increased plant health, achieved through direct and/or plant-mediated control of plant pests and pathogens. A few studies reported that root-associated microbes, particularly mycorrhizae and/or rhizobacteria, might influence and change plant physiology such that the aboveground parts are less prone to attack by phytophagous insects [14]. Plant defense is then achieved by priming for enhanced expression of sequences regulated by the production of jasmonic acid, ethylene, or salicylic acid. In other cases, beneficial microbes, such as root-colonizing pseudomonads, may directly act against plant-feeding insects by producing volatile organic compounds (VOCs) that have insecticidal properties [15]. In various studies, most of the antagonistic relationships between beneficial microbes and pathogens have been successful in explaining efficient biocontrol activity against many fungal diseases [16]. In a number of studies, researchers have found that endophytic microorganisms may have a symbiotic association with their host plants. According to Benhamou et al. [17], the endophytic *Bacillus pumilus* efficiently protected pea plants from *Fusarium oxysporum* f. sp. *pisi*, the causal agent of *Fusarium* root rot. Similarly, Varma et al. [18] demonstrated the growth-promoting activity in various plants elicited by the endophytic fungus *Piriformospora indica*. These endophytic microorganisms provide real advantages to the host plants, for example, by enhancing the physiological activity of the plant or facilitating the uptake of nutrients from the soil. Thus, they may serve as biocontrol agents or plant growth promoters [19]. Among other microorganisms, a variety of actinomycetes inhabits a wide range of plants as endophytes [20–25]; therefore, such actinobacteria may have both the potential to serve as effective biocontrol agents and to be considered as efficient plant growth promoters [26–28]. The genus *Streptomyces* has been extensively studied and used for biocontrol of soilborne fungal pathogens because of its intense antagonistic activity through the production of various antifungal metabolites [29–31].

2. *Streptomyces* spp. as Endophytes

Streptomycetes are Gram-positive bacteria belonging to the order *Actinomycetales* and the family *Streptomycetaceae*; roughly, streptomycetes are represented by more than 570 different species [32]. Streptomycetes are aerobic and filamentous bacteria able to produce vegetative hyphae that eventually form a complex mycelium and are able to grow and colonize different substrates. They are spore-forming bacteria and their spores may aid the dispersion and dissemination of the microorganism [33]. The genus *Streptomyces* includes ten plant pathogenic species, most of which are causal agents of the common scab of potatoes [34]. In nature, streptomycetes have a quite widespread distribution and are found in soils of very different structure and chemistry, in surface waters, and in plants as rhizosphere colonizers or true endophytes. As endophytic microorganisms, they colonize the internal part of plants, mainly the root system and the xylem tissues of the stem, causing no apparent change to their host's morphology and physiology [35,36]. In different natural environments, they often play a major role in nutrient cycling. They may also have a strong influence in the population structure of environmental microbial communities due to their ability to produce a large set of secondary metabolites, many of which are of clinical and biotechnological importance [37,38].

From the medical point of view, *Streptomyces* is the largest antibiotic-producing genus against clinical microorganisms (fungi and bacteria) and parasites. They also produce other clinically important bioactive compounds such as immunosuppressants [39]. Only very recently streptomycetes has been considered as a prospective biocontrol agent in agriculture. Indeed, their ability to produce antibiotics may be used to control plant pathogenic bacteria and fungi [40]. Interference competition, an important strategy in interspecific interactions, is the production of growth inhibitory secondary metabolites (for example, antibiotics, toxins, biosurfactants, volatiles, and others) that can suppress or kill microbial opponents [41,42]. This feature is particularly present in streptomycetes, thus suggesting their use in excluding plant pathogens from their crop plants.

Interestingly, in a few cases, their interactions with plants may lead to suppression of the innate plant responses to phytopathogens. Therefore, it is of great importance to choose and characterize single *Streptomyces* strains for possible use as microbial antagonists. This is conveniently done through extensive in vitro and in planta studies on the roles of their antibiotics and possible production of VOCs [43]. One of the most common metabolites in streptomycetes communities is geosmin, a bicyclic alcohol derivative of decalin that confers the typical "earthy" flavor to the substrates they colonize [44]. Geosmin may be regarded as a volatile organic compound of microbial origin to which the human nose is extremely sensitive [45]. Although geosmin has no known antibiotic activity and its adaptive significance is not yet known, this metabolite might have an important role in the biology of streptomycetes [46]; indeed, it is a well-conserved trait and the gene responsible is highly conserved among *Streptomyces* spp. [47]. Geosmin enables bacteria to adapt to various environments, such as microbial communities or the host, ultimately influencing bacterial competition and cooperation [48]. It also has the ability to induce selective growth of geosmin-utilizing bacteria [49].

Microbial endophytes that efficiently and stably colonize different plant tissues, from roots to all aerial parts, have been long known, although their pivotal importance in agriculture has become evident only in recent decades. The main roles of endophytic microorganisms were discussed around 20 years ago, when several authors focused on symbiotic microorganisms and their possible plant–microbe interactions from a systematic, ecological, and physiological point of view [50–53]. Later discovery of the metabolic potential of such endophytes *in planta*, their ability to efficiently compete with other endophytes (included plant pathogens), and their role in stimulating the expression or overexpression of plant genomic sequences involved in tolerance/resistance to plant stresses (abiotic and biotic) indicated that selected endophytes may be considered as very promising agents to control plant pests and diseases.

Actinobacteria, and streptomycetes in particular, are known to constitute a large part of the rhizosoil microbiota. They may live saprophytically and endophytically in both natural and agricultural environments where they may colonize the rhizosphere and different morphological parts of plant roots [54]. Therefore, considering their plant growth-promoting activity, streptomycetes represent an excellent alternative for improving nutrient availability to crop plants and promoting innovation and sustainability in agricultural systems [55]. Plant growth-promoting streptomycetes (PGPS) stimulate and enhance several direct and indirect biosynthetic pathways in plants, for example, inorganic phosphate solubilisation, biosynthesis of chelating compounds, phytohormones production, inhibition of plant pathogens, and alleviation of various abiotic stresses (Figure 1) [56].

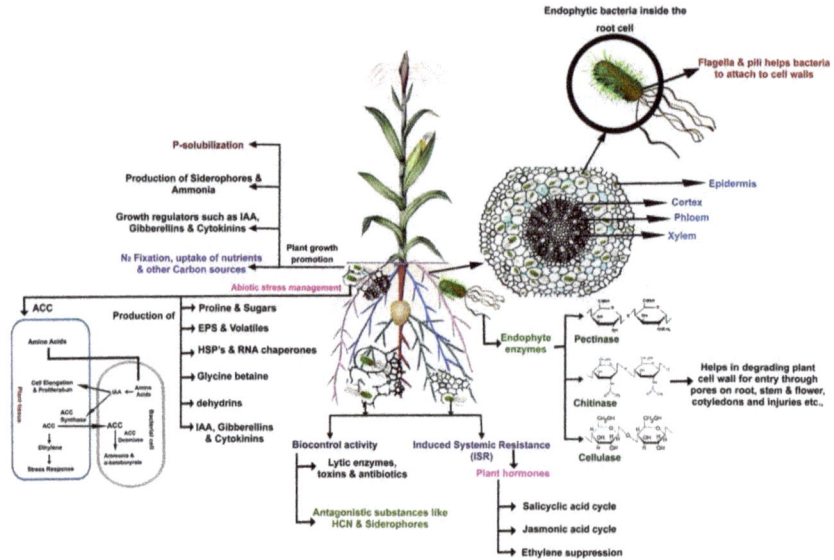

Figure 1. Representation of possible plant–microbe interactions favouring plant growth and/or biocontrol of phytopathogens by streptomycetes as rhizosphere competent microorganisms and/or endophytes (adapted from [57]).

The isolation of actinomycetes in pure culture is an important step for screening the production of bioactive compounds. The most studied actinomycetes are species from the genus *Frankia*, a nitrogen-fixing bacterium of non-leguminous plants [58], and a few species of the genus *Streptomyces* that are phytopathogens [59]. Mundt and Hinckle [60] were able to isolate different species of *Streptomyces* and *Nocardia* from 27 different plant species, finding these actinobacteria present as endophytes in different plant tissues such as seeds and ovules. Sardi et al. [20] isolated and observed, through direct microscope examination, endophytic actinomycetes from the roots of 28 plant species from Northwestern Italy, finding actinomycetes belonging to the genus *Streptomyces* and other common genera, namely *Streptoverticillium*, *Nocardia*, *Micromonospora*, and *Streptosporangium*.

3. *Streptomyces* spp. as Plant Growth Promoters and Improvement of Plant Nutrition

Actinobacteria may have, in general, a positive role in plant mineral nutrition. This is correlated to both nitrogen fixation and metal mobilizing ability involving mineral nutrients such as Fe, Zn, and Se. Nonetheless, metagenomic analyses have not proven that streptomycetes are involved in such beneficial processes [61]. Metagenomic analyses of bacterial microbiota in plants have shown that the phylogenetic and taxonomic composition of such microbial communities is limited to few bacterial phyla, including actinobacteria.

More recently, Viaene et al. [7] highlighted the contribution of streptomycetes to plant growth and health. The plant has an important role in shaping its root microbiome through root exudate composition (chemotaxis) and nutritional interactions [62–64]. Plant root exudates are a source of metabolic signals (such as flavonoids, strigolactones, and terpenoids) that have the ability to shape the microbial communities in the rhizosphere. The signals that attract streptomycetes into the rhizosphere are still unknown. From the rhizosphere, streptomycetes are able to enter roots and colonize root tissues and vessels from where they can be isolated [24] and purified to identify them and describe their physiology and their microbe–microbe interactions.

Actinobacteria, such as *Streptomyces* spp., influence soil fertility through the involvement of many components and serve as nutrient enhancers. Besides producing siderophores and solubilizing phosphate, they are known to produce various enzymes—including amylase, chitinase, cellulase, invertase, lipase, keratinase, peroxidase, pectinase, protease, phytase, and xylanase—which make the complex nutrients into simple mineral forms. This nutrient cycling capacity makes them ideal candidates for natural fertilizers [65]. The relationship between PGPS and their host plant and the biochemical processes involved deserve deeper investigation. This knowledge would allow manipulation of those interactions, particularly the biochemical mechanisms leading to a compatible relationship between the host plant and its endophytes. Most streptomycetes are free-living in the soil as saprophytes and are able to colonize the rhizosphere and rhizoplane of the host plant. For instance, some PGPS, initially known as soil-dwelling microorganisms, were found to efficiently colonize the inner tissues of selected host plants as endophytes, therefore proving their ability to fully or partly conduct their life cycle inside plant tissues [66]. Additionally, a wide variety of *Streptomyces* species may establish beneficial plant–microbe interactions [67–69]. Table 1 summarizes the plant growth-promoting activity of *Streptomyces* species—many of them not fully identified—that gain access to root tissues from the rhizosoil. These species thus acquire an endophytic status without causing any visible harm or symptoms in the host plant. Such streptomycetes, although not always identified at the species level, are reported to have marked plant growth-promoting activity in their host plants. These species are most likely present in the apoplast of different parts of the plant (that is, roots, stems, leaves, flowers, fruits, and seeds) [69]. Coombs and Franco [70] demonstrated the endophytic colonization of wheat embryos, endosperm, and emerging radicles by tagging *Streptomyces* spp. strain EN27 with green fluorescent protein.

The endophytic streptomycetes can also be a source of metabolites that promote or improve host plant fitness and growth, as well as reduce disease symptoms that are caused by plant pathogens or various environmental stresses [71].

Table 1. List of streptomycetes isolated from plants or the rhizosphere showing plant growth-promoting (PGP) activity.

Species	Host Plant	PGP Traits/Observed Effects in Plants	Reference
Streptomyces sp.	Clover	Nutrient uptake	[72]
Streptomyces sp.	Rice, chickpea	Nutrient uptake and plant growth	[73,74]
Streptomyces lydicus	Pea	Nodulation	[75]
Streptomyces sp.	Mung bean	Enhanced plant growth	[68]
Streptomyces sp.	Soybean	Nutrient uptake and plant growth	[76]
Streptomyces atrovirens, S. griseoviridis, S. lydicus, S. olivaceoviridis, S. rimosus, S. rochei, S. viridis	Rhizosphere of different plants	Auxin/IAA production	[77–81]
Streptomyces sp.	-	Gibberellin biosynthesis	[82]
Streptomyces igroscopicus	-	ACC deaminase	[83]
Streptomyces sp.	*Terfezia leonis* Tul.	Siderophore production, IAA, and gibberellic acid production	[84]
Streptomyces sp.	Marine environments	Gibberellic acid, IAA, abscisic acid, kinetin, and benzyladenine	[85]
Streptomyces aurantiogriseus	Rice	IAA production	[86,87]
Streptomyces sp.	Soil	Synthesis of IAA and siderophore production	[88]
Streptomyces spp.	-	B-1,3-Glucanase, IAA, and HCN synthesis	[73,89]
Streptomyces sp.	-	Siderophore production	[90]

Table 1. *Cont.*

Species	Host Plant	PGP Traits/Observed Effects in Plants	Reference
Streptomyces rochei, *S. carpinensis,* *S. thermolilacinus*	Wheat rhizosphere	Production of siderophore, IAA synthesis, and phosphate solubilization	[91]
Streptomyces sp.	Soil	Siderophore production, phosphate solubilization, and N_2 fixation	[72]
Streptomyces spp.	*Alnus glutinosa, Casuarina glauca, Eleagnus angustifolia*	Production of zeatin, gibberellic acid, and IAA	[92]
Streptomyces olivaceoviridis, S. rochei	Wheat	Auxin, gibberellin, and cytokinin synthesis	[93]
Streptomyces hygroscopicus	Kidney beans	Formation of adventitious roots in hypocotyls	[94]
Streptomyces sp.	Rhododendron	Accelerated emergence and elongation of adventitious roots in tissue-cultured seedlings	[28]
Streptomyces filipinensis, *S. atrovirens*	Tomato	Plant growth promotion	[78]
Streptomyces spiralis	Cucumber	Plant growth promotion	[95]
Streptomyces spp.	Sorghum	Enhanced agronomic traits of sorghum	[96]
	Rice	Enhanced stover yield, grain yield, total dry matter, and root biomass	

Note: IAA: Indole-3-acetic acid; ACC: 1-amino cyclopropane-1-carboxylic acid; HCN: Hydrogen cyanide.

Plant Hormone Production by Streptomycetes

Many scientific reports have explained the ability of endophytic actinobacteria to stimulate the secretion of plant growth hormones and enhance their growth-promoting activity. A study by Dochhil et al. [97] described the evidence of plant growth-promoting activity and a higher percentage of seed germination due to the synthesis of higher concentrations (71 g/mL and 197 g/mL) of the plant growth hormone indole acetic acid (IAA) by two *Streptomyces* spp. strains isolated from *Centella asiatica*. In field trials, increased growth promotion and yield of cucumber was achieved by the application of *Streptomyces spiralis* alone, or in combination with other microbial "activators" such as *Actinoplanes campanulatus* or *Micromonospora chalcea*. Such experiments highlight the role of multiple microbes (or a microbial consortium) in very productive crop systems [98,99].

In soil, most of the known actinomycetes belong to genus *Streptomyces* and have been used for various agricultural purposes, mainly due to their production of antifungal and antibacterial metabolites and a number of plant growth-promoting (PGP) traits [100,101]. Indeed, more than 60% of known compounds with antimicrobial or plant growth-promoting activity originate from this genus [102]. In agricultural environments, *Streptomyces* species are an important group of soil bacteria because of their ample capacity to produce PGP substances, secondary metabolites (such as antibiotics), and enzymes [77,103].

Indole-3-acetic acid (IAA) is a common plant hormone belonging to the class of auxins. It has an important role in plant growth and development since it induces cell elongation and division. Manulis et al. [104] studied the production of IAA and the pathways of its synthesis by various *Streptomyces* spp., including *Streptomyces violaceus, Streptomyces griseus, Streptomyces exfoliates, Streptomyces coelicolor,* and *Streptomyces lividans*. Reddy et al. [105] isolated *Streptomyces atrovirens* from groundnut roots. This bacteria has shown excellent growth-promoting activity not only on groundnut but also on a number of other crops. These results are particularly interesting since they show the ability of a single streptomycete to promote growth in multiple different plants. El-Sayed et al. [106] and El-Shanshoury [107] reported IAA production in plants stimulated by *Streptomyces* sp. in greenhouse experiments while El-Tarabily [78] was successful in comparing different *Streptomyces* spp. strains. In these experiments, remarkably efficient growth promotion was stimulated by *Streptomyces filipinensis* due to its production of IAA.

1-aminocyclopropane-1-carboxylate (ACC) is a derived amino acid that is required for the endogenous biosynthesis of ethylene in plants. Comparing different streptomycetes, El-Tarabily [78] noted that the increased growth promoted by *Streptomyces filipinensis*, when compared to *S. atrovirens*, was due to the production of both IAA and ACC, whereas *S. atrovirens* produced only ACC deaminase. Therefore, a single streptomycetes was shown to produce more than one plant hormone. These results are of great interest for the possible exploitation of streptomycetes as plant growth stimulants.

The endophytic colonization of streptomycetes connected with their influence on plant nutrition is poorly studied so far. However, nitrogen-fixing actinobacteria and their correlation with plant nutrition and productivity have been recently described [108]. Among such actinobacterial communities, a few *Streptomyces* spp. with nitrogen-fixing capacity have been identified. Siderophore production has also been described. In particular, isolates of *Streptomyces* spp. were able to produce and excrete an enterobactin, an iron-chelating compound characteristic of some *Enterobacteriaceae* [90,109].

4. Streptomycetes in Plant Protection against Biotic Stresses

Microbial biocontrol agents have the ability to perform antibiosis, parasitism, or competition with the pathogen for nutrients and space. They may also induce disease resistance in the host plant that they colonize, acting along different steps of the infection process. Therefore, protection of plants from biotic stresses may be the result of one or more microbe–microbe or plant–microbe interactions [110]. Actinomycetes, and particularly *Streptomyces* species, are well known for their production of a wide spectrum of antibiotics. These are often species specific and allow them to develop symbiotic interactions with plants by protecting them from various pathogens; at the same time, plant exudates promote *Streptomyces* growth [111]. In the last two decades, there has been an increasing interest in antibiosis by PGPB and such biocontrol mechanisms are now better understood [112]. Several metabolites with antibiotic nature produced by pseudomonads have been studied and characterized so far, e.g., the cyclic lipopeptide amphysin, 2,4-diacetylphloroglucinol (DAPG), oomycin A, the aromatic polyketide pyoluteorin, pyrrolnitrin, the antibacterial compound tropolone [113,114]. Other bacterial genera, such as *Bacillus*, *Streptomyces*, *Stenotrophomonas* spp., produce the macrolide oligomycin A, kanosamine, the linear aminopolyol zwittermicin A, and xanthobactin [115,116]. They also synthesize several enzymes that are able to disrupt fungal cell walls [39]. Early studies performed during the 1950s described the production by streptomycetes of a set of antibiotics suitable for controlling foliage diseases caused by phytopathogenic fungi [117,118]. Later, several other authors reported excellent biocontrol activity of some phytopathogenic soilborne fungi such as *Pythium* spp., *Fusarium* spp. [119], *Rhizoctonia solani* [120], and *Phytophthora* spp. [121] (Table 2).

Table 2. Biocontrol activity of several *Streptomyces* spp. against different fungi.

Species/Strain	Plant	Disease	Target Pathogens	References
Streptomyces viridodiasticus	Lettuce	Basal drop disease	*Sclerotinia minor*	[122]
S. violaceusniger G10	Banana	Wilt	*Fusarium oxysporum* f. sp. *cubense* race 4	[123]
Streptomyces sp. KH-614	Rice	Blast	*Pyricularia oryzae*	[124]
Streptomyces sp. AP77	Porphyra	Red rot	*Pythium porphyrae*	[125]
Streptomyces sp. S30	Tomato	Damping off	*Rhizoctonia solani*	[126]
S. halstedii	Red pepper	Blight	*Phytophthora capsica*	[127]
Streptomyces spp. 47W08, 47W10	Pepper	Blight	*Phytophthora capsica*	[128]
S. violaceusniger XL-2	Many	Wood rot	*Phanerochaete chrysosporium, Postia placenta, Coriolus versicolor, Gloeophyllum trabeum*	[129]
S. ambofaciens S2	Red chili fruits	Anthracnose	*Colletotrichum gloeosporioides*	[130]
Streptomyces spp.	Sugar beet	Damping off	*Sclerotium rolfsii*	[131]

Table 2. *Cont.*

Species/Strain	Plant	Disease	Target Pathogens	References
S. hygroscopicus	Many	Anthracnose and leaf blight	*Colletotrichum gloeosporioides* and *Sclerotium rolfsii*	[132]
Streptomyces spp.	Sunflower	Head and stem rot	*Sclerotinia sclerotiorum*	[133]
Streptomyces sp.	Sweet pea	Powdery mildew	*Oidium* sp.	[134]
S. vinaceusdrappus	Rice	Blast	*Curvularia oryzae, Pyricularia oryzae, Bipolaris oryzae, Fusarium oxysporum*	[135]
Streptomyces sp. RO3	Lemon fruit	Green mold and sour rot	*Penicillium digitatum, Geotrichum candidum*	[136]
S. spororaveus RDS28	Many	Collar or root rot, stalk rot, leaf spots, and gray mold rot or botrytis blight	*Rhizoctonia solani, Fusarium solani, Fusarium verticillioides, Alternaria alternata, Botrytis cinerea*	[137]
S. toxytricini vh6	Tomato	Root rot	*Rhizoctonia solani*	[138]
Streptomyces spp.	Sugar beet	Root rot	*Rhizoctonia solani, Phytophthora drechsleri*	[139]
Streptomyces sp.	Chili	Root rot, blight, and fruit rot	*Alternaria brassicae, Colletotrichum gloeosporioides, Rhizoctonia solani, Phytophthora capsica*	[140]
Streptomyces sp.	Chili	Wilt	*Fusarium oxysporum* f. sp. *capsici*	[141]
Streptomyces sp.	Ginger	Rhizome rot	*Fusarium oxysporum* f. sp. *zingiberi*	[142]
Streptomyces sp. CBE	Groundnut	Stem rot	*Sclerotium rolfsii*	[143]
Streptomyces sp.	Tomato	Damping off	*Rhizoctonia solani*	[144]
Streptomyces sp.	Tobacco	Brown spot	*Alternaria* spp.	[145]
S. aurantiogriseus VSMGT1014	Rice	Sheath blight	*Rhizoctonia solani*	[87]
S. felleus YJ1	Oilseed rape	Stem rot	*Scleotinia sclerotiorum*	[146]
S. vinaceusdrappus S5MW2	Tomato	Root rot	*Rhizoctonia solani*	[147]
Streptomyces sp. CACIS-1.16CA	Many	-	*Curvularia* sp., *Aspergillus niger, Helminthosporium* sp., *Fusarium* sp. *Alternaria* sp., *Phytophthora capsici, Colletotrichum* sp., and *Rhizoctonia* sp.	[148]
S. griseus	Tomato	Wilt	*Fusarium* sp.	[149]
Streptomyces sp.	Potato	Silver scurf	*Helminthosporium solani*	[150]
S. rochei	Pepper	Root rot	*Phytophthora capsica*	[151]
Streptomyces sp.	Maize	Seed fungi	*Aspergillus* sp.	[152]
S. lydicus WYEC108	Many	Foliar and root fungal diseases	-	[153,154]
S. griseoviridis K61	Many	Root rot and wilt pathogenic fungi	-	
Streptomyces sp. YCED9 and WYEC108	Lettuce	Damping off	*Pythium ultimum, Sclerotinia homeocarpa, Rhizoctonia solani*	[153,155]
Streptomyces G10	Banana	Wilt	*Fusarium oxysporum* f. sp. *cubense*	[156]
S. violaceusniger YCED9	Turfgrass	Crown/foliar disease	*Rhizoctonia solani*	[157]
Streptomyces sp.	Cucurbit	Anthracnose	*Colletotrichum orbiculare*	[19]
Streptomyces sp. A1022	Pepper and cherry tomato	Anthracnose	*Colletotrichum gloeosporioides*	[158]
S. halstedii K122	Many	-	*Aspergillus fumigatus, Mucor hiemalis, Penicillium roqueforti, Paecilomyces variotii*	[159]
Streptomyces sp. MT17		Wood rotting	Different fungi	[160]
S. lavendulae HHFA1 *S. coelicolor* HHFA2	Onion	Bacterial rot	*Erwinia carotovora* subsp. *carotovora, Burkholderia cepacia*	[161]
Streptomyces sp. 5406	Cotton	Soilborne diseases	Soilborne plant pathogens	[162]
Streptomyces sp.	Raspberry	Root rot	*Phytophthora fragariae* var. *rubi*	[163]
S. albidoflavus	Tomato	Many	*Alternaria solani, A. alternata, Colletotrichum gloeosporioides, Fusarium oxysporum, Fusarium solani, Rhizoctonia solani, Botrytis cinerea*	[164]
Streptomyces sp.	Soybean	Bacterial blight	*Xanthomonas campestris* pv. *glycines*	[165]
Streptomyces sp. BSA25 and WRAI	Chickpea	-	*Phytophthora medicaginis*	[166]
Streptomyces sp.	Chickpea	Fusarium wilt	*Fusarium oxysporum* f. sp. *ciceri*	[167]
Streptomyces sp.	Chickpea	Basal rot	*Macrophomina phaseolina*	[73]
Streptomyces sp.	Cucumber	Fusarium wilt	*Fusarium oxysporum*	[168]

As shown in Table 2, streptomycetes are promising microbial biocontrol organisms that are able to antagonize and/or kill fungal and bacterial plant pathogens. Their biocontrol activity is often performed before the pathogens completely infect their respective host. Recently, these organisms have been the focus of different approaches toward the development of biocontrol strategies against soilborne pathogens [120,169]. For instance, by treating seeds with endophytic *Streptomyces* spp. and *Micromonospora* spp. prior to sowing, *Arabidopsis thaliana* was protected from infection by *Erwinia carotovora* and *F. oxysporum*. Streptomycetes were observed antagonizing pathogens by inducing the expression of defense pathways in the plant [170]. This observation implies that the microbial antagonists penetrated the seeds during their germination and colonized the seedlings. Bacon and Hinton [171] reported that varying levels of disease suppression in the field were positively correlated with similar results obtained from in vitro experiments. In other experiments, a significant pathogen inhibition in vitro was not always correlated with disease protection *in planta*. Growth inhibition of plant pathogens by endophytic bacteria indicates the presence of antagonistic activities between them, which may act directly (by mechanisms of antibiosis, competition, and lysis) or indirectly (by inducing plant defense or by growth-promoting substances) [31,172] (Figure 2a,b).

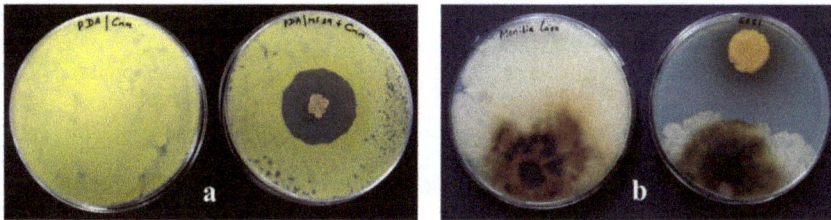

Figure 2. In vitro biocontrol activity of *Streptomyces* spp.: (**a**) antimicrobial activity against *Clavibacter michiganensis* subsp. *michiganensis*, the causal agent of the tomato bacterial canker and (**b**) antifungal activity against *Monilinia laxa*, the causal agent of the brown rot of stone fruits.

Production of chitinolytic enzymes and siderophores (iron-chelating compounds) is a known additional mode of action for fungal growth inhibition by endophytic actinobacteria. Endophytic actinobacteria can produce enzymes that degrade fungal cell walls, especially by the production of chitinases. Over 90% of chitinolytic microorganisms are actinomycetes. These have been extensively studied during the last two decades, starting in the mid-1990s [173]. The production of chitinases by actinomycetes and by streptomycetes in particular makes these organisms very promising microbial biocontrol agents. In *Streptomyces plicatus*, chitinases are encoded in a region of *chi65*, the expression of which is induced by N,N'-diacetylchitobiose and activated by allosamidin [174]. In fungi, chitinase is necessary for fungal development, such as hyphal growth and branching [175]. Several bacteria, and streptomycetes in particular, also produce a set of chitinases to obtain nutrients through degradation of environmental chitin, including the cell wall of soil fungi. Therefore, this ability may be exploited in the selection and exploitation of chitinolytic microbial agents for the biocontrol of phytopathogenic fungi [176,177]. Allosamidin is an important secondary metabolite of streptomycetes and was initially reported as a chitinase inhibitor [178]. Later, they further investigated the role of allosamidin in its producing *Streptomyces* and showed that allosamidin inhibits all family 18 chitinases, but can dramatically promote chitinase production and growth of its producer *Streptomyces* [174,179]. This appears particularly important for the bacterial growth in soil where chitin, mainly originating from insect cuticle and fungal cell walls, is a major nutrient source. Therefore, allosamidin is listed as a potent secondary metabolite with antifungal activity [180].

Although both crude and purified chitinases have great potential for cell wall lysis of fungal pathogens, in common agricultural systems the use of selected streptomycetes as microbial biocontrol agents targeting important phytopathogens appears to be a more effective strategy.

This is due to the high cost of purified antimicrobial molecules, making them more suitable as pharmaceuticals against clinical and animal pathogens. The ability of siderophores to promote plant growth and enhance antagonism to phytopathogens has gained more significance [69,181,182]. El-Shatoury et al. [183] reported actinobacteria from *Achillea fragrantissima* that were capable of producing both chitinases and siderophores; they also showed remarkable inhibitory activity against phytopathogenic fungi. These reports were strongly supported and further explained by Gangwar et al. [184] studying actinobacteria from *Aloe vera*, *Mentha arvensis*, and *Ocimum sanctum*. The latter authors provided quantitative data for different types of siderophore compounds: the hydroxamate-type of siderophore ranged between 5.9 and 64.9 $\mu g \cdot mL^{-1}$ and the catechol-type of siderophore occurred in a range of 11.2–23.1 $\mu g \cdot mL^{-1}$. In another investigation, El-Tarabily et al. [185] applied endophytic *Streptomyces spiralis* together with *Actinoplanes campanulatus* and *Micromonospora chalcea* to cucumber seedlings. Since this group of microorganisms, applied as a microbial consortium, showed more effectiveness in the reduction of seedling damping off and root- and crown-rot diseases by *Pythium aphanidermatum* than the single actinobacterium, this study recommended their use as very effective biocontrol agents. Therefore, as for other PGPR, several streptomycetes produce siderophores to sequester iron in the rhizosphere, making iron unavailable to certain rhizoplane microorganisms, in particular to some phytopathogens. These pathogenic microorganisms are often unable to obtain essential quantities of iron for their growth because they do not produce siderophores, produce comparatively less siderophores than PGPR, and/or produce siderophores that have less affinity for iron than those of PGPR [186].

Igarashi studied the new bioactive compound 6-prenylindole produced by a *Streptomyces* spp. [187]. In the beginning, it was reported as a component of liverwort (*Hepaticae*) and it showed significant antifungal activity against *Alternaria brassicicola*. Interestingly, this molecule was isolated from both plants and microorganisms [187]. Similar reports by Zhang et al. [188] explained the inhibition of the phytopathogenic fungi *Colletotrichum orbiculare*, *Phytophthora capsici*, *Corynespora cassiicola*, and *Fusarium oxysporum* by a new prenylated compound and three known hybrid isoprenoids with IC_{50} in the range 30.55–89.62. Another study by Lu and Shen [189,190] reported inhibition of *Penicillium avellaneum* UC-4376 by naphthomycins A and K produced by *Streptomyces* spp. The synthesis of fistupyrone, a metabolite produced by *Streptomyces* spp. isolated from leaves of spring onion (*Allium fistulosum*), was found by Igarashi [187] to inhibit *Alternaria brassicicola*, the causal agent of the black leaf spot in *Brassica* plant. This study reported that fistupyrone was able to inhibit the fungal infection process by pre-treating the seedlings with such compound at a concentration of 100 ppm. In support of this statement, experiments by Igarashi et al. [191] evidenced that fistupyrone did not inhibit the growing hyphae but suppressed spore germination of fungi at 0.1 ppm concentration.

Streptomyces spp. have the capacity to produce cellulolytic enzymes and various secondary metabolites, which directly act on herbivorous insects and show toxic activity on phytopathogens and/or insect pests [192,193]. A set of different molecules from *Streptomyces* spp. that act against insect pests have been found and characterized; these are, for instance, flavensomycin [194], antimycin A [195], piericidins [196], macrotetralides [197] and prasinons [198]. *Streptomyces avermitilis*, a common soil inhabitant, was shown to produce avermectins, molecules with potent activity against arthropods and nematodes [199]. These compounds derive from lactones and are macrocyclic in nature; they mainly act on the insect peripheral nervous system by targeting the γ-aminobutyric acid (GABA) receptors, leading to paralysis of the neuromuscular system [200]. Commercial insecticides based on avermectin mixtures are known as abamectin and they act on phytophagous arthropods directly by contact and ingestion. They are not systemic in plants, showing just a limited translaminar activity. Similar molecules produced by *Streptomyces* spp. are emamectin—particularly toxic to Lepidoptera, and milbemectin—and are specifically isolated from *S. hygroscopicus*.

5. Commercialization, Environmental Effects, and Biosafety of *Streptomyces* Products

Streptomycete producing antimicrobial secondary metabolites present an attractive alternative to chemical fertilizers, pesticides, and supplements, which may result in a significant increase in agricultural plant growth and pest and disease control [201]. While increasing our knowledge of the mechanisms triggered by actinomycetes for suppressing plant diseases, improving nutrient uptake by plants, and stimulating and/or increasing the production of phytohormones in planta, a great deal of research is being carried out worldwide for the development of correct formulations containing actinomycete inoculants as their active ingredients. Nevertheless, very few actinomycete-based products are currently commercialized. Although biocontrol with PGPR is an acceptable green approach, the proportion of registration of *Streptomyces* spp. as biocontrol agents for commercial availability is very low. Mycostop (Verdera Oy, Finland) is the only *Streptomyces*-based plant protection product registered in the EU; it is also registered in Canada and the USA. Actofit and Astur, based on *Streptomyces avermitilis*, are registered as insecticides in the Ukraine. Table 3 lists the microbial pesticides that are registered in particular countries worldwide. In most cases, metabolites produced by *Streptomyces* spp. are registered as active substances in plant-protection products, as shown in Table 4.

Table 3. List of *Streptomyces* spp.-based products available in the market worldwide (data collected and modified into a table from [202]).

Commercial Product Name	Organism as Active Substance	Registered as Microbial Pesticide	Targeted Pest/Pathogen/Disease
Actinovate, Novozymes BioAg Inc., USA	*S. lydicus* WYEC 108	Canada, USA	Soilborne diseases, viz. *Pythium, Fusarium, Phytophthora, Rhizoctonia,* and *Verticillium;* foliar diseases such as powdery and downy mildew, *Botrytis* and *Alternaria, Postia, Geotrichum,* and *Sclerotinia*
Mycostop, Verdera Oy, Finland	*Streptomyces K61*	EU, Canada, USA	Damping off caused by *Alternaria* and *R. solani* and *Fusarium, Phytophthora,* and *Pythium* wilt and root diseases
Mykocide KIBC Co. Ltd. South Korea	*S. colombiensis*	South Korea	Powdery mildews, grey mold, brown patch
Safegrow KIBC Co. Ltd. South Korea	*S. kasugaensis*	South Korea	Sheath blight, large patch
Actofit, Astur	*S. avermitilis*	Ukraine	Colorado potato beetle, web mites, other phytophags
Bactophil	*Streptomyces albus*	Ukraine	Seed germination diseases
Bialaphos, Toku-E, USA	*S. hygroscopicus, S. viridochromogenes*	USA	Herbicide
Incide SP, Sri Biotech Laboratories India Ltd., India	*S. atrovirens*	India	Insecticide
Actin, Sri Biotech Laboratories India Ltd., India	*S. atrovirens*	India	Fungicide

Table 4. List of active substances derived from *Streptomyces* spp. (in bold) and registered as commercial products in different geographical areas (data collected and modified into a table from [193,203–205]).

Biocontrol Metabolite (Bold) and Commercial Names	Organism	Country	Targeted Pathogen/Disease
Blasticidin-S BLA-S	*S. griseochromogenes*	USA	Rice blast (*Pyricularia oryzae*)
Kasugamycin Kasumin, Kasurab	*S. kasugaensis*	Ukraine	Leaf spot in sugar beet and celery (*Cercospora* spp.) and scab in pears and apples (*Venturia* spp.), soybean root rot (*Phytophthora sojae*)
Streptomycin Agrimycin, Paushak, Cuprimicin 17, AAstrepto 17, AS-50, Dustret, Cuprimic 100 and 500	*S. griseus*	India, USA, New Zealand, China, Ukraine, Canada	Bacterial rots, canker, and other bacterial diseases, *Xanthomonas oryzae, Xanthomonas citri,* and *Pseudomonas tabaci* of pome fruit, stone fruit, citrus, olives, vegetables, potatoes, tobacco, cotton, and ornamentals

Table 4. *Cont.*

Biocontrol Metabolite (Bold) and Commercial Names	Organism	Country	Targeted Pathogen/Disease
Phytomycin Mycoshield, Cuprimic 100 and 500, Mycoject	*S. rimosus*	-	Fire blight (*Erwinia amylovora*) and diseases caused by *Pseudomonas* and *Xanthomonas* sp. and mycoplasma-like organisms
Validamycin Validacin, Valimun, Dantotsupadanvalida, Mycin Hustler, Valida, Sheathmar	*S. hygroscopicus*	-	*Rhizoctonia solani* and other *Rhizoctonia* in rice, potatoes, vegetables, strawberries, tobacco, ginger, cotton, rice, sugar beet, etc.
Polyoxorim Endorse, PolyoxinZ, Stopit, Polyoxin AL and Z, Polybelin	*S. cacaoi* var. *asoensis*	-	Plant pathogenic fungi, *Sphaerotheca* spp. and other powdery mildews, *Botrytis cinerea*, *Sclerotinia sclerotiorum*, *Corynespora melonis*, *Cochliobolus miyabeanus*, *Alternaria alternate* and other species in vines, apples, pears, vegetables, and ornamentals. Rice sheath blight (*R. solani*), apple, pear canker, and *Helminthosporium* in rice; also inhibits cell wall biosynthesis and causes abnormal germ tube swelling of spores and hyphal tips, rendering fungus nonpathogenic
Natamycin Delvolan	*S. natalensis* and *S. chattanoogensis*	-	Basal rots on daffodils and ornamentals caused by *Fusarium oxysporum*
Abamectin (Avermectins) Agri-Meck Avid, Clinch, Dynamec, Vertimec, Abacide, Abamex, Vapcomic, Vibamec, Agromec, Belpromec, Vamectin 1.8 EC, Vivid and many others	*S. avermitilis*	European Union, Worldwide	Mites, leaf miners, suckers, beetles, fire ants, and other insects in ornamentals, cotton, citrus, pome and nut fruit, vegetables
Polynactin Mitecidin	*S. aureus*	Japan	Spider mites (*Tetranychus cinnabarinus*), two-spotted mite (*Tetranychus urticae*), European red mite (*Panonychus ulmi*) in orchard fruit trees
Milbemycine Milbeknock, Koromite, Mesa, Ultiflora and Matsuguard	*S. hygroscopicus* subsp. *aureolacrimosus*	-	Citrus red mites, Kanzawa spider mites, and leaf miners in citrus, tea, eggplant

Any formulation with an increased shelf life and a broad spectrum of actions, such as plant growth promotion and/or disease suppression under field conditions, could open the way for technological exploitation and marketing. Many reports suggest that commercial biocontrol agents are easy to deliver, induce plant growth and stress resistance and, eventually, increase plant biomass and yield. As very promising and rich sources of agro-active compounds and biocontrol tools, actinomycetes have gained increasing interest in several agricultural sectors [206,207]. In fact, in the last 30 years, about 60% of new insecticides and herbicides reported have originated from *Streptomyces* [206]; this is because three-quarters of all *Streptomyces* spp. are able to produce some class of antibiotics [208]. As a single example among many, we mention the production of polyoxin B and D by *Streptomyces cacaoi* var. *asoensis* as a new class of natural fungicides [209]. *Kasugamycin*, registered in several countries as a bactericidal and fungicidal agrochemical, was discovered in *Streptomyces kasugaensis* [210]. More recently, Siddique et al. [211] reported that avermectin B1b, a component of commercially available abamectin, was obtained as a fermentation product of *Streptomyces avermitilis*, which has frequently been used as an insecticidal agent. Very few *Streptomyces*-based commercial formulations are available in the worldwide market, compared with products based on *Streptomyces* metabolites; the former are mainly indicated for pest and disease control. Additionally, few products have been specifically commercialized for plant growth promotion, although significant research has been carried out on actinomycete production of growth-promoting substances [212]. One reason for this gap may reside in the difficulty of preparing a commercial product formulation with one or more streptomycetes as an active substance. Ideally, the industrial process used should not affect the bioproduct's plant growth-promoting activity and/or antimicrobial characteristics for 18–24 months.

In keeping with current quality and safety standards, ideal microbial biocontrol agents should be univocally identified as a taxonomical unit; be effective against target plant pathogens or pests; show no clinical or animal toxicity; should not persist in the agro-environment (included surface water), having a short growing period, a level-off, and a final lining to the background microorganisms; and should not transfer genetic material to other taxonomically related microorganisms. Therefore, the antagonistic potential, environmental fate, and behavioural features of a putative microbial biocontrol agent must be thoroughly addressed by the industry to allow its registration as a bio-pesticide and approval for use in plant protection. All this may hinder the transfer of an effective biocontrol agent from the research lab into a commercially available product. Indeed, the BCB Manual of Biocontrol Agents, 5th Edition, lists over 120 microorganisms with potential use in agriculture, 54 of which have been approved in the EU for use in plant protection; however, only a few are commercially available. A complex regulatory landscape must be navigated by applicants applying for authorization to release a biocontrol agent. This is particularly true if the biocontrol agent is not indigenous [213,214].

Environmental risks associated with the inoculation of streptomycetes in agricultural environments as organisms beneficial to plants are associated with the lack of available data concerning the use of genetically modified organisms and their impact on the natural microbial communities [215]. In addition, the release of not genetically modified microbials may pose a risk related to the possible horizontal transfer of entire or partial gene clusters; this might be particularly risky in the case of antibiotic resistance. This was initially reported by Egan et al. [216] and later confirmed by Egan et al. [217]. This risk might be minimized during the search and study of prospective microbial inoculants, which should be accurately tested and characterized prior to their registration for the lack of known antibiotic-resistant genes or gene clusters.

To ensure food security for an increasing worldwide human population, most agricultural systems presently depend on the use of chemical fertilizers and pesticides [218]. Industrially produced chemical fertilizers are rich in nitrogen, phosphorous, and potassium, the repeated use of which leads to pollution of the soil, air, and groundwater [219]. Given these known problems, beneficial agricultural microorganisms used as microbial inoculants will be an important focus in pursuing sustainable agriculture and the provision of safe food without depleting natural resources in the coming decades [220]. The application of these naturally occurring beneficial microorganisms to soil ecosystems improves the soil's physical–chemical properties, fitness and stability, and microbial development along with promoting plant growth promotion and crop yield [221].

The microbial agents with the greatest agricultural prospects are rhizobacteria (as plant growth promoters), nitrogen-fixing cyanobacteria, mycorrhizal fungi, bacterial antagonists to plant pathogens, different biotic and abiotic stress-tolerance endophytes, and biodegrading bacteria [222,223]. Current registration and authorization procedures for microbe-based products to be used in agriculture as "fertilizers" are much less demanding; several microbe-based products or microbial consortia are, therefore, commercially available to farmers worldwide. Among them, a few contain streptomycetes and other *Actinomycetales*. Examples include Micosat F® (CCS Aosta srl, Aosta, Italy), containing three different *Streptomyces* spp.; Forge SP® (Blacksmith Bioscience, Spring, TX, USA), containing *Streptomyces nigrescens*; and Mykorrhyza soluble 30G (Glückspilze, Innsbruck, Austria), containing *Streptomyces griseus* and *S. lydicus*.

6. Formulations and Inoculation Methods

For any agro-pharma industry, the major challenge for the success or failure of a commercial product developed from an experimentally efficient biocontrol agent is its "formulation." Formulations should include the novel microorganism(s) in a calibrated quantity as an active ingredient and a set of other inert ingredients (frequently not specified in detail by the manufacturer) to produce a commercial product suitable for use in field conditions. Commercial formulations should comply with local and national legislation on agrochemicals (specifically legislation on microbial inoculants, where applicable), as well as with growers' requirements: repeated positive results, reasonable pricing, and easy handling. With regard to these considerations, microbial inoculants have a major problem specific to microorganisms: loss of viability during storage complicates the need for a long shelf life and stability over a range of −5 to 30 °C, which are typical growers' storage conditions [224]. Commercial microbial formulations can be prepared and made available on the market in four types: powder, liquid suspension, granules, and slurry [225,226]. The physical formulation of *Streptomyces*-based products also indicates how these microbials should be inoculated in agricultural systems, prior or during cropping (Table 5). Different types of low-cost raw materials are used to prepare different types of commercial formulations: peat, perlite, charcoal, vermicompost, inorganic soil fractions, and many others [227].

Moreover, biocontrol agents may not show the same results in both in vitro and in vivo experiments; this is regarded as the crucial challenge in the industrial development of bio-inoculants. The efficacy of biocontrol agents, among them streptomycetes, is affected by soil organic matter, pH, nutrient levels, and moisture level. Variations in agri-environmental conditions tend to affect the results of biocontrol agents that perform well in vitro: an experimentally excellent biocontrol agent might fail in greenhouse or field experiments. Thus, environmental variables should always be taken into proper consideration when selecting an appropriate biocontrol agent for a precise location. Ideally, the most active and prospective microbes should be isolated from the same agricultural area [228].

Additionally, any physical formulation, method, or procedure of inoculation (for example, soil, seed, seedling, or vegetative part) should be thoroughly screened since each method may play an important role in obtaining satisfactory results during field experiments [228,229]. For instance, in inoculating soil with a biocontrol agent, microbes are mixed with soil or sowing furrows or are spread in the field by dripping systems [230]. Seed inoculation methods commonly involve soaking seeds in a suspension containing the selected biocontrol agent(s); alternatively, they are mixed with suitable wetting agents [228–232]. The inoculation of vegetative parts is done by spraying a suspension of the biocontrol agent to aerial parts of the plant or dipping the roots of seedlings into a microbial suspension prior to transplantation [230,232]. Each method of microbial application may contribute to achieving effective results in commercial agriculture (open field, nurseries, glasshouse production, etc.) [228,233]. (Table 5).

Table 5. Technical formulations of a set of *Streptomyces*-based products available on international markets, their indications of use, and their inoculation methods. The table has been prepared according to the information given on the labels of the respective products.

Technical Formulation	Commercial Example	Microbial Biocontrol Agent(s)	Indications	Inoculation Method
Granules (G)	Micosat F UNO, CCS Aosta Srl	*Streptomyces* sp. strain SB14	Transplants mortality, plant growth promoter	Soil application as dry granules
Wettable Granules (WG)	Micosat F MO, CCS Aosta Srl	*Streptomyces* spp. strains SA51; SB14 and SL81	Soil bioremediation in viticulture and vegetable production	Soil application as microbial suspension in water
Wettable Powder (WP)	Mykostop, Verdera Oy	*Streptomyces griseoviridis* strain K61	Damping-off fungi, *Phytophthora* spp.	Drip irrigation, cutting/bulb soaking.
Soluble Powder (SP)	Actinovate, Novozymes BioAg Inc.	*Streptomyces lydicus* WYEC 108	Soilborne fungi, powdery mildews, gray moulds	Soil drench, transplants root dipping, foliar sprays, bulb soaking.
Slurry (SE)	-	*Streptomyces* spp. consortium	Bioremediation of organic and inorganic soil pollutants	Soil application as diluted slurry

As described in previous chapters, most streptomycetes are soil inhabitants, commonly colonizing the rhizosphere and frequently showing the ability to enter plants, thus efficiently colonizing plant tissues as endophytes. Since pathogens necessarily require an endophytic state to initiate an infection, a successful endophytic biocontrol agent, such as a selected *Streptomyces* sp., should be able to rapidly move from the rhizosphere into the roots and/or other plant parts. Therefore, its antagonistic activity may be both preventive (competitive exclusion of plant pathogens in the rhizosphere and *in planta*) and curative (killing plant pathogens post-infection). From the industrial point of view, the selection and choice of prospective *Streptomyces* spp. as candidates for development and implementation of innovative and sustainable biopesticides should necessarily consider that goal.

Industrial exploitation of research results is not necessarily easy. Most papers published worldwide demonstrate the excellent biocontrol activity of streptomycetes during experiments in vitro or, when in planta, under strictly standardized conditions. Frequently, no field research results are presented to support the applicability of experimental data in commercial fields/greenhouses. Clearly this is a weak point, hindering the commercial development of successful biopesticides. Development is also hindered because the production of antimicrobial molecules by actinobacteria, and particularly by streptomycetes, is strictly dependent on the substrate where they grow [234] and on the natural microbial community around them.

7. Future Aspects and Challenges

Streptomyces spp. have great potential to become an essential constituent of modern agricultural practice as biofertilizers and biocontrol agents, with the capacity to dominate agrimarkets in coming decades. Actinomycetes, particularly abundant *Streptomyces* as filamentous spore-forming bacteria with superior biocontrol and nutrient-cycling activity, are among the most promising PGPR to increase overall soil health and boost agricultural productivity. Nevertheless, some unresolved problems need to be addressed in order to reproduce results from the controlled laboratory environment into large-scale field trials and commercial marketing. More focus is still needed to develop novel formulations that could increase the shelf life of streptomycetes, thus ensuring their long-term viability, their sporulation activity, and their efficacy as microbial-based agrochemicals. Additionally, the potential of streptomycetes to control post-harvest bacterial and fungal diseases of fruits and vegetables is totally unexplored. Further extensive studies on the complex *Streptomyces*–rhizosphere environment and the mechanisms of PGP action are needed. Shedding light on the symbiotic association of *Streptomyces* with other PGPR might lead to developing highly effective and efficient bioinoculants across different soil types and environmental conditions. The knowledge of various aspects such as interactions between rhizosphere PGPS and native microbiota and infection processes by endophytic PGPS are still not sufficiently explained, even though many reports have indicated that PGPS can promote plant growth by colonizing their host plants epiphytically and/or endophytically. Metagenomics and molecular biology studies, such as tagging green fluorescent protein (GFP) markers to microorganisms, will be necessary to understand the fate of PGPS microbial populations in plants, their endophytic distribution, and their pattern of colonization. Much more focus is still needed to design and implement industrial processes that are able to produce effective formulations with one or more microbial agents, using different additives, carriers, and with various methods of field inoculations.

8. Conclusions

Many studies have been conducted on actinomycetes, highlighting the ability of these microorganisms to promote plant growth and their additive/synergistic effects on plant growth and protection. As the above discussion makes clear, actinomycetes, and especially *Streptomyces* as helper bacteria, are truly prospective for use as plant coinoculants: this to improve plant–microbe symbiosis in a way that could lead to an increased sustainable production of agriculture products under diverse conditions. This promise is mainly based on the use of eco-friendly microorganisms that control pests and improve plant growth. The use of biofertilizers, biopesticides, or consortiums of plant beneficial microbes in correct formulations provides a potential solution for a more sustainable agricultural future. The studies mentioned in this review support the belief that designing new formulations with cooperative microbes might contribute to growth improvement and plant protection of several crops. However, these studies also highlight the importance of continuing research on this subject, especially focusing on actinomycetes, which up to now have been little used as inoculants to enhance agricultural production and ensuring food security, despite the excellent potential shown in a large number of scientific publications so far.

Acknowledgments: The authors are thankful to CCS Aosta Srl, Quart (AO) for financial support of a Ph.D. grant devoted to study the role and use of Actinobacteria in agricultural systems (symbiotic agriculture). The authors kindly acknowledge the EU COST Action FA1103: "Endophytes in Biotechnology and Agriculture" (www.endophytes.eu) for providing an international network of scientists devoted to increase knowledge and use of endophytes in plant production.

Author Contributions: Sai Shiva Krishna Prasad Vurukonda, Davide Giovanardi, Emilio Stefani contributed equally to the manuscript.

Conflicts of Interest: The authors declare no conflict of interest.

References

1. Smith, S.; Read, D. *Mycorrhizal Symbiosis*; Academic Press: London, UK, 1997; pp. 453–469.
2. Smith, S.E.; Smith, A.F.; Jakobsen, I. Mycorrhizal fungi can dominate phosphate supply to plants irrespective of growth responses. *Plant Physiol.* **2003**, *133*, 16–20. [CrossRef] [PubMed]
3. Brader, G.; Compant, S.; Mitter, B.; Trognitz, F.; Sessitsch, A. Metabolic potential of endophytic bacteria. *Curr. Opin. Biotechnol.* **2014**, *27*, 30–37. [CrossRef] [PubMed]
4. Okon, Y.; Labandera-González, C. Agronomic applications of *Azospirillum*: An evaluation of 20 years of worldwide field inoculation. *Soil Biol. Biochem.* **1994**, *26*, 1591–1601. [CrossRef]
5. Glick, B.R.; Patten, C.L.; Holguin, G.; Penrose, D.M. *Biochemical and Genetic Mechanisms Used by Plant Growth Promoting Bacteria*; Imperial College Press: London, UK, 1999; pp. 1–13.
6. Dias, M.P.; Bastos, M.S.; Xavier, V.B.; Cassel, E.; Astarita, L.V.; Santarém, E.R. Plant growth and resistance promoted by *Streptomyces* spp. in tomato. *Plant Physiol. Biochem.* **2017**, *118*, 479–493. [CrossRef] [PubMed]
7. Viaene, T.; Langendries, S.; Beirinckx, S.; Maes, M.; Goormachtig, S. *Streptomyces* as a plant's best friend? *FEMS Microbiol. Ecol.* **2016**, *92*, fiw119. [CrossRef] [PubMed]
8. Sturz, A.V.; Nowak, J. Endophytic communities of rhizobacteria and the strategies required to create yield enhancing associations with crops. *Appl. Soil Ecol.* **2000**, *15*, 183–190. [CrossRef]
9. Smith, S.E.; Read, D. *Mycorrhizal Symbiosis*, 3rd ed.; Academic Press: London, UK, 2008; pp. 1–9.
10. Bertram, R.; Schlicht, M.; Mahr, K.; Nothaft, H.; Saier, M.H., Jr.; Titgemeyer, F. In silico and transcriptional analysis of carbohydrate uptake systems of *Streptomyces coelicolor* A3(2). *J. Bacteriol.* **2004**, *186*, 1362–1373. [CrossRef] [PubMed]
11. Chater, K.F.; Biró, S.; Lee, K.J.; Palmer, T.; Schrempf, H. The complex extracellular biology of *Streptomyces*. *FEMS Microbiol. Rev.* **2010**, *34*, 171–198. [CrossRef] [PubMed]
12. Thompson, B.J.; Widdick, D.A.; Hicks, M.G.; Chandra, G.; Sutcliffe, I.C.; Palmer, T.; Hutchings, M.I. Investigating lipoprotein biogenesis and function in the model Gram-positive bacterium *Streptomyces coelicolor*. *Mol. Microbiol.* **2010**, *77*, 943–956. [CrossRef] [PubMed]
13. Seipke, R.F.; Kaltenpoth, M.; Hutchings, M.I. *Streptomyces* as symbionts: An emerging and widespread theme? *FEMS Microbiol. Rev.* **2012**, *36*, 862–876. [CrossRef] [PubMed]
14. Pangesti, N.; Pineda, A.; Pieterse, C.; Dicke, M.; Van Loon, J.J.A. Two-way plant mediated interactions between root-associated microbes and insects: From ecology to mechanisms. *Front. Plant Sci.* **2013**, *4*, 414. [CrossRef] [PubMed]
15. Kupferschmied, P.; Maurhofer, M.; Keel, C. Promise for plant pest control: Root-associated pseudomonads with insecticidal activities. *Front. Plant Sci.* **2013**, *4*, 287. [CrossRef] [PubMed]
16. Baker, R. Diversity in biological control. *Crop Prot.* **1991**, *10*, 85–94. [CrossRef]
17. Benhamou, N.; Kloepper, J.W.; Quadt-Hallman, A.; Tuzun, S. Induction of defense-related ultrastructural modifications in pea root tissues inoculated with endophytic bacteria. *Plant Physiol.* **1996**, *112*, 919–929. [CrossRef] [PubMed]
18. Varma, A.; Verma, S.; Sahay, N.; Bütehorn, B.; Franken, P. *Piriformospora indica*, a cultivable plant-growth-promoting root endophyte. *Appl. Environ. Microbiol.* **1999**, *65*, 2741–2744. [PubMed]
19. Shimizu, M.; Yazawa, S.; Ushijima, Y.A. Promising strain of endophytic *Streptomyces* sp. for biological control of cucumber anthracnose. *J. Gen. Plant. Pathol.* **2009**, *75*, 27–36. [CrossRef]

20. Sardi, P.; Saracchi, M.; Quaroni, S.; Petrolini, B.; Borgonovi, G.E.; Merli, S. Isolation of endophytic *Streptomyces* strains from surface-sterilized roots. *Appl. Environ. Microbiol.* **1992**, *58*, 2691–2693. [PubMed]

21. Shimizu, M.; Nakagawa, Y.; Sato, Y.; Furumai, T.; Igarashi, Y.; Onaka, H.; Yoshida, R.; Kunoh, H. Studies on endophytic actinomycetes (I) *Streptomyces* sp. isolated from rhododendron and its antifungal activity. *J. Gen. Plant Pathol.* **2000**, *66*, 360–366. [CrossRef]

22. Nishimura, T.; Meguro, A.; Hasegawa, S.; Nakagawa, Y.; Shimizu, M.; Kunoh, H. An endophytic actinomycetes, *Streptomyces* sp. AOK-30, isolated from mountain laurel and its antifungal activity. *J. Gen. Plant Pathol.* **2002**, *68*, 390–397. [CrossRef]

23. Castillo, U.; Harper, J.K.; Strobel, G.A.; Sears, J.; Alesi, K.; Ford, E.; Lin, J.; Hunter, M.; Maranta, M.; Ge, H.; et al. Kakadumycins, novel antibiotics from *Streptomyces* sp. NRRL 30566, an endophyte of *Grevillea pteridifolia*. *FEMS Microbiol. Lett.* **2003**, *224*, 183–190. [CrossRef]

24. Coombs, J.T.; Franco, C.M.M. Isolation and identification of actinobacteria from surface-sterilized wheat roots. *Appl. Environ. Microbiol.* **2003**, *69*, 5603–5608. [CrossRef] [PubMed]

25. Tian, X.L.; Cao, L.X.; Tan, H.M.; Zeng, Q.G.; Jia, Y.Y.; Han, W.Q.; Zhou, S.N. Study on the communities of endophytic fungi and endophytic actinomycetes from rice and their antipathogenic activities in vitro. *World J. Microbiol. Biotechnol.* **2004**, *20*, 303–309. [CrossRef]

26. Kunoh, H. Endophytic actinomycetes: Attractive biocontrol agents. *J. Gen. Plant Pathol.* **2002**, *68*, 249–252. [CrossRef]

27. Coombs, J.T.; Michelsen, P.P.; Franco, C.M.M. Evaluation of endophytic actinobacteria as antagonists of *Gaeumannomyces graminis* var. *tritici* in wheat. *Biol. Control.* **2004**, *29*, 359–366. [CrossRef]

28. Meguro, A.; Ohmura, Y.; Hasegawa, S.; Shimizu, M.; Nishimura, T.; Kunoh, H. An endophytic actinomycete, *Streptomyces* sp. MBR-52, that accelerates emergence and elongation of plant adventitious roots. *Actinomycetologica* **2006**, *20*, 1–9. [CrossRef]

29. Rothrock, C.S.; Gottlieb, D. Role of antibiosis in antagonism of *Streptomyces hygroscopicus* var. *geldanus* to *Rhizoctonia solani* in soil. *Can. J. Microbiol.* **1984**, *30*, 1440–1447. [CrossRef]

30. Xiao, K.; Kinkel, L.L.; Samac, D.A. Biological control of *Phytophthora* root rots on *alfalfa* and soybean with *Streptomyces*. *Biol. Control.* **2002**, *23*, 285–295. [CrossRef]

31. El-Tarabily, K.A.; Sivasithamparam, K. Non-streptomycete actinomycetes as biocontrol agents of soil-borne fungal plant pathogens and as plant growth promoters. *Soil Biol. Biochem.* **2006**, *38*, 1505–1520. [CrossRef]

32. Kämpfer, P. The family Streptomycetaceae, part I: Taxonomy. In *The Prokaryotes: A Handbook on the Biology of Bacteria*; Dworkin, M., Falkow, S., Rosenberg, E., Schleifer, K.H., Stackebrandt, E., Eds.; Springer: New York, NY, USA, 2006; pp. 538–604.

33. Chater, K.F. Morphological physiological differentiation in *Streptomyces*. In *Microbial Development*; Losick, R., Shapiro, L., Eds.; Cold Spring Harbor Laboratory: Cold Spring Harbor, NY, USA, 1984; pp. 89–115.

34. Labeda, D.P. Multilocus sequence analysis of phytopathogenic species of the genus *Streptomyces*. *Int. J. Syst. Evol. Microbiol.* **2011**, *61*, 2525–2531. [CrossRef] [PubMed]

35. Kumar, V.; Kumar, A.; Pandey, K.D.; Roy, B.K. Isolation and characterization of bacterial endophytes from the roots of *Cassia tora* L. *Ann. Microbiol.* **2014**, *65*, 1391–1399. [CrossRef]

36. Marella, S. Bacterial endophytes in sustainable crop production: Applications, recent developments and challenges ahead. *Int. J. Life Sci. Res.* **2014**, *2*, 46–56.

37. Strobel, G.A.; Hess, W.M.; Ford, E.; Sidhu, R.S.W.; Yang, X. Taxol from fungal endophytes and the issue of biodiversity. *J. Ind. Microbiol.* **1996**, *17*, 417–423. [CrossRef]

38. Strobel, G.A.; Long, D.M. Endophytic microbes embody pharmaceutical potential. *Am. Soc. Microbiol. News* **1998**, *64*, 263–268.

39. Watve, M.; Tickoo, R.; Jog, M.; Bhole, B. How many antibiotics are produced by the genus *Streptomyces*? *Arch. Microbiol.* **2001**, *176*, 386–390. [CrossRef] [PubMed]

40. Schrey, S.D.; Tarkka, M.T. Friends and foes: Streptomycetes as modulators of plant disease and symbiosis. *Antonie Leeuwenhoek* **2008**, *94*, 11–19. [CrossRef] [PubMed]

41. Hibbing, M.; Fuqua, C.; Parsek, M.R.; Peterson, S.B. Bacterial competition: Surviving and thriving in the microbial jungle. *Nat. Rev. Microbiol.* **2010**, *8*, 15–25. [CrossRef] [PubMed]

42. Cornforth, D.M.; Foster, K.R. Competition sensing: The social side of bacterial stress responses. *Nat. Rev. Microbiol.* **2013**, *11*, 285–293. [CrossRef] [PubMed]

43. Citron, C.A.; Barra, L.; Wink, J.; Dickschat, J.S. Volatiles from nineteen recently genome sequenced actinomycetes. *Org. Biomol. Chem.* **2015**, *13*, 2673–2783. [CrossRef] [PubMed]

44. Gerber, N.N.; Lechevalier, H.A. Geosmin, an earthly-smelling substance isolated from actinomycetes. *Appl. Microbiol.* **1965**, *13*, 935–938. [PubMed]

45. Polak, E.H.; Provasi, J. Odor sensitivity to geosmin enantiomers. *Chem. Sens.* **1992**, *17*, 23–26. [CrossRef]

46. Rosenzweig, N. The importance and application of bacterial diversity in sustainable agricultural crop production systems. In *Bacterial Diversity in Sustainable Agriculture*; Maheshwari, D.K., Ed.; Springer International Publishing: Cham, Switzerland, 2014; pp. 341–366.

47. Hopwood, D. *Streptomyces in Nature and Medicine: The Antibiotic Makers*; Oxford University Press: Oxford, UK, 2007; p. 260.

48. Audrain, B.; Farag, M.A.; Choong-Min, R.; Jean-Marc, G. Role of bacterial volatile compounds in bacterial biology. *FEMS Microbiol. Rev.* **2015**, *39*, 222–233. [CrossRef] [PubMed]

49. Ngoc, D.P.; Eun-Hee, L.; Seon-Ha, C.; Yongdeok, C.; Hyejin, S.; Ahjeong, S. Bacterial Community Structure Shifted by Geosmin in Granular Activated Carbon System of Water Treatment Plants. *J. Microbiol. Biotechnol.* **2016**, *26*, 99–109.

50. Isaac, S. *Fungal-Plant Interactions*; Chapman & Hall: London, UK, 1992; p. 418, ISBN 0412353903.

51. Redlin, S.C.; Carris, L.M. Endophytic Fungi in Grasses and Woody Plants: Systemics, ecology and evolution. In *Isolation and Analysis of Endophytic Fungal Communities from Woody Plants*; Bills, G.F., Ed.; American Phytopathological Society Press: St Paul, MN, USA, 1996; pp. 31–65.

52. Baldani, J.I.; Olivares, F.L.; Hemerly, A.S.; Reis, F.B., Jr.; Oliveira, A.L.M.; Baldani, V.D.L.; Goi, S.R.; Reis, V.M.; Dobereiner, J. Nitrogen-fixing endophytes: Recent advances in the association with graminaceous plants grown in the tropics. In *Biological Nitrogen Fixation for the 21st Century*; Elmerich, C., Konderosi, A., Newton, W.E., Eds.; Springer: Dordrecht, The Netherlands, 1997; pp. 203–296.

53. Azevedo, J.L. Microrganismos endofíticos. In *Ecologia Microbiana*; Meloj, I.S., Azevedo, L., Eds.; Editora Embrapa: Jaguariuna, Brasil, 1998; pp. 117–137.

54. Saleem, M.; Law, A.D.; Moe, L.A. Nicotiana roots recruit rare rhizosphere taxa as major root-inhabiting microbes. *Microb. Ecol.* **2016**, *71*, 469–472. [CrossRef] [PubMed]

55. Figueiredo, M.V.B.; Seldin, L.; de Araujo, F.F.; Mariano, R.L.R. Plant Growth Promoting Rhizobacteria: Fundamentals and Applications. In *Plant Growth and Health Promoting Bacteria*; Maheshwari, D., Ed.; Microbiology Monographs; Springer: Berlin/Heidelberg, Germany, 2010; pp. 2–43.

56. Sousa, J.A.A.; Olivares, F.L. Plant growth promotion by streptomycetes: Ecophysiology, mechanisms and applications. *Chem. Biol. Technol. Agric.* **2016**, *3*, 24. [CrossRef]

57. Vardharajula, S.; Shaik Zulfikar, A.; Vurukonda, S.S.K.P.; Shrivastava, M. Plant growth promoting endophytes and their interaction with plants to alleviate abiotic stress. *Curr. Biotechnol.* **2017**, *6*, 252–263. [CrossRef]

58. Benson, D.R.; Silvester, W.B. Biology of *Frankia* strains, actinomycete symbionts of actinorhizal plants. *Microbiol. Rev.* **1993**, *28*, 293–319.

59. Bradbury, J.F. *Guide to Plant Pathogenic Bacteria*; CAB International Mycological Institute Publishers: London, UK, 1986; p. 332, ISBN 0851985572.

60. Mundt, J.O.; Hinckle, N.F. Bacteria within ovules and seeds. *Appl. Environ. Microbiol.* **1976**, *32*, 694–698. [PubMed]

61. Sathya, A.; Vijayabharathi, R.; Gopalakrishnan, S. Plant growth-promoting actinobacteria: A new strategy for enhancing sustainable production and protection of grain legumes. *3 Biotech* **2017**, *7*, 102. [CrossRef] [PubMed]

62. Bulgarelli, D.; Schlaeppi, K.; Spaepen, S.; Ver, E.; Van Themaat, L.; Schulze-Lefert, P. Structure and functions of the bacterial microbiota of plants. *Annu. Rev. Plant Biol.* **2013**, *64*, 807–838. [CrossRef] [PubMed]

63. Massalha, H.; Korenblum, E.; Tholl, D.; Aharoni, A. Small molecules below-ground: The role of specialized metabolites in the rhizosphere. *Plant J.* **2017**, *90*, 788–807. [CrossRef] [PubMed]

64. Bais, H.P.; Weir, T.L.; Perry, L.G.; Gilroy, S.; Vivanco, J.M. The role of root exudates in rhizosphere interactions with plants and other organisms. *Ann. Rev. Plant Biol.* **2006**, *57*, 233–266. [CrossRef] [PubMed]

65. Jog, R.; Nareshkumar, G.; Rajkumar, S. Enhancing soil health and plant growth promotion by actinomycetes. In *Plant Growth Promoting Actinobacteria*; Gopalakrishnan, S., Sathya, A., Vijayabharathi, R., Eds.; Springer: Singapore, 2016; pp. 33–45.

66. Meschke, H.; Schrempf, H. *Streptomyces lividans* inhibits the proliferation of the fungus *Verticillium dahliae* on seeds and roots of *Arabidopsis thaliana*. *Microb. Biotechnol.* **2010**, *3*, 428–443. [CrossRef] [PubMed]

67. Palaniyandi, S.A.; Yang, S.H.; Damodharan, K.; Suh, J.W. Genetic and functional characterization of culturable plant-beneficial actinobacteria associated with yam rhizosphere. *J. Basic Microbiol.* **2013**, *53*, 985–995. [CrossRef] [PubMed]

68. Rungin, S.; Indananda, C.; Suttiviriya, P.; Kruasuwan, W.; Jaemsaeng, R.; Thamchaipenet, A. Plant growth enhancing effects by a siderophore producing endophytic streptomycete isolated from a Thai jasmine rice plant (*Oryza sativa* L. cv. KDML105). *Antonie Leeuwenhoek* **2012**, *102*, 463–472. [CrossRef] [PubMed]

69. Qin, S.; Xing, K.; Jiang, J.H.; Xu, L.H.; Li, W.J. Biodiversity, bioactive natural products and biotechnological potential of plant-associated endophytic actinobacteria. *Appl. Microbiol. Biotechnol.* **2011**, *89*, 457–473. [CrossRef] [PubMed]

70. Coombs, J.T.; Franco, C.M. Visualization of an endophytic *Streptomyces* species in wheat seed. *Appl. Environ. Microbiol.* **2003**, *69*, 4260–4262. [CrossRef] [PubMed]

71. Shimizu, M. Endophytic actinomycetes: Biocontrol agents and growth promoters. In *Bacteria in Agrobiology: Plant Growth Responses*; Maheshwari, D.K., Ed.; Elsevier Academic Press: San Diego, CA, USA, 2011; pp. 201–220.

72. Franco-Correa, M.; Quintana, A.; Duque, C.; Suarez, C.; Rodriguez, M.X.; Barea, J.M. Evaluation of actinomycete strains for key traits related with plant growth-promotion and mycorrhiza helping activities. *Appl. Soil Ecol.* **2010**, *45*, 209–217. [CrossRef]

73. Gopalakrishnan, S.; Vadlamudi, S.; Bandikinda, P.; Sathya, A.; Vijayabharathi, R.; Rupela, O.; Kudapa, H.; Katta, K.; Varshney, R.K. Evaluation of *Streptomyces* strains isolated from herbal vermicompost for their plant growth-promotion traits in rice. *Microbiol. Res.* **2014**, *169*, 40–48. [CrossRef] [PubMed]

74. Gopalakrishnan, S.; Srinivas, V.; Alekhya, G.; Prakash, B.; Kudapa, H.; Varshney, R.K. Evaluation of *Streptomyces* sp. obtained from herbal vermicompost for broad spectrum of plant growth-promoting activities in chickpea. *Org. Agric.* **2015**, *5*, 123–133. [CrossRef]

75. Tokala, R.; Strap, J.; Jung, C.M.; Crawford, D.L.; Salove, M.H.; Deobald, L.A.; Bailey, J.F.; Morra, M.J. Novel plant microbe rhizosphere interaction involving *Streptomyces lydicus* WYEC 108 and the pea plant (*Pisum sativum*). *Appl. Environ. Microbiol.* **2002**, *68*, 2161–2171. [CrossRef] [PubMed]

76. Nimnoi, P.; Pongsilp, N.; Lumyong, S. Co-inoculation of soybean (*Glycine max*) with actinomycetes and *Bradyrhizobium japonicum* enhances plant growth, nitrogenase activity and plant nutrition. *J. Plant Nutr.* **2014**, *37*, 432–446. [CrossRef]

77. Abd-Alla, M.H.; El-Sayed, E.A.; Rasmey, A.M. Indole-3-acetic acid (IAA) production by *Streptomyces atrovirens* isolated from rhizospheric soil in Egypt. *J. Biol. Earth Sci.* **2013**, *3*, 182–193.

78. El-Tarabily, K.A. Promotion of tomato (*Lycopersicon esculentum* Mill.) plant growth by rhizosphere competent 1-aminocyclopropane-1-carboxylic acid deaminase-producing streptomycete actinomycetes. *Plant Soil* **2008**, *308*, 161–174. [CrossRef]

79. Khamna, S.; Yokota, A.; Peberdy, J.F.; Lumyong, S. Indole-3-acetic acid production by *Streptomyces* sp. isolated from some Thai medicinal plant rhizosphere soils. *Eur. Asia J. Biosci.* **2010**, *4*, 23–32. [CrossRef]

80. Verma, V.C.; Singh, S.K.; Prakash, S. Bio-control and plant growth-promotion potential of siderophore producing endophytic *Streptomyces* from *Azadirachta indica* A. Juss. *J. Basic Microbiol.* **2011**, *51*, 550–556. [CrossRef] [PubMed]

81. Lin, L.; Xu, X. Indole-3-acetic acid production by endophytic *Streptomyces* sp. En-1 isolated from medicinal plants. *Curr. Microbiol.* **2013**, *67*, 209–217. [CrossRef] [PubMed]

82. Tsavkelova, E.A.; Klimova, S.Y.; Cherdyntseva, T.A.; Netrusov, A.I. Microbial producers of plant growth-stimulators and their practical use: A review. *Appl. Biochem. Microbiol.* **2006**, *42*, 117–126. [CrossRef]

83. Nascimento, F.X.; Rossi, M.J.; Soares, C.R.; McConkey, B.J.; Glick, B.R. New insights into 1-aminocyclopropane-1-carboxylate (ACC) deaminase phylogeny, evolution and ecological significance. *PLoS ONE* **2014**, *6*, e99168. [CrossRef] [PubMed]

84. Goudjal, Y.; Zamoum, M.; Meklat, A.; Sabaou, N.; Mathieu, F.; Zitouni, A. Plant growth-promoting potential of endosymbiotic actinobacteria isolated from sand truffles (*Terfezia leonis* Tul.) of the Algerian Sahara. *Ann. Microbiol.* **2015**, *66*, 91–100. [CrossRef]

85. Rashad, F.M.; Fathya, H.M.; El-Zayata, A.S.; Elghonaimy, A.M. Isolation and characterization of multifunctional *Streptomyces* species with antimicrobial, nematicidal and phytohormone activities from marine environments in Egypt. *Microbiol. Res.* **2015**, *175*, 34–47. [CrossRef] [PubMed]

86. Harikrishnan, H.; Shanmugaiah, V.; Balasubramanian, N. Optimization for production of indole acetic acid (IAA) by plant growth-promoting *Streptomyces* sp. VSMGT1014 isolated from rice rhizosphere. *Int. J. Curr. Microbiol. Appl. Sci.* **2014**, *3*, 158–171.

87. Harikrishnan, H.; Shanmugaiah, V.; Balasubramanian, N.; Sharma, M.P.; Kotchoni, S.O. Antagonistic potential of native strain *Streptomyces aurantiogriseus* VSMGT1014 against Sheath Blight of rice disease. *World J. Microbiol. Biotechnol.* **2014**, *30*, 3149–3161. [CrossRef] [PubMed]

88. Rafik, E.; Rahal, E.; Ahmed, L. Isolation and screening of actinomycetes strains producing substances plant growth-promoting. *Indo-Am. J. Agric. Vet. Sci.* **2014**, *2*, 1–12.

89. Gopalakrishnan, S.; Vadlamudi, S.; Apparla, S.; Bandikinda, P.; Vijayabharathi, R.; Bhimineni, R.K.; Rupela, O. Evaluation of *Streptomyces* spp. for their plant growth-promotion traits in rice. *Can. J. Microbiol.* **2013**, *59*, 534–539. [CrossRef] [PubMed]

90. Lee, J.; Postmaster, A.; Soon, H.P.; Keast, D.; Carson, K.C. Siderophore production by Actinomycetes isolates from two soil sites in Western Australia. *Biometals* **2012**, *25*, 285–296. [CrossRef] [PubMed]

91. Jog, R.; Nareshkumar, G.; Rajkumar, S. Plant growth promoting potential and soil enzyme production of the most abundant *Streptomyces* spp. from wheat rhizosphere. *J. Appl. Microbiol.* **2012**, *113*, 1154–1164. [CrossRef] [PubMed]

92. Ghodhbane-Gtari, F.; Essoussi, I.; Chattaoui, M.; Chouaia, B.; Jaouani, A.; Daffonchio, D.; Boudabous, A.; Gtari, M. Isolation and characterization of non-*Frankia* actinobacteria from root nodules of *Alnus glutinosa*, *Casuarina glauca* and *Elaeagnus angustifolia*. *Symbiosis* **2010**, *50*, 51–57. [CrossRef]

93. Aldesuquy, H.S.; Mansour, F.A.; Abo-Hamed, S.A. Effect of the culture filtrates of *Streptomyces* on growth and productivity of wheat plants. *Folia Microbiol.* **1998**, *43*, 465–470. [CrossRef]

94. Igarashi, Y.; Iida, T.; Yoshida, R.; Furumai, T. Pteridic acids A and B, novel plant growth-promoters with auxin-like activity from *Streptomyces hygroscopicus* TP-A0451. *J. Antibiot.* **2002**, *55*, 764–767. [CrossRef] [PubMed]

95. El-Tarabily, K.A.; Nassar, A.H.; Hardy, G.E.S.J.; Sivasithamparam, K. Plant growth-promotion and biological control of *Pythium aphanidermatum*, a pathogen of cucumber, by endophytic actinomycetes. *J. Appl. Microbiol.* **2009**, *106*, 13–26. [CrossRef] [PubMed]

96. Gopalakrishnan, S.; Srinivas, V.; Vidya, M.S.; Rathore, A. Plant growth-promoting activities of *Streptomyces* spp. in sorghum and rice. *SpringerPlus* **2013**, *2*, 574. [CrossRef] [PubMed]

97. Dochhil, H.; Dkhar, M.S.; Barman, D. Seed germination enhancing activity of endophytic Streptomyces isolated from indigenous ethno-medicinal plant Centella asiatica. *Int. J. Pharm. Biol. Sci.* **2013**, *4*, 256–262.

98. El-Tarabily, K.A.; Hardy, G.E.S.J.; Sivasithamparam, K. Performance of three endophytic actinomycetes in relation to plant growth promotion and biological control of *Pythium aphanidermatum*, a pathogen of cucumber under commercial field production conditions in the United Arab Emirates. *Eur. J. Plant Pathol.* **2010**, *128*, 527–539. [CrossRef]

99. Golinska, P.; Magdalena, W.; Gauravi, A.; Dnyaneshwar, R.; Hanna, D.; Mahendra, R. Endophytic actinobacteria of medicinal plants: Diversity and bioactivity. *Antonie Leeuwenhoek* **2015**, *108*, 267–289. [CrossRef] [PubMed]

100. Goodfellow, M.; Simpson, K.E. Ecology of streptomycetes. *Front. Appl. Microbiol.* **1987**, *2*, 97–125.

101. Suzuki, S.; Yamamoto, K.; Okuda, T.; Nishio, M.; Nakanishi, N.; Komatsubara, S. Selective isolation and distribution of *Actinomadura rugatobispora* strains in soil. *Actinomycetologica* **2000**, *14*, 27–33. [CrossRef]

102. Alam, M.; Dahrni, S.; Khaliq, A.; Srivastava, S.K.; Samad, A.; Gupta, M.K. A promising strain of *Streptomyces* sp. with agricultural traits for growth-promotion and disease management. *Indian J. Exp. Microbiol.* **2012**, *50*, 559–568.

103. Inbar, E.; Green, S.J.; Hadar, Y.; Minz, D. Competing factors of compost concentration and proximity to root affect the distribution of Streptomycetes. *Microb. Ecol.* **2005**, *50*, 73–81. [CrossRef] [PubMed]

104. Manulis, S.; Epstein, E.; Shafrir, H.; Lichter, A.; Barash, I. Biosynthesis of indole-3-acetic acid via the indole-3-acetamide pathway in *Streptomyces* spp. *Microbiology* **1994**, *140*, 1045–1050. [CrossRef] [PubMed]

105. Reddy, K.R.K.; Jyothi, G.; Sowjanya, Ch.; Kusumanjali, K.; Malathi, N.; Reddy, K.R.N. Plant Growth-Promoting Actinomycetes: Mass Production, Delivery systems, and commercialization. In *Plant Growth Promoting Actinobacteria*; Subramaniam, G., Ed.; Springer: Singapore, 2016; pp. 287–298, ISBN 978-981-10-0705-7.

106. El-Sayed, M.A.; Valadon, L.R.G.; El-Shanshoury, A. Biosynthesis and metabolism of indole-3-acetic acid in Streptomyces mutabilis and Streptomyces atroolivaceus. *Microbiol. Lett.* **1987**, *36*, 85–95.

107. El-Shanshoury, A.R. Biosynthesis of indole-3-acetic acid in Streptomyces atroolivaceus and its changes during spore germination and mycelial growth. *Microbiol. Lett.* **1991**, *67*, 159–164.

108. Sellstedt, A.; Richau, K.H. Aspects of nitrogen-fixing Actinobacteria, in particular free-living and symbiotic *Frankia*. *FEMS Microbiol. Lett.* **2013**, *342*, 179–186. [CrossRef] [PubMed]

109. Fiedler, H.; Krastel, P.; Müller, J.; Gebhardt, K.; Zeeck, A. Enterobactin: The characteristic catecholate siderophore of Enterobacteriaceae is produced by *Streptomyces* species. *FEMS Microbiol. Lett.* **2001**, *196*, 147–151. [CrossRef] [PubMed]

110. Van Driesche, R.G.; Bellows, T.S., Jr. *Biological Control*; Chapman & Hall: New York, NY, USA, 1996; p. 539.

111. Procópio, R.E.; Silva, I.R.; Martins, M.K.; Azevedo, J.L.; Araújo, J.M. Antibiotics produced by *Streptomyces*. *Braz. J. Infect. Dis.* **2012**, *16*, 466–471. [CrossRef] [PubMed]

112. Whipps, J.M. Microbial interactions and biocontrol in the rhizosphere. *J. Exp. Bot.* **2001**, *52*, 487–511. [CrossRef] [PubMed]

113. Défago, G. 2,4-Diacetylphloroglucinol, a promising compound in biocontrol. *Plant Pathol.* **1993**, *42*, 311–312. [CrossRef]

114. De Souza, J.T.; de Boer, M.; de Waard, P.; van Beek, T.A.; Raaijmakers, J.M. Biochemical, genetic, and zoosporicidal properties of cyclic lipopeptide surfactants produced by *Pseudomonas fluorescens*. *Appl. Environ. Microbiol.* **2003**, *69*, 7161–7172. [CrossRef] [PubMed]

115. Kim, B.S.; Moon, S.S.; Hwang, B.K. Isolation, identification and antifungal activity of a macrolide antibiotic, oligomycin A, produced by *Streptomyces libani*. *Can. J. Bot.* **1999**, *77*, 850–858.

116. Compant, S.; Duffy, B.; Nowak, J.; Clément, C.; Barka, E.A. Use of plant growth-promoting bacteria for biocontrol of plant diseases: Principles, mechanisms of action and future prospects. *Appl. Environ. Microbiol.* **2005**, *71*, 4951–4959. [CrossRef] [PubMed]

117. Pridham, T.G.; Lindenfesler, L.A.; Shotwell, O.L.; Stodola, F.H.; Bendict, R.G.; Foley, C.; Jackson, R.W.; Zaumeyr, J.W.; Preston, W.H., Jr.; Mitchell, J.W. Antibiotics against plant disease, I. Laboratory and green house survey. *Phytopathology* **1956**, *46*, 568–575.

118. Pridham, T.G.; Shotwell, O.L.; Stodola, F.H.; Lindenfesler, L.S.; Bendict, R.G.; Jackson, R.W. Antibiotics against plant disease. II. Effective agents produced by *Streptomyces cinnamomeous forma azacoluta* F. Nov. *Phytopathol.* **1956**, *46*, 575–581.

119. Reddy, G.S.; Rao, A.S. Antagonism of soil actinomycetes to some soil borne plant pathogenic fungi. *Indian Phytopathol.* **1971**, *24*, 649–657.

120. Rothrock, C.S.; Gottlieb, D. Importance of antibiotic production in antagonism of selected *Streptomyces* species to two soil-borne plant pathogens. *J. Antibiot.* **1981**, *34*, 830–835. [CrossRef] [PubMed]

121. Papavizas, G.C.; Sutherland, E.D. Evaluation of oospore hyper-Parasites for the control of *Phytophthora* crown rot of pepper. *J. Phytopathol.* **1991**, *131*, 33–39.

122. El–Tarabily, K.A.; Soliman, M.H.; Nassar, A.H.; Al–Hassani, H.A.; Sivasithamparam, K.; McKenna, F.; Hardy, G.E.S.J. Biocontrol of *Sclerotinia* minor using a chitinolytic bacterium and actinomycetes. *Plant Pathol.* **2000**, *49*, 573–583. [CrossRef]

123. Getha, K.; Vikinesward, S. Antagonistic effects of *Streptomyces violaceusniger* strain G10 on *Fusarium oxysporum* f. sp. *cubense* race 4: Indirect evidence for the role of antibiosis in the antagonistic process. *J. Ind. Microbiol. Biotechnol.* **2002**, *28*, 303–310. [CrossRef] [PubMed]

124. Rhee, K.H. Purification and identification of an antifungal agent from *Streptomyces* sp. KH-614 antagonistic to rice blast fungus, *Pyricularia oryzae*. *J. Microbiol. Biotechnol.* **2003**, *13*, 984–988.

125. Woo, J.H.; Kamei, Y. Antifungal mechanism of an anti–pythium protein (SAP) from the marine bacterium *Streptomyces* sp. strain AP77 is specific for *P. porphyrae*, a causative agent of red rot disease in *Porhyra* spp. *Appl. Microbiol. Biotechnol.* **2003**, *62*, 407–413. [CrossRef] [PubMed]

126. Cao, L.; Qiu, Z.; You, J.; Tan, H.; Zhou, S. Isolation and characterization of endophytic *Streptomyces* strains from surface–sterilized tomato (*Lycopersicon esculentum*) roots. *Lett. Appl. Microbiol.* **2004**, *39*, 425–430. [CrossRef] [PubMed]

127. Joo, G.J. Production of an anti-fungal substance for biological control of *Phytophthora capsici* causing phytophthora blight in red–peppers by *Streptomyces halstedii*. *Biotechnol. Lett.* **2005**, *27*, 201–205. [CrossRef] [PubMed]

128. Liang, J.F.; Xue, Q.H.; Niu, X.L.; Li, Z.B. Root colonization and effects of seven strains of actinomycetes on leaf PAL and PPO activities of capsicum. *Acta Bot. Boreal-Occident Sin.* **2005**, *25*, 2118–2123.

129. Shekhar, N.; Bhattacharya, D.; Kumar, D.; Gupta, K.R. Biocontrol of wood-rotting fungi with *Streptomyces violaceusniger* XL-2. *Can. J. Microbiol.* **2006**, *52*, 805–808. [CrossRef] [PubMed]

130. Heng, J.L.S.; Md Shah, U.K.; Abdul Rahman, N.A.; Shaari, K.; Halizah, H. *Streptomyces ambofaciens* S2—A potential biological control agent for *Colletotrichum gleosporioides* the causal agent for anthracnose in red chilli fruits. *J. Plant Pathol. Microbiol.* **2006**, *1*, 2. [CrossRef]

131. Errakhi, R.; Bouteau, F.; Lebrihi, A.; Barakate, M. Evidences of biological control capacities of *Streptomyces* spp. against *Sclerotium rolfsii* responsible for damping-off disease in sugar beet (*Beta vulgaris* L.). *World J. Microbiol. Biotechnol.* **2007**, *23*, 1503–1509. [CrossRef]

132. Prapagdee, B.; Kuekulvong, C.; Mongkolsuk, S. Antifungal potential of extracellular metabolites produced by *Streptomyces hygroscopicus* against phytopathogenic fungi. *Int. J. Biol. Sci.* **2008**, *4*, 330–337. [CrossRef] [PubMed]

133. Baniasadi, F.; Bonjar, G.H.S.; Baghizadeh, A.; Nick, A.K.; Jorjandi, M.; Aghighi, S.; Farokhi, P.R. Biological control of *Sclerotinia sclerotiorum*, causal agent of sunflower head and stem rot disease, by use of soil borne actinomycetes isolates. *Am. J. Agric. Biol. Sci.* **2009**, *4*, 146–151. [CrossRef]

134. Sangmanee, P.; Bhromsiri, A.; Akarapisan, A. The potential of endophytic actinomycetes, (*Streptomyces* sp.) for the biocontrol of powdery mildew disease in sweet pea (*Pisum sativum*). *Asian J. Food Agric. Ind.* **2009**, *93*, e8.

135. Ningthoujam, D.S.; Sanasam, S.; Tamreihao, K.; Nimaichand, S. Antagonistic activities of local actinomycete isolates against rice fungal pathogens. *Afr. J. Microbiol. Res.* **2009**, *3*, 737–742.

136. Maldonado, M.C.; Orosco, C.E.; Gordillo, M.A.; Navarro, A.R. In vivo and in vitro antagonism of *Streptomyces* sp. RO3 against *Penicillium digitatum* and *Geotrichum candidum*. *Afr. J. Microbiol. Res.* **2010**, *4*, 2451–2456.

137. Al–Askar, A.A.; Abdul Khair, W.M.; Rashad, Y.M. In vitro antifungal activity of *Streptomyces spororaveus* RDS28 against some phytopathogenic fungi. *Afr. J. Agric. Res.* **2011**, *6*, 2835–2842.

138. Patil, H.J.; Srivastava, A.K.; Singh, D.P.; Chaudhari, B.L.; Arora, D.K. Actinomycetes mediated biochemical responses in tomato (*Solanum lycopersicum*) enhances bioprotection against *Rhizoctonia solani*. *Crop Prot.* **2011**, *30*, 1269–1273. [CrossRef]

139. Karimi, E.; Sadeghi, A.; Dehaji, P.A.; Dalvand, Y.; Omidvari, M.; Nezhad, M.K. Biocontrol activity of salt tolerant *Streptomyces* isolates against phytopathogens causing root rot of sugar beet. *Biocontrol. Sci. Technol.* **2012**, *22*, 333–349. [CrossRef]

140. Srividya, S.; Thapa, A.; Bhat, D.V.; Golmei, K.; Nilanjan, D. *Streptomyces* sp. 9p as effective biocontrol against chili soilborne fungal phytopathogens. *Eur. J. Exp. Biol.* **2012**, *2*, 163–173.

141. Saengnak, V.; Chaisiri, C.; Nalumpang, S. Antagonistic *Streptomyces* species can protect chili plants against wilt disease caused by *Fusarium*. *J. Agric. Technol.* **2013**, *9*, 1895–1908.

142. Manasa, M.; Yashoda, K.; Pallavi, S.; Vivek, M.N.; Onkarappa, R.; Kekuda, T.R.P. Biocontrol potential of *Streptomyces* species against *Fusarium oxysporum* f. sp. *zingiberi* (causal agent of rhizome rot of ginger). *J. Adv. Sci. Res.* **2013**, *4*, 1–3.

143. Adhilakshmi, M.; Latha, P.; Paranidharan, V.; Balachandar, D.; Ganesamurthy, K.; Velazhahan, R. Biological control of stem rot of groundnut (*Arachis hypogaea* L.) caused by *Sclerotium rolfsii* Sacc. with actinomycetes. *Arch. Phytopathol. Plant Protect.* **2014**, *47*, 298–311. [CrossRef]

144. Goudjal, Y.; Toumatiaa, O.; Yekkour, A.; Sabaoua, N.; Mathieuc, F.; Zitouni, A. Biocontrol of *Rhizoctonia solani* damping–off and promotion of tomato plant growth by endophytic actinomycetes isolated from native plants of Algerian Sahara. *Microbiol. Res.* **2014**, *169*, 59–65. [CrossRef] [PubMed]

145. Gao, F.; Wu, U.; Wang, M. Identification and antifungal activity of an actinomycete strain against *Alternaria* spp. *Span. J. Agric. Res.* **2014**, *12*, 1158–1165. [CrossRef]

146. Cheng, G.; Huang, Y.; Yang, Y.; Liu, F. *Streptomyces felleus* YJ1: Potential biocontrol agents against the sclerotinia stem rot (*Sclerotinia sclerotiorum*) of oilseed rape. *J. Agric. Sci.* **2014**, *6*, 91–98. [CrossRef]

147. Yandigeri, M.S.; Malviya, N.; Solanki, M.K.; Shrivastava, P.; Sivakumar, G. Chitinolytic *Streptomyces vinaceusdrappus* S5MW2 isolated from Chilika lake, India enhances plant-growth and biocontrol efficacy through chitin supplementation against *Rhizoctonia solani*. *World J. Microbiol. Biotechnol.* **2015**, *31*, 1217–1225. [CrossRef] [PubMed]

148. Zahaed, E.M. Isolation and characterization of soil *Streptomyces* species as potential biological control agents against fungal plant pathogens. *World J. Microbiol. Biotechnol.* **2014**, *30*, 1639–1647.

149. Anitha, A.; Rabeeth, M. Control of *Fusarium* wilt of tomato by bioformulation of *Streptomyces griseus* in green house condition. *Afr. J. Basic Appl. Sci.* **2009**, *1*, 9–14.
150. Elson, M.K. Selection of microorganisms for biological control of silver scurf (*Helminthosporium solani*) of potato tubers. *Plant Dis.* **1997**, *81*, 647–652. [CrossRef]
151. Ezziyyani, M.; Requena, M.E.; Egea-Gilabert, C.; Candela, M.E. Biological control of *Phytophthora* root rot of pepper using *Trichoderma harzianum* and *Streptomyces rochei* in combination. *J. Phytopathol.* **2007**, *155*, 342–349. [CrossRef]
152. Bressan, W. Biological control of maize seed pathogenic fungi by use of actinomycetes. *BioControl* **2003**, *48*, 233–240. [CrossRef]
153. Crawford, D.L.; Lynch, J.M.; Whipps, J.M.; Ousley, M.A. Isolation and characterization of actinomycete antagonists of a fungal root pathogen. *Appl. Environ. Microbiol.* **1993**, *59*, 3899–3905. [PubMed]
154. Lahdenpera, M.L. The control of *Fusarium* wilt on carnation with a *Streptomyces* preparation. *Acta Horticult.* **1987**, *216*, 85–92. [CrossRef]
155. Crawford, D.L. Use of *Streptomyces* Bacteria to Control Plant Pathogens. U.S. Patent No. 5527526, 18 June 1996.
156. Getha, K.; Vikineswary, S.; Wong, W.H.; Seki, T.; Ward, A.; Goodfellow, M. Evaluation of *Streptomyces* sp. strain g10 for suppression of *Fusarium* wilt and rhizosphere colonization in pot-grown banana plantlets. *J. Ind. Microbiol. Biotechnol.* **2005**, *32*, 24–32. [CrossRef] [PubMed]
157. Trejo-Estrada, S.R.; Sepulveda, I.R.; Crawford, D.L. In vitro and in vivo antagonism of *Streptomyces violaceusniger* YCED9 against fungal pathogens of turfgrass. *World J. Microbiol. Biotechnol.* **1998**, *14*, 865–872. [CrossRef]
158. Kim, H.J.; Lee, E.J.; Park, S.H.; Lee, H.S.; Chung, N. Biological control of anthracnose (*Colletotrichum gloeosporioides*) in pepper and cherry tomato by *Streptomyces* sp. A1022. *J. Agric. Sci.* **2014**, *2*, 54–62. [CrossRef]
159. Frändberg, E.; Petersson, C.; Lundgern, L.N.; Schnürer, J. *Streptomyces halstedii* K122 produces the antifungal compounds bafilomycin B1 and C1. *Can. J. Microbiol.* **2000**, *46*, 753–758. [CrossRef] [PubMed]
160. Nagpure, A.; Choudhary, B.; Shanti, K.; Gupta, R.K. Isolation and characterization of chitinolytic *Streptomyces* sp. MT7 and its antagonism towards wood rotting fungi. *Ann. Microbiol.* **2014**, *64*, 531–541. [CrossRef]
161. Abdallah, M.E.; Haroun, S.A.; Gomah, A.A.; El-Naggar, N.E.; Badr, H.H. Application of actinomycetes as biocontrol agents in the management of onion bacterial rot diseases. *Arch. Phytopathol. Plant Protect.* **2013**, *46*, 1797–1808. [CrossRef]
162. Yin, S.Y.; Chang, J.K.; Xun, P.C. Studies in the mechanisms of antagonistic fertilizer "5406". IV. The distribution of the antagonist in soil and its influence on the rhizosphere. *Acta Microbiol. Sin.* **1965**, *11*, 259–288.
163. Valois, D.; Fayad, K.; Barasubiye, T.; Garon, M.; Dery, C.; Brzezinski, R.; Beaulieu, C. Glucanolytic actinomycetes antagonistic to *Phytophthora fragariae* var. *rubi*, the causal agent of raspberry root rot. *Appl. Environ. Microbiol.* **1996**, *62*, 1630–1635. [PubMed]
164. Haggag, W.M.; Singer, S.M.; Mohamed, D.E.H.A. Application of broad-spectrum of marine *Streptomyces albidoflavus* a biofungicide and plant promoting of tomato diseases. *Res. J. Pharm. Biol. Chem.* **2014**, *5*, 142–148.
165. Mingma, R.; Pathom-aree, W.; Trakulnaleamsai, S.; Thamchaipenet, A.; Duangmal, K. Isolation of rhizospheric and roots endophytic actinomycetes from *Leguminosae* plant and their activities to inhibit soybean pathogen, *Xanthomonas campestris* pv. *glycine*. *World J. Microbiol. Biotechnol.* **2014**, *30*, 271–280. [CrossRef] [PubMed]
166. Misk, A.; Franco, C. Biocontrol of chickpea root rot using endophytic actinobacteria. *BioControl* **2011**, *56*, 811–822. [CrossRef]
167. Gopalakrishnan, S.; Pandey, S.; Sharma, M.; Humayun, P.; Kiran, B.K.; Sandeep, D.; Vidya, M.S.; Deepthi, K.; Rupela, O. Evaluation of actinomycete isolates obtained from herbal vermicompost for the biological control of *Fusarium* wilt of chickpea. *Crop Prot.* **2011**, *30*, 1070–1078. [CrossRef]
168. Singh, P.P.; Shin, Y.C.; Park, C.S.; Chung, Y.R. Biological control of *fusarium* wilt of cucumber by chitinolytic bacteria. *Phytopathology* **1999**, *89*, 92–99. [CrossRef] [PubMed]
169. Samac, D.A.; Willert, A.M.; McBride, M.J.; Kinkel, L.L. Effects of antibiotic-producing *Streptomyces* on nodulation and leaf spot in *Alfalfa*. *Appl. Soil. Ecol.* **2003**, *22*, 55–66. [CrossRef]
170. Conn, V.M.; Walker, A.R.; Franco, C.M. Endophytic actinobacteria induce defense pathways in *Arabidopsis thaliana*. *Mol. Plant Microbe Interact.* **2008**, *21*, 208–218. [CrossRef] [PubMed]
171. Bacon, C.W.; Hinton, D.M. Bacterial endophytes: The endophytic niche, its occupants, and its utility. In *Plant Associated Bacteria*; Gnanamanickam, S.S., Ed.; Springer: Dordrech, The Netherlands, 2006; pp. 154–194.

172. Hastuti, R.D.; Yulin, L.; Antonius, S.; Rasti, S. Endophytic *Streptomyces* sp. as Biocontrol Agents of Rice Bacterial Leaf Blight Pathogen (*Xanthomonas oryzae* pv. *oryzae*). *HAYATI J. Biosci.* **2012**, *19*, 155–162. [CrossRef]

173. Gupta, R.; Saxena, R.K.; Chaturvedi, P.; Virdi, J.S. Chitinase production by *Streptomyces viridificans*: Its potential in fungal cell wall lysis. *J. Appl. Bacteriol.* **1995**, *78*, 378–383. [CrossRef] [PubMed]

174. Suzuki, S.; Nakanishi, E.; Ohira, T.; Kawachi, R.; Ohnishi, Y.; Horinouchi, S.; Nagasawa, H.; Sakuda, S. Chitinase inhibitor allosamidin is a signal molecule for chitinase production in its producing *Streptomyces*. II. Mechanism for regulation of chitinase production by allosamidin through a two-component regulatory system. *J. Antibiot.* **2006**, *59*, 410–417. [CrossRef] [PubMed]

175. Adams, D.J. Fungal cell wall chitinases and glucanases. *Microbiology* **2004**, *150*, 2029–2035. [CrossRef] [PubMed]

176. Chernin, L.; Ismailov, Z.; Haran, S.; Chet, I. Chitinolytic *Enterobacter agglomerans* antagonistic to fungal plant pathogens. *Appl. Environ. Microbiol.* **1995**, *61*, 1720–1726. [PubMed]

177. De Boer, W.; Klein Gunnewiek, P.J.A.; Kowalchuk, G.A.; Van Veen, J.A. Growth of Chitinolytic Dune Soil β-Subclass *Proteobacteria* in Response to Invading Fungal Hyphae. *Appl. Environ. Microbiol.* **2001**, *67*, 3358–3362. [CrossRef] [PubMed]

178. Sakuda, S.; Isogai, A.; Matsumoto, S.; Suzuki, A. Search for microbial insect growth regulators II. Allosamidin, a novel insect chitinase inhibitor. *J. Antibiot.* **1987**, *40*, 296–300. [PubMed]

179. Suzuki, S.; Nakanishi, E.; Ohira, T.; Kawachi, R.; Nagasawa, H.; Sakuda, S. Chitinase inhibitor allosamidin is a signal molecule for chitinase production in its producing *Streptomyces*. I. Analysis of the chitinase whose production is promoted by allosamidin and growth accelerating activity of allosamidin. *J. Antibiot.* **2006**, *59*, 402–409. [CrossRef] [PubMed]

180. Sakuda, S.; Inoue, H.; Nagasawa, H. Novel Biological Activities of Allosamidins. *Molecules* **2013**, *18*, 6952–6968. [CrossRef] [PubMed]

181. Cao, L.; Qiu, Z.; You, J.; Tan, H.; Zhou, S. Isolation and characterization of endophytic *Streptomyces* antagonists of *Fusarium* wilt pathogen from surface sterilized banana roots. *FEMS Microbiol. Lett.* **2005**, *247*, 147–152. [CrossRef] [PubMed]

182. Tan, H.M.; Cao, L.X.; He, Z.F.; Su, G.J.; Lin, B.; Zhou, S.N. Isolation of endophytic actinomycetes from different cultivars of tomato and their activities against *Ralstonia solanacearum* in vitro. *World J. Microbiol. Biotechnol.* **2006**, *22*, 1275–1280. [CrossRef]

183. El-Shatoury, S.; El-Kraly, O.; El-Kazzaz, W.; Dewedar, A. Antimicrobial activities of Actinomycetes inhabiting *Achillea fragrantissima* (Family: Compositae). *Egypt. J. Nat. Toxins* **2009**, *6*, 1–15.

184. Gangwar, M.; Dogra, S.; Gupta, U.P.; Kharwar, R.N. Diversity and biopotential of endophytic actinomycetes from three medicinal plants in India. *Afr. J. Microbiol. Res.* **2014**, *8*, 184–191.

185. El-Tarabily, K.A. An endophytic chitinase-producing isolate of Actinoplanes missouriensis, with potential for biological control of root rot of lupine caused by *Plectosporium tabacinum*. *Aust. J. Bot.* **2003**, *51*, 257–266. [CrossRef]

186. Kloepper, J.W.; Leong, J.; Teintze, M.; Schiroth, M.N. Enhanced plant growth by siderophores produced by plant growth promoting rhizobacteria. *Nature* **1980**, *286*, 885–886. [CrossRef]

187. Igarashi, Y. Screening of novel bioactive compounds from plant-associated actinomycetes. *Actinomycetolog* **2004**, *18*, 63–66. [CrossRef]

188. Zhang, J.; Wang, J.D.; Liu, C.X.; Yuan, J.H.; Wang, X.J.; Xiang, W.S. A new prenylated indole derivative from endophytic actinobacteria *Streptomyces* sp. neau-D50. *Nat. Prod. Res.* **2014**, *28*, 431–437. [CrossRef] [PubMed]

189. Lu, C.; Shen, Y.A. New macrolide antibiotic with antitumor activity produced by *Streptomyces* sp. CS, a commensal microbe of Maytenus hookeri. *J. Antibiot.* **2003**, *56*, 415–418. [CrossRef] [PubMed]

190. Lu, C.; Shen, Y. A novel ansamycin, naphthomycin k from *Streptomyces* sp. *J. Antibiot.* **2007**, *60*, 649–653. [CrossRef] [PubMed]

191. Igarashi, Y.; Iida, T.; Sasaki, T.; Saito, N.; Yoshida, R.; Furumai, T. Isolation of actinomycetes from live plants and evaluation of anti-phytopathogenic activity of their metabolites. *Actinomycetolog* **2002**, *16*, 9–13. [CrossRef]

192. Book, A.J.; Lewin, G.R.; McDonald, B.R.; Takasuka, T.E.; Doering, D.T.; Adams, A.S.; Blodgett, J.A.V.; Clardy, J.; Raffa, K.F.; Fox, B.G.; et al. Cellulolytic *Streptomyces* strains associated with herbivorous insects share a phylogenetically linked capacity to degrade lignocelluloses. *Appl. Environ. Microbiol.* **2014**, *80*, 4692–4701. [CrossRef] [PubMed]

193. Copping, G.L.; Menn, J.J. Biopesticides: A review of their action, applications and efficacy. *Pest Manag. Sci.* **2000**, *56*, 651–676. [CrossRef]

194. Craveri, R.; Giolitti, G. An antibiotic with fungicidal and insecticidal activity produced by *Streptomyces*. *Nature* **1957**, *179*, 1307. [CrossRef] [PubMed]

195. Kido, G.S.; Spyhalski, E. Antimycin A, an antibiotic with insecticidal and miticidal properties. *Science* **1950**, *112*, 172–173. [CrossRef] [PubMed]

196. Takahaski, N.; Suzuki, A.; Kimura, Y.; Miyamoto, S.; Tamura, S.; Mitsui, T.; Fukami, J. Isolation, structure and physiological activities of piericidin B, natural insecticide produced by a *Streptomyces*. *Agric. Biol. Chem.* **1968**, *32*, 1115–1122.

197. Oishi, H.; Sugawa, T.; Okutomi, T.; Suzuki, K.; Hayashi, T.; Sawada, M.; Ando, K. Insecticidal activity of macrotetrolide antibiotics. *J. Antibiot.* **1970**, *23*, 105–106. [CrossRef] [PubMed]

198. Box, S.J.; Cole, M.; Yeoman, G.H. Prasinons A and B: Potent insecticides from *Streptomyces prasinus*. *Appl. Microbiol.* **1973**, *29*, 699–704.

199. Turner, M.J.; Schaeffer, J.M. Mode of action of ivermectin. In *Ivermectin and Abamectin*; Cambell, W.C., Ed.; Springer: New York, NY, USA, 1989; pp. 73–88.

200. Bloomquist, J.R. Ion Channels as Targets for Insecticides. *Annu. Rev. Entomol.* **1996**, *41*, 163–190. [CrossRef] [PubMed]

201. Sousa, S.C.; Soares, F.A.C.; Garrido, S.M. Characterization of Streptomycetes with potential to promote plant growth and biocontrol. *Sci. Agric.* **2008**, *65*, 50–55. [CrossRef]

202. Kabaluk, J.T.; Svircev, A.M.; Goettel, M.S.; Woo, S.G. *The Use and Regulation of Microbial Pesticides in Representative Jurisdiction Worldwide*; IOBC Global: Hong Kong, China, 2010; p. 99.

203. Copping, L.G.; Duke, S.O. Review-natural products that have been used commercially as crop protection agents. *Pest Manag. Sci.* **2007**, *63*, 524–554. [CrossRef] [PubMed]

204. Saxena, S. Microbial metabolites for development of ecofriendly agrochemical. *Allelopathy J.* **2014**, *33*, 1–24.

205. Aggarwal, N.; Thind, S.K.; Sharma, S. Role of secondary metabolites of Actinomycetes in crop protection. In *Plant Growth Promoting Actinobacteria*; Subramaniam, G., Ed.; Springer: Singapore, 2016; pp. 99–121, ISBN 978-981-10-0705-7.

206. Tanaka, Y.; Omura, S. Agro-active compounds of microbial origin. *Annu. Rev. Microbiol.* **1993**, *47*, 57–87. [CrossRef] [PubMed]

207. Behal, V. Bioactive products from *Streptomyces*. *Adv. Appl. Microbiol.* **2000**, *47*, 113–157. [PubMed]

208. Alexander, M. *Introduction to Soil Microbiology*; Krieger Publishing Company: Malabar, India, 1977; p. 467.

209. Isono, K.; Nagatsu, J.; Kobinata, K.; Sasaki, K.; Suzuki, S. Studies on polyoxins, antifungal antibiotics. Part I. Isolation and characterization of polyoxins A and B. *Agric. Biol. Chem.* **1965**, *29*, 848–854.

210. Umezawa, H.; Okami, Y.; Hashimoto, T.; Suhara, Y.; Hamada, M.; Takeuchi, T. A new antibiotic, kasugamycin. *J. Antibiot.* **1965**, *18*, 101–103. [PubMed]

211. Siddique, S.; Syed, Q.; Adnan, A.; Nadeem, M.; Irfan, M.; Qureshi, F.A. Production of avermectin B1b from Streptomyces avermitilis 41445 by batch submerged fermentation. *Jundishapur J. Microbiol.* **2013**, *6*, e7198. [CrossRef]

212. Reddy, K.R.K.; Jyothi, G.; Sowjanya, Ch.; Kusumanjali, K.; Malathi, N.; Reddy, K.R.N. Plant Growth-Promoting Actinomycetes: Mass Production, Delivery systems, and commercialization. In *Plant Growth Promoting Actinobacteria*; Subramaniam, G., Ed.; Springer: Singapore, 2016; ISBN 978-981-10-0705-7.

213. Subramaniam, G.; Arumugam, S.; Rajendran, V.; Vadlamudi, S. Plant Growth Promoting Micorbes for Field Applications. In *Microbial Inoculants in Sustainable Agriculture Productivity*; Singh, D.P., Singh, H.B., Prabha, R., Eds.; Springer: New Delhi, India, 2016; pp. 239–251, ISBN 978-81-322-2644-4.

214. FAO. International Standards for Phytosanitary Measures. Guidelines for the export, shipment, import and release of biological control agents and other beneficial organisms. In *Second International Symposium on Biological Control of Arthropods*; Hoddle, M., Ed.; USDA Forest Service: Davos, Switzerland, 2005; pp. 726–734.

215. Rafii, F.; Crawford, D.L.; Bleakley, B.H.; Wang, Z. Assessing the risks of releasing recombinant *Streptomyces* in soil. *Microbiol. Sci.* **1988**, *12*, 358–362.

216. Egan, S.; Wiener, P.; Kallifidas, D.; Wellington, E.M. Transfer of streptomycin biosynthesis gene clusters within streptomycetes isolated from soil. *Appl. Environ. Microbiol.* **1998**, *12*, 5061–5063.

217. Egan, S.; Wiener, P.; Kallifidas, D.; Wellington, E.M. Phylogeny of *Streptomyces* species and evidence for horizontal transfer of entire and partial antibiotic gene clusters. *Antoine Leeuwenhoek* **2001**, *79*, 127–133. [CrossRef]

218. Ward, M.G. The Regulatory Landscape for Biological Control Agents. *EPPO Bull* **2016**, *46*, 249–253. [CrossRef]

219. Santos, V.B.; Araujo, S.F.; Leite, L.F.; Nunes, L.A.; Melo, J.W. Soil microbial biomass and organic matter fractions during transition from conventional to organic farming systems. *Geoderma* **2012**, *170*, 227–231. [CrossRef]

220. Youssef, M.M.A.; Eissa, M.F.M. Biofertilizers and their role in management of plant parasitic nematodes. A review. *E3 J. Biotechnol. Pharm. Res.* **2014**, *5*, 1–6.

221. Nina, K.; Thomas, W.K.; Prem, S.B. *Beneficial Organisms for Nutrient Uptake*; VFRC Report 2014/1; Virtual Fertilizer Research Centre, Wageningen Academic Publishers: Washington, DC, USA, 2014; p. 63.

222. Sahoo, R.K.; Ansari, M.W.; Dangar, T.K.; Mohanty, S.; Tuteja, N. Phenotypic and molecular characterisation of efficient nitrogen-fixing *Azotobacter* strains from rice fields for crop improvement. *Protoplasma* **2014**, *251*, 511–523. [CrossRef] [PubMed]

223. Singh, J.S.; Pandey, V.C.; Singh, D.P. Efficient soil microorganisms: A new dimension for sustainable agriculture and environmental development. *Agric. Ecosyst. Environ.* **2011**, *140*, 339–353. [CrossRef]

224. Bhardwaj, D.; Ansari, M.W.; Sahoo, R.K.; Tuteja, N. Biofertilizers function as key player in sustainable agriculture by improving soil fertility, plant tolerance and crop productivity. *Microb. Cell Fact.* **2014**, *13*, 66. [CrossRef] [PubMed]

225. Bashan, Y.; de-Bashan, L.E.; Prabhu, S.R.; Juan-Pablo, H. Advances in plant growth promoting bacterial inoculant technology: Formulations and practical perspectives (1998–2013). *Plant Soil* **2014**, *378*, 1–33. [CrossRef]

226. Bashan, Y. Inoculants of plant growth-promoting bacteria for use in agriculture. *Biotechnol. Adv.* **1998**, *16*, 729–770. [CrossRef]

227. Catroux, G.; Hartmann, A.; Revellin, C. Trends in rhizobial inoculant production and use. *Plant Soil* **2001**, *230*, 21–30. [CrossRef]

228. Suprapta, D.N. Potential of microbial antagonists as biocontrol agents against plant fungal pathogens. *J. ISSAAS* **2012**, *18*, 1–8.

229. Ou, S.H. A look at worldwide rice blast disease control. *Plant Dis.* **1980**, *64*, 439–445. [CrossRef]

230. Vasudevan, P.; Kavitha, S.; Priyadarisini, V.B.; Babujee, L.; Gnanamanickam, S.S. Biological control of rice diseases. In *Biological Control of Crop Diseases*; Gnanamanickam, S.S., Ed.; Marcel Dekker Inc.: New York, NY, USA, 2002; pp. 11–32.

231. Dubey, R.C. Biopesticides: Biological control of plant pathogens, pests and weeds. In *A Textbook of Biotechnology*; Dubey, R.C., Ed.; S. Chand Publishing: Ram Nagar, India, 1993; pp. 457–475, ISBN 978812926089.

232. Yang, J.H.; Liu, H.X.; Zhu, G.M.; Pan, Y.L.; Xu, L.P.; Guo, J.H. Diversity analysis of antagonists from rice-associated bacteria and their application in biocontrol of rice diseases. *J. Appl. Microbiol.* **2008**, *104*, 91–104. [CrossRef] [PubMed]

233. Law, J.W.-F.; Ser, H.-L.; Khan, T.M.; Chuah, L.-H.; Pusparajah, P.; Chan, K.-G.; Goh, B.-H.; Lee, L.-H. The Potential of *Streptomyces* as Biocontrol Agents against the Rice Blast Fungus, *Magnaporthe oryzae* (*Pyricularia oryzae*). *Front. Microbiol.* **2017**, *8*. [CrossRef] [PubMed]

234. Bibb, M.J. Understanding and manipulating antibiotic production in actinomycetes. *Biochem. Soc. Trans.* **2013**, *41*, 1355–1364. [CrossRef] [PubMed]

International Journal of
Molecular Sciences

MDPI

Article

Cultivar and Metal-Specific Effects of Endophytic Bacteria in *Helianthus tuberosus* Exposed to Cd and Zn

Blanca Montalbán [1,2,*], Sofie Thijs [2], Mª Carmen Lobo [1], Nele Weyens [2], Marcel Ameloot [3], Jaco Vangronsveld [2] and Araceli Pérez-Sanz [1,4]

[1]   Departamento de Investigación Agroambiental, IMIDRA, Finca "El Encín",
     Autovía del Noreste A-2 Km 38.2, 28800 Alcalá de Henares, Madrid, Spain;
     carmen.lobo@madrid.org (Mª.C.L.); a.perez@nhm.ac.uk (A.P.-S.)
[2]   Environmental Biology, Centre for Environmental Sciences, Hasselt University, Agoralaan Building D,
     BE3590 Diepenbeek, Belgium; sofie.thijs@uhasselt.be (S.T.); nele.weyens@uhasselt.be (N.W.);
     jaco.vangronsveld@uhasselt.be (J.V.)
[3]   Biomedical Research Department, Hasselt University, Agoralaan building D, BE3590 Diepenbeek, Belgium;
     marcel.ameloot@uhasselt.be
[4]   Department of Earth Sciences, Natural History Museum, Cromwell Road, London SW7 5BD, UK
*   Correspondence: blanca.montalban@madrid.org; Tel.: +34-91-887-94-73; Fax: +34-91-887-94-74

Received: 18 August 2017; Accepted: 12 September 2017; Published: 21 September 2017

Abstract: Plant growth promoting endophytic bacteria (PGPB) isolated from *Brassica napus* were inoculated in two cultivars of *Helianthus tuberosus* (VR and D19) growing on sand supplemented with 0.1 mM Cd or 1 mM Zn. Plant growth, concentrations of metals and thiobarbituric acid (TBA) reactive compounds were determined. Colonization of roots of *H. tuberosus* D19 by *Pseudomonas* sp. 262 was evaluated using confocal laser scanning microscopy. *Pseudomonas* sp. 228, *Serratia* sp. 246 and *Pseudomonas* sp. 262 significantly enhanced growth of *H. tuberosus* D19 exposed to Cd or Zn. *Pseudomonas* sp. 228 significantly increased Cd concentrations in roots. *Serratia* sp. 246, and *Pseudomonas* sp. 256 and 228 resulted in significantly decreased contents of TBA reactive compounds in roots of Zn exposed D19 plants. Growth improvement and decrease of metal-induced stress were more pronounced in D19 than in VR. *Pseudomonas* sp. 262-green fluorescent protein (GFP) colonized the root epidermis/exodermis and also inside root hairs, indicating that an endophytic interaction was established. *H. tuberosus* D19 inoculated with *Pseudomonas* sp. 228, *Serratia* sp. 246 and *Pseudomonas* sp. 262 holds promise for sustainable biomass production in combination with phytoremediation on Cd and Zn contaminated soils.

Keywords: metal contaminated soil; *Helianthus tuberosus*; phytoremediation; high biomass crop; green fluorescent protein; plant growth promoting bacteria

1. Introduction

During the last two decades, the potential use of plants to remediate metal contaminated soils has been intensively investigated. For application of phytotechnologies on metal contaminated soils, and especially in the case of phytoextraction, metal availability, uptake and phytotoxicity are the main limiting factors [1–5]. The interactions between plants and beneficial bacteria may increase the efficiency of phytoextraction because of increased biomass, metal uptake and plant tolerance to toxic metals [6–9]. Plant growth can be enhanced: (1) indirectly, by preventing growth and activity of plant pathogens through the production of antibiotics or through competition for space and nutrients [10]; and (2) directly, by increasing available nutrients through different mechanisms such as nitrogen fixation [11], solubilization of minerals such as phosphorous and iron [12,13], and production of phytohormones (as IAA, indole-3-acetic acid) [14] and 1-aminocyclopropane-1-carboxylate (ACC) deaminase [15,16]. Metal and nutrient availability can be enhanced by excreting organic acids that decrease pH in the rhizosphere or by enhancing the Fe(III) mobility and other cations through production of siderophores [9,17,18]. Some microorganisms are equipped with metal-resistance/sequestration systems that can contribute to metal detoxification [19]. Plant-associated bacteria can also adsorb metals by binding them to anionic functional groups or to extracellular polymeric substances of the cell wall [20–22]. This leads to a reduced metal uptake and translocation inside the plant, improving its growth through decreasing phytotoxicity [8,23]. In a previous study, we showed that bacterial strains isolated from a Zn contaminated soil increased root length of *Brassica napus* seedlings in the presence of Cd and Zn under in vitro conditions [24].

Many studies have evaluated the interactions between plants and their associated bacteria for the removal or stabilization of metals in contaminated soils [25]. In some cases, bacteria isolated from metal tolerant plants promoted the growth of plants from different taxonomic groups [23,26–28] and demonstrated high levels of colonization in plant species different from the original host. Several studies have been performed under hydroponic conditions to evaluate the effects of bacteria on growth, metal uptake and production of thiobarbituric acid (TBA) reactive compounds for different plants and metals [29–31]. However, most studies tested the use of single inocula on the same plant species. In the light of bacteria-stimulated phytoremediation, it is however important to assess the bacterial colonization of more than one plant cultivar under different metal pollution contexts.

The strategy of bacterial inoculation is one of the most critical steps in phytotechnology applications [32]. The colonization must be effective in order to achieve beneficial effects on plant growth and metal uptake [33]. A profound knowledge about plant growth promoting endophytic bacteria (PGPB) colonization routes and plant–bacteria interactions is essential to develop an effective method of inoculation [34]. The use of fluorescent proteins in non-invasive microscopy is a well-established and valuable tool in biology and biotechnology [35]. Labeling with enhanced green fluorescent protein (EGFP) can be adopted to observe the colonization patterns of bacteria [36–38]. GFP has been described to be a good marker for studying bacterial behavior in the rhizosphere and the endosphere [39,40]. Recently, Ma et al. [3] pointed out that endophytes could be a more reliable source of natural biocenosis than rhizobacteria because of their intimate association with plants, although their effects in phytotechnologies still should be investigated more in depth.

Helianthus tuberosus L. (Asteraceae) is a high biomass crop used for bio-ethanol production. It is vegetatively propagated by tubers [41] with low production costs and negligible pests and disease problems [42,43]. Several studies have demonstrated the tolerance of this crop to metals such as Cd, Pb and Zn [44–48]. All these characteristics make *H. tuberosus* a promising candidate for phytoremediation of metal contaminated soils, as well as to produce renewable energy. Therefore, the aim of this work was to evaluate the effects of PGPB strains, isolated from *B. napus* growing on a metal contaminated soil, on growth, metal uptake and TBA reactive compounds, in two cultivars of *H. tuberosus* (VR and D19) exposed to Cd and Zn. Of one particular interesting endophytic strain, *Pseudomonas* sp. 262, the colonization of the roots of *H. tuberosus* was studied using confocal laser scanning microscopy.

2. Results and Discussion

2.1. Plant Growth and Metal Uptake

Exposure to Cd and Zn significantly decreased the weight of *H. tuberosus* in comparison to the non-exposed control plants (Figure 1). In particular, the shoot weight decreased by 57% and the root weight by 67% when plants were exposed to 0.1 mM Cd; the reductions reached 70% and 50% in shoot and root weights in the case plants were grown in presence of 1 mM Zn. Some of the inoculated bacterial strains significantly improved growth of metal exposed plants. In the presence of Zn, inoculation of *Pseudomonas* sp. 228 significantly increased both shoot and root weights of the

D19 cultivar, by 145% and 263% respectively (Figure 1). *Serratia* sp. 246 increased the shoot weight of the VR cultivar under Zn exposure by 78%. In Cd exposed plants of the D19 cultivar, inoculation of *Pseudomonas* sp. 262 and *Serratia* sp. 246 significantly increased the shoot weight by 68% and 46%, respectively. These beneficial effects on weight are in line with earlier studies in which positive effects of inoculation with PGPB on growth of plants exposed to metals were reported [49–51]. However, these positive effects of the endophytes *Pseudomonas* sp. and *Serratia* sp. have been not described before in a tuberous plant exposed to metals.

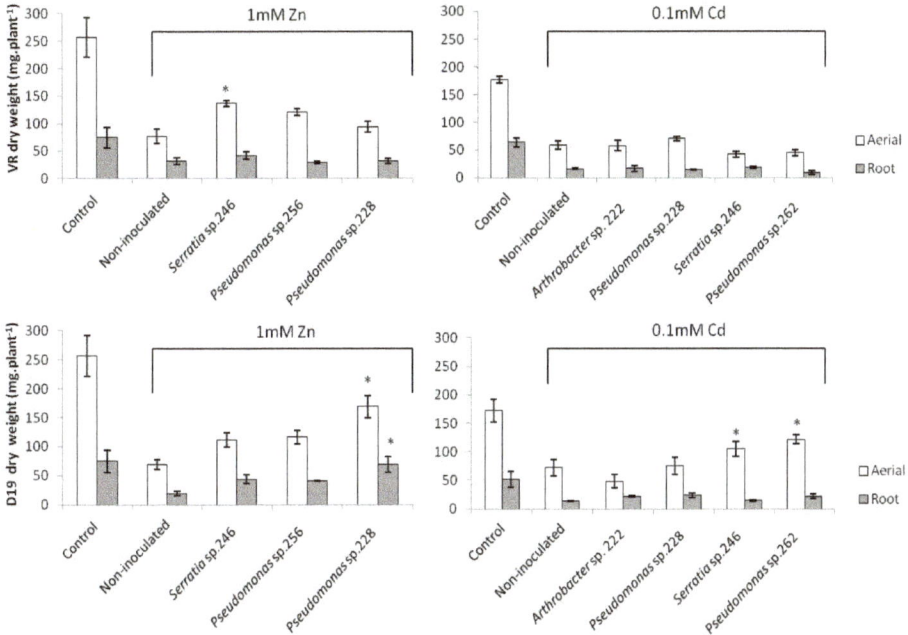

Figure 1. Dry weight (mg·plant^{-1}) of the *H. tuberosus* cultivars VR and D19 after three weeks of growth in presence of 1 mM Zn or 0.1 mM Cd. * Significant differences between inoculated and non-inoculated after Tukey's test, $p < 0.05$; mean values ± SE; $n = 4$.

In vitro, the inoculated bacterial strains demonstrated plant growth-promoting characteristics like production of IAA, acetoin and ACC deaminase activity that can improve the growth of their host plant (Table 1). Production of IAA and acetoin can stimulate root formation [52,53], and thereby increase the nutrient absorption capacity of the plant. ACC deaminase activity can reduce the ethylene levels generated due to stress, improving the growth of plants in presence of toxic concentrations of metals [16]. It is important to mention that the endophytic bacterial strains that increased the growth of *H. tuberosus* also increased the length of the roots of *Brassica napus* seedlings in vertical agar plates containing toxic concentrations of Cd and Zn [24]. This suggests that these endophytes are beneficial for plants from different families.

Table 1. Metal tolerance and plant growth promoting (PGP) characteristics of selected bacterial strains for inoculation in *H. tuberosus* under hydroponic conditions with Cd and Zn, modified from [24].

Comp.[1]	Strain	Identification	Accesion	Zn	Cd	Fe 0 µM	Fe 0.25 µM	OA	ACC	IAA	Ace	Psol	N fix
Soil	222	*Arthrobacter* sp.	KT461847	+++	+++	−	−	++	+++	−	−	−	+
Root	228	*Pseudomonas* sp.	KT461831	++	++	+	+	+	++	+	−	++	−
Root	246	*Serratia* sp.	KT461863	+++	+++	+	+	++	+++	++	−	−	−
Root	256	*Pseudomonas* sp.	KT461831	+++	+	+	+	−	+	++	+	+++	++
Root	262	*Pseudomonas* sp.	KT461831	+	+	++	−	+	+++	++	+	−	−

[1] Compartment (Comp.), growth in the presence of Zn (1 mM) and Cd (0.8 mM), siderophores (Fe 0 µM and Fe 0.25 µM), Organic acids (OA), ACC (ACC deaminase activity), IAA (indole-3-acetic acid), Ace (Acetoin), phosphate solubilization (Psol), nitrogen fixation (N fix). + low, ++ medium, +++ high production, − absence of production.

The bacterial inoculation also affected the metal concentrations in both cultivars of *H. tuberosus* (Table 2). The Zn concentration significantly decreased in roots of the VR cultivar inoculated with *Pseudomonas* sp. 228. No significant differences were found between inoculated and non-inoculated plants of the D19 cultivar. In the case of Cd exposure, the effects were different. Inoculation of *Pseudomonas* sp. 228 significantly increased the Cd concentration in roots of the D19 cultivar in comparison to non-inoculated plants. In contrast, inoculation of *Pseudomonas* sp. 262 and *Arthrobacter* sp. 222 decreased the Cd concentration in roots of the VR cultivar. *Serratia* sp. 246 and *Pseudomonas* sp. 262 also decreased the concentrations of Cd in the shoots of, respectively, the VR and the D19 cultivar. In metal contaminated nutrient solutions, metals are almost entirely available to plants. Therefore, the effects of the bacteria on the plant uptake could be masked because of the high metal uptake that usually occurs in these cases. Wan et al. [30] did not observe significant differences in Cd uptake by hydroponically grown *Solanum nigrum* after inoculation of *Serratia nematodiphila* LRE07 in the presence of high Cd concentrations. These authors concluded that the effect of the strain was more significant at lower concentrations (10 µM of Cd). Moreover, the decreases in Cd and Zn concentrations in inoculated plants could be due to the capacity of some bacteria to adsorb and immobilize toxic ions from the solution through the production of extracellular polysaccharides and proteins that can bind and precipitate metals [54]. It this way, bacteria can reduce the phytotoxic effects of the metals improving the growth of the host plant [21,23,29]. Several authors have reported such effect in different plant species and diverse growth conditions. Marques et al. [55] observed that the Cd and Zn concentrations in roots of *Helianthus annuus* decreased after inoculation with *Chrysiobacterium humi*, isolated from a Cd-Zn contaminated soil. They attributed this effect to the fact that some bacteria can share the metal load with the plant, thereby decreasing the metal uptake in the plant. Tripathi et al. [56] described that the growth of *Phaseolus vulgaris* improved after inoculation of *Pseudomonas putida* KNP9 in a soil spiked with Cd and Pb. They suggested that the improved growth was possibly due to a decreased metal uptake by the plant. Vivas et al. [22] reported that the inoculation of *Brevibacillus* sp. alleviated the toxicity of Zn in *Trifolium repens* by reducing the metal uptake by plants growing on a Zn contaminated soil. Inoculation with *Serratia* sp. MSMC541 decreased the metal translocation of *Lupinus luteus* when growing in a soil spiked with As, Cd, Pb and Zn [57]. They concluded that this strain protects the plants against metal toxicity by reducing their uptake and, in this way promoting plant growth. In our work, the weight of D19 plants increased in presence of *Pseudomonas* sp. 228 under Zn exposure, and after addition of *Pseudomonas* sp. 262 and *Serratia* sp. 246 in presence of Cd. In the case of the VR cultivar, the weight also increased in plants inoculated with *Serratia* sp. 246 in presence of Zn. The concentration of metals tended to decrease in roots of plants inoculated with *Pseudomonas* sp. 228 and *Serratia* sp. 246, although this decrease was only significant in the case of the D19 cultivar, after inoculation with *Pseudomonas* sp. 262 in presence of Cd. Taking this into account, our data support the hypothesis that bacteria have cultivar-dependent and metal-specific effects on plant growth. *Pseudomonas* sp. 262 is a promising endophyte because it can lower metal uptake in presence of Cd, decrease phytotoxicity, and improve plant growth.

Table 2. Total metal concentrations (mg·kg^{-1} dry matter) in two cultivars of *H. tuberosus* grown in absence (control) and in presence of 1 mM Zn or 0.1 mM Cd.

		VR		D19	
	Treatments	Zn			
		Aerial	Root	Aerial	Root
	Control	58 ± 14a	40 ± 10a	75 ± 23a	41 ± 5a
	Non-inoculated	1533 ± 149b	4533 ± 945c	1097 ± 175b	3862 ± 1063bc
Zn	*Serratia* sp. 246	1155 ± 23b	4195 ± 355bc	1283 ± 207b	3455 ± 1767b
	Pseudomonas sp. 256	1349 ± 183b	4368 ± 442bc	1554 ± 299b	3484 ± 651b
	Pseudomonas sp. 228	975 ± 154b	2237 ± 368b	1317 ± 177b	3504 ± 1167b
		Cd			
	Control	0.43 ± 0.09a	1.2 ± 0.2a	0.6 ± 0.1a	0.5 ± 0.2a
	Non-inoculated	152 ± 10c	1118 ± 177def	106 ± 44bc	889 ± 196cde
Cd	*Arthrobacter* sp. 222	83 ± 6bc	492 ± 85bc	58 ± 8b	631 ± 140bc
	Pseudomonas sp. 228	112 ± 23bc	1250 ± 320ef	106 ± 18bc	1365 ± 145f
	Serratia sp. 246	24 ± 4b	908 ± 314cde	129 ± 45c	798 ± 65bcd
	Pseudomonas sp. 262	145 ± 29c	487 ± 57bc	81 ± 5bc	383 ± 107b

Different letters represent significant differences per column, cultivar and metal after Tukey's test, $p < 0.05$; mean values ± SE; $n = 4$.

2.2. Nutrient Status

In general, inoculation of bacterial strains did not have clear effects on the nutrient concentrations in both cultivars (Tables S1–S4). Macronutrients as Na and Ca were significantly lower in roots of, respectively, VR and D19 plants when plants were inoculated with *Serratia* sp. 246, *Pseudomonas* sp. 228 and 256 in presence of 1 mM of Zn (Table S1). In the case of exposure of the plants to 0.1 mM Cd, inoculation of *Arthrobacter* sp. 222, *Serratia* sp. 246, *Pseudomonas* sp. 228 and 262 led to lower K concentrations in shoots of the VR cultivar (Table S2).

Micronutrient concentrations changed in some cases after bacterial inoculation. When plants were grown in the presence of Zn, inoculation of *Serratia* sp. 246, *Pseudomonas* sp. 228 and 256 significantly decreased the Cu and Fe concentrations in roots of respectively the VR and D19 cultivars (Table S3). The concentrations of Cu in roots of Cd exposed plants of the VR cultivar were also lower after inoculation with *Arthrobacter* sp. 222, *Serratia* sp. 246 and *Pseudomonas* sp. 262 (Table S4). However, *Serratia* sp. 246 increased the Fe content in the shoots of the VR cultivar in presence of Zn (Table S3). The latter strain also increased the weight of VR plants exposed to Zn.

The lower Cu and Fe concentrations in roots of both cultivars when inoculated with *Serratia* sp. 246 and *Pseudomonas* (262 and 256) can be due to the above-mentioned bacterial mechanisms of metal sequestration and/or biosorption. Microorganisms indeed have developed complex mechanisms of metal resistance that can affect the availability of metals and nutrients [58,59]. PGPB can sequestrate elements through extracellular production of polysaccharides, by fixing elements such as Fe or Cu on the membrane or cell wall or they can precipitate them in the form of hydroxides or other insoluble metal salts [23,60,61]. Bacterial surfaces hold polar functional groups that can interact with cations [62]. In our work, the excess of Cd and Zn might induce the bacterial mechanisms of metal resistance that lower the availability of metals and also the solubility of other nutrients that might be precipitated on the cell surface.

The synthesis of siderophores is stimulated in presence of toxic metals in order to supply the appropriate amounts of ions to the plant and diminish the phytotoxicity symptoms [17,29]. This PGP characteristic plays an important role under soil conditions in which the nutrients are mainly present for the plants in unavailable chemical forms. Therefore, it can be expected that, under the growth conditions used in this study (sand moistened with half-strength Hoagland whether or not supplemented with Zn or Cd), the effects of the bacteria on nutrient uptake are less pronounced, since Fe is supplied in an appropriate concentration with the nutrient solution. Thus, the differences observed in the nutrient concentrations in the plants might also be due to the imbalance of nutrients generated by the presence of metals in the solution.

2.3. Lipid Peroxidation

TBA reactive compounds are produced as a result of peroxidation of membrane lipids. This process is initiated by excess of free radicals in consequence of oxidative stress. Increased levels of TBA reactive compounds are an indicator for physiological stress [63]. Many studies reported that levels of TBA reactive compounds increased in plants exposed to toxic concentrations of metals such as Cd, Zn, and Pb [64–67].

In the present work, exposure to 1 mM Zn and 0.1 mM Cd significantly increased the levels of TBA reactive compounds in roots of both cultivars of *H. tuberosus* (Figure 2). No significant differences in TBA-levels were found in leaves of metal-exposed plants compared to non-exposed plants. Nouairi et al. [68] obtained similar results for leaves of *Brassica juncea* exposed to 50 μM Cd. According to them, this result could be related with a tolerance mechanism of the plant to avoid oxidative stress generated by the presence of metals in the leaves. A reduction of the concentrations of TBA reactive compounds has been reported to result from increased activities of anti-oxidative enzymes, which limit H_2O_2 levels and membrane damage [69].

Interestingly, the inoculation of *Serratia* sp. 246, *Pseudomonas* sp. 256 and 228 significantly decreased the amounts of TBA reactive compounds in roots of the D19 cultivar grown in the presence of Zn (Figure 2a). The roots of the VR cultivar also contained lower levels of TBA reactive compounds when plants were inoculated with *Pseudomonas* sp. 228. In Cd exposed plants, no significant differences in TBA reactive compounds were observed between inoculated and non-inoculated plants (Figure 2b). Decreases of TBA reactive compounds after inoculation of PGPB were reported by several authors in different plant species. Pandey et al. [31] described that inoculation of *Ochrobactrum* strain CdSP9 lowered the content of TBA reactive compounds in hydroponically grown *Oryza sativa* exposed to Cd. Wan et al. [30] also observed that the inoculation of *Serratia nematodiphila* LRE07 decreased the concentration of TBA reactive compounds in *Solanum nigrum* exposed to Cd under hydroponic conditions.

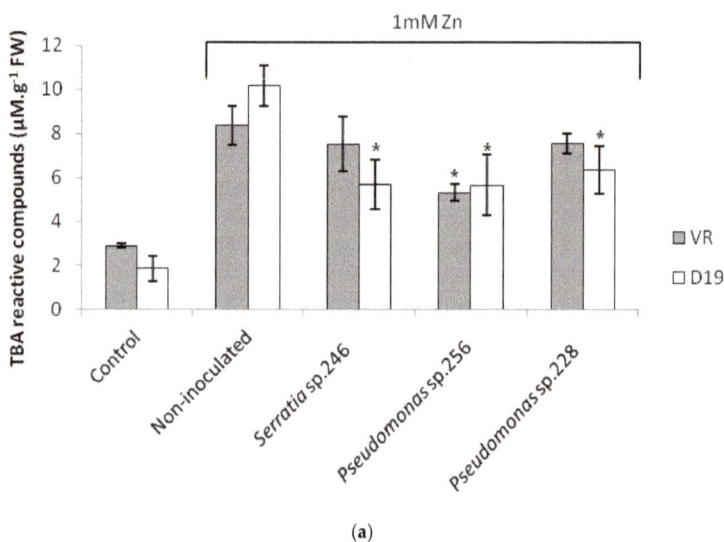

(a)

Figure 2. *Cont.*

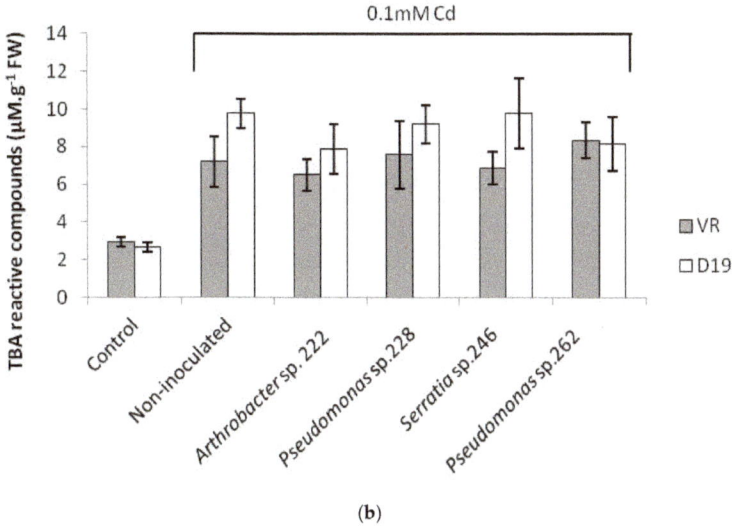

(b)

Figure 2. Thiobarbituric acid reactive compounds ($\mu M \cdot g^{-1}$ fresh weight) in roots of *H. tuberosus* cultivars VR and D19 after three weeks of exposure to: 1 mM of Zn (**a**); and 0.1 mM Cd (**b**). * Significant differences between inoculated and non-inoculated after Tukey's test, $p < 0.05$; mean values $\pm$ SE; $n = 4$.

These results suggest that the inoculated bacteria can assist metal exposed plants to keep the oxidative stress under control. In the inoculated plants, the Zn concentrations tended to decrease which could at least partially explain the lowering of TBA reactive compounds in inoculated plants.

2.4. Colonization of Enhanced Green Fluorescent Protein (EGFP): Tetracycline® Pseudomonas sp. 262 in the Roots of H. tuberosus

Pseudomonas sp. 262 was able to grow in the presence of 0.8 mM Cd and showed in vitro the capacity to produce siderophores (in absence of iron), organic acids, indole acetic acid, acetoin and ACC deaminase (Table 1). Moreover, this bacterial strain increased the shoot weight of the D19 cultivar of *H. tuberosus* exposed to 0.1 mM Cd. Taking this into account, *Pseudomonas* sp. 262 was selected to be labeled with the EGFP: tetracycline® plasmid to study the bacterial colonization of the roots of the *H. tuberosus* D19 cultivar.

Figure 3a demonstrates that the conjugation was effective, since *Pseudomonas* sp. 262 showed fluorescence after blue light (488 nm) excitation, and was able to grow in presence of tetracycline (20 $\mu g \cdot mL^{-1}$). In Figure 3b, EGFP-*Pseudomonas* sp. 262 can be seen as single cells attached to the surfaces of root hairs. Two days after inoculation, bacterial cells were also found inside the root hair which can be observed from the orthogonal plot (Figure 3c). These results support that *Pseudomonas* sp. 262 is an endophytic strain.

Ma et al. [70] reported that another *Pseudomonas* sp. A3R3 isolated from roots of *Alyssum serpyllifolium* showed a high level of colonization in root and shoot interior of *Brassica juncea*. He et al. [27] also observed that *Rahnella* sp. JN6, originally isolated from *Polygonum pubescens*, could colonize the root, stem and leaf tissues of *Brassica napus*. *Rahnella aquatilis* SPb, an endophytic bacterial strain from *Ipomoea batatas* was inoculated in hybrid poplar and increased the growth of the cuttings in comparison with non-inoculated conditions, illustrating the beneficial effects of the strains in growth of another plant species not related with the initial host plant [71].

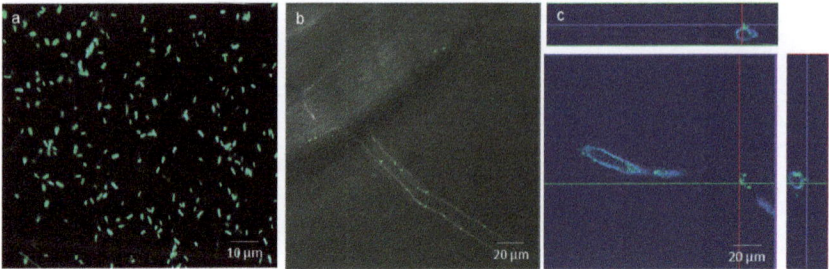

Figure 3. Confocal images of EGFP-labeled *Pseudomonas* sp. 262 colonising the root hairs of one-week-old seedlings of *H. tuberosus* D19 cultivar: (**a**) solution with EGFP-labeled *Pseudomonas* sp. strain 262 with blue light (488 nm) excitation; (**b**) single cells attached to a root hair; and (**c**) ortho-image of the root hair, showing bacterial cells (green) inside plant cells (in blue).

In our study, the EGFP-labeled bacterial strain was found in the root interior of *H. tuberosus* in the studied conditions. Since the inoculated bacterial cells were also found attached to the root hair surface, we suggest this one of the entry routes of the bacterial cells to the plant. Moreover, after inoculation of this strain, the growth of the *H. tuberosus* D19 cultivar improved significantly when exposed to Cd. This beneficial effect on plant growth, together with the visualization of the bacteria on the root hair surfaces and inside roots, indicates that a beneficial plant–microbe interaction was established.

3. Materials and Methods

3.1. Plant Material

Tubers of two cultivars of *H. tuberosus* (Violet de Rennes abbreviated as VR, and Blanc Précoce commonly named D19) were collected in spring in the field collection of IMIDRA (Instituto Madrileño de Investigación y Desarrollo Rural, Agrario y Alimentario; Madrid, Spain) to perform the experiments. The tubers were kept during two weeks at 4 °C for vernalization. After this period and before starting the experiments, the tubers were vigorously washed in tap water to remove the adhered soil.

3.2. PGPB Strains

Cultivable bacteria were isolated from soil, rhizosphere and plant-endosphere of *Brassica napus* growing on a Zn-contaminated site in Belgium [24]. Based on their PGP characteristics (Table 1), 3 Zn-tolerant strains (*Serratia* sp. strain 246, *Pseudomonas* sp. strain 228, and *Pseudomonas* sp. strain 256) and 4 Cd-tolerant strains (*Arthrobacter* sp. strain 222, *Pseudomonas* sp. strain 228, *Pseudomonas* sp. strain 262, and *Serratia* sp. strain 246) were selected to inoculate *H. tuberosus*. *Serratia* sp. strain 246 and *Pseudomonas* sp. strain 228 were inoculated in the presence of Zn and Cd because both strains showed high tolerance to grow with both metals. The strains were grown in 869 liquid medium [72] at 30 °C under shaking conditions.

3.3. Inoculation of PGPB Strains in H. tuberosus

Tuber slices with buds were incubated in 1 L plastic pots filled with moist quartz sand that were placed in a growth chamber at 25/12 °C, 14/12 h of photoperiod. The following conditions were established after one week of growth: (i) control plants grown in sand without metal and bacteria; (ii) non-inoculated, metal-exposed plants grown in the presence of metals (Cd or Zn), but without bacteria; (iii) plants inoculated with bacterial strains (*Arthrobacter* sp. 222, *Pseudomonas* sp. 228, 262 and *Serratia* sp. 246) and grown in presence of Cd; and (iv) plants inoculated with bacterial strains (*Pseudomonas* sp. 228, 256 and *Serratia* sp. 246) and grown in presence of Zn.

Half strength modified Hoagland's solution (1 mM Ca $(NO_3)_2 \cdot 4H_2O$, 1.5 mM KNO_3, 0.5 mM $NH_4H_2PO_4$, 0.25 mM $MgSO_4 \cdot 7H_2O$, 1 µM $MnSO_4 \cdot H_2O$, 12.5 µM H_3BO_3, 0.25 µM $(NH_4)_6Mo_7O_4$,

0.05 μM $CuSO_4 \cdot 5H_2O$, 1 μM $ZnSO_4 \cdot 7H_2O$, 10 μM $NaFe^{III}$-EDTA, and demineralized water buffered with 1 mM of 2-(*N*-morpholino) ethanesulfonic acid, at pH 5.5 ± 0.5) was added to the sand until saturation. Metal exposures were performed by adding 0.1 mM of Cd (added as $CdSO_4 \cdot 8H_2O$) or 1 mM of Zn (added as $ZnSO_4 \cdot 7H_2O$) to the nutrient solution. Plants were watered every two days with the nutrient solution supplemented with metals. The metal concentrations used in this experiment were chosen based on a former study [73]. Two plants were put per pot with four independent replicates per treatment. The bacterial suspension (10^8 cfu·mL^{-1}) in buffer (10 mM $MgSO_4$) was added into the pots. Buffer (10 mM $MgSO_4$) without bacteria was added to the controls. After 3 weeks of growth, plants were harvested.

3.4. Plant Analysis

After harvest, the roots were rinsed in 10 mM sodium ethylenediaminetetraacetic acid (Na_2EDTA) to remove the adhering metal-containing particles, and subsequently washed in distilled water. Plants were subdivided into leaves, stems and roots, weighed and dried in a forced air oven for 48 h at 60 °C to determine the dry weights. Subsequently, the dried tissues were individually ground and digested (30 mg) according to [47]. Total concentrations of metals and macro/micronutrients were determined by flame atomic absorption spectrometry (Fast Sequential Model AA240FS, Varian, Santa Clara, CA, USA). The quality of the digestion and analytical methods was verified by including blanks and certified reference materials (NCS DC73348 Brush Branches and Leaves, China National Analysis Center for Iron and Steel, and CTA-VTL-2 Virginia Tobacco Leaves, Polish Academy of Sciences and Institute of Nuclear Chemistry and Technology) with every set of samples. The recovery percentages for metals were: Cd (~95%) and Zn (~101%).

The membrane lipid peroxidation in the plant tissues was estimated in terms of the content of thiobarbituric acid reactive (TBA) compounds according to the method of [74], modified by [75]. The calibration curve was carried out with every set of samples, using 1,1,3,3-Tetraethoxypropane (TEP) as precursor of malondialdehyde (MDA). Absorbances were determined with a UV–Vis light spectrophotometer (Thermo Spectronic Helios Alpha, Thermo Fisher Scientific, Madison, WI, USA).

3.5. Evaluation of the Colonization Process: Localization of Inoculated EGFP Labeled Pseudomonas sp. 262

3.5.1. Bacterial Strains and Growth Conditions

The receptor, *Pseudomonas* sp. 262 was grown in 284 minimal medium [76] supplemented with 0.4 mM of Cd (added as $CdSO_4 \cdot 8H_2O$) at 30 °C. The donor, *Escherichia coli* strain dH5a, carrying the EGFP pMP4655 plasmid, was grown in 869 medium [72] supplemented with 20 μg·mL^{-1} tetracycline at 30 °C. The helper, *E. coli* strain dH5a, carrying the pRK2013 plasmid, was grown in 869 medium at 30 °C. Donor and helper were constructed in the Institute of Biology Leiden, Leiden University (The Netherlands) [39].

3.5.2. Introduction of the EGFP: Tetracycline into Pseudomonas sp. 262

Triparental mating was carried out to label *Pseudomonas* sp. 262 with the EGFP: tetracycline®plasmid. The strains were grown in 869 medium at 30 °C under shaking conditions. Growth curves were obtained by diluting an overnight culture in order to verify the time needed to reach the appropriate optical density (OD) for conjugation (donor and helper OD 0.3–0.4, and receptor OD 0.7). The OD was measured at 660 nm every 30 min using a Visible Diode Array Spectrophotometer, Novaspec Plus, Amersham Biosciences, Piscataway, New York, United States. Once the appropriate OD was reached, the bacterial strains were centrifuged at 3000 rpm during 10 min, and then added to the mating filter in a Petri dish with 869 medium. After the conjugation, 284 minimal medium supplemented with 0.4 mM of Cd (added as $CdSO_4 \cdot 8H_2O$) and tetracycline (20 μg·mL^{-1}) was used to isolate the receptor labeled strains. Fluorescence of the strains was checked using a Nikon 80i fluorescence microscope (High-pressure Mercury Lamp;

Excitation filters: 465–495 nm, dichroic mirror 505 nm, emission filter 515–555 nm. Objectives used: 40×/0.95 Air Plan Apo WD 0.14 mm and 100×/1.25 Oil Plan Apo WD 0.17 mm).

3.5.3. Inoculation of EGFP *Pseudomonas* sp. 262 on Roots of *H. tuberosus*

Tuber slices with buds of *H. tuberosus* cultivar D19 were grown on coarse perlite moistened with a half strength modified Hoagland's solution (see above) under greenhouse conditions (25–30 °C temperature and 70–90% relative humidity). The bacterial suspension (10^8 cfu·mL^{-1}) was added to the pots (0.2 L) after appearance of the first roots (at 5 days). Four repetitions were used.

3.5.4. Confocal Laser Scanning Microscopy

After 48 h of incubation, one-week-old plant roots were washed to remove weakly adhered bacterial cells and, subsequently, intact root preparations (at 25 °C) were observed with a Zeiss LSM510 confocal laser scanning microscope (Carl Zeiss, Jena, Germany) mounted on an Axiovert 200M. The objective used was 40×/1.1 water immersion (Zeiss LD C-Apochomat 40×/1.1 WKorr UV–VIS-IR, Carl Zeiss).

Excitation was performed at 488 nm using an Argon laser source. Backward GFP signal was filtered using a 500–550 nm band pass filter. Images were edited using the software Zen 2009 Light Edition (Carl Zeiss MicroImaging GmbH, Jena, Germany).

3.6. Statistical Analysis

Statistical analysis of data was performed using the IBM SPSS Statistics 19.0 software (Armonk, NY, USA). Two-way analysis of variance (ANOVA) and Tukey's test were applied. Differences at $p < 0.05$ levels were considered significant.

4. Conclusions

The effects of the bacterial strains on the growth of *H. tuberosus* differed in function of the metal, the inoculated bacterial strain and the plant cultivar. The improvement of growth and the decrease of the metal-induced stress were more pronounced in the D19 cultivar than in the VR cultivar. Three endophytes of *Brassica napus* enhanced the growth of the D19 cultivar exposed Cd or Zn. Only *Pseudomonas* sp. 228 increased Cd uptake. Using confocal microscopy, we observed that, two days after inoculation, EGFP-labeled *Pseudomonas* sp. 262 colonized the root surface and interior of *H. tuberosus*. In combination with the growth promotion that was observed after inoculation, this demonstrates an established plant–microbe interaction. Therefore, use of the D19 cultivar in combination with *Pseudomonas* sp. 228, *Serratia* sp. 246 and *Pseudomonas* sp. 262 holds promise for application in phytoremediation strategies on Cd-Zn contaminated soils.

Supplementary Materials: Supplementary materials can be found at www.mdpi.com/1422-0067/18/10/2026/s1.

Acknowledgments: We would like to acknowledge Rik Paesen from the Biomedical Research Institute of Hasselt University for his assistance in the confocal laser microscope and Marta Letón Rojo from Agro-environmental department of Instituto Madrileño de Investigación y Desarrollo Rural, Agrario y Alimentario (IMIDRA) for her help in Flame Atomic Absorption measurements. We would like also to thank to Instituto Nacional de Investigación y Tecnología Agraria y Alimentaria (INIA) to support the grant Formación de Personal Investigador FPI-INIA 2010 of B. Montalbán, EIADES Project S2009/AMB-1478 (Comunidad de Madrid) and the UHasselt Methusalem project 08M03VGRJ. S. Thijs and N. Weyens are grateful to the FWO (Fund for Scientific Research Flanders) for, respectively, a PhD and post-doc grant.

Author Contributions: Blanca Montalbán, Araceli Pérez-Sanz, Mª Carmen Lobo, Jaco Vangronsveld and Nele Weyens conceived and designed the plant experiments. Blanca Montalbán, Araceli Pérez-Sanz and Mª Carmen Lobo analyzed the data. Sofie Thijs, Blanca Montalbán and Marcel Ameloot designed, performed and analyzed the colonization assays and localization of inoculated EGFP labeled bacteria with confocal laser scanning microscopy. Blanca Montalbán wrote the paper.

Conflicts of Interest: The authors declare no conflict of interest. The founding sponsors had no role in the design of the study; in the collection, analyses, or interpretation of data; in the writing of the manuscript, and in the decision to publish the results.

Abbreviations

PGPB	Plant Growth Promoting Bacteria
TBA	Thiobarbituric Acid
EGFP	Enhanced Green Fluorescent Protein
EDTA	Ethylenediaminetetraacetic Acid
MDA	Malondialdehyde

References

1. Mulligan, C.N.; Yong, R.N.; Gibbs, B.F. Heavy metal removal from sediments by biosurfactants. *J. Hazard Mater.* **2001**, *85*, 111–125. [CrossRef]
2. Weyens, N.; van der Lelie, D.; Taghavi, S.; Newman, L.; Vangronsveld, J. Exploiting plant-microbe partnerships to improve biomass production and remediation. *Trends Biotechnol.* **2009**, *27*, 591–598. [CrossRef] [PubMed]
3. Ma, Y.; Rajkumar, M.; Zhang, C.H.; Freitas, H. Beneficial role of bacterial endophytes in heavy metal phytoremediation. *J. Environ. Manag.* **2016**, *174*, 14–25. [CrossRef] [PubMed]
4. Coninx, L.; Martinova, V.; Rineau, F. Mycorrhiza-Assisted Phytoremediation. In *Phytoremediation. Advances in Botanical Research*; Cuypers, A., Vangronsveld, J., Eds.; Elsevier: Amsterdam, The Netherlands, 2017; Volume 83, pp. 127–188.
5. Kidd, P.S.; Álvarez-López, V.; Becerra-Castro, C.; Cabello-Conejo, M.; Prieto-Fernández, Á. Potential role of plant-associated bacteria in plant metal uptake and implications in phytotechnologies. In *Phytoremediation. Advances in Botanical Research*; Cuypers, A., Vangronsveld, J., Eds.; Elsevier: Amsterdam, The Netherlands, 2017; Volume 83, pp. 87–126.
6. Germida, J.J.; Siciliano, S.D.; Renato de Freitas, J.; Seib, A.M. Diversity of root-associated bacteria associated with field-grown canola (*Brassica napus* L.) and wheat (*Triticuma estivum* L.). *FEMS Microbiol. Ecol.* **1998**, *26*, 43–50. [CrossRef]
7. Genrich, I.; Burd, D.; George, D.; Glick, B.R. Plant growth promoting bacteria that decrease heavy metal toxicity in plants. *Can. J. Microbiol.* **2000**, *46*, 237–245.
8. Rajkumar, M.; Sandhya, S.; Prasad, M.N.V.; Freitas, H. Perspectives of plant-associated microbes in heavy metal phytoremediation. *Biotechnol. Adv.* **2012**, *30*, 1562–1574. [CrossRef] [PubMed]
9. Ullah, A.; Heng, S.; Munis, M.F.H.; Fahads, S.; Yang, X. Phytoremediation of heavy metals assisted by plant growth promoting (PGP) bacteria: A review. *Environ. Exp. Bot.* **2015**, *117*, 28–40. [CrossRef]
10. Lugtenberg, B.; Kamilova, F. Plant-Growth-Promoting Rhizobacteria. *Annu. Rev. Microbiol.* **2009**, *63*, 541–556. [CrossRef] [PubMed]
11. Roper, M.M.; Ladha, J.K. Biological N_2-fixation by heterotrophic and phototrophic bacteria in association with straw. *Plant Soil* **1995**, *174*, 211–224. [CrossRef]
12. Kim, K.Y.; Jordan, D.; McDonald, G.A. Enterobacter agglomerans, phosphate solubilizing bacteria, and microbial activity in soil: Effect of carbon sources. *Soil Biol. Biochem.* **1998**, *30*, 995–1003. [CrossRef]
13. Ryan, R.P.; Germaine, K.; Franks, A.; Ryan, D.J.; Dowling, D.N. Bacterial endophytes: Recent developments and applications. *FEMS Microbiol. Lett.* **2008**, *278*, 1–9. [CrossRef] [PubMed]
14. Dobbelaere, S.; Croonenborghs, A.; Thys, A.; Vandebroek, A.; Vanderleyden, J. Phytostimulatory effect of *Azospirillum brasilense* wild type and mutant strains altered in IAA production on wheat. *Plant Soil* **1999**, *212*, 155–164. [CrossRef]
15. Glick, B.R.; Penrose, D.M.; Li, J. A model for the lowering of plant ethylene concentrations by plant growth promoting bacteria. *J. Theor. Biol.* **1998**, *190*, 63–68. [CrossRef] [PubMed]
16. Glick, B.R. Phytoremediation: Synergistic use of plants and bacteria to clean up the environment. *Biotechnol. Adv.* **2003**, *21*, 383–393. [CrossRef]
17. Glick, B.R.; Bashan, Y. Genetic manipulation of plant growth-promoting bacteria to enhance biocontrol of phytopathogens. *Biotechnol. Adv.* **1997**, *15*, 353–378. [CrossRef]
18. Fasim, F.; Ahmed, N.; Parsons, R.; Gadd, G.M. Solubilization of zinc salts by a bacterium isolated from the air environment of a tannery. *FEMS Microbiol. Lett.* **2002**, *213*, 1–6. [CrossRef] [PubMed]
19. Diels, L.; De Smet, M.; Hooyberghs, L.; Corbisier, P. Heavy metals bioremediation of soil. *Mol. Biotechnol.* **1999**, *12*, 149–158. [CrossRef]

20. Rouch, D.A.; Lee, T.O.B.; Morby, A.P. Understanding cellular responses to toxic agents: A model for mechanisms-choice in bacterial resistance. *J. Ind. Microbiol.* **1995**, *14*, 132–141. [CrossRef] [PubMed]

21. Madhaiyan, M.; Poonguzhali, S.; Sa, T. Influence of plant species and environmental conditions on epiphytic and endophytic pink-pigmented facultative methylotrophic bacterial populations associated with field-grown rice cultivars. *J. Microbiol. Biotechnol.* **2007**, *17*, 1645–1654. [PubMed]

22. Vivas, A.; Biró, B.; Ruíz-Lozano, J.M.; Barea, J.M.; Azcón, R. Two bacterial strains isolated from a Zn-polluted soil enhance plant growth and mycorrhizal efficiency under Zn-toxicity. *Chemosphere* **2006**, *62*, 1523–1533. [CrossRef] [PubMed]

23. Ma, Y.; Prasad, M.N.V.; Rajkumar, M.; Freitas, H. Plant growth promoting rhizobacteria and endophytes accelerate phytoremediation of metalliferous soils. *Biotechnol. Adv.* **2011**, *29*, 248–258. [CrossRef] [PubMed]

24. Montalbán, B.; Croes, S.; Weyens, N.; Lobo, M.C.; Pérez-Sanz, A.; Vangronsveld, J. Characterization of bacterial communities associated with *Brassica napus* L. growing on a Zn-contaminated soil and their effects on root growth. *Int. J. Phytoremediat.* **2016**, *18*, 985–993.

25. Chen, B.; Shen, J.; Zhang, X.; Pan, F.; Yang, X.; Feng, Y. The endophytic bacterium, *Sphingomonas* SaMR12, improves the potential for zinc phytoremediation by its host, *Sedum alfredii*. *PLoS ONE* **2014**, *9*, e106826. [CrossRef] [PubMed]

26. Sheng, X.; Sun, L.; Huang, Z.; He, L.; Zhang, W.; Chen, Z. Promotion of growth and Cu accumulation of bio-energy crop (*Zea mays*) by bacteria: Implications for energy plant biomass production and phytoremediation. *J. Environ. Manag.* **2012**, *103*, 58–64. [CrossRef] [PubMed]

27. He, H.; Ye, Z.; Yang, D.; Yan, J.; Xiao, L.; Zhong, T.; Yuan, M.; Cai, X.; Fang, Z.; Jing, Y. Characterization of endophytic *Rahnella* sp. JN6 from *Polygonum pubescens* and its potential in promoting growth and Cd, Pb, Zn uptake by *Brassica napus*. *Chemosphere* **2013**, *90*, 1960–1965. [CrossRef] [PubMed]

28. Sessitsch, A.; Kuffner, M.; Kidd, P.; Vangronsveld, J.; Wenzel, W.W.; Fallmann, K.; Puschenreiter, M. The role of plant-associated bacteria in the mobilization and phytoextraction of trace elements in contaminated soils. *Soil Biol. Biochem.* **2013**, *60*, 182–194. [CrossRef] [PubMed]

29. Rajkumar, M.; Ae, N.; Freitas, H. Endophytic bacteria and their potential to enhance heavy metal phytoextraction. *Chemosphere* **2009**, *77*, 153–160. [CrossRef] [PubMed]

30. Wan, Y.; Luo, S.; Chen, J.; Xiao, X.; Chen, L.; Zeng, G.; Liu, C.; He, Y. Effect of endophyte-infection on growth parameters and Cd-induced phytotoxicity of Cd-hyperaccumulator *Solanum nigrum* L. *Chemosphere* **2012**, *89*, 743–750. [CrossRef] [PubMed]

31. Pandey, S.; Ghosh, P.K.; Ghosh, S.; Kumar, D.T.; Maiti, T.K. Role of heavy metal resistant *Ochrobactrum* sp. and *Bacillus* spp. strains in bioremediation of a rice cultivar and their PGPR like activities. *J. Microbiol.* **2013**, *5*, 7–11. [CrossRef] [PubMed]

32. Weyens, N.; van der Lelie, D.; Taghavi, S.; Vangronsveld, J. Phytoremediation: Plant-endophyte partnerships take the challenge. *Curr. Opin. Biotechnol.* **2009**, *20*, 248–254. [CrossRef] [PubMed]

33. Lugtenberg, B.J.J.; Dekkers, L.; Bloemberg, G.V. Molecular determinants of rhizosphere colonization by *Pseudomonas*. *Annu. Rev. Phytopathol.* **2001**, *39*, 461–490. [CrossRef] [PubMed]

34. Compant, S.; Clément, C.; Sessitsch, A. Plant growth-promoting bacteria in the rhizo- and endosphere of plants: Their role, colonization, mechanisms involved and prospects for utilization. *Soil Biol. Biochem.* **2010**, *42*, 669–678. [CrossRef]

35. Lagendijk, E.L.; Validov, S.; Lamers, G.E.M.; de Weert, S.; Bloemberg, G.V. Genetic tools for tagging Gram-negative bacteria with mCherry for visualization in vitro and in natural habitats, biofilm and pathogenicity studies. *FEMS Microbiol. Lett.* **2010**, *305*, 81–90. [CrossRef] [PubMed]

36. Bloemberg, G.V.; Lugtenberg, B.J.J. Molecular basis of plant growth promotion and biocontrol by rhizobacteria. *Curr. Opin. Plant Biol.* **2001**, *4*, 343–350. [CrossRef]

37. Germaine, K.; Keogh, E.; Borremans, B.; van der Lelie, D.; Barac, T.; Oeyen, L.; Vangronsveld, J.; Moore, F.P.; Moore, E.R.B.; Campbel, C.D.; et al. Colonisation of poplar trees by gfp expressing endophytes. *FEMS Microbiol. Ecol.* **2004**, *48*, 109–118. [CrossRef] [PubMed]

38. Weyens, N.; Boulet, J.; Adriaensen, D.; Timmermans, J.P.; Prinsen, E.; Van Oevelen, S.; D'Haen, J.; Smeets, K.; van der Lelie, D.; Taghavi, S.; et al. Contrasting colonization and plant growth promoting capacity between wild type and a gfp-derative of the endophyte *Pseudomonas putida* W619 in hybrid poplar. *Plant Soil* **2012**, *356*, 217–230. [CrossRef]

39. Bloemberg, G.V.; Wijfjes, A.H.; Lamers, G.E.; Stuurman, N.; Lugtenberg, B.J. Simultaneous imaging of *Pseudomonas fluorescens* WCS365 populations expressing three different autofluorescent proteins in the rhizosphere: New perspectives for studying microbial communities. *Mol. Plant Microbe Interact.* **2000**, *13*, 1170–1176. [CrossRef] [PubMed]

40. Newman, K.L.; Almeida, R.P.P.; Purcell, A.H.; Lindow, S.E. Use of a green fluorescent strain for analysis of *Xylella fastidiosa* colonization of *Vitis vinifera*. *Appl. Environ. Microbiol.* **2003**, *69*, 7319–7327. [CrossRef] [PubMed]

41. Serieys, H.; Souyris, I.; Gil, A.; Poinso, B.; Berville, A. Diversity of Jerusalem artichoke clones (*Helianthus tuberosus* L.) from the INRA-Montpellier collection. *Genet. Resour. Crop Evol.* **2010**, *57*, 1207–1215. [CrossRef]

42. Denoroy, P. The crop physiology of *Helianthus tuberosus* L.: A model oriented view. *Biomass Bioenergy* **1996**, *11*, 11–32. [CrossRef]

43. Kays, S.J.; Nottingham, S.F. *Biology and Chemistry of Jerusalem Artichoke Helianthus tuberosus L.*; Taylor and Francis Group: Boca Raton, FL, USA; Abingdon, UK; New York, NY, USA, 2007; pp. 1–496, ISBN 9781420044959.

44. Cui, S.; Zhou, Q.; Chao, L. Potential hyperaccumulation of Pb, Zn, Cu and Cd in endurant plants distributed in an old smeltery, northeast China. *Environ. Geol.* **2007**, *51*, 1043–1048. [CrossRef]

45. Chen, L.; Long, X.; Zhang, Z.; Zheng, X.; Rengel, Z.; Liu, Z. Cadmium accumulation and translocation in two Jerusalem Artichoke (*Helianthus tuberosus* L.) Cultivars. *Pedosphere* **2011**, *21*, 573–580. [CrossRef]

46. Long, X.; Ni, N.; Wang, L.; Wang, X.; Wang, Y.; Zhang, Z.; Zed, R.; Liu, Z.; Shao, H. Phytoremediation of cadmium-contaminated soil by two Jerusalem Artichoke (*Helianthus tuberosus* L.) genotypes. *CLEAN Soil Air Water* **2013**, *41*, 202–209. [CrossRef]

47. Montalbán, B.; Lobo, M.C.; Alonso, J.; Pérez-Sanz, A. Metal(loid)s uptake and effects on the growth of *Helianthus tuberosus* cultivar-clones under multi-polluted hydroponic cultures. *CLEAN Soil Air Water* **2016**, *44*, 1368–1374. [CrossRef]

48. Willscher, S.; Jablonski, L.; Fona, Z.; Rahmi, R.; Wittig, J. Phytoremediation experiments with *Helianthus tuberosus* under different pH and heavy metal soil concentrations. *Hydrometallurgy* **2016**, *168*, 153–158. [CrossRef]

49. Zaidi, S.; Usmani, S.; Singh, B.R.; Musarrat, J. Significance of *Bacillus subtilis* strain SJ-101 as a bioinoculant for concurrent plant growth promotion and nickel accumulation in *Brassica juncea*. *Chemosphere* **2006**, *64*, 991–997. [CrossRef] [PubMed]

50. Sheng, X.F.; Xia, J.J. Improvement of rape (*Brassica napus*) plant growth and cadmium uptake by cadmium-resistant bacteria. *Chemosphere* **2006**, *64*, 1036–1042. [CrossRef] [PubMed]

51. Sheng, X.F.; Xia, J.J.; Jiang, C.Y.; He, L.Y.; Qian, M. Characterization of heavy metal-resistant endophytic bacteria from rape (*Brassica napus*) roots and their potential in promoting the growth and lead accumulation of rape. *Environ. Pollut.* **2008**, *156*, 1164–1170. [CrossRef] [PubMed]

52. Duan, J.; Jiang, W.; Cheng, Z.; Heikkila, J.J.; Glick, B.R. The Complete Genome Sequence of the Plant Growth-Promoting Bacterium *Pseudomonas* sp. UW4. *PLoS ONE* **2013**, *8*, e58640. [CrossRef] [PubMed]

53. Glick, B.R. Using soil bacteria to facilitate phytoremediation. *Biotechnol. Adv.* **2010**, *28*, 367–374. [CrossRef] [PubMed]

54. Burd, G.I.; Dixon, D.G.; Glick, B.R. A plant growth-promoting bacterium that decreases nickel toxicity in seedlings. *Appl. Environ. Microbiol.* **1998**, *64*, 3663–3668. [PubMed]

55. Marques, A.P.G.C.; Moreira, H.; Franco, A.R.; Rangel, A.O.S.S.; Castro, P.M.L. Inoculating *Helianthus annuus* (sunflower) grown in zinc and cadmium contaminated soils with plant growth promoting bacteria—Effects on phytoremediation strategies. *Chemosphere* **2013**, *92*, 74–83. [CrossRef] [PubMed]

56. Tripathi, M.; Munot, H.P.; Shouche, Y.; Meyer, J.M.; Goel, R. Isolation and functional characterization of siderophore-producing lead- and cadmium-resistant *Pseudomonas putida* KNP9. *Curr. Microbiol.* **2005**, *50*, 233–237. [CrossRef] [PubMed]

57. Aafi, N.E.; Brhada, F.; Dary, M.; Maltouf, A.F.; Pajuelo, E. Rhizostabilization of metals in soils using *Lupinus luteus* inoculated with the metal resistant rhizobacterium *Serratia* sp. MSMC 541. *Int. J. Phytoremediat.* **2012**, *14*, 261–274. [CrossRef] [PubMed]

58. Nies, D.H. Microbial heavy-metal resistance. *Appl. Microbiol. Biotechnol.* **1999**, *51*, 730–750. [CrossRef] [PubMed]

59. Bruins, M.R.; Kapil, S.; Oehme, F.W. Microbial resistance to metals in the environment. *Ecotoxicol. Environ. Saf.* **2000**, *45*, 198–207. [CrossRef] [PubMed]

60. Chen, H.; Cutright, T.J. Preliminary evaluation of microbiallymediated precipitation of cadmium, chromium, and nickel by rhizosphere consortium. *J. Environ. Eng.* **2003**, *129*, 4–9. [CrossRef]

61. Kidd, P.; Barceló, J.; Bernal, M.P.; Navari-Izzo, F.; Poschenrieder, C.; Shilev, S.; Clemente, R.; Monterroso, C. Trace element behaviour at the root–soil interface: Implications in Phytoremediation. *Environ. Exp. Bot.* **2009**, *67*, 243–259. [CrossRef]

62. Vecchio, A.; Finoli, C.; Di Simine, D.; Andreoni, V. Heavy metal biosorption by bacterial cells. *Fresen. J. Anal. Chem.* **1998**, *361*, 338–342. [CrossRef]

63. Li, Y.; Zhang, S.; Jiang, W.; Liu, D. Cadmium accumulation, activities of antioxidant enzymes, and malondialdehyde (MDA) content in *Pistiastratiotes* L. *Environ. Sci. Pollut. Res.* **2013**, *20*, 1117–1123. [CrossRef] [PubMed]

64. Smeets, K.; Cuypers, A.; Lambrechts, A.; Semane, B.; Hoet, P.; Van Laere, A.; Vangronsveld, J. Induction of oxidative stress and antioxidative mechanisms in *Phaseolus vulgaris* after Cd application. *Plant Physiol. Biochem.* **2005**, *43*, 437–444. [CrossRef] [PubMed]

65. Wang, C.; Zhang, S.H.; Wang, P.F.; Hou, J.; Zhang, W.J.; Li, W.; Lin, Z.P. The effect of excess Zn on mineral nutrition and antioxidative response in rapeseed seedlings. *Chemosphere* **2009**, *75*, 1468–1476. [CrossRef] [PubMed]

66. Bauddh, K.; Singh, R.P. Growth, tolerance, efficiency and phytoremediation potential of *Ricinuscommunis* (L.) and *Brassica juncea* (L.) in salinity and drought affected cadmium contaminated soil. *Ecotoxicol. Environ. Saf.* **2012**, *85*, 13–22. [CrossRef] [PubMed]

67. Jozefczak, M.; Keunen, E.; Schat, H.; Bliek, M.; Hernández, L.E.; Carleer, R.; Remans, T.; Bohler, S.; Vangronsveld, J.; Cuypers, A. Differential response of *Arabidopsis* leaves and roots to cadmium: Glutathione-related chelating capacity vs antioxidant capacity. *Plant Physiol. Biochem.* **2014**, *83*, 1–9. [CrossRef] [PubMed]

68. Nouairi, I.; Ammar, W.B.; Youssef, N.B.; Miled, D.D.B.; Ghorbal, M.H.; Zarrouk, M. Antioxidant defense system in leaves of Indian mustard (*Brassica juncea*) and rape (*Brassica napus*) under cadmium stress. *Acta Physiol. Plant.* **2009**, *31*, 237–247. [CrossRef]

69. Zhang, F.Q.; Wang, Y.S.; Lou, Z.P.; Dong, J.D. Effect of heavy metal stress on antioxidative enzymes and lipid peroxidation in leaves and roots of two mangrove plant seedlings (*Kandelia candel* and *Bruguiera gymnorrhiza*). *Chemosphere* **2007**, *67*, 44–50. [CrossRef] [PubMed]

70. Ma, Y.; Rajkumar, M.; Luo, Y.M.; Freitas, H. Inoculation of endophytic bacteria on host and non-host plants—Effects on plant growth and Ni uptake. *J. Hazard. Mater.* **2011**, *195*, 230–237. [CrossRef] [PubMed]

71. Khan, Z.; Doty, S.L. Characterization of bacterial endophytes of sweet potato plants. *Plant Soil* **2009**, *322*, 197–207. [CrossRef]

72. Mergeay, M.; Nies, D.; Schlegel, H.G.; Gerits, J.; Charles, P.; Van Gijsegem, F. *Alcaligenes eutrophus* CH34 is a facultative chemolithotroph with plasmid-bound resistgance to heavy metals. *J. Bacteriol.* **1985**, *162*, 328–334. [PubMed]

73. Montalbán, B.; García-Gonzalo, P.; Pradas del Real, A.E.; Alonso, J.; Lobo, M.C.; Pérez-Sanz, A. *Brachypodium distachyon* tolerance to metals under in vitro conditions: A comparison with two metal-tolerant energy crops. *Fresen. Environ. Bull.* **2014**, *23*, 2086–2092.

74. Reilly, C.A.; Aust, S.D. Measurement of lipid peroxidation. *Curr. Protoc. Toxicol.* **2001**, *28*, 659–671.

75. Catalá, M.; Gasulla, F.; Pradas del Real, A.E.; Garcia-Breijo, F.; Reig-Arminana, J.; Barreno, E. Fungal-associated NO is involved in the regulation of oxidative stress during rehydration in lichen symbiosis. *BMC Microbiol.* **2010**, *10*, 297. [CrossRef] [PubMed]

76. Schlegel, H.G.; Kaltwasser, H.; Gottschalk, G. Ein Submersverfahren zur Kultur wassers toffoxy dieren der Bakterien: Wachstums physiologische Untersuchungen. *Arch. Mikrobiol.* **1961**, *38*, 209–222. [CrossRef] [PubMed]

International Journal of
Molecular Sciences

MDPI

Article

Synergistic Effects of *Bacillus amyloliquefaciens* (GB03) and Water Retaining Agent on Drought Tolerance of Perennial Ryegrass

An-Yu Su [1,†], Shu-Qi Niu [1,†], Yuan-Zheng Liu [1,†], Ao-Lei He [1], Qi Zhao [1], Paul W. Paré [2], Meng-Fei Li [3], Qing-Qing Han [1], Sardar Ali Khan [1] and Jin-Lin Zhang [1,3,*]

[1]   State Key Laboratory of Grassland Agro-ecosystems, College of Pastoral Agriculture Science and Technology, Lanzhou University, Lanzhou 730020, China; suay13@lzu.edu.cn (A.-Y.S.); niushq14@lzu.edu.cn (S.-Q.N.); liuyzh13@lzu.edu.cn (Y.-Z.L.); heal15@lzu.edu.cn (A.-L.H.); qzhao@lzu.edu.cn (Q.Z.); hanqq16@lzu.edu.cn (Q.-Q.H.); ali.khan13@lzu.edu.cn (S.A.K.)
[2]   Department of Chemistry and Biochemistry, Texas Tech University, Lubbock, TX 79409, USA; paul.pare@ttu.edu
[3]   College of Life Science and Technology, Gansu Agricultural University, Lanzhou 730070, China; lmf@gsau.edu.cn
*   Correspondence: jlzhang@lzu.edu.cn; Tel.: +86-931-8912357
†   These authors contributed equally to this work.

Received: 11 October 2017; Accepted: 4 December 2017; Published: 11 December 2017

Abstract: Water retaining agent (WRA) is widely used for soil erosion control and agricultural water saving. Here, we evaluated the effects of the combination of beneficial soil bacterium *Bacillus amyloliquefaciens* strain GB03 and WRA (the compound is super absorbent hydrogels) on drought tolerance of perennial ryegrass (*Lolium perenne* L.). Seedlings were subjected to natural drought for maximum 20 days by stopping watering and then rewatered for seven days. Plant survival rate, biomass, photosynthesis, water status and leaf cell membrane integrity were measured. The results showed that under severe drought stress (20-day natural drought), compared to control, GB03, WRA and GB03+WRA all significantly improved shoot fresh weight, dry weight, relative water content (RWC) and chlorophyll content and decreased leaf relative electric conductivity (REC) and leaf malondialdehyde (MDA) content; GB03+WRA significantly enhanced chlorophyll content compared to control and other two treatments. Seven days after rewatering, GB03, WRA and GB03+WRA all significantly enhanced plant survival rate, biomass, RWC and maintained chlorophyll content compared to control; GB03+WRA significantly enhanced plant survival rate, biomass and chlorophyll content compared to control and other two treatments. The results established that GB03 together with water retaining agent promotes ryegrass growth under drought conditions by improving survival rate and maintaining chlorophyll content.

Keywords: *Bacillus amyloliquefaciens*; perennial ryegrass; water retaining agent; synergistic effects; drought tolerance

1. Introduction

Drought as one of the major abiotic stresses has been weighing heavily against the agricultural productivity worldwide [1,2], since most of crops and forage plants grown to feed the global population are highly sensitive to drought [3]. Drought also induces severe desertification, with a progressive reduction of the vegetation cover coupled with rapid soil erosion in arid and semi-arid climatic regions [4,5]. Drought affects water potential and turgor in plants, resulting in the changes of physiological and morphological traits. Fresh weight and relative water content are two parameters commonly adopted to measure the impact of drought stress on plant growth [6]. Drought decreases plant chlorophyll content, which is directly related to photosynthesis rate [7]. Drought is also known to increase the reactive oxygen species (ROS) in plant cells, which are well recognized for lipid peroxidation and cell membrane deterioration, resulting in secondary oxidative stress [8,9]. Among various impacts of drought stress on plant growth, nutrient and water availability are mainly discussed [10,11].

Plant growth promoting rhizobacteria (PGPRs) are microorganisms associated with plant roots and can confer beneficial effects on the host plants [12]. Early research reported that the PGPR *Paenibacillus polymyxa* enhanced drought tolerance of *Arabidopsis thaliana* [13]. *Bacillus amyloliquefaciens* had been applied to several commercial crops and shown the remarkable effects on the increase in plant growth, disease resistance as well as salt and drought tolerance [14–16]. *B. amyloliquefaciens* strain GB03 enhanced growth and abiotic stress tolerance in *Arabidopsis* by emitting a complex blend of volatile organic compounds (VOCs) [17–21]. These VOCs activated differential expression of approximately 600 transcripts including genes related to cell wall modifications, primary and secondary metabolism, hormone regulation and stress response [19]. Recent studies reported that GB03 also promoted growth and salt tolerance in wheat (*Triticum aestivum*) [22], white clover (*Trifolium repens* L.) [23] and a halophytic grass *Puccinellia tenuiflora* [24].

Super absorbent hydrogels used as water retaining agents (WRA) in agriculture were formed from highly hydrophilic cross-linked polymers, which possess high water absorption capacity [25]. It was found that the hydrogels can mitigate soil erosion by reducing sediment and nutrient losses [26–28]. The hydrogels can also absorb water and nutrients and subsequently release them gradually [29–31]. In addition, Sojka et al. found that hydrogel promoted soil colonization of microorganism, including bacteria and mycorrhiza [32]. It increases the plant available water in the soil, which prolongs plant survival time under drought stress [33–36]. Hayat et al. proved that the administration of WRA induced substantial changes in soil physical properties by increasing saturation percentage while decreasing particle density and bulk density, so as to promote crop productivity [37].

Perennial ryegrass (*Lolium perenne* L.) is an important grass species for pasture, forage and turf in the world [38]. This cool-season grass species is native to Northern Europe, Asia and Africa and widely distributed in many temperate regions all over the world [39]. It has good turf quality with quick establishment [40]. Like many other turf grasses, however, it is also a drought sensitive species [41].

Although either *B. amyloliquefaciens* GB03 or WRA can enhance the plant performance under drought conditions, the combination effects of both of them have not been reported to date. The objective of this research was to investigate the synergistic effects of the beneficial soil bacterium strain *Bacillus amyloliquefaciens* GB03 and water retaining agent on drought tolerance of perennial ryegrass. Plants were subjected to drought treatments by stopping watering for 20 days. Plant survival rate and parameters related to plant biomass, photosynthesis, water status and leaf cell membrane integrity were assessed.

2. Results

2.1. B. amyloliquefaciens GB03 and WRA Promoted Ryegrass Growth under Drought Condition

Table 1 showed that the soil water contents of WRA and GB03+WRA treatments were higher than those of control and GB03 treatment. GB03 and WRA significantly increased leaf growth and plant density of ryegrass compared to control during all phases ($p < 0.05$) (Figure 1). No significant difference was observed among all treatments for 20-day-old (before drought treatment, BT) and 10-day natural drought seedlings. However, after 20 days natural drought and then seven days after rewatering, seedlings treated with GB03+WRA grew well and exhibited the highest survival rate (approximately 94%), compared with those treated with WRA (76%) that was significantly higher than that of seedlings treated with GB03 (55%) or control (10%) (Figure 2).

Table 1. Soil water content after 20 days of natural drought. Values are means with SEs ($n = 12$). Different letters indicate significant differences among treatments at $p < 0.05$ (ANOVA and Duncan's multiple range test).

Treatments	Control	GB03	WRA	GB03+WRA
Soil water content (mg/g·DW)	37.7 ± 3.9 b	33.3 ± 04.7 b	61.8 ± 2.0 a	62.4 ± 5.9 a

Figure 1. Ryegrass growth status of different treatments in four vegetation phases. From left to right: control, GB03, WRA and GB03+WRA treatments; from top to bottom: 20-day-old seedlings (before treatment, BT), 10-day natural drought (10DD), 20-day natural drought (20DD) and 7 days after rewatering (7DR).

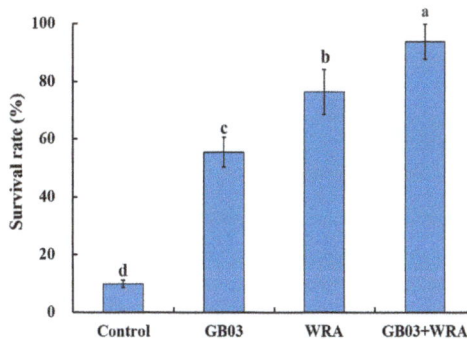

Figure 2. The survival rate of ryegrass seedlings with 20 days natural drought and then seven days after rewatering. Values are means and bars indicate standard errors (SEs) ($n = 12$). Columns with different letters indicate significant differences among treatments at $p < 0.05$ (ANOVA and Duncan's post-hoc multiple comparison test).

After growing for 20 days (BT), GB03 and WRA significantly enhanced shoot fresh weight by 96.3% and 59.7% compared to control, respectively; moreover, the shoot fresh weight of GB03+WRA treatment was 116% and 35.0% significantly higher than that of control and WRA treatment, respectively (Figure 3A). After 10-days of drought treatment, three treatments (GB03, WRA and GB03+WRA) had significantly 28.5%, 32.8%, and 38.6% higher fresh weight than control, respectively. Drought stress triggered significant decreases in shoot fresh weight when seedlings were withheld water for 20 days: all treatments (control, GB03, WRA and GB03+WRA, respectively) decreased by 77.8%, 65.1%, 35.3% and 34.2%. However, in this phase, the fresh weights of GB03 and WRA treatments were significantly 1.02-fold and 2.87-fold higher than that of control, respectively; GB03+WRA treatment was 1.04-fold and 3.11-fold significantly higher in shoot fresh weight than GB03 treatment and control, respectively. Seven days after rewatering, shoot fresh weights for all treatments (except for control) increased significantly than that in 20-day drought. In this phase, three treatments (GB03, WRA and GB03+WRA) were significantly 5.20-fold, 6.52-fold and 8.14-fold higher in fresh weight than control, respectively; the shoot fresh weight in GB03+WRA and WRA treatments were 47.4% and 21.3% significantly higher than that in GB03 treatment, respectively, and GB03+WRA increased shoot fresh weight significantly compared to control and other two treatments (Figure 3A).

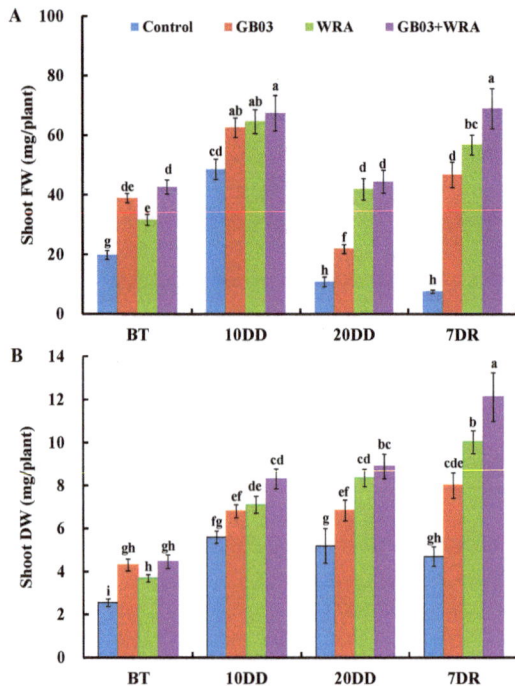

Figure 3. Effects of GB03, water retaining agent (WRA), or the combination of GB03+WRA on shoot fresh weight (FW) (**A**) and dry weight (DW) (**B**) of ryegrass. BT, before treatment (20-day-old seedling); 10DD, 10-day natural drought; 20DD, 20-day natural drought; 7DR, 7 days after rewatering. Values are means and bars indicate SEs ($n = 12$). Columns with different letters indicate significant differences among treatments at $p < 0.05$ (ANOVA and Duncan's post-hoc multiple comparison test).

Three treatments (GB03, WRA and GB03+WRA) significantly increased shoot dry weight by 68.7%, 44.5% and 74.7% compared to control, respectively, after growing for 20 days (Figure 3B). After 10-day drought, WRA and GB03+WRA treatments increased shoot dry weight significantly by 26.8% and 48.2%, respectively, compared to control. After 10-day drought, three treatments (GB03, WRA and GB03+WRA) were significantly 31.5%, 60.5% and 70.9% higher in dry weight than control, respectively.

Seven days after rewatering, WRA and GB03+WRA treatments increased shoot dry weight significantly by 20.1% and 36.4%, respectively, compared to those in 20-day drought treatment. In this phase, three treatments (GB03, WRA and GB03+WRA) were significantly 0.72-fold, 1.13-fold and 1.57-fold higher in shoot dry weight, respectively, compared to control, and GB03+WRA increased shoot dry weight significantly compared to control and other two treatments (Figure 3B).

2.2. GB03 and WRA Maintained the Relative Water Content (RWC) under Drought Condition

To probe the plant water status, RWC was assayed in ryegrass leaves. As shown in plant biomass (shoot fresh weight and dry weight), 10-day drought treatment was tolerable for ryegrass with all treatments no significant difference in RWC. After stopping watering for 20 days, all treatments decreased RWC significantly by 56.7%, 36.6%, 27.4% and 20.0%, respectively, compared to those in 10-day drought; the three treatments (GB03, WRA and GB03+WRA) were 51.4%, 74.8% and 95.5% significantly higher in RWC, respectively, compared to control; GB03+WRA treatment was also 29.1% significantly higher than GB03 treatment, indicating that GB03 and WRA together could effectively maintain the RWC in ryegrass. Seven days after rewatering, three treatments (GB03, WRA and GB03+WRA) were significantly 1.19-, 1.15- and 1.27-folds higher than control in RWC, respectively (Figure 4).

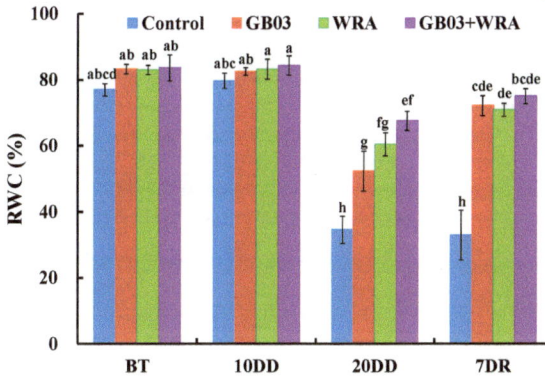

Figure 4. Effects of GB03, water retaining agent (WRA), or the combination of GB03+WRA on leaf relative water content (RWC) of ryegrass. BT, before treatment (20-day-old seedling); 10DD, 10-day natural drought; 20DD, 20-day natural drought; 7DR, 7 days after rewatering. Values are means and bars indicate SEs ($n = 12$). Columns with different letters indicate significant differences among treatments at $p < 0.05$ (ANOVA and Duncan's post-hoc multiple comparison test).

2.3. GB03 and WRA Maintained Chlorophyll Content

After growing for 20 days (BT), chlorophyll content for GB03 and GB03+WRA treatments was 15.1% and 24.8% significantly higher than that in control, respectively; chlorophyll content in GB03+WRA was significantly 14.1% higher than that in WRA. After 10-day drought treatment, chlorophyll content was 38.8%, 31.4% and 50.8% significantly higher in GB03, WRA and GB03+WRA treatments than in control, respectively (Figure 5). After 20-day drought treatment, the chlorophyll content in WRA was 21.4% significantly lower than that in GB03+WRA, however, seedlings with GB03 treatment and control became too wilt and chlorophyll content was unmeasurable. Seven days after rewatering, the chlorophyll content in all three treatments returned to the BT level, while that in control was still beyond measurement. The chlorophyll content in GB03+WRA treatment was 17.1% and 11.7% significantly higher than those in GB03 and WRA treatments, respectively. Therefore, WRA together with GB03 effectively maintained ryegrass chlorophyll content, especially when seedlings were under severe drought stress (20-day drought) (Figure 5).

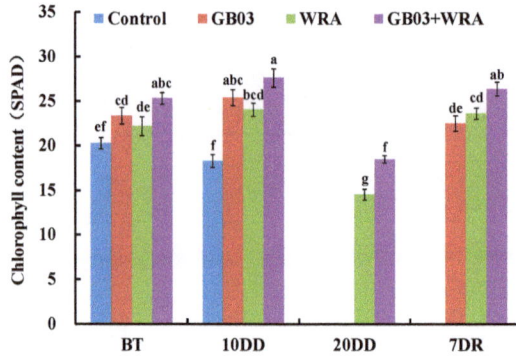

Figure 5. Effects of GB03, water retaining agent (WRA), or the combination of GB03+WRA on leaf chlorophyll content of ryegrass. BT, before treatment (20-day-old seedling); 10DD, 10-day natural drought; 20DD, 20-day natural drought; 7DR, 7 days after rewatering. Values are means and bars indicate SEs ($n = 12$). Columns with different letters indicate significant differences among treatments at $p < 0.05$ (ANOVA and Duncan's post-hoc multiple comparison test).

2.4. GB03 and WRA Reduced Relative Electric Conductivity (REC) and MDA Content under Drought Stress

After stopping watering for 20 days, a significant increase in REC was observed compared to 10 days of drought in control and three treatments; REC in control increased drastically by 6.80-fold. However, GB03 and WRA helped seedlings to maintain relatively lower REC and reduced REC significantly by 56.6% and 48.4% compared to control, respectively; moreover, REC in GB03+WRA treatment was 62.1% significantly lower than that in control (Figure 6). These results suggested that GB03 and WRA can ensured the relative low level of REC in ryegrass under severe drought conditions. Seven days after rewatering, REC in WRA and GB03+WRA significantly decreased by 59.8% and 51.5% compared to 20-day drought, respectively; and REC in GB03+WRA was significantly 48.2% lower than GB03. While REC in control plants was beyond measurement (Figure 6).

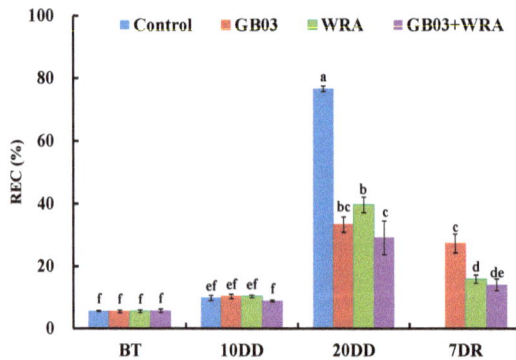

Figure 6. Effects of GB03, water retaining agent (WRA), or the combination of GB03+WRA on relative electric conductivity (REC) of ryegrass. BT, before treatment (20-day-old seedling); 10DD, 10-day natural drought; 20DD, 20-day natural drought; 7DR, 7 days after rewatering Values are means and bars indicate SEs ($n = 12$). Columns with different letters indicate significant differences among treatments at $p < 0.05$ (ANOVA and Duncan's post-hoc multiple comparison test).

After 10-day drought, MDA contents were increased significantly by 90.9%, 57.5%, 57.0% and 65.3% compared with their corresponding BT levels. After 20-day drought, MDA content only in control plants was increased significantly by 1.07-fold compared to 10-day drought, whereas those in the three treatments were still maintained the previous phase level; GB03+WRA treatments was 55.9% significantly lower than control in MDA content (Figure 7).

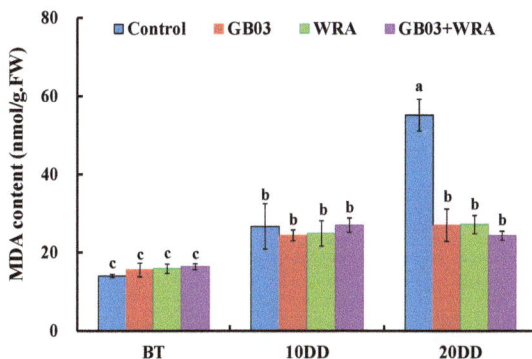

Figure 7. Effects of GB03, water retaining agent (WRA), or the combination of GB03+WRA on leaf malondialdehyde (MDA) content of ryegrass. BT, before treatment (20-day-old seedling); 10DD, 10-day natural drought; 20DD, 20-day natural drought; 7DR, 7 days after rewatering. Values are means and bars indicate SEs (n = 12). Columns with different letters indicate significant differences among treatments at $p < 0.05$ (ANOVA and Duncan's post-hoc multiple comparison test).

3. Discussion

3.1. Synergistic Effects of GB03 and WRA on Plant Growth under Drought Condition

Considerable progress has been made in fathoming mechanisms underlying *Bacillus*-mediated plant growth promotion and crop yield increase; these mechanisms include increased nutrient availability, synthesizing plant hormones and the production of volatile organic compounds [15–20]. Plant growth promotion meditated by *Bacillus amyloliquefaciens* has been reported in many species including *Arabidopsis* [15,20,42], maize (*Zea mays* L.) [43], tomato (*Lycopersicon esculentum*) [44], wheat (*Triticum aestivum*) [22], white clover (*Trifolium repens* L. cultivar Huia) [23] and *Puccinellia tenuiflora* [24]. Gagné-Bourque et al. also proved that *B. subtilis* enhanced *Brachypodium distachyon* growth under drought stress [45].

Consistent with the recent study of Galeş et al. [34], the soil moisture at the end of the drought treatment showed that the WRA's property in retaining the water and releasing it afterward (Table 1). Because of this property, treatments amended with WRA in this study positively affected plant growth and the more severe the drought was, the better they performed in comparison to controls (Figures 1 and 2). Similar results for tomato and cucumber were reported by El-Hady and Wanas [46]. The property of WRA, however, is hard to evaluate due to its depending largely on temperature, humidity, the particle size of hydrogel and the properties of the soil [47]. Interestingly, here we found that the combined effect of GB03 and WRA on promoting plant growth was greater than that of either of them.

3.2. GB03 and WRA Maintained Relatively Higher RWC Level in Ryegrass Leaves

As one of the best criteria for measuring the water status in plants, RWC indicates the water metabolic activity in tissues as drought-resistant species usually have higher RWC in their leaves [48]. Hence, RWC could also be used as an ideal parameter to probe the PGPR-mediated plant drought tolerance. Indeed, many researchers have reported that under drought stress, plants with PGPR inoculation maintained higher RWC as compared to those without, suggesting that PGPR strains could effectively prolong the plant survival under drought conditions [49–51]. In this work, GB03 effectively enhanced the RWC by 51.4% over control under severe drought treatment (20-day drought) (Figure 4), which was consistent with previous research in sorghum [52]. Dodd et al. claimed that the increase in RWC might be a consequence of changes of the sensitivity in stomatal closure [53]. Despite the progress in the recent decade, the mechanisms behind increased RWC with PGPR treatment remain to be elucidated. After stopping watering for 20 days, WRA also elevated the level of RWC in ryegrass by 74.8% over control. This was understandable because WRA retained soil water available for plant. GB03 together with WRA maintained relatively higher RWC level than GB03 or WRA in ryegrass leaves although not significantly.

3.3. Synergistic Effects of GB03 and WRA in Ryegrass Leaf Chlorophyll Content under Severe Drought Condition

Leaf chlorophyll content is also an important physiological parameter positively affecting plant photosynthesis rate [7]. As one of the symptoms of photo-oxidation, drought-triggered decrease in chlorophyll content has been observed in maize [54], sorghum [52] and white clover [23]. Zhang et al. found that GB03 increased photosynthetic capacity in *Arabidopsis* by raising photosystem II photosynthetic activity and chlorophyll content [20]. In ryegrass, the increase and maintenance in chlorophyll content by GB03 inoculation were observed under normal condition (BT) and moderate drought condition (10-day drought) (Figure 5). It was observed that WRA enhanced chlorophyll content in maize and soybean crop [33]. With the effect of WRA, the chlorophyll content in this study maintained relatively high level even under severe drought stress (20-day drought). GB03 alone failed to function under severe drought; however, the combination of t GB03 and WRA showed significantly better effect than WRA alone (Figure 5).

3.4. GB03 and WRA Alleviated Cell Membrane Damage under Drought Conditions

Drought incurs oxidative stress in plants by increasing ROS that is able to effectively degrade membrane lipids and to exacerbate lipid peroxidation [55,56]. The degradation of cell membrane results in the increase of REC. Besides, with the increase of ROS, the content of MDA follows rapidly. Thus the content of MDA has also been considered as a suitable index of oxidative damage [50,57]. Vardharajula et al. reported *Bacillus* spp. HYD-B17 decreased REC in maize by 26.3% under drought conditions [50]. Similar results were reported by Sandhya et al. [49] and Naveed et al. [51]. In the current study, GB03 and WRA reduced REC significantly in ryegrass under severe drought stress (20-day drought) and prolonged the seedlings' survivability (Figure 6). As for using MDA as a marker to probe cell damage, Han et al. found GB03 significantly decreased the MDA content in white clover under stress [23]. Here we found similar result in ryegrass under severe drought stress (20-day drought); in addition, we also found that WRA together with GB03 decreased MDA content to the same statistical level as GB03 inoculation and WRA treatments (Figure 7).

4. Materials and Methods

4.1. Bacterial Suspension Culture

Bacillus amyloliquefaciens strain GB03 (presented by Professor Paul W. Paré at Texas Tech University, Lubbock, TX, USA) was streaked onto Luria broth (LB) agar plates and incubated at 28 °C in darkness for 24 h. Bacterial cells were then transferred from LB agar plates to liquid LB medium and cultured under 28 °C and 250 rpm to yield 10^9 colony forming units (CFU)/mL, as measured by optical density and a series of dilutions [24,58].

4.2. Plant Growth and Treatments

Perennial ryegrass (*Lolium perenne* L. cv. Esquire) seeds (Beijing Top Green Seed Co., Ltd., Beijing, China) were surface sterilized with 2% NaClO (sodium hypochlorite) for 1 min followed by 70% ethanol for 10 min, and rinsed by sterilized water five times. Seeds were then grown in pre-sterilized plastic pots (diameter 20 cm, depth 15 cm, 1.5 g seeds/pot with 12 replicates) filled with 1800 g of heat-sterilized (95 °C, 48 h) vermiculite, sand and field top soil mixture with their volume ratio of 1:1:1. Each pot was inoculated with 10 mL GB03 suspension culture or 10 mL liquid LB medium. Then, half of each group of above two treatments was applied WRA (5 g/pot, the compound is super absorbent hydrogels provided by Zhuhai Nongshen Biotechnology Co., Ltd., Zhuhai, China). Thus, four treatments were set: control, GB03, WRA and GB03+WRA. Each treatment contained 12 replications (12 pots) and all 48 pots were daily rearranged to avoid any border effect or light heterogeneity. Each pot was irrigated with tap water (250 mL) every 3 days until drought treatment started. Plants were grown in greenhouse under 28 °C/23 °C (day/night), the photoperiod was 16/8 h (light/dark) and the relative humidity was about 70%.

Twenty-day-old plants were subjected to drought stress by stopping watering for 20 days, and then regular watering was continued. Seedlings were harvested on the 0th (BT), 10th and 20th day after stopping watering and 7th day after rewatering for plant biomass and physiological measurements. Soil samples were taken to measure soil moisture on the 20th day after stopping watering (Table 1).

4.3. Plant Survival Rate, Biomass and Physiological Measurements

The numbers of survived plants and dead plants in each pot were counted after 20 days natural drought and then and survival rate (%) was calculated (*n* = 12).

Two plants from each of the 12 pots (12 replications) at each stage were sampled to measure biomass and physiological indexes. Shoot fresh weight was measured at once after harvest. Turgid weights of leaves were measured after they were soaked in distilled water in test tubes at 4 °C overnight in the dark. Finally, shoots and leaves were dried in an oven under 80 °C for 48 h and weighed again to get shoot and leaf dry weight. Leaf relative water content (RWC) was estimated according to the method described by Barrs and Weatherley [59] and then calculated using the following equation, where FW represents the leaf fresh weight, TW the leaf turgid weight and DW the leaf dry weight.

$$\text{RWC (\%)} = \frac{FW - DW}{TW - DW} \times 100$$

Total chlorophyll content was measured by using chlorophyll meter (SPAD 502, Konica Minolta Sensing, Inc., Osaka, Japan).

Leaf relative electric conductivity (REC) was measured to estimate leaf cell membrane damage using an electric conductivity meter (EC215, Hanna Corporation, Italy) as described by Peever and Higgins [60] and Niu et al. [61] with slight modifications. REC (%) was calculated using the following equation, where S1 and S2 refer to conductivity of ryegrass live leaves and boiled leaves, respectively.

$$\text{REC (\%)} = \frac{S1}{S2} \times 100$$

To probe leaf oxidative damage, the biomarker malondialdehyde (MDA) was extracted and determined spectrophotometrically using a thiobarbituric acid (TBA) protocol [23]. Reagent kit was supplied by Suzhou Comin Biotechnology Co., Ltd. (Suzhou, China) Absorbance was determined at 532 and 600 nm using a UV spectrophotometer (UV-2102C, Unico Instrument Co., Ltd., Shanghai, China).

4.4. Data Analysis

Results of the growth and physiological parameters were presented as means with standard errors (*n* = 12). All the data were subjected to one-way analysis of variance (ANOVA) and Duncan's post-hoc multiple comparison tests were used to detect significant differences among means at a significance level of $p < 0.05$ using SPSS 17.0 (SPSS Inc., Chicago, IL, USA).

5. Conclusions

This study demonstrated that either GB03 or WRA promotes ryegrass survival and growth under drought conditions by directly or indirectly maintaining survival rate, biomass, relative water content, leaf chlorophyll content and cell membrane integrity. Furthermore, the synergistic effect of GB03 and WRA on drought tolerance promotion of ryegrass were greater than that of either of them by improving survival rate, biomass and chlorophyll content. This work could be helpful to develop a fresh and excellent approach for turf to cope with the challenge of global fresh water insufficiency.

Acknowledgments: This research was supported by National Natural Science Foundation of China (grant No. 31222053 and 31172256), Hui-Chun Chin and Tsung-Dao Lee Chinese Undergraduate Research Endowment (JZH0089) and Science and Technology Support Program of Gansu Province, China (1604NKCA077).

Author Contributions: Jin-Lin Zhang conceived and designed the experiments. An-Yu Su, Shu-Qi Niu and Yuan-Zheng Liu performed the experiments and Ao-Lei He, Qi Zhao, Qing-Qing Han and Sardar Ali Khan gave helps. An-Yu Su and Jin-Lin Zhang analyzed the data and wrote the paper. Paul W. Paré and Meng-Fei Li revised the paper.

Conflicts of Interest: The authors declare no conflict of interest.

Abbreviations

BT	Before drought treatment
CFU	Colony Forming Units
LB	Luria Broth
MDA	Malondialdehyde
PGPR	Plant Growth Promoting Rhizobacteria
REC	Relative electric conductivity
RWC	Relative Water Content
SEs	Standard errors
VOCs	Volatile Organic Compounds
WRA	Water Retaining Agent

References

1. Kramer, P.J.; Boyer, J.S. *Water Relations of Plants and Soils*; Academic Press: Cambridge, MA, USA, 1995.
2. Hassine, A.B.; Bouzid, S.; Lutts, S. Does habitat of *Atriplex halimus* L. affect plant strategy for osmotic adjustment? *Acta Physiol. Plant.* **2010**, *32*, 325–331. [CrossRef]
3. Wu, G.Q.; Xi, J.J.; Wang, Q.; Bao, A.K.; Ma, Q.; Zhang, J.L.; Wang, S.M. The *ZxNHX* gene encoding tonoplast Na$^+$/H$^+$ antiporter from the xerophyte *Zygophyllum xanthoxylum* plays important roles in response to salt and drought. *J. Plant Physiol.* **2011**, *168*, 758–767. [CrossRef] [PubMed]
4. Martínez, J.P.; Kinet, J.M.; Bajji, M.; Lutts, S. NaCl alleviates polyethylene glycol-induced water stress in the halophyte species *Atriplex halimus* L. *J. Exp. Bot.* **2005**, *56*, 2421–2431. [CrossRef] [PubMed]
5. Slama, I.; Ghnaya, T.; Messedi, D.; Hessini, K.; Labidi, N.; Savoure, A.; Abdelly, C. Effect of sodium chloride on the response of the halophyte species *Sesuvium portulacastrum* grown in mannitol-induced water stress. *J. Plant Res.* **2007**, *120*, 291–299. [CrossRef] [PubMed]
6. Jaleel, C.A.; Manivannan, P.; Wahid, A.; Farooq, M.; Al-Juburi, H.J.; Somasundaram, R.; Panneerselvam, R. Drought Stress in Plants: A review on morphological characteristics and pigments composition. *Int. J. Agric. Biol.* **2009**, *11*, 100–105.
7. Ma, Q.; Yue, L.J.; Zhang, J.L.; Wu, G.Q.; Bao, A.K.; Wang, S.M. Sodium chloride improves photosynthesis and water status in the succulent xerophyte *Zygophyllum xanthoxylum*. *Tree Physiol.* **2012**, *32*, 4–13. [CrossRef] [PubMed]
8. Hendry, G.A.F. Oxygen, free radical processes and seed longevity. *Seed Sci. Res.* **1993**, *3*, 141–153. [CrossRef]
9. Sgherri, C.L.M.; Maffei, M.; Navari-Izzo, F. Antioxidative enzymes in wheat subjected to increasing water deficit and rewatering. *J. Plant Physiol.* **2000**, *157*, 273–279. [CrossRef]

10. Gholamhoseini, M.; Ghalavand, A.; Dolatabadian, A.; Jamshidi, E.; Khodaei-Joghan, A. Effects of arbuscular mycorrhizal inoculation on growth, yield, nutrient uptake and irrigation water productivity of sunflowers grown under drought stress. *Agric. Water Manag.* **2013**, *117*, 106–114. [CrossRef]

11. Agnew, C.; Warren, A. A framework for tackling drought and land degradation. *J. Arid Environ.* **1996**, *33*, 309–320. [CrossRef]

12. Egamberdieva, D.; Kucharova, Z. Selection for root colonising bacteria stimulating wheat growth in saline soils. *Biol. Fertil. Soils* **2009**, *45*, 563–571. [CrossRef]

13. Timmusk, S.; Wagner, E.G. The plant-growth-promoting rhizobacterium *Paenibacillus polymyxa* induces changes in *Arabidopsis thaliana* gene expression: A possible connection between biotic and abiotic stress responses. *Mol. Plant Microbe Interact.* **1999**, *12*, 951–959. [CrossRef] [PubMed]

14. Kloepper, J.W. Induced systemic resistance and promotion of plant growth by *Bacillus* spp. *Phytopathology* **2004**, *94*, 1259–1266. [CrossRef] [PubMed]

15. Paré, P.W.; Zhang, H.; Aziz, M.; Xie, X.; Kim, M.-S.; Shen, X.; Zhang, J. Beneficial rhizobacteria induce plant growth: Mapping signaling networks in *Arabidopsis*. In *Biocommunication in Soil Microorganisms*; Springer: Berlin, Germany, 2011; pp. 403–412.

16. Gao, S.; Wu, H.; Wang, W.; Yang, Y.; Xie, S.; Xie, Y.; Gao, X. Efficient colonization and harpins mediated enhancement in growth and biocontrol of wilt disease in tomato by *Bacillus subtilis*. *Lett. Appl. Microbiol.* **2013**, *57*, 526–533. [CrossRef] [PubMed]

17. Ryu, C.-M.; Farag, M.A.; Hu, C.; Reddy, M.S.; Wei, H.; Paré, P.W.; Kloepper, J.W. Bacterial volatiles promote growth in *Arabidopsis*. *Proc. Natl. Acad. Sci. USA* **2003**, *100*, 4927–4932. [CrossRef] [PubMed]

18. Ryu, C.-M.; Farag, M.A.; Hu, C.-H.; Reddy, M.S.; Kloepper, J.W.; Paré, P.W. Bacterial volatiles induce systemic resistance in *Arabidopsis*. *Plant Physiol.* **2004**, *134*, 1017–1026. [CrossRef] [PubMed]

19. Zhang, H.; Kim, M.S.; Krishnamachari, V.; Payton, P.; Sun, Y.; Grimson, M.; Farag, M.A.; Ryu, C.M.; Allen, R.; Melo, I.S.; et al. Rhizobacterial volatile emissions regulate auxin homeostasis and cell expansion in *Arabidopsis*. *Planta* **2007**, *226*, 839–851. [CrossRef] [PubMed]

20. Zhang, H.; Xie, X.; Kim, M.S.; Kornyeyev, D.A.; Holaday, S.; Paré, P.W. Soil bacteria augment *Arabidopsis* photosynthesis by decreasing glucose sensing and abscisic acid levels in planta. *Plant J.* **2008**, *56*, 264–273. [CrossRef] [PubMed]

21. Zhang, H.; Sun, Y.; Xie, X.; Kim, M.S.; Dowd, S.E.; Paré, P.W. A soil bacterium regulates plant acquisition of iron via deficiency-inducible mechanisms. *Plant J.* **2009**, *58*, 568–577. [CrossRef] [PubMed]

22. Zhang, J.L.; Aziz, M.; Qiao, Y.; Han, Q.Q.; Li, J.; Wang, Y.Q.; Shen, X.; Wang, S.M.; Paré, P.W. Soil microbe *Bacillus subtilis* (GB03) induces biomass accumulation and salt tolerance with lower sodium accumulation in wheat. *Crop Pasture Sci.* **2014**, *65*, 423–427. [CrossRef]

23. Han, Q.-Q.; Lü, X.-P.; Bai, J.-P.; Qiao, Y.; Paré, P.W.; Wang, S.-M.; Zhang, J.-L.; Wu, Y.-N.; Pang, X.-P.; Xu, W.-B.; et al. Beneficial soil bacterium *Bacillus subtilis* (GB03) augments salt tolerance of white clover. *Front. Plant Sci.* **2014**, *5*, 525. [CrossRef] [PubMed]

24. Niu, S.-Q.; Li, H.-R.; Paré, P.W.; Aziz, M.; Wang, S.-M.; Shi, H.; Li, J.; Han, Q.-Q.; Guo, S.-Q.; Li, J.; et al. Induced growth promotion and higher salt tolerance in the halophyte grass *Puccinellia tenuiflora* by beneficial rhizobacteria. *Plant Soil* **2015**, *407*, 217–230. [CrossRef]

25. Karadağ, E.; Uzum, O.B.; Saraydin, D. Swelling equilibria and dye adsorption studies of chemically crosslinked superabsorbent acrylamide/maleic acid hydrogels. *Eur. Polym. J.* **2002**, *38*, 2133–2141. [CrossRef]

26. Sojka, R.E.; Bjorneberg, D.L.; Entry, J.A.; Lentz, R.D.; Orts, W.J. Polyacrylamide in agriculture and environmental land management. *Adv. Agron.* **2007**, *92*, 75–162.

27. Szögi, A.A.; Leib, B.G.; Redulla, C.A.; Stevens, R.G.; Mathews, G.R.; Strausz, D.A. Erosion control practices integrated with polyacrylamide for nutrient reduction in rill irrigation runoff. *Agric. Water Manag.* **2007**, *91*, 43–50. [CrossRef]

28. Sepaskhah, A.R.; Shahabizad, V. Effects of water quality and PAM application rate on the control of soil erosion, water infiltration and runoff for different soil textures measured in a rainfall simulator. *Biosyst. Eng.* **2010**, *106*, 513–520. [CrossRef]

29. Sepaskhah, A.R.; Bazrafshan-Jahromi, A.R. Controlling runoff and erosion in sloping land with polyacrylamide under a rainfall simulator. *Biosyst. Eng.* **2006**, *93*, 469–474. [CrossRef]

30. El-Hamshary, H. Synthesis and water sorption studies of pH sensitive poly (acrylamide-*co*-itaconic acid) hydrogels. *Eur. Polym. J.* **2007**, *43*, 4830–4838. [CrossRef]

31. Farrell, C.; Ang, X.Q.; Rayner, J.P. Water-retention additives increase plant available water in green roof substrates. *Ecol. Eng.* **2013**, *52*, 112–118. [CrossRef]

32. Sojka, R.E.; Entry, J.A.; Fuhrmann, J.J. The influence of high application rates of polyacrylamide on microbial metabolic potential in an agricultural soil. *Appl. Soil Ecol.* **2006**, *32*, 243–252. [CrossRef]

33. Galeş, D.C.; Trincă, L.C.; Cazacu, A.; Peptu, C.A.; Jităreanu, G. Effects of a hydrogel on the cambic chernozem soil's hydrophysic indicators and plant morphophysiological parameters. *Geoderma* **2016**, *267*, 102–111. [CrossRef]

34. Jobin, P.; Caron, J.; Bernier, P.-Y.; Dansereau, B. Impact of Two hydrophilic acrylic-based polymers on the physical properties of three substrates and the growth of *Petunia* ×*hybrida* 'Brilliant Pink'. *J. Am. Soc. Hortic. Sci.* **2004**, *129*, 449–457.

35. Hüttermann, A.; Orikiriza, L.J.B.; Agaba, H. Application of superabsorbent polymers for improving the ecological chemistry of degraded or polluted lands. *Clean Soil Air Water* **2009**, *37*, 517–526. [CrossRef]

36. Agaba, H.; Orikiriza, L.J.B.; Obua, J.; Kabasa, J.D.; Worbes, M.; Hüttermann, A. Hydrogel amendment to sandy soil reduces irrigation frequency and improves the biomass of *Agrostis stolonifera*. *Agric. Sci.* **2011**, *2*, 544–550. [CrossRef]

37. Hayat, R.; Ali, S. Water absorption by synthetic polymer (Aquasorb) and its effect on soil properties and tomato yield. *Int. J. Agric. Biol.* **2004**, *6*, 998–1200.

38. Kane, K.H. Effects of endophyte infection on drought stress tolerance of *Lolium perenne* accessions from the Mediterranean region. *Environ. Exp. Bot.* **2011**, *71*, 337–344. [CrossRef]

39. Buxton, D.R.; Mertens, D.R.; Fisher, D.S. *Forage Quality and Ruminant Utilization*; American Society of Agronomy, Crop Science Society of America, Soil Science Society of America: Madison, WI, USA, 1996.

40. Wang, J.P.; Bughrara, S.S. Evaluation of drought tolerance for *Atlas fescue*, perennial ryegrass, and their progeny. *Euphytica* **2008**, *164*, 113–122. [CrossRef]

41. Norris, I.B. Relationships between growth and measured weather factors among contrasting varieties of *Lolium*, *Dactylis* and *Festuca* species. *Grass Forage Sci.* **1985**, *40*, 151–159. [CrossRef]

42. Xie, X.; Zhang, H.; Paré, P.W. Sustained growth promotion in *Arabidopsis* with long-term exposure to the beneficial soil bacterium *Bacillus subtilis* (GB03). *Plant Signal. Behav.* **2009**, *4*, 948–953. [CrossRef] [PubMed]

43. Cavaglieri, L.; Orlando, J.; Rodríguez, M.I.; Chulze, S.; Etcheverry, M. Biocontrol of *Bacillus subtilis* against *Fusarium verticillioides* in vitro and at the maize root level. *Res. Microbiol.* **2005**, *156*, 748–754. [CrossRef] [PubMed]

44. Chowdappa, P.; Mohan Kumar, S.P.; Jyothi Lakshmi, M.; Upreti, K.K. Growth stimulation and induction of systemic resistance in tomato against early and late blight by *Bacillus subtilis* OTPB1 or *Trichoderma harzianum* OTPB3. *Biol. Control* **2013**, *65*, 109–117. [CrossRef]

45. Gagné-Bourque, F.; Mayer, B.F.; Charron, J.B.; Vali, H.; Bertrand, A.; Jabaji, S. Accelerated growth rate and increased drought stress resilience of the model grass *Brachypodium distachyon* colonized by *Bacillus subtilis* B26. *PLoS ONE* **2015**, *10*, 1–23. [CrossRef] [PubMed]

46. El-Hady, O.A.; Wanas, S.A. Water and fertilizer use efficiency by cucumber grown under stress on sandy soil treated with acrylamide hydrogels. *J. Appl. Sci. Res.* **2006**, *2*, 1293–1297.

47. Kim, S.; Iyer, G.; Nadarajah, A.; Frantz, J.M.; Spongberg, A.L. Polyacrylamide hydrogel properties for horticultural applications. *Int. J. Polym. Anal. Charact.* **2010**, *15*, 307–318. [CrossRef]

48. Jarvis, P.G.; Jarvis, M.S. The water relations of tree seedlings. IV. Some aspects of the tissue water relations and drought resistance. *Physiol. Plant* **1963**, *16*, 215–235. [CrossRef]

49. Sandhya, V.; Ali, S.Z.; Grover, M.; Reddy, G.; Venkateswarlu, B. Effect of plant growth promoting *Pseudomonas* spp. on compatible solutes, antioxidant status and plant growth of maize under drought stress. *Plant Growth Regul.* **2010**, *62*, 21–30. [CrossRef]

50. Vardharajula, S.; Zulfikar Ali, S.; Grover, M.; Reddy, G.; Bandi, V. Drought-tolerant plant growth promoting Bacillus spp.: Effect on growth, osmolytes, and antioxidant status of maize under drought stress. *J. Plant Interact.* **2011**, *6*, 1–14. [CrossRef]

51. Naveed, M.; Mitter, B.; Reichenauer, T.G.; Wieczorek, K.; Sessitsch, A. Increased drought stress resilience of maize through endophytic colonization by *Burkholderia phytofirmans* PsJN and *Enterobacter* sp. FD17. *Environ. Exp. Bot.* **2014**, *97*, 30–39. [CrossRef]

52. Grover, M.; Madhubala, R.; Ali, S.Z.; Yadav, S.K.; Venkateswarlu, B. Influence of *Bacillus* spp. strains on seedling growth and physiological parameters of sorghum under moisture stress conditions. *J. Basic Microbiol.* **2014**, *54*, 951–961. [CrossRef] [PubMed]

53. Dodd, I.C.; Zinovkina, N.Y.; Safronova, V.I.; Belimov, A.A. Rhizobacterial mediation of plant hormone status. *Ann. Appl. Biol.* **2010**, *157*, 361–379. [CrossRef]

54. Nelson, D.E.; Repetti, P.P.; Adams, T.R.; Creelman, R.A.; Wu, J.; Warner, D.C.; Anstrom, D.C.; Bensen, R.J.; Castiglioni, P.P.; Donnarummo, M.G.; et al. Plant nuclear factor Y (NF-Y) B subunits confer drought tolerance and lead to improved corn yields on water-limited acres. *Proc. Natl. Acad. Sci. USA* **2007**, *104*, 16450–16455. [CrossRef] [PubMed]

55. Mittler, R. Oxidative stress, antioxidants and stress tolerance. *Trends Plant Sci.* **2002**, *7*, 405–410. [CrossRef]

56. Farooq, M.; Wahid, A.; Kobayashi, N.; Fujita, D.; Basra, S.M.A. Plant drought stress: Effects, mechanisms and management. *Agron. Sustain. Dev.* **2009**, *29*, 185–212. [CrossRef]

57. Møller, I.M.; Jensen, P.E.; Hansson, A. Oxidative modifications to cellular components in plants. *Annu. Rev. Plant Biol.* **2007**, *58*, 459–481. [CrossRef] [PubMed]

58. Zhang, H.; Kim, M.-S.; Sun, Y.; Dowd, S.E.; Shi, H.; Paré, P.W. Soil bacteria confer plant salt tolerance by tissue-specific regulation of the sodium transporter *HKT1*. *Mol. Plant Microbe Interact.* **2008**, *21*, 737–744. [CrossRef] [PubMed]

59. Barrs, H.D.; Weatherley, P.E. A re-examination of the relative turgidity technique for estimating water deficits in leaves. *Aust. J. Biol. Sci.* **1962**, *15*, 413–428. [CrossRef]

60. Peever, T.L.; Higgins, V.J. Electrolyte leakage, lipoxygenase, and lipid peroxidation induced in tomato leaf tissue by specific and nonspecific elicitors from *Cladosporium fulvum*. *Plant Physiol.* **1989**, *90*, 867–875. [CrossRef] [PubMed]

61. Niu, M.; Huang, Y.; Sun, S.; Sun, J.; Cao, H.; Shabala, S.; Bie, Z. Root respiratory burst oxidase homologue-dependent H_2O_2 production confers salt tolerance on a grafted cucumber by controlling Na^+ exclusion and stomatal closure. *J. Exp. Bot.* **2017**. [CrossRef] [PubMed]

International Journal of
Molecular Sciences

MDPI

Article

Induced Salt Tolerance of Perennial Ryegrass by a Novel Bacterium Strain from the Rhizosphere of a Desert Shrub *Haloxylon ammodendron*

Ao-Lei He [†], Shu-Qi Niu [†], Qi Zhao, Yong-Sheng Li, Jing-Yi Gou, Hui-Juan Gao, Sheng-Zhou Suo and Jin-Lin Zhang *

State Key Laboratory of Grassland Agro-Ecosystems, College of Pastoral Agriculture Science and Technology, Lanzhou University, Lanzhou 730000, China; heal15@lzu.edu.cn (A.-L.H.); niushq14@lzu.edu.cn (S.-Q.N.); qzhao@lzu.edu.cn (Q.Z.); liys@lzu.edu.cn (Y.-S.L.); goujy16@lzu.edu.cn (J.-Y.G.); gaohj15@lzu.edu.cn (H.-J.G.); suoshzh16@lzu.edu.cn (S.-Z.S.)
* Correspondence: jlzhang@lzu.edu.cn; Tel.: +86-931-8912357
† These authors contributed equally to this work.

Received: 25 December 2017; Accepted: 1 February 2018; Published: 5 February 2018

Abstract: Drought and soil salinity reduce agricultural output worldwide. Plant-growth-promoting rhizobacteria (PGPR) can enhance plant growth and augment plant tolerance to biotic and abiotic stresses. *Haloxylon ammodendron*, a C4 perennial succulent xerohalophyte shrub with excellent drought and salt tolerance, is naturally distributed in the desert area of northwest China. In our previous work, a bacterium strain numbered as M30-35 was isolated from the rhizosphere of *H. ammodendron* in Tengger desert, Gansu province, northwest China. In current work, the effects of M30-35 inoculation on salt tolerance of perennial ryegrass were evaluated and its genome was sequenced to identify genes associated with plant growth promotion. Results showed that M30-35 significantly enhanced growth and salt tolerance of perennial ryegrass by increasing shoot fresh and dry weights, chlorophyll content, root volume, root activity, leaf catalase activity, soluble sugar and proline contents that contributed to reduced osmotic potential, tissue K^+ content and K^+/Na^+ ratio, while decreasing malondialdehyde (MDA) content and relative electric conductivity (REC), especially under higher salinity. The genome of M30-35 contains 4421 protein encoding genes, 12 rRNA, 63 tRNA-encoding genes and four rRNA operons. M30-35 was initially classified as a new species in *Pseudomonas* and named as *Pseudomonas* sp. M30-35. Thirty-four genes showing homology to genes associated with PGPR traits and abiotic stress tolerance were identified in *Pseudomonas* sp. M30-35 genome, including 12 related to insoluble phosphorus solubilization, four to auxin biosynthesis, four to other process of growth promotion, seven to oxidative stress alleviation, four to salt and drought tolerance and three to cold and heat tolerance. Further study is needed to clarify the correlation between these genes from M30-35 and the salt stress alleviation of inoculated plants under salt stress. Overall, our research indicated that desert shrubs appear rich in PGPRs that can help important crops tolerate abiotic stress.

Keywords: *Haloxylon ammodendron*; rhizobacteria; perennial ryegrass; salt tolerance; complete genome sequence; *Pseudomonas* sp.

1. Introduction

Drought and soil salinity limit crop productivity worldwide [1,2]. Osmotic stress from high salinity exposure triggers imbalance of ions, ion toxicity-induced metabolism imbalances and a series of metabolic responses and water deficiency induced by osmotic stress in plants [3,4]. PGPR can help plants tolerate abiotic stresses [5,6]. In the last decade, researchers reported that bacteria belonging to various genera including *Rhizobium*, *Bacillus*, *Pseudomonas*, *Pantoea*, *Paenibacillus*, *Burkholderia*, *Achromobacter*, *Azospirillum*, *Microbacterium*, *Methylobacterium*, *Variovorax*, *Enterobacter*, etc. provided tolerance to host plants under different abiotic stress environments [7–9]. These bacteria mediated salt tolerance in plants through modulating of reactive oxygen species (ROS) scavenging enzyme expression [10], altering the selectivity of Na$^+$, K$^+$, and Ca^{2+} and sustaining a higher K$^+$/Na$^+$ ratio in plants [11]. Our previous work showed that *Bacillus subtilis* GB03 triggered upregulation of *PtHKT1;5* and *PtSOS1* but downregulation of *PtHKT2;1* in roots reduced Na$^+$ transport from root to shoot as well as Na$^+$ uptake in roots [12]; GB03 promoted growth in *Codonopsis pilosula* (Franch.) [13], drought tolerance in ryegrass [14] and salt tolerance in wheat [15], white clover [16] and *Codonopsis pilosula* [17].

Among the many known PGPR genera, *Pseudomonas* has received much research attention because it is widely distributed in various environments and is easy to culture under laboratory conditions [18]. *Pseudomonas* can survive and prosper in a wide range of environments with many strains isolated from various environments, such as soil [19], plant [20], straw [21], animal [22] and fresh and saline water [23,24]. By now, 255 species with validly published names have been described (http://www.bacterio.net). *Pseudomonas* is beneficial for plant growth promotion [25–28]. The use of these beneficial microorganisms was considered as one of the most promising methods for safe crop-management practices [29]. In order to further exploit the relevant genetic PGPR traits, complete genome sequence technology of *Pseudomonas* was widely used and many species from the genus were genetically studied [30–36]. *Haloxylon ammodendron*, a C4 perennial shrub, is a succulent xerohalophyte that dominantly colonizes in arid areas. It is mainly distributed in Junggar Basin, northeast of Tarim basin, Badan Jaran desert, Tengger desert and Ulanbuh desert, China, where the average annual rainfall is around 100 mm, while the average annual evaporation is over 3500 mm [37].

Turfgrasses are increasingly subjected to soil salinity in many areas due to the accelerated soil salinization and increasing effluent water use for irrigating turfgrass landscapes [38]. Perennial ryegrass (*Lolium perenne* L.) is one of the most popular cool-season perennial grass species with high yield and superior quality in temperate regions around the world [39,40]. Perennial ryegrass has good turf quality such as dense root system, superior tillering, and regeneration ability [41]. However, the salinity tolerance of perennial ryegrass is ranked as moderate for commercial cultivars [42]. Therefore, enhancing the perennial ryegrass to better counter salt stress is very essential for improving its growth and production.

Therefore, exploring novel *Pseudomonas* strains is becoming more and more important.

Now that *H. ammodendron* can survive in harsh environmental conditions with strong roots, we proposed that the root system of *H. ammodendron* could provide a unique habitat for beneficial bacteria and these bacteria could help *H. ammodendron* itself and crops adapt to the extreme environment. In our previous work, we isolated over 290 bacterium strains from the rhizosphere of *H. ammodendron* in Tengger desert, Gansu province, northwestern China (unpublished data). Among these strains, we found that a strain numbered M30-35 had the ability to promote plant growth of model plant *Arabidopsis thaliana* under salt stress conditions in our preliminary experiment (unpublished data). The aims of this work were to evaluate the effects of M30-35 inoculation on salt tolerance on Perennial ryegrass and explore its genetic property as a PGPR strain by complete genome sequencing technology.

2. Results

2.1. M30-35 Promoted Ryegrass Growth under Salinity Conditions

Statistically significant growth differences in the whole plant level were observed between M30-35 treatment and the other two treatments, *Escherichia coli* strain DH5α and Luria broth (LB) medium, after two week treatments. Plants inoculated with M30-35 had larger size than those inoculated with DH5α and LB medium (Figure 1A). Strain M30-35 enhanced shoot fresh weight, dry weight and root volume of ryegrass under both non-saline and saline stress (150 and 300 mM NaCl) (Figure 1B). Compared to LB medium (control), shoot fresh weight was significantly increased by 35%, 21% and

38%, and dry weight by 42%, 42% and 32% ($p < 0.05$) under 0, 150 and 300 mM NaCl treatments, respectively. Root volume was increased significantly by 17%, 30% and 42% ($p < 0.05$) compared to corresponding controls, respectively, under 0, 150 and 300 mM NaCl treatments.

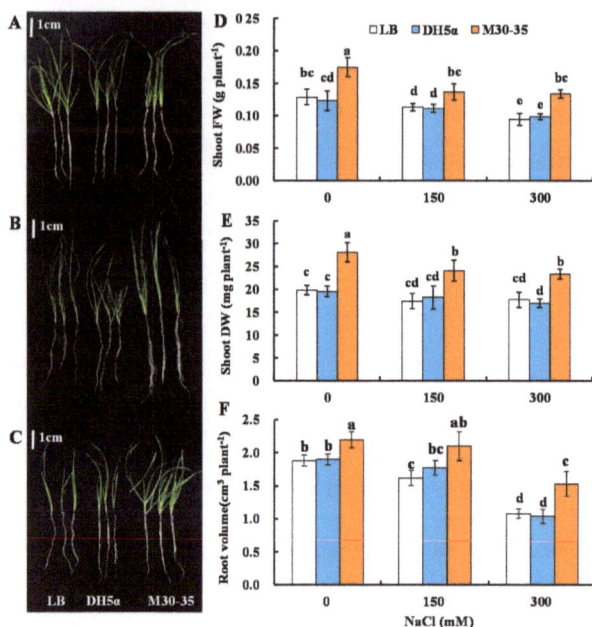

Figure 1. Effects of M30-35 on whole plant growth (**A–C**), shoot fresh weight (FW) (**D**), shoot dry weight (DW) (**E**) and root volume (**F**) of ryegrass under various salt treatments (0, 150 and 300 mM NaCl). For (**A–C**), from left to right: Luria broth (LB) medium, DH5α and M30-35 treatments; from top to bottom: 0, 150 and 300 mM NaCl treatments. Values are means and bars indicate standard errors (SEs) (*n* = 12). Columns with different letters indicate significant differences among treatments at $p < 0.05$ (ANOVA and Duncan's multiple comparison test).

2.2. Effects of M30-35 on Chlorophyll Content under Salinity Conditions

After two-week treatments, in addition to promoting shoot growth, M30-35 increased both leaf chlorophyll a and chlorophyll b contents under both non-saline and salinity. Compared to corresponding controls, chlorophyll a content in plants inoculated with M30-35 was increased significantly by 28%, 30% and 28% (Figure 2A) and chlorophyll b content by 32%, 35% and 28% ($p < 0.05$) under 0, 150 and 300 mM NaCl treatments, respectively (Figure 2B).

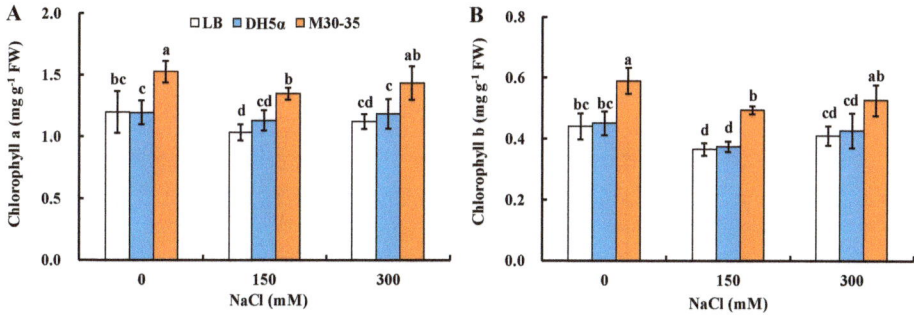

Figure 2. Effects of M30-35 on leaf chlorophyll a content (**A**) and leaf chlorophyll b content (**B**) of ryegrass under various salt stress (0, 150 and 300 mM NaCl). Values are means and bars indicate SEs (*n* = 12). Columns with different letters indicate significant differences among treatments at *p* < 0.05 (ANOVA and Duncan's post hoc multiple comparison test).

2.3. Effects of M30-35 on Root Activity of Ryegrass

To test if M30-35 could maintain root vigor, root activity of plants inoculated with M30-35 was significantly enhanced by 54% and 60% (*p* < 0.05) under 150 and 300 mM NaCl conditions, respectively, compared to corresponding controls (Figure 3). The above results demonstrated that M30-35 can maintain stable root vigor of ryegrass under salt stress.

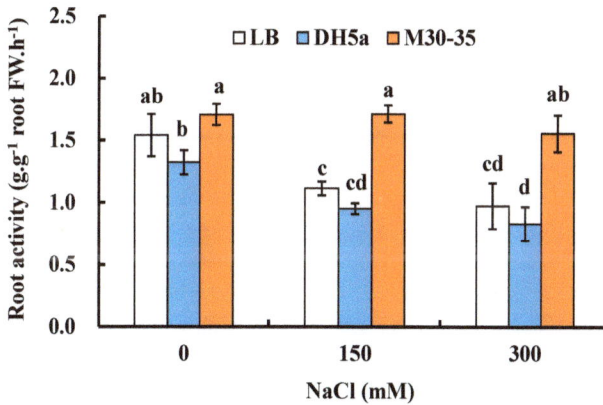

Figure 3. Effects of M30-35 on root activity of ryegrass under salt treatment (0, 150 and 300 mM NaCl). Values are means and bars indicate SEs (*n* = 12). Columns with different letters indicate significant differences among treatments at *p* < 0.05 (ANOVA and Duncan's multiple comparison test).

2.4. Effects of M30-35 on Leaf Cell Membrane Integrity of Ryegrass under Salinity Conditions

M30-35 alleviated oxidative stress of ryegrass under salt stress. After two week treatments, application of strain M30-35 significantly improved catalase (CAT) activity by 46% and 63% under salt conditions (150 and 300 mM NaCl), respectively, compared to corresponding controls (Figure 4).

To test leaf cell membrane integrity under salt stress, malonyldialdehyde (MDA) content and relative electric conductivity (REC) were measured. After two-week treatments, MDA content in the leaves of plants inoculated with M30-35 was 21%, 84% and 34% significantly (*p* < 0.05) lower than corresponding controls under 0, 150 and 300 mM NaCl treatments, respectively (Figure 5A). Similarly,

REC was 37% and 15% significantly ($p < 0.05$) lower than corresponding controls under 0 and 300 mM NaCl treatments, respectively (Figure 5B).

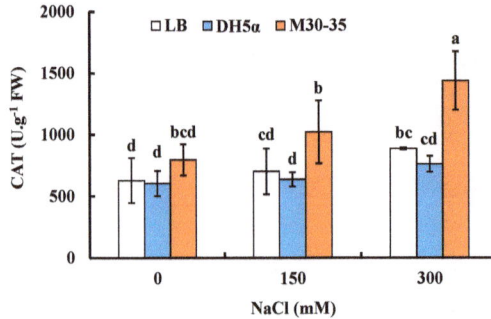

Figure 4. Effects of M30-35 on leaf catalase (CAT) activity of ryegrass under various salt treatment (0, 150 and 300 mM NaCl). Values are means and bars indicate SEs ($n = 12$). Columns with different letters indicate significant differences among treatments at $p < 0.05$ (ANOVA and Duncan's multiple comparison test).

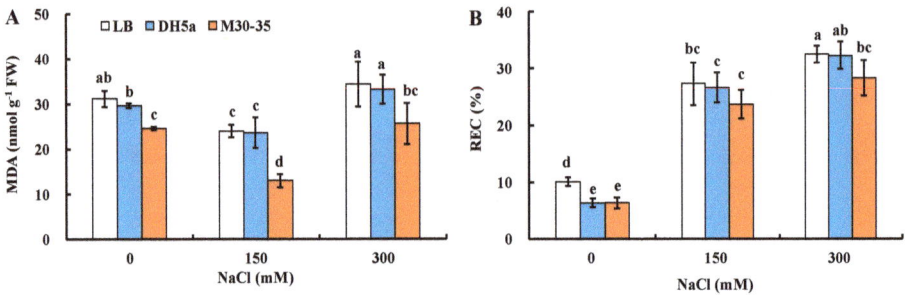

Figure 5. Effects of M30-35 on leaf malonyldialdehyde (MDA) content (**A**) and relative electric conductivity (REC) (**B**) of ryegrass under various salt treatments (0, 150 and 300 mM NaCl). Values are means and bars indicate SEs ($n = 12$). Columns with different letters indicate significant differences among treatments at $p < 0.05$ (ANOVA and Duncan's multiple comparison test).

2.5. Effects of M30-35 on Leaf Osmotic Adjustment Capability of Ryegrass under Higher Salt Stress

Leaf soluble sugar and proline are two major osmotic adjustment substances and their contents can reflect leaf osmotic adjustment capability. Under 0 and 150 mM NaCl, M30-35 had no significant effects on leaf soluble sugar content; however, it significantly enhanced leaf soluble sugar content by 1.2-fold ($p < 0.05$) compared to control under 300 mM NaCl (Figure 6A).

Leaf proline content was significantly increased by 62%, 1.1 and 1.1-fold ($p < 0.05$) by M30-35 under 0, 150 and 300 mM NaCl treatments, respectively, compared to corresponding controls (Figure 6B).

M30-35 had no significant effects on leaf osmotic potential under 0 and 150 mM NaCl conditions; however, osmotic potential was significantly decreased by 35% ($p < 0.05$) by M30-35 under 300 mM NaCl compared to control (Figure 6C). Therefore, leaf osmotic adjustment capability of ryegrass under higher salt stress was enhanced by M30-35.

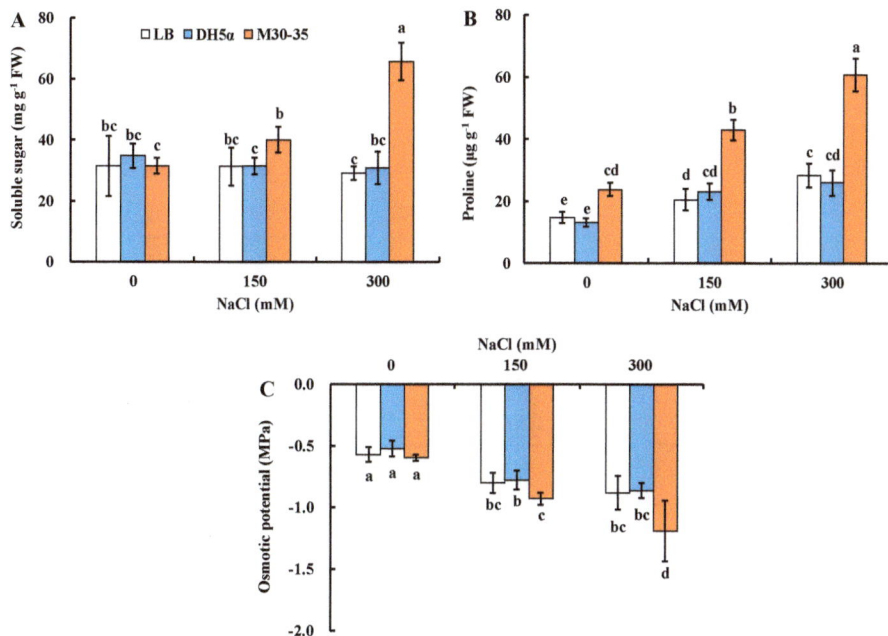

Figure 6. Effects of M30-35 on soluble sugar content (**A**), proline content (**B**) and osmotic potential (**C**) of ryegrass under salt treatment (0, 150 and 300 mM NaCl). Values are means and bars indicate SEs (n = 12). Columns with different letters indicate significant differences among treatments at $p < 0.05$ (ANOVA and Duncan's multiple comparison test).

2.6. M30-35 Maintained K^+/Na^+ Ratio in Ryegrass under Salinity Conditions

To test whether M30-35 inoculation could alter ion accumulation in ryegrass under various salinity conditions, endogenous Na^+ and K^+ contents were measured, and K^+/Na^+ ratio was calculated accordingly.

Compared to corresponding controls, M30-35 significantly decreased shoot Na^+ content by 35% and 32% and root Na^+ content by 37% and 31% ($p < 0.05$) with 150 and 300 mM NaCl treatments, respectively (Figure 7A,D). M30-35 had no significant effects on K^+ content in both shoot and root under 0 and 150 mM NaCl; however, K^+ content in shoot and root was significantly improved by 28% and 63% under 300 mM NaCl ($p < 0.05$) (Figure 7B,E).

Tissue K^+/Na^+ ratio was also significantly improved by M30-35. Compared to corresponding controls, shoot K^+/Na^+ ratio was significantly increased by 18%, 63% and 76% under 0, 150 and 300 mM NaCl and root K^+/Na^+ ratios by 63% and 1.08-fold ($p < 0.05$) under 150 and 300 mM NaCl, respectively (Figure 7C,F).

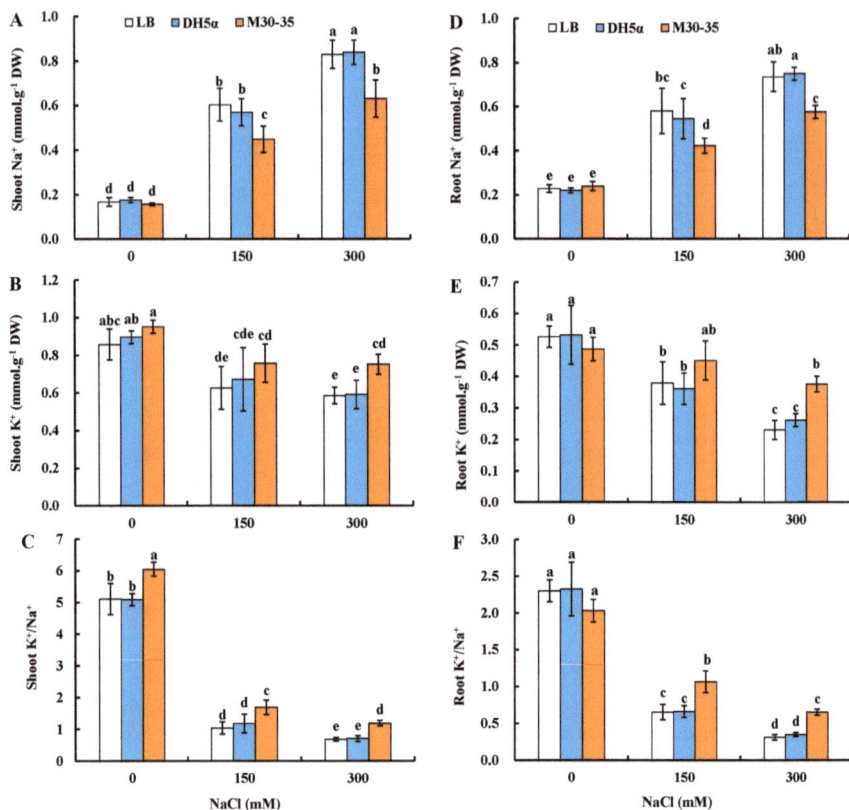

Figure 7. Effects of M30-35 on tissue Na$^+$ (**A,D**) and K$^+$ (**B,E**) contents and K$^+$/Na$^+$ (**C,F**) ratio of ryegrass under various salt conditions (0, 150 and 300 mM NaCl). Values are means and bars indicate SEs (n = 12). Columns with different letters indicate significant differences among treatments at $p < 0.05$ (ANOVA and Duncan's multiple comparison test).

2.7. Complete Genome Sequence of M30-35 and Identification of Potential Genes Responsible for Plant Growth Promotion

Classification analysis and identification based on 16S rRNA gene sequence in complete genome indicated that strain M30-35 was closely related to members of the genus *Pseudomonas*. The complete genome sequence of *Pseudomonas* sp. M30-35 was deposited in GenBank under the accession number CP020892. We initially classified the strain M30-35 as a new species in *Pseudomonas* and named it as *Pseudomonas* sp. M30-35.

The genome of M30-35 was composed a circular chromosome with a size of 4,926,954 bp and an overall guanine (G) and cytosine (C) content of 54.3% with the total number of genes 4500. The chromosome contains 63 tRNA, 12 rRNA and 4 rRNA operons (Table 1).

We performed clusters of orthologous genes (COG) function classification according to Shen et al. [31] and Powell et al. [43]. Based on COG, in detail, there is one gene for RNA processing and modification, two genes for chromatin structure and dynamics, 250 genes for energy production and conversion, 31 genes for cell cycle control, cell division, chromosome partitioning, 436 genes for amino acid transport and metabolism, 78 genes for nucleotide transport and metabolism, 200 genes for carbohydrate transport and metabolism, 140 genes for coenzyme transport and metabolism, 177 genes for lipid transport and metabolism, 184 genes for translation, ribosomal structure and biogenesis,

332 genes for transcription, 136 genes for replication, recombination and repair, 211 genes for cell wall, cell membrane, cell envelope biogenesis, 94 genes for cell motility, 156 genes for posttranslational modification, protein turnover, chaperones, 221 genes for inorganic ion transport and metabolism, 120 genes for secondary metabolites biosynthesis, transport and catabolism, 528 genes for general function prediction only, 224 genes for signal transduction mechanisms, 106 genes for intracellular trafficking, secretion, and vesicular transport, 47 genes for defense mechanisms, and 337 genes for unknown function (Figure 8).

We identified 34 genes that were responsible for plant growth promotion and abiotic stress tolerance in *Pseudomonas* sp. M30-35 genome (Table 2). These included 12 genes related to insoluble phosphorus solubilization, four to auxin biosynthesis, four to other process of growth promotion, seven to oxidative stress alleviation, four to salt and drought tolerance and three to cold and heat tolerance.

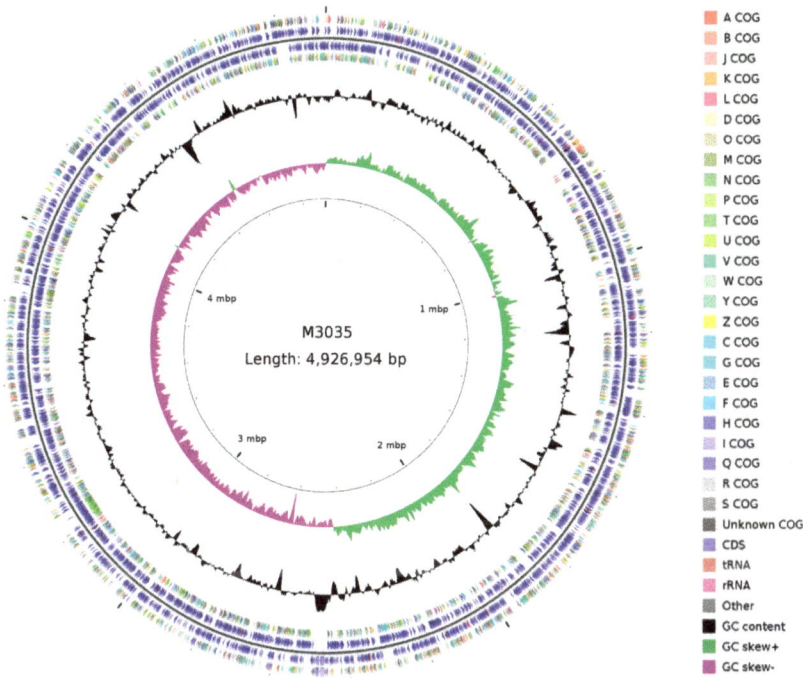

Figure 8. Circular representation of M30-35 genome. From the inner to outer circle: the first circle shows the scale; the second and third circles show G + C skew and G + C content, respectively; the fourth and seventh circle shows the distribution of genes related to COG categories; the fifth and sixth circle shows CDS, tRNA and rRNA on the location of the complete genome. A: RNA processing and modification; B: Chromatin structure and dynamics; J: Translation, ribosomal structure and biogenesis; K: Transcription; L: Replication, recombination and repair; D: Cell cycle control, cell division, chromosome partitioning; O: Posttranslational modification, protein turnover, chaperones; M: Cell wall, cell membrane, cell envelope biogenesis; N: Cell motility; P: Inorganic ion transport and metabolism; T: Signal transduction mechanisms; U: Intracellular trafficking, secretion, and vesicular transport; V: Defense mechanisms; W: Extracellular structures; Y: Nuclear structure; Z: Cytoskeleton; C: Energy production and conversion; G: Carbohydrate transport and metabolism; E: Amino acid transport and metabolism; F: Nucleotide transport and metabolism; H: Coenzyme transport and metabolism; I: Lipid transport and metabolism; Q: Secondary metabolites biosynthesis, transport and catabolism; R: General function prediction only; S: Unknown function.

Table 1. Genome features of *Pseudomonas* sp. M30-35.

Features	Values
Genome size (bp)	4,926,954
Guanine (G) + cytosine (C) content (%)	54.3
Total number of genes	4500
Number of proteins	4364
Protein encoding genes	4421
rRNAs (5S, 16S, 23S)	12
tRNAs	63
rRNA operons	4
ncRNA	4
Pseudogenes	57

Table 2. Potential genes responsible for plant growth promotion and stress tolerance in *Pseudomonas* sp. M30-35 Genome.

Categories	Gene ID	Gene Annotation
		Insoluble phosphorus solubilization
	ORF03861	Pyruvate kinase
	ORF00725	Malate synthase
	ORF01978	Phosphoenolpyruvate carboxylase
	ORF04759	Acetate kinase
	ORF03358	Citrate synthase
	ORF05557	Shikimate kinase
	ORF00482	Llactate dehydrogenase
	ORF02892	2-methylcitrate synthase
	ORF05912	Exopolyphosphatase
	ORF01099	Inorganic pyrophosphatase
Plant growth promotion	ORF05696	Alkaline phosphatase
	ORF02576	Nicotinamide adenine dinucleotide (NADH) pyrophosphatase
		Auxin biosynthesis
	ORF00180	Tryptophan synthase α chain (*trpA*)
	ORF00181	Tryptophan synthase β chain (*trpB*)
	ORF02417	Tryptophan-tRNA ligase (*trpS*)
	ORF05537	Tryptophan 2-halogenase (*cmdE*)
		Others related to plant growth promotion
	ORF03057	Nitrogen fixation protein (*fixG* and *anfA*)
	ORF01526	Acetolactate synthase 3 small subunit (*ilvH*)
	ORF05343	Biosynthetic arginine decarboxylase (*speA*)
	ORF00855	S-adenosylmethionine decarboxylase proenzyme (*speD*)
		Oxidative stress alleviation
	ORF00943	Catalase
	ORF04857	Superoxide dismutase
	ORF03162	Glutathione S-transferase
	ORF02859	Glutathione peroxidase
	ORF03063	Glutathione reductase
	ORF04403	S-(hydroxymethyl) glutathione dehydrogenase
	ORF00537	Glutathione synthetase
Stress tolerance		**Salt and drought tolerance**
	ORF04529	Na$^+$/H$^+$ antiporter (*nhaC*)
	ORF05136	Glycine betaine transporter (*opuD*)
	ORF00154	Trehalose/maltose-binding protein
	ORF04015	1-aminocyclopropane-1-carboxylate (ACC) deaminase
		Cold and heat shock protein
	ORF01759	Cold shock protein (*capB*)
	ORF02091	Cold shock protein (*cspA*)
	ORF00761	Heat shock protein (*hs1R*)

3. Discussion

3.1. Pseudomonas sp. M30-35 Promoted Ryegrass Growth under Salinity Conditions

Beneficial rhizobacteria promote plant growth as well as alleviate various abiotic stresses including salinity [44] and drought [14]. PGPR species like *Azospirillum* sp. and *Pseudomonas* sp. increased the growth and biomass of canola plants by regulating the oxidative stress enzymes and essential nutrient under salinity stress [45]. The plants inoculated with *P. mendocina* had significantly greater shoot biomass than the controls, suggesting that the inoculation with selected PGPRs could be an effective tool for alleviating salinity stress in salt sensitive plants [25,46]. It was suggested that varying responses of rhizobacteria on plant root architecture indicated the specificity of plant-rhizobacterium associations [47]. Similarly, *Pseudomonas* sp. M30-35, isolated directly from the rhizosphere of *H. ammodendron*, promotes the growth of ryegrass under both control condition and salt stress through improving its root volume and root activity (Figures 1F and 3).

Leaf chlorophyll content reflects plant growth [48]. Salinity inhibited photosynthesis mainly through reducing chlorophyll content, leaf area and photosystem II efficiency [16,49,50]. The mitigating effect of *P. putida* and *Bacillus subtilis* against salt stress and the increase in chlorophyll content were observed in canola and white clover seedlings, respectively [16,51]. In agreement with those observations, in our research, the significant increase in chlorophyll content by M30-35 was also observed under salt stress (0, 150, 300 mM NaCl) (Figure 2). These results implied that *Pseudomonas* sp. M30-35 plays a positive regulatory role in improving photosynthesis of ryegrass under salt treatment.

3.2. Pseudomonas sp. M30-35 Maintained Cell Membrane Integrity and Improved Osmotic Adjustment Capability of Ryegrass under Salinity Conditions

Abiotic stressors increase cellular levels of ROS like superoxide radical, hydroxyl radicals and hydrogen peroxide, leading to lipid peroxidation of membranes [52] and the increases in the content of the biomarker, malondialdehyde (MDA) [4,16,53] and the relative electric conductivity (REC) [14,54]. Plants have various enzymes that reduce ROS level and alleviate oxidative stress [4,53,55]. It was reported that *Bacillus subtilis* GB03 reduced MDA content in white clover under salinity [16]; root endophyte *Piriformospora indica* induced salt tolerance of barley through a strong increase in antioxidants [56]; and PGPRs enhanced abiotic stress tolerance in *Solanum tuberosum* through inducing changes in the expression of ROS-scavenging enzymes [10]. Our data showed that M30-35 inoculation significantly increased leaf CAT activity and decreased in MDA content and REC of ryegrass under higher salinity (300 mM NaCl), therefore maintaining leaf cell membrane integrity. These observations suggested that *Pseudomonas* sp. M30-35 contributed to the reduction of lipid peroxidation and the maintenance of membrane functions when plants were subjected to salinity.

Soluble sugars (such as glucose, fructose and sucrose) have been reported to be pivotal components for osmotic adaptation when plants respond to abiotic stress [4,57], especially increasing soluble sugar content could decrease osmotic potential in cells and maintain normal physiological function of plant cells in abiotic stress conditions [58,59]. Proline is a major plant osmotic regulator with multiple functions in adjusting osmosis, stabilizing the structure of proteins and scavenging ROS under salt stress [60–62]. Our results showed that both soluble sugar and proline contents were significantly enhanced by M30-35 inoculation in ryegrass under salt stress, especially higher salinity (300 mM NaCl) (Figure 6); therefore, leaf osmotic potential was significantly reduced under higher salinity. These results suggested that the elevated soluble sugar and proline accumulation could contribute to the osmotic balance in perennial ryegrass against salt stress.

3.3. Pseudomonas sp. M30-35 Maintained Ionic Homeostasis of Ryegrass under Salinity Conditions

Salt stress is commonly caused by high concentrations of Na^+ in soil [2]. Salts taken up by roots were then transported into shoots, and eventually accumulated in leaves through the transpiration stream [63]. High concentration of Na^+ in the cytoplasm disrupt the uptake of K^+ into plant cell, which is important for the catalytic activities of many enzymes [4]. PGPRs could help plants maintain ion homeostasis and high K^+/Na^+ ratios in shoots by reducing Na^+ and Cl^+ accumulation in leaves, increasing Na^+ exclusion via roots, and boosting the activity of high affinity K^+ transporters [64]. Inoculation of *Bacillus subtilis* GB03 in white clover, wheat and *Codonopsis pilosula* (Franch.) Nannf. under salt stress improved their salt tolerance by reducing Na^+ accumulation [13,15,16]. *Acinetobacter calcoaceticus* applied *Cucumis sativus* plants had reduced sodium concentration, while potassium was abundantly present under salt stress [65]. GB03-triggered upregulation of *PtHKT1;5* and *PtSOS1* and downregulation of *PtHKT2;1* in roots restricted Na^+ transport from root into shoot as well as Na^+ uptake into roots of the halophyte grass *Puccinellia tenuiflora* [12]. In current research, M30-35 inoculation decreased shoot and root Na^+ concentration, but enhanced shoot and root K^+ concentration, and thereby increased K^+/Na^+ ratio in both shoot and root of ryegrass under salt stress, especially under higher salinity (300 Mm NaCl) (Figure 7). The mechanism for M30-35 to regulate ion accumulation remained to be further explored.

3.4. Genetic Property of Pseudomonas sp. M30-35 as a PGPR Strain

Complete genome information of endophyte *Bacillus flexus* KLBMP 4941; *Pseudomonas azotoformans* S4, *P. Antarctica* PAMC 27494, *P. putida* BIRD-1 and *P. aurantiaca* strain JD37 contains genes related to plant growth promotion, biocontrol and salt tolerance [30,32,35,66]. In agreement with those observations, we also found genes that are responsible for plant growth promotion and abiotic stress tolerance the complete genome of *Pseudomonas* sp. M30-35. The specific genes and their functions are described in our research (Table 2).

Phosphorous (P) ranks second among essential plant nutrients and a major component of vital molecules [67]. Phosphorous plays an important role in metabolic processes [68]. However, the high sorption capacity of phosphate to soil particles results in a very low mobility and availability for uptake by plants [69]. In order to avoid phosphate deficiency, the environmentally unfriendly phosphate fertilizer has been widely used in agriculture worldwide, resulting in inevitable severe environmental pollution caused by P runoff [70]. Phosphate-solubilizing bacteria (PSB) exists in soil and rhizosphere and are able to release soluble phosphate from insoluble mineral phosphate [71]. The major group of phosphate solubilizing bacteria is distributed in the genus *Pseudomonas*, *Bacillus* and *Acinetobacter* [72]. Through various mechanisms of solubilization and mineralization, these microorganisms can convert inorganic and organic P into available form and ultimately contribute to plant growth [68]. *Acinetobacter calcoaceticus* applied plants had abundant phosphorus as compared to control in *Cucumis sativus* under salt stress [65]. In this study, 12 genes responsible for insoluble phosphorus solubilization were identified in the complete genome of *Pseudomonas* sp. M30-35, including pyruvate kinase, malate synthase, phosphoenolpyruvate carboxylase, acetate kinase, citrate synthase, Shikimate kinase, L-lactate dehydrogenase, 2-methylcitrate synthase, exopolyphosphatase, inorganic pyrophosphatase, alkaline phosphatase and NADH pyrophosphatase, respectively (Table 2). Four genes responsible for tryptophan synthesis, including tryptophan synthase α chain (*trpA*), tryptophan synthase β chain (*trpB*) and tryptophan—tRNA ligase (*trpS*) and Tryptophan 2-halogenase (*cmdE*) were also identified. These genes may play important roles in the synthesis of tryptophan and, therefore, participate in auxin indole-3-acetic acid (IAA) biosynthesis and enhance the growth of plants [73]. In addition, four other genes responsible for nitrogen fixation protein, acetolactate synthase 3 small subunit, biosynthetic arginine decarboxylase and S-adenosylmethionine decarboxylase proenzyme were identified, which may also contribute to growth promotion in ryegrass inoculated with M30-35.

Several studies indicated PGPRs could alleviate oxidative stress induced by salinity. Oxidative stress was mitigated by *Acinetobacter calcoaceticus* through reducing activities of catalase, peroxidase, polyphenol oxidase and total polyphenol in *Cucumis sativus* [65]. However, *Pseudomonas aeruginosa* inoculation ameliorated adverse effects of Zn stress by enhancing antioxidative enzyme activities, superoxide dismutase (SOD), peroxidase (POD) and CAT, in wheat [74]. Also in wheat, *Dietzia natronolimnaea* inoculation enhanced gene expression of various antioxidant enzymes such as ascorbate peroxidase (APX), SOD, CAT, POD, glutathione peroxidase (GPX) and glutathione reductase (GR) and higher proline content and contributed to increased tolerance to salinity stress [75]. In our research, seven genes responsible for oxidative stress alleviation were identified in the complete genome of *Pseudomonas* sp. M30-35, including catalase, superoxide dismutase, glutathione *S*-transferase, glutathione peroxidase, glutathione reductase, *S*-(hydroxymethyl) glutathione dehydrogenase and glutathione synthetase. Further study is needed to clarify the correlation between these genes and the oxidative stress alleviation in ryegrass inoculated with *Pseudomonas* sp. M30-35 under salt stress.

Seven genes responsible for abiotic stress in the complete genome of *Pseudomonas* sp. M30-35 were also identified, including Na^+/H^+ antiporter (*nhaC*), glycine betaine transporter (*opuD*), trehalose/maltose-binding protein, 1-aminocyclopropane-1-carboxylate (ACC) deaminase, cold shock protein (*capB*), cold shock protein (*cspA*) and heat shock protein (*hs1R*) (Table 2). Trehalose/maltose-binding protein can alleviate different stress conditions (dehydration, heat and cold) of plant growth environment [76]. ACC deaminase could lower the concentration of plant ethylene and decrease the deleterious effect of abiotic stress [77].

4. Materials and Methods

4.1. Bacterial Suspension Culture

Pseudomonas sp. strain M30-35 and *Escherichia coli* strain DH5α as a positive control bacterium strain were grown in liquid Luria broth (LB) medium without light for 24 h at 28 °C with 250 rpm rotation to yield 10^9 colony forming units (CFU) mL^{-1}, as determined by optical density and serial dilutions. Strain M30-35 was in our laboratory at Lanzhou University, Lanzhou, China, and *E. coli* strain DH5α was purchased from Takara Biotechnology (Dalian) Co., Ltd., Dalian, China.

4.2. Plant Growth and Treatments

Perennial ryegrass (*Lolium perenne* L. cv. Esquire) seeds (from Beijing Top Green Seed Co., Ltd., Beijing, China) were surface sterilized with 2% NaClO for 1 min followed by 70% ethanol for 10 min, and rinsed with sterile water five times. Seeds were then sown in pre-sterilized plastic pots (diameter 20 cm, depth 15 cm, 1.0 g seeds/pot) containing $1800\times g$ of heat-sterilized (95 °C, 72 h) vermiculite and sand mix (volume ratio 1:1) and watered with half strength Hoagland's nutrient solution (including 2 mM KNO_3, 0.5 mM $NH_4H_2PO_4$, 0.1 mM $Ca(NO_3)_2 \cdot 4H_2O$, 0.25 mM $MgSO_4 \cdot 7H_2O$, 0.5 mM Fe citrate, 92 μM H_3BO_3, 18 μM $MnCl_2 \cdot 4H_2O$, 1.6 μM $ZnSO_4 \cdot 7H_2O$, 0.6 μM $CuSO_4 \cdot 5H_2O$, and 0.7 μM $(NH_4)_6Mo_7O_{24} \cdot 4H_2O$) once per week. After seed germination, each pot was inoculated with 10 mL bacterial suspension culture of M30-35 or DH5α, or 10 mL liquid LB medium as control. Plants were grown in a glasshouse at the temperature regulated to around 28 °C during the day and 23 °C at night. Relative humidity averaged 65% and 75% for day and night periods, respectively. The photoperiod was 16 h/8 h (light/dark).

Twenty days after germination, seedlings were watered with 0, 150 or 300 mM NaCl as salt treatments. Two weeks after salt treatments, plants were harvested for biomass and physiological index measurements (12 replications for each treatment).

4.3. Plant Biomass and Physiological Measurements

Two weeks after salt treatments, plants were removed from the pots and roots were water rinsed to remove attached soil. Root and shoot were separated and blotted gently. Shoot fresh weights and root volume were determined immediately and samples were oven dried at 80 °C for 3 days to obtain dry weights.

Leaf chlorophyll content was measured using acetone and alcohol method with slight modification [16]. Briefly, fresh leaf sample was ground thoroughly with 80% acetone (5 mL) and 95% alcohol (5 mL) as a solvent in the dark and centrifuged at $9000\times g$ for 10 min at 4 °C. Absorbance reading (UV-2102C Spectrophotometer, Unico Instrument Co., Ltd., Shanghai, China) at 645 and 663 nm for collected supernatant was used to estimate chlorophyll a and chlorophyll b contents, respectively. Leaf chlorophylls content was calculated according to the formula: Chlorophyll a content = $(12.72A_{663} - 2.59A_{645}) \times 10 \div 1000 \div W$; Chlorophyll b content = $(22.88 A_{645} - 4.67A_{663}) \times 10 \div 1000 \div W$. W: fresh weight of leaves; A_{663} represented the absorbance value at 663 nm; A_{645} represented the absorbance value at 645 nm.

Root activity was measured according to the triphenyltetrazolium chloride (TTC) reduction method [78]. Fresh root tissue (0.1 g) was soaked in 10 mL reaction solution (5 mL 0.4% TTC + 5 mL phosphate buffer, pH = 7.0) at 37 °C for 3 h, and then 2 mL of 1 mol L^{-1} sulfuric acid was added to stop the reaction. Roots were taken out from the reaction solution, blotted and grounded with 3 mL ethyl acetate. Then, the supernatant was collected by ethyl acetate to the final volume of 10. Absorbance reading (UV-2102C Spectrophotometer, Unico Instrument Co., Ltd., Shanghai, China) at 485 nm of collected supernatant was used to estimate root activity. The standard curve for reduction value of TTC (*y*): $y = 0.0015x + 0.0053$, $R^2 = 0.9914$ and *y* represented absorbance reading. Root activity = reduction value of TTC (*x*) $\div$ W $\times$ time. W represented fresh weight of root and time 3 h.

Catalase (CAT) activity was measured according to [55]. Briefly, 0.5 g fresh leaf sample was grounded with 3 mL cold (4 °C) phosphate buffer (pH = 7.0) and enzyme was extracted in a total volume of 10 mL. CAT activity was estimated by the decrease in H_2O_2 according to the absorbance at 240 nm. CAT = $678 \times \Delta A \div W$, $\Delta A = A_1 - A_2$, A_1 represented the initial absorbance value at 240 nm, A_2 absorbance value after one minute and W fresh weight of leaves.

The level of lipid peroxidation in leaves was assessed by measuring the content of malondialdehyde (MDA) using the thiobarbituric acid reaction method [12,14]. Absorbance was determined at 532 (A_{532}) and 600 (A_{600}) nm using a UV spectrophotometer (UV-2102C, Unico Instrument Co., Ltd., Shanghai, China) to estimate MDA content. MDA = $25.8 \times \Delta A \div W$. $\Delta A = A_{532} - A_{600}$ and W represented fresh weight of leaves.

To estimate leaf cell membrane damage, leaf relative electric conductivity (REC) was measured using an electric conductivity meter EC215 (Hanna Instruments Romania Srl, Nusfalau, Romania) [12]. REC (%) was calculated the following equation, REC (%) = $(S1/S2) \times 100$, where S1 and S2 refer to electric conductivity of live leaves and boiled leaves, respectively.

The soluble sugar content was measured according to the method of anthrone colorimetry [79]. In addition, 1 g fresh leaf sample was grounded thoroughly with water and centrifuged at $8000 \times g$ for 10 min at 25 °C to extract soluble sugar in 10 mL. Then, 1 mL supernatant, 0.5 mL anthrone and 5 mL H_2SO_4 were mixed slowly and kept for 10 min. After cooling down to room temperature, absorbance was determined at 620 nm using a UV spectrophotometer (UV-2102C, Unico Instrument Co., Ltd., Shanghai, China) to estimate soluble sugar content. Soluble sugar content = $1.17 \times (\Delta A + 0.07) \div W$. $\Delta A = A_1 - A_0$, A_1 represented the absorbance value of sample at 620 nm, A_0 represented the absorbance value of blank solution at 620 nm and W fresh weight of leaves.

Leaf proline content was determined by the sulfosalicylic acid method described by [80]. Proline was extracted from 0.2 g of fresh leaf tissue into 10 mL of 3% sulfosalicylic acid and filtered through filter paper and absorbance was determined at 520 nm (A_{520}) using a spectrophotometer (UV-2102C Spectrophotometer, Unico Instrument Co., Ltd., Shanghai, China). Proline content = $19.2 \times (A_{520} + 0.0021) \div W$ and W represented fresh weight of leaves.

CAT, MDA, the soluble sugar and proline content were measured using reagent kit (Suzhou Comin Biotechnology Co., Ltd., Suzhou, China), and the specific steps of all methods were operated according to the instructions provided by the company.

Leaf osmotic potential (ψ_s) was measured according to Ma et al. [48]. Fresh leaf samples were frozen in liquid nitrogen. Cell sap was collected by thawing slowly and then ψ_s was determined using a cryoscopic osmometer (Osmomat-030, Gonotec GmbH, Berlin, Germany) at 25 °C. The readings (mol L^{-1}) were used to calculate the solute potential (ψ_s) in MPa with the formula $\psi_s = -$The readings $\times R \times T$, here $R = 0.008314$ MPa L mol^{-1} K^{-1} and $T = 298.8$ K.

Na$^+$ and K$^+$ contents were measured according to the method described by Han et al. [16]. For ion content analysis, plants were harvested two weeks after bacterial inoculation. Roots were washed twice for 8 min in ice-cold 20 mM CaCl$_2$ to exchange cell-wall-bound K$^+$ and Na$^+$, and the shoots were rinsed in deionised water to remove surface salts. Roots and shoots were separated and samples were oven dried at 80 °C for 3 days. Ions were extracted from dried tissues with 10 mL 100 mM acetic acid at 90 °C for 2 h. Ion analysis was conducted by an atomic absorption spectrophotometry r (2655-00, Cole-Parmer Instrument Co., Vernon Hills, IL, USA).

4.4. Data Analysis

Results of the growth and physiological parameters were presented as means with standard errors ($n = 12$). All the data were subjected to one-way analysis of variance (ANOVA) and Duncan's multiple comparison tests were used to detect significant differences among means at a significance level of $p < 0.05$ by SPSS 19.0 (SPSS Inc., Chicago, IL, USA).

4.5. Complete Genome Sequencing and Analysis of M30-35

Complete genome of strain M30-35 was sequenced by Illumin Miseq, Pacific Biosciences (PacBio) RS II and Single Molecule Real Time (SMRT) sequencing technology. The complete genome sequencing work was in cooperation with Shanghai Personal Biotechnology Co., Ltd., Shanghai, China.

The genome annotation was performed using NCBI Prokaryotic Genome Annotation Pipeline (PGAP) version 4.10 (National Center for Biotechnology Information, National Library of Medicine, Bethesda, MD, USA). The analysis of complete genome was conducted at https://www.ncbi.nlm.nih.gov/genome/microbes/ and sequence blasting at https://blast.ncbi.nlm.nih.gov/Blast.cgi. The open reading frames (ORFs) were predicted using Glimmer version 3.0 (currently supported by National Library of Medicine, Bethesda, MD, USA) [31,32], tRNA and rRNA genes were identified by tRNAscan-SE version 1.3.1 (The Lowe Lab, Biomolecular Engineering, University of California Santa Cruz, Santa Cruz, CA, USA) [33,35] and rRNAmmer version 1.2 (Department of Bio. and Health Informatics, Technical University of Denmark Bioinformatics, Lyngby, Denmark) [81], respectively. We also performed functional analysis and classification of the clusters of orthologous genes (COG) [43].

5. Conclusions

Pseudomonas sp. M30-35 live naturally in the rhizosphere desert shrub and xerohalophyte *Haloxylon ammodendron*. Our study firstly demonstrated that M30-35 had the feature as a PGPR strain. The soil inoculation of M30-35 significantly enhanced growth and salt tolerance of perennial ryegrass through increasing chlorophyll content, root activity, catalase activity, soluble sugar content, proline content, tissue K$^+$ content and K$^+$/Na$^+$ ratio, while decreasing malondialdehyde content and relative electric conductivity and osmotic potential under salt condition, especially higher salinity. M30-35 was initially classified as a new species in *Pseudomonas* and named it as *Pseudomonas* sp. M30-35. Thirty-four genes, which were responsible for plant growth promotion and abiotic stress tolerance in *Pseudomonas* sp. M30-35 genome, were identified through complete genome sequencing technology. Further study is needed to clarify the correlation between these genes from M30-35 and the salt stress alleviation in inoculated plants under salt stress. Genome sequencing and physiological analysis in our study indicated that *Pseudomonas* sp. M30-35 is a novel PGPR strain. Moreover, this work will be useful to explore novel PGPR strains from rhizosphere of plants under extreme environments and the potential interaction mechanism between PGPRs strains and crop plants.

Acknowledgments: Suo-Min Wang from Lanzhou University was appreciated for his advice and laboratory support to this work. This work was financially supported by the National Basic Research Program of China (Grant No. 2014CB138701), the National Natural Science Foundation of China (Grant No. 31222053), the Science and Technology Support Program of Gansu Province, China (Grant No. 1604NKCA077), the 111 project (Grant No. B12002) and the Fundamental Research Funds for the Central Universities (Grant No. lzujbky-2017-1).

Author Contributions: Jin-Lin Zhang conceived and designed the experiments and wrote the manuscript; Ao-Lei He performed the experiments and contributed to the manuscript writing; Shu-Qi Niu performed the experiments and analyzed the data; Qi Zhao, Yong-Sheng Li, Jing-Yi Gou, Hui-Juan Gao and Sheng-Zhou Suo contributed to data analysis and manuscript editing.

Conflicts of Interest: The authors declare no conflict of interest.

Abbreviations

ACC	1-aminocyclopropane-1-carboxylate
APX	Ascorbate peroxidase
CAT	Catalase
CFU	Colony forming unit
COG	Clusters of orthologous gene
G + C	Guanine and Cytosine
GDH	Glucose dehydrogenase
GPX	Glutathione peroxidase
GR	Glutathione reductase
LB	Luria broth
MDA	Malondialdehyde
NADH	Nicotinamide adenine dinucleotide
PGAP	Prokaryotic genome annotation pipeline
PGPR	Plant growth promoting rhizobacterium
POD	Peroxidase
PSB	Phosphate solubilizing bacterium
REC	Relative electric conductivity
ROS	Reactive oxygen species
SEs	Standard errors
SMRT	Single molecule real time
SOD	Superoxide dismutase
TBA	Thiobarbituric acid
TTC	Triphenyltetrazolium chloride

References

1. Hu, H.; Xiong, L. Genetic engineering and breeding of drought-resistant crops. *Annu. Rev. Plant Biol.* **2014**, *65*, 715–741. [CrossRef] [PubMed]
2. Zhang, J.L.; Shi, H. Physiological and molecular mechanisms of plant salt tolerance. *Photosynth. Res.* **2013**, *115*, 1–22. [CrossRef] [PubMed]
3. Zhang, J.L.; Flowers, T.J.; Wang, S.M. Mechanisms of sodium uptake by roots of higher plant. *Plant Soil* **2010**, *326*, 45–60. [CrossRef]
4. Yang, Y.; Guo, Y. Elucidating the molecular mechanisms mediating plant salt-stress responses. *New Phytol.* **2017**. [CrossRef] [PubMed]
5. Hayat, R.; Ali, S.; Amara, U.; Khalid, R.; Ahmed, I. Soil beneficial bacteria and their role in plant growth promotion: A review. *Ann. Microbiol.* **2010**, *60*, 579–598. [CrossRef]
6. Lugtenberg, B.; Kamilova, F. Plant-growth-promoting rhizobacteria. *Annu. Rev. Microbiol.* **2009**, *63*, 541–556. [CrossRef]
7. Yang, J.; Kloepper, J.W.; Ryu, C.M. Rhizosphere bacteria help plants tolerate abiotic stress. *Trends Plant Sci.* **2009**, *14*, 1–4. [CrossRef] [PubMed]
8. Grover, M.; Ali, S.Z.; Sandhya, V.; Rasul, A.; Venkateswarlu, B. Role of microorganisms in adaptation of agriculture crops to abiotic stresses. *World J. Microb. Biotechnol.* **2011**, *27*, 1231–1240. [CrossRef]
9. Bharti, N.; Yadav, D.; Barnawal, D.; Maji, D.; Kalra, A. *Exiguobacterium oxidotolerans*, a halotolerant plant growth promoting rhizobacteria, improves yield and content of secondary metabolites in *Bacopa monnieri* (L.) Pennell under primary and secondary salt stress. *World J. Microb. Biotechnol.* **2013**, *29*, 379–387. [CrossRef]

10. Gururani, M.A.; Upadhyaya, C.P.; Baskar, V.; Venkatesh, J.; Nookaraju, A.; Park, S.W. Plant growth-promoting rhizobacteria enhance abiotic stress tolerance in *Solanum tuberosum* through inducing changes in the expression of ROS-scavenging enzymes and improved photosynthetic performance. *J. Plant Growth Regul.* **2013**, *32*, 245–258. [CrossRef]

11. Hamdia, E.S.; Shaddad, M.A.K.; Doaa, M.M. Mechanisms of salt tolerance and interactive effects of *Azospirillum brasilense* inoculation on maize cultivars grown under salt stress conditions. *Plant Growth Regul.* **2004**, *44*, 165–174. [CrossRef]

12. Niu, S.Q.; Li, H.R.; Paré, P.W.; Aziz, M.; Wang, S.M.; Shi, H.Z.; Li, J.; Han, Q.Q.; Guo, S.Q.; Li, J.; et al. Induced growth promotion and higher salt tolerance in the halophyte grass *Puccinellia tenuiflora* by beneficial rhizobacteria. *Plant Soil* **2016**, *407*, 217–230. [CrossRef]

13. Zhao, Q.; Wu, Y.N.; Fan, Q.; Han, Q.Q.; Paré, P.W.; Xu, R.; Wang, Y.Q.; Wang, S.M.; Zhang, J.L. Improved growth and metabolite accumulation in *Codonopsis pilosula* (Franch.) Nannf. by inoculation of *Bacillus amyloliquefaciens* GB03. *J. Agric. Food Chem.* **2016**, *64*, 8103–8108. [CrossRef] [PubMed]

14. Su, A.Y.; Niu, S.Q.; Liu, Y.Z.; He, A.L.; Zhao, Q.; Paré, P.W.; Li, M.F.; Han, Q.Q.; Ali Khan, S.; Zhang, J.L. Synergistic effects of *Bacillus amyloliquefaciens* (GB03) and water retaining agent on drought tolerance of perennial ryegrass. *Int. J. Mol. Sci.* **2017**, *18*, 2651. [CrossRef] [PubMed]

15. Zhang, J.L.; Aziz, M.; Qiao, Y.; Han, Q.Q.; Li, J.; Wang, Y.Q.; Shen, X.; Wang, S.M.; Paré, P.W. Soil microbe *Bacillus subtilis* (GB03) induces biomass accumulation and salt tolerance with lower sodium accumulation in wheat. *Crop Pasture Sci.* **2014**, *65*, 423–427. [CrossRef]

16. Han, Q.Q.; Lü, X.P.; Bai, J.P.; Qiao, Y.; Paré, P.W.; Wang, S.M.; Zhang, J.L.; Wu, Y.N.; Pang, X.P.; Xu, W.B.; et al. Beneficial soil bacterium *Bacillus subtilis* (GB03) augments salt tolerance of white clover. *Front. Plant Sci.* **2014**, *5*, 525. [CrossRef] [PubMed]

17. Han, Q.Q.; Wu, Y.N.; Xu, R.; Paré, P.W.; Shi, H.; Zhao, Q.; Li, H.R.; Sardar, A.K.; Wang, Y.Q.; Wang, S.M.; et al. Improved salt tolerance of medicinal plant *Codonopsis pilosula* by *Bacillus amyloliquefaciens* GB03. *Acta Physiol. Plant.* **2017**, *39*, 35. [CrossRef]

18. Santoro, M.V.; Bogino, P.C.; Nocelli, N.; Rosario, C.L.D.; Giordano, W.F.; Erika, B. Analysis of plant growth-promoting effects of *fluorescent Pseudomonas* strains isolated from *Mmentha piperita* rhizosphere and effects of their volatile organic compounds on essential oil composition. *Front. Microbiol.* **2016**, *7*, 1085. [CrossRef] [PubMed]

19. Zhang, L.; Pan, Y.; Wang, K.; Zhang, X.; Zhang, C.; Zhang, S.; Fu, X.; Jiang, J. *Pseudomonas zhaodongensis* sp. nov., isolated from saline and alkaline soils. *Int. J. Syst. Evol. Microbiol.* **2015**, *65*, 1022–1030. [CrossRef] [PubMed]

20. Anwar, N.; Abaydulla, G.; Zayadan, B.; Abdurahman, M.; Hamood, B.; Erkin, R.; Ismayil, N.; Mamtimin, H.; Rahman, E. *Pseudomonas populi* sp. nov., an endophytic bacterium isolated from *Populus euphratica*. *Int. J. Syst. Evol. Microbiol.* **2016**, *66*, 1419–1425. [CrossRef] [PubMed]

21. Ramos, E.; Ramírez-Bahena, M.H.; Valverde, A.; Velázquez, E.; Zúñiga, D.; Velezmoro, C.; Peix, A. *Pseudomonas punonensis* sp. nov., isolated from straw. *Int. J. Syst. Evol. Microbiol.* **2013**, *63*, 1834–1839. [CrossRef] [PubMed]

22. Menéndez, E.; Ramírez-Bahena, M.H.; Fabryová, A.; Igual, J.M.; Benada, O.; Mateos, P.F.; Peix, A.; Kolarik, M.; García-Fraile, P. *Pseudomonas coleopterorum* sp. nov., a cellulase-producing bacterium isolated from the bark beetle *Hylesinus fraxini*. *Int. J. Syst. Evol. Microbiol.* **2015**, *65*, 2852–2858. [CrossRef] [PubMed]

23. Pascual, J.; Lucena, T.; Ruvira, M.A.; Giordano, A.; Gambacorta, A.; Garay, E.; Arahal, D.R.; Pujalte, M.J.; Macián, M.C. *Pseudomonas litoralis* sp. nov., isolated from Mediterranean seawater. *Int. J. Syst. Evol. Microbiol.* **2012**, *62*, 438–444. [CrossRef] [PubMed]

24. Zhong, Z.P.; Liu, Y.; Hou, T.T.; Liu, H.C.; Zhou, Y.G.; Wang, F.; Liu, Z.P. *Pseudomonas salina* sp. nov., isolated from a salt lake. *Int. J. Syst. Evol. Microbiol.* **2015**, *65*, 2846–2851. [CrossRef] [PubMed]

25. Kohler, J.; Caravaca, F.; Carrasco, L.; Roldán, A. Contribution of *Pseudomonas mendocina* and *Glomus intraradices* to aggregate stabilization and promotion of biological fertility in rhizosphere soil of lettuce plants under field conditions. *Soil Use Manag.* **2006**, *22*, 298–304. [CrossRef]

26. Kang, S.M.; Radhakrishnan, R.; Khan, A.L.; Kim, M.J.; Park, J.M.; Kim, B.R.; Shin, D.H.; Lee, I.J. Gibberellin secreting rhizobacterium, *Pseudomonas putida* H-2-3 modulates the hormonal and stress physiology of soybean to improve the plant growth under saline and drought conditions. *Plant Physiol. Biochem.* **2014**, *84*, 115–124. [CrossRef] [PubMed]

27. Zerrouk, I.Z.; Benchabane, M.; Khelifi, L.; Yokawa, K.; Ludwig-Müller, J.; Baluska, F. A *Pseudomonas* strain isolated from date-palm rhizospheres improves root growth and promotes root formation in maize exposed to salt and aluminum stress. *J. Plant Physiol.* **2016**, *191*, 111–119. [CrossRef] [PubMed]

28. Liu, R.; Yu, Z.; Ping, C.; Lin, H.; Ye, G.; Wang, Z.; Ge, C.; Bo, Z.; Ren, D. Genomic and phenotypic analyses of *Pseudomonas psychrotolerans* PRS08-11306 reveal a turnerbactin biosynthesis gene cluster that contributes to nitrogen fixation. *J. Biotechnol.* **2017**, *253*, 10–13. [CrossRef] [PubMed]

29. Tiwari, S.; Prasad, V.; Chauhan, P.S.; Lata, C. *Bacillus amyloliquefaciens* confers tolerance to various abiotic stresses and modulates plant response to phytohormones through osmoprotection and gene expression regulation in rice. *Front. Plant Sci.* **2017**, *8*, 1510. [CrossRef] [PubMed]

30. Roca, A.; Pizarro-Tobias, P.; Udaondo, Z.; Fernandez, M.; Matilla, M.A.; Molina-Henares, M.A.; Molina, L.; Segura, A.; Duque, E.; Ramos, J.L. Analysis of the plant growth-promoting properties encoded by the genome of the rhizobacterium *Pseudomonas putida* BIRD-1. *Environ. Microbiol.* **2013**, *15*, 780–794. [CrossRef] [PubMed]

31. Shen, X.; Hu, H.; Peng, H.; Wang, W.; Zhang, X. Comparative genomic analysis of four representative plant growth-promoting rhizobacteria in *Pseudomonas*. *BMC Genom.* **2013**, *14*, 271. [CrossRef] [PubMed]

32. Lee, J.; Cho, Y.J.; Yang, J.Y.; Jung, Y.J.; Hong, S.G.; Kim, O.S. Complete genome sequence of *Pseudomonas antarctica* PAMC 27494, a bacteriocin-producing psychrophile isolated from Antarctica. *J. Biotechnol.* **2017**, *259*, 15–18. [CrossRef] [PubMed]

33. Li, X.; Li, C.Z.; Mao, L.Q.; Yan, D.Z.; Zhou, N.Y. Complete genome sequence of the cyclohexylamine-degrading *Pseudomonas plecoglossicida* NyZ12. *J. Biotechnol.* **2015**, *199*, 29–30. [CrossRef] [PubMed]

34. Fang, Y.; Wu, L.; Chen, G.; Feng, G. Complete genome sequence of *Pseudomonas azotoformans* S4, a potential biocontrol bacterium. *J. Biotechnol.* **2016**, *227*, 25–26. [CrossRef] [PubMed]

35. Jiang, Q.; Xiao, J.; Zhou, C.; Mu, Y.; Xu, B.; He, Q.; Xiao, M. Complete genome sequence of the plant growth-promoting rhizobacterium *Pseudomonas aurantiaca* strain JD37. *J. Biotechnol.* **2014**, *192*, 85–86. [CrossRef] [PubMed]

36. Yanzhen, M.; Yang, L.; Xiangting, X.; Wei, H.; Chuanchao, D. Complete genome sequence of a bacterium *Pseudomonas fragi* P121, a strain with degradation of toxic compounds. *J. Biotechnol.* **2016**, *224*, 68–69. [CrossRef] [PubMed]

37. Huang, Z.; Zhang, X.; Zheng, G.; Gutterman, Y. Influence of light, temperature, salinity and storage on seed germination of *Haloxylon ammodendron*. *J. Arid Environ.* **2003**, *55*, 453–464. [CrossRef]

38. Huang, B.; DaCosta, M.; Jiang, Y. Research advances in mechanisms of grass tolerance to abiotic stress from physiology to molecular biology. *Crit. Rev. Plant Sci.* **2014**, *33*, 141–189. [CrossRef]

39. Kane, K.H. Effects of endophyte infection on drought stress tolerance of *Lolium perenne* accessions from the Mediterranean region. *Environ. Exp. Bot.* **2011**, *71*, 337–344. [CrossRef]

40. Li, X.; Han, S.; Wang, G.; Liu, X.; Amombo, E.; Xie, Y.; Fu, J. The fungus *Aspergillus aculeatus* enhances salt-stress tolerance, metabolite accumulation, and improves forage quality in perennial ryegrass. *Front. Microbiol.* **2017**, *8*, 1664. [CrossRef] [PubMed]

41. Wang, J.; Suleimans, B. Evaluation of drought tolerance for Atlas fescue, perennial ryegrass, and their progeny. *Euphytica* **2008**, *164*, 113–122. [CrossRef]

42. Tang, J.; Yu, X.; Luo, N.; Xiao, F.; Camberato, J.J.; Jiang, Y. Natural variation of salinity response, population structure and candidate genes associated with salinity tolerance in perennial ryegrass accessions. *Plant Cell Environ.* **2013**, *36*, 2021–2033. [CrossRef] [PubMed]

43. Powell, S.; Forslund, K.; Szklarczyk, D.; Trachana, K.; Roth, A.; Huertacepas, J.; Gabaldon, T.; Rattei, T.; Creevey, C.; Kuhn, M.; et al. eggNOG v4.0: Nested orthology inference across 3686 organisms. *Nucleic Acids Res.* **2014**, *42*, 231–239. [CrossRef] [PubMed]

44. Dimkpa, C.; Weinand, T.; Asch, F. Plant–rhizobacteria interactions alleviate abiotic stress conditions. *Plant Cell Environ.* **2009**, *32*, 1682–1694. [CrossRef] [PubMed]

45. Baniaghil, N.; Arzanesh, M.H.; Ghorbanli, M.; Shahbazi, M. The effect of plant growth promoting rhizobacteria on growth parameters, antioxidant enzymes and microelements of canola under salt stress. *J. Appl. Environ. Biol. Sci.* **2013**, *3*, 17–27.

46. Kohler, J.; Hernández, J.A.; Caravaca, F.; Roldán, A. Induction of antioxidant enzymes is involved in the greater effectiveness of a PGPR versus AM fungi with respect to increasing the tolerance of lettuce to severe salt stress. *Environ. Exp. Bot.* **2009**, *65*, 245–252. [CrossRef]

47. Spaepen, S.; Bossuyt, S.; Engelen, K.; Marchal, K.; Vanderleyden, J. Phenotypical and molecular responses of *Arabidopsis thaliana* roots as a result of inoculation with the auxin-producing bacterium *Azospirillum brasilense*. *New Phytol.* **2014**, *201*, 850–861. [CrossRef] [PubMed]

48. Ma, Q.; Yue, L.J.; Zhang, J.L.; Wu, G.Q.; Bao, A.K.; Wang, S.M. Sodium chloride improves photosynthesis and water status in the succulent xerophyte *Zygophyllum xanthoxylum*. *Tree Physiol.* **2012**, *32*, 4–13. [CrossRef] [PubMed]

49. Zhang, H.M.; Xie, X.T.; Kim, M.S.; Kornyeyev, D.A.; Holaday, S.; Paré, P.W. Soil bacteria augment *Arabidopsis* photosynthesis by decreasing glucose sensing and abscisic acid levels in *planta*. *Plant J.* **2008**, *56*, 264–273. [CrossRef] [PubMed]

50. Kalaji, H.M.; Bosa, K.; Kościelniak, J.; Żuk-Gołaszewska, K. Effects of salt stress on photosystem II efficiency and CO_2 assimilation of two Syrian barley landraces. *Environ. Exp. Bot.* **2011**, *73*, 64–72. [CrossRef]

51. Glick, B.R.; Liu, C.; Ghosh, S.; Dumbroff, E.B. Early development of canola seedlings in the presence of the plant growth-promoting rhizobacterium *Pseudomonas putida* GR12-2. *Soil Biol. Biochem.* **1997**, *29*, 1233–1239. [CrossRef]

52. Gill, S.S.; Tuteja, N. Reactive oxygen species and antioxidant machinery in abiotic stress tolerance in crop plants. *Plant Physiol. Biochem.* **2010**, *48*, 909–930. [CrossRef] [PubMed]

53. Das, K.; Roychoudhury, A. Reactive oxygen species (ROS) and response of antioxidants as ROS-scavengers during environmental stress in plants. *Front. Environ. Sci.* **2014**, *2*, 53. [CrossRef]

54. Niu, M.; Huang, Y.; Sun, S.; Sun, J.; Cao, H.; Shabala, S.; Bie, Z. Root respiratory burst oxidase homologue-dependent H_2O_2 production confers salt tolerance on a grafted cucumber by controlling Na^+ exclusion and stomatal closure. *J. Exp. Bot.* **2017**. [CrossRef] [PubMed]

55. Li, J.; Liu, J.; Wang, G.; Cha, J.Y.; Li, G.; Chen, S.; Li, Z.; Guo, J.; Zhang, C.; Yang, Y. A chaperone function of NO CATALASE ACTIVITY1 is required to maintain catalase activity and for multiple stress responses in *Arabidopsis*. *Plant Cell* **2015**, *27*, 908–925. [CrossRef] [PubMed]

56. Baltruschat, H.; Fodor, J.; Harrach, B.D.; Niemczyk, E.; Balazs, B.; Gullner, G.; Janeczko, A.; Kogel, K.H.; Schafer, P.; Schwarczinger, I.; et al. Salt tolerance of barley induced by the root endophyte *Piriformospora indica* is associated with a strong increase in antioxidants. *New Phytol.* **2008**, *180*, 501–510. [CrossRef] [PubMed]

57. Krasensky, J.; Jonak, C. Drought, salt, and temperature stress-induced metabolic rearrangements and regulatory networks. *J. Exp. Bot.* **2012**, *63*, 1593–1608. [CrossRef] [PubMed]

58. Bohnert, H.J.; Shen, B. Transformation and compatible solutes. *Sci. Hortic.* **1998**, *78*, 237–260. [CrossRef]

59. Yin, Y.G.; Kobayashi, Y.; Sanuki, A.; Kondo, S.; Fukuda, N.; Ezura, H.; Sugaya, S.; Matsukura, C. Salinity induces carbohydrate accumulation and sugar-regulated starch biosynthetic genes in tomato (*Solanum lycopersicum* L. cv. 'Micro-Tom') fruits in an ABA- and osmotic stress-independent manner. *J. Exp. Bot.* **2010**, *61*, 563–574. [CrossRef] [PubMed]

60. Verbruggen, N.; Hermans, C. Proline accumulation in plants: A review. *Amino Acids* **2008**, *35*, 753–759. [CrossRef] [PubMed]

61. Hayat, S.; Hayat, Q.; Alyemeni, M.N.; Wani, A.S.; Pichtel, J.; Ahmad, A. Role of proline under changing environments: A review. *Plant Signal. Behav.* **2012**, *7*, 1456–1466. [CrossRef] [PubMed]

62. Gharsallah, C.; Fakhfakh, H.; Grubb, D.; Gorsane, F. Effect of salt stress on ion concentration, proline content, antioxidant enzyme activities and gene expression in tomato cultivars. *Aob Plants* **2016**, *8*, plw055. [CrossRef] [PubMed]

63. Munns, R.; Tester, M. Mechanisms of salinity tolerance. *Annu. Rev. Plant Biol.* **2008**, *59*, 651–681. [CrossRef] [PubMed]

64. Ilangumaran, G.; Smith, D.L. Plant growth promoting rhizobacteria in amelioration of salinity stress: A systems biology perspective. *Front. Plant Sci.* **2017**, *8*, 1768. [CrossRef] [PubMed]

65. Kang, S.M.; Khan, A.L.; Waqas, M.; You, Y.H.; Kim, J.H.; Kim, J.G.; Hamayun, M.; Lee, I.J. Plant growth-promoting rhizobacteria reduce adverse effects of salinity and osmotic stress by regulating phytohormones and antioxidants in *Cucumis sativus*. *J. Plant Interact.* **2014**, *9*, 673–682. [CrossRef]

66. Wang, T.T.; Ding, P.; Chen, P.; Xing, K.; Bai, J.L.; Wan, W.; Jiang, J.H.; Qin, S. Complete genome sequence of endophyte *Bacillus flexus* KLBMP 4941 reveals its plant growth promotion mechanism and genetic basis for salt tolerance. *J. Biotechnol.* **2017**, *260*, 38–41. [CrossRef] [PubMed]

67. Hanif, M.K.; Hameed, S.; Imran, A.; Naqqash, T.; Shahid, M.; Elsas, J.D.V. Isolation and characterization of a β-propeller gene containing phosphobacterium *Bacillus subtilis* strain KPS-11 for growth promotion of potato (*Solanum tuberosum* L.). *Front. Microbiol.* **2015**, *6*, 583. [CrossRef] [PubMed]

68. Khan, A.A.; Jilani, G.; Akhtar, M.S.; Naqvi, S.M.S.; Rasheed, M. Phosphorus solubilizing bacteria: Occurrence, mechanisms and their role in crop production. *J. Agric. Biol. Sci.* **2009**, *1*, 48–58.

69. Péret, B.; Desnos, T.; Jost, R.; Kanno, S.; Berkowitz, O.; Nussaume, L. Root architecture responses: In search of phosphate. *Plant Physiol.* **2014**, *166*, 1713–1723. [CrossRef] [PubMed]

70. Hart, M.R.; Quin, B.F.; Nguyen, M.L. Phosphorus runoff from agricultural land and direct fertilizer effects: A review. *J. Environ. Qual.* **2004**, *33*, 1954–1972. [CrossRef] [PubMed]

71. Vassilev, N.; Eichlerlöbermann, B.; Vassileva, M. Stress-tolerant P-solubilizing microorganisms. *Appl. Microbiol. Biot.* **2012**, *95*, 851–859. [CrossRef] [PubMed]

72. Behera, B.C.; Singdevsachan, S.K.; Mishra, R.R.; Dutta, S.K.; Thatoi, H.N. Diversity, mechanism and biotechnology of phosphate solubilising microorganism in mangrove—A review. *Biocatal. Agric. Biotechnol.* **2014**, *3*, 97–110. [CrossRef]

73. Vessey, J.K. Plant growth promoting rhizobacteria as biofertilizers. *Plant Soil.* **2003**, *255*, 571–586. [CrossRef]

74. Islam, F.; Yasmeen, T.; Ali, Q.; Ali, S.; Arif, M.S.; Hussain, S.; Rizvi, H. Influence of *Pseudomonas aeruginosa* as PGPR on oxidative stress tolerance in wheat under Zn stress. *Ecotoxicol. Environ. Saf.* **2014**, *104*, 285–293. [CrossRef] [PubMed]

75. Bharti, N.; Pandey, S.S.; Barnawal, D.; Patel, V.K.; Kalra, A. Plant growth promoting rhizobacteria *Dietzia natronolimnaea* modulates the expression of stress responsive genes providing protection of wheat from salinity stress. *Sci. Rep.* **2016**, *6*, 34768. [CrossRef] [PubMed]

76. Crowe, J.H.; Crowe, L.M.; Chapman, D. Preservation of membranes in anhydrobiotic organisms: The role of trehalose. *Science* **1984**, *223*, 701–703. [CrossRef] [PubMed]

77. Glick, B.R.; Penrose, D.M.; Li, J. A model for the lowering of plant ethylene concentrations by plant growth-promoting bacteria. *J. Theor. Biol.* **1998**, *190*, 63–68. [CrossRef] [PubMed]

78. Stūrīte, I.; Henriksen, T.M.; Breland, T.A. Distinguishing between metabolically active and inactive roots by combined staining with 2,3,5-triphenyltetrazolium chloride and image colour analysis. *Plant Soil* **2005**, *271*, 75–82. [CrossRef]

79. Leakey, A.D.; Ainsworth, E.A.; Bernacchi, C.J.; Rogers, A.; Long, S.P.; Ort, D.R. Elevated CO_2 effects on plant carbon, nitrogen, and water relations: Six important lessons from FACE. *J. Exp. Bot.* **2009**, *60*, 2859–2876. [CrossRef] [PubMed]

80. Tiwari, S.; Lata, C.; Chauhan, P.S.; Nautiyal, C.S. *Pseudomonas putida* attunes morphophysiological, biochemical and molecular responses in *Cicer arietinum* L. during drought stress and recovery. *Plant Physiol. Biochem.* **2016**, *99*, 108–117. [CrossRef] [PubMed]

81. Lagesen, K.; Hallin, P.; Rodland, E.A.; Staerfeldt, H.H.; Rognes, T.; Ussery, D.W. RNAmmer: Consistent and rapid annotation of ribosomal RNA genes. *Nucleic Acids Res.* **2007**, *35*, 3100–3108. [CrossRef] [PubMed]

International Journal of
Molecular Sciences

MDPI

Article

Improvement of *Verticillium* Wilt Resistance by Applying Arbuscular Mycorrhizal Fungi to a Cotton Variety with High Symbiotic Efficiency under Field Conditions

Qiang Zhang [1,2], Xinpeng Gao [1], Yanyun Ren [3], Xinhua Ding [1,2], Jiajia Qiu [1], Ning Li [1], Fanchang Zeng [1,*] and Zhaohui Chu [1,*]

[1]   State Key Laboratory of Crop Biology, College of Agronomy, Shandong Agricultural University, Tai'an 271018, China; zqsdau@163.com (Q.Z.); gaoxinpeng1990@foxmail.com (X.G.); xhding@sdau.edu.cn (X.D.); chunshuijiaren@163.com (J.Q.); nli@sdau.edu.cn (N.L.)
[2]   Shandong Provincial Key Laboratory of Vegetable Disease and Insect Pests, College of Plant Protection, Shandong Agricultural University, Tai'an 271018, China
[3]   Jining Academy of Agricultural Sciences, Jining 272031, China; renyanyun@126.com
*   Correspondence: fczeng@sdau.edu.cn (F.Z.); zchu@sdau.edu.cn (Z.C.); Tel.: +86-538-8241-828 (F.Z.); +86-538-8249-913 (Z.C.)

Received: 7 December 2017; Accepted: 10 January 2018; Published: 13 January 2018

Abstract: Arbuscular mycorrhizal fungi (AMF) play an important role in nutrient cycling processes and plant stress resistance. To evaluate the effect of *Rhizophagus irregularis* CD1 on plant growth promotion (PGP) and *Verticillium* wilt disease, the symbiotic efficiency of AMF (SEA) was first investigated over a range of 3% to 94% in 17 cotton varieties. The high-SEA subgroup had significant PGP effects in a greenhouse. From these results, the highest-SEA variety of Lumian 1 was selected for a two-year field assay. Consistent with the performance from the greenhouse, the AMF-mediated PGP of Lumian 1 also produced significant results, including an increased plant height, stem diameter, number of petioles, and phosphorus content. Compared with the mock treatment, AMF colonization obviously inhibited the symptom development of *Verticillium dahliae* and more strongly elevated the expression of pathogenesis-related genes and lignin synthesis-related genes. These results suggest that AMF colonization could lead to the mycorrhiza-induced resistance (MIR) of Lumian 1 to *V. dahliae*. Interestingly, our results indicated that the AMF endosymbiont could directly inhibit the growth of phytopathogenic fungi including *V. dahliae* by releasing undefined volatiles. In summary, our results suggest that stronger effects of AMF application result from the high-SEA.

Keywords: mycorrhizal colonization; *Gossypium hirsutum*; *Verticillium* wilt; symbiotic efficiency; plant growth promotion; resistance; antifungal activity

1. Introduction

Cotton (*Gossypium* spp.) is an essential resource for thousands of consumables and industrial products manufactured across the world. It continues to grow in importance in the fiber and oil industries. However, *Verticillium* wilt caused by the soil-borne fungus *Verticillium dahliae*, also known as the "cancer" of cotton crops, has become one of the most devastating diseases in cotton-growing areas, resulting in significant losses of plant biomass, lint yield, and fiber quality worldwide [1,2]. This pathogenic fungus can persist alone in the soil for up to 15 years by forming microsclerotia as resting structures [3]. Because of its highly variable pathogenicity and strong vitality, it is extremely difficult to control *Verticillium* wilt disease [4].

In China, approximately 2.5 million hectares of the cotton are affected by *Verticillium* wilt, which may cause direct economic loss of 250–310 million US dollars annually [4]. More seriously, in some regions, the losses of lint cotton yield may be as high as 80% [5]. Unfortunately, to date no effective fungicides are available to control this pathogen because of the strong viability of its microsclerotia along with its wide host range and highly variable pathogenicity [6]. Although many island cotton varieties (*Gossypium barbadense*) possess resistance to *Verticillium* wilt, upland cotton (*Gossypium hirsutum*), the commercial cultivar that accounts for more than 95% of the annual global cotton crop, lacks germplasm resources resistant to this disease [7]. Traditional cross-breeding between the two species has not been successful because of hybrid dysgenesis, linkage drag, or abnormal separations in the progeny [8,9].

Histochemical analyses have revealed that intrinsic defenses play a dominant role in the disease resistance of island cotton, while such induced resistance is severely limited in upland cotton [10]. During the past decades, soil solarization and fumigation have been used as primary management strategies to reduce microsclerotia. However, in addition to playing only a limited role, they pose costs to the economy, ecological balance, and public health [11]. Currently, the molecular mechanisms underlying cotton resistance to *Verticillium* wilt are still poorly understood [12,13]. With the rapid development of genetic engineering, several genes have been identified to improve the cotton *Verticillium* wilt resistance, such as *GAFP4* [4] and a series of *Ve1* gene homologs including *Gbvdr5* [14], *Gbvdr3* [15], *GbaVd1*, and *GbaVd2* [16]. Alternatively, transgenic cotton that produced the RNAi construct of *VdH1* (VdH1i)-derived siRNAs showed efficient protection against *V. dahliae* [17]. They are potential sources to introduce into integrated pest management in the future.

Many reports emphasize that an induced resistance mechanism using mycorrhiza-induced resistance (MIR) rather than increased tolerance or other effects plays a major role in plant defense against a broad spectrum of pathogens [18–21]. MIR, a mild but efficient activation of the plant immune response, not only locally but also systemically shares some characteristics with systemic acquired resistance (SAR) after pathogen invasion and induced systemic resistance (ISR) triggered by non-pathogenic rhizobacteria [22]. At the precontact phase of mycorrhizal symbiosis, plant roots release strigolactones (which may also include an *N*-acetylglucosamine-based molecule) to induce the germination of fungal spores and stimulate hyphal branching [23,24]. Subsequently, arbuscular mycorrhizal fungi (AMF) produce lipochitooligosaccharides and chitooligosaccharides, which can be recognized by the host roots and then activate the symbiosis signaling pathway [25]. During this period, the innate immune system of host plant recognizes the microbe-associated molecular patterns (MAMPs) from AMF and initiates the transient expression of MAMP-triggered immunity (PTI), as well as the accumulation of the plant defense hormone salicylic acid (SA) in the vascular tissues [26]. Although the initial SA accumulation is restrained during successive stages of AMF infection, the primed defense state of SA-dependent defense and SAR can be sustained for long periods [27,28]. To establish a successful infection, AMF secrete specific effectors to suppress PTI and transiently induce the production of abscisic acid (ABA) in the roots [29]. ABA can be transported through the xylem to the phloem, by which it primes cell wall defense [30]. With the development of symbiosis and the modulation of plant immunity, the root exudation chemistry changes and then results in the delivery of ISR-eliciting signals by the mycorrhizosphere bacteria, which can be perceived by the host plant. Subsequently, the host plant generates long-distance signals that prime jasmonate- and ethylene-dependent plant defenses and cause ISR [18,31]. In summary, MIR can induce a primed state of the mycorrhizal plants that allows a more effective activation of defense mechanisms to address environmental challenges [22].

AMF are generally endomycorrhizal fungi of the phylum *Glomeromycota*, which can establish symbionts with most types of terrestrial plants [32]. The growth and development of AMF rely on the colonization of host roots to obtain their sugar and lipids [33,34]. In return, the AMF benefit the host plants, such as by helping to absorb and deliver phosphorus, enhancing resistance to biotic and abiotic stresses, and stimulating growth promotion [35–37]. In the past, research on AMF primarily focused on its beneficial effects on plant growth and nutrition, while recent studies have paid more attention to its biocontrol potential, in particular for soil-borne pathogens. AMF have exhibited inhibitory effects on several soil-borne pathogens including species of *Rhizoctonia* [38], *Verticillium* [39], *Fusarium* [40], *Phytophthora* [41], *Macrophomina* [42], and *Aphanomyces* [43]. Considering its unique advantages in food security, environmental protection and low risk of antimicrobial resistance, AMF should provide new avenues to protect cotton from the fungal pathogen *V. dahliae* in sustainable and organic agriculture.

Although it has been confirmed that AMF can benefit cotton growth and may behave as a biocontrol agent against *Verticillium* wilt [39,44], studies of symbiotic efficiency between AMF and different cotton cultivars have not yet been reported, and the effects of plant growth promotion (PGP) or resistance towards *V. dahliae* by AMF on high-SEA cotton varieties have never been investigated. The main objective of this study was to assess the potential of AMF *Rhizophagus irregularis* CD1 in controlling *Verticillium* wilt in high-SEA cotton cultivars directly in the field.

2. Results

2.1. Symbiotic Efficiency of Seventeen Cotton Varieties Inoculated by Arbuscular Mycorrhizal Fungi (AMF) R. irregularis

Sufficient mycorrhizal colonization is one of the key factors to initiate the symbiotic system. To select an appropriate cotton variety for this study, we first investigated the SEA of 17 cotton varieties at 40 days post inoculation (dpi) under greenhouse conditions. A variety of infection structures could be observed in the roots, including vesicles, hyphae, spores and arbuscules (Figure S1). Based on two repeated experiments, we found that the total SEA ranged dramatically from 3% to 94% with the 17 cotton varieties tested. Lumian 1 is one of the highest SEA (SEA = 94%) varieties. In addition, the colonization ratio of hyphae and vesicles was inconsistent in different varieties, and Lumian 1 had the highest ratio, reaching 41% and 55%, respectively (Figure 1).

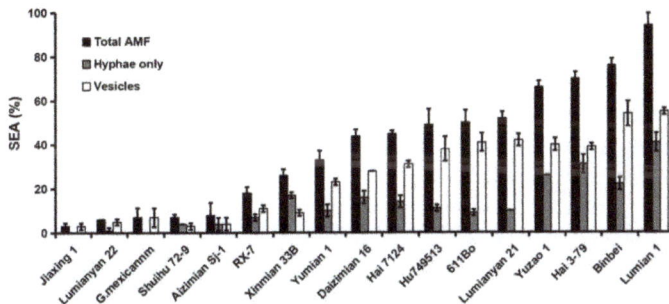

Figure 1. Evaluation of the symbiotic efficiency of AMF (SEA) among seventeen cotton cultivars. One hundred 1-cm root fragments were investigated from each plant after 40 days of growth with *Rhizophagus irregularis* CD1 during greenhouse conditions. The parameters Hyphae only (%), Vesicles (%), and Total AMF (%) denote the frequency of internal hyphae (without vesicles), vesicles, and total mycorrhizal colonization, respectively. The experiment was repeated twice under greenhouse conditions. Error bars represent ± SD.

2.2. Effect of Plant Growth Promotion (PGP) among Cotton Varieties with Different Symbiotic Efficiency of AMF (SEA)

To test the effect of PGP among different SEA cotton varieties, two subgroups that contained four cotton varieties, each with high-SEA and low-SEA, were inoculated with *R. irregularis* CD1 or water (control) under greenhouse conditions, respectively. As expected, AMF-mediated PGP was dependent on the SEA. In the high-SEA subgroup (SEA ≥ 66%), AMF colonization resulted in a significant increase in plant growth parameters including shoot fresh weight and root fresh weight. In addition, treatment of Yuzao 1 and Hai 3-79 resulted in a significant level of promotion in plant height (Figure 2A–D). In contrast, the low-SEA subgroup (SEA ≤ 26%) did not show any positive growth responses in their plant height and shoot and root fresh weights (Figure 2A–D). This suggests that the beneficial effects of AMF on cottons depend on a highly defined symbiotic relationship.

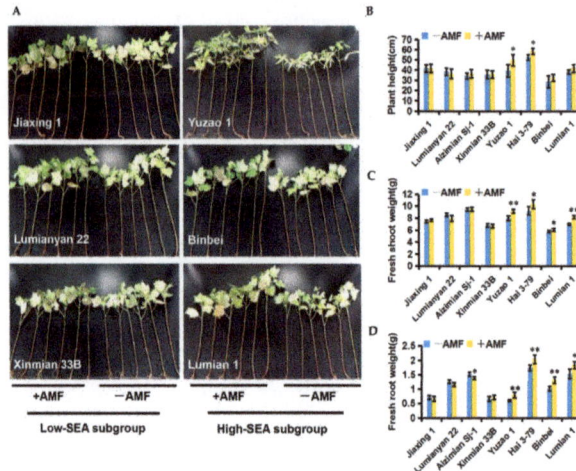

Figure 2. Effects of different symbiotic efficiency of AMF (SEA) on cotton growth promotion at 60 days post inoculation (dpi) under greenhouse conditions. +AMF: mycorrhizal; −AMF: nonmycorrhizal. (**A**) The growth phenotypes of two cotton subgroups. The low-SEA subgroup (SEA ≤ 26%) was comprised of Jiaxing 1, Lumianyan 22, Aizimian Sj-1, and Xinmian 33B. The high-SEA subgroup (SEA ≥ 66%) was comprised of Lumian 1, Binbei, Hai 3-79, and Yuzao 1; (**B–D**) Biomass statistics of two cotton subgroups including plant height (**B**), shoot fresh weight (**C**), and root fresh weight (**D**). Error bars represent ± SD. * $p < 0.05$; ** $p < 0.01$.

2.3. Field Evaluation of PGP of Lumian 1 by Applying for AMF

Given the highest SEA of Lumian 1 and its effective growth promotion associated with AMF under greenhouse conditions, we expected that it could be extended into the field. Lumian 1 is one of the leading cultivars in China because of its high yield and outstanding quality. However, it is not resistant to *Verticillium* wilt [45]. Therefore, it is a good candidate to use to appraise the potential of AMF application to improve plant growth and *Verticillium* wilt resistance.

We first conducted a field trial for AMF-mediated PGP in 2015. At 40 dpi, mycorrhiza-treated cotton grew significantly better than the control (Figure 3A). Statistical analyses revealed that the AMF applications significantly enhanced almost all of the growth parameters including plant height, stem diameter, number of petioles, and the area of the largest functional leaf (Figure 3B–E). Obvious effects of PGP could sequentially be observed by applying the AMF at 55 dpi and 72 dpi in both 2015 and 2016 (Table S1). Compared with the control, the mycorrhiza-treated cotton exhibited a higher inorganic phosphorus (Pi) content in both the roots and leaves (Figure 4A). Correspondingly, the transcription levels of several phosphate transport genes, including Gh_A02G0202, Gh_A02G0203, Gh_D02G0263, and Gh_D10G1372, which shared 78%, 77%, 77%, and 78% similarity with the phosphate transporter 1–5 gene (At2G32830) from *Arabidopsis thaliana*, respectively, were significantly induced in both the roots and leaves by AMF colonization (Figure 4B–D). This implies that these genes may contribute to the Pi transport from AMF to cotton during mycorrhizal symbiosis. Consistent with the benefits from AMF in a greenhouse assay, AMF is also highly effective at improving the plant growth of high SEA cotton varieties under field conditions.

Figure 3. Effects of AMF on the growth of Lumian 1 at 40 dpi under field conditions. +AMF: mycorrhizal; −AMF: nonmycorrhizal. (**A**) The growth phenotypes of Lumian 1. (**B–E**) Biomass statistics of Lumian 1 including plant height (**B**), stem diameter (**C**), the number of petioles (**D**), and the area of the largest true leaf (**E**). Error bars represent ± SD. * $p < 0.05$; ** $p < 0.01$.

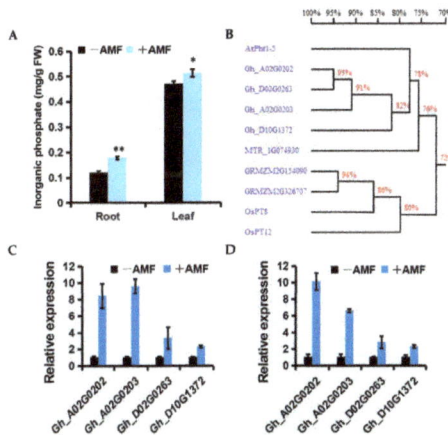

Figure 4. Effects of AMF on inorganic phosphorus (Pi) transport of Lumian 1 at 60 dpi under field conditions. (**A**) Pi content of both the root and leaf. +AMF: mycorrhizal; −AMF: nonmycorrhizal. Error bars represent ±SD. * $p < 0.05$; ** $p < 0.01$. (**B**) Homology tree for amino acid sequences of AtPht 1–5 (At2G32830) and its homologs within *Gossypium hirsutum* (Gh_A02G0202, Gh_A02G0203, Gh_D02G0263 and Gh_D10G1372), *Oryza sativa* (OsPT8, AAN39049; OsPT12, AAN39053), *Zea mays* (GRMZM2G326707, GRMZM2G154090) and *Medicago truncatula* (MTR_1g074930). This was generated by DNAMAN version 5.2.2.0 (Lynnon Biosoft, San Ramon, CA, USA); (**C,D**) Expression level of cotton homologs of *AtPht 1–5* in root (**C**) and leaf (**D**). The test performed by quantitative RT-PCR analysis. Transcript abundance of genes was normalized to that of the reference gene *UBQ7* (GenBank Accession Number: DQ116441). Three biological replicates were used for each reaction with three technical replicates each. Mean values and standard errors were calculated from three biological replicates.

2.4. Field Performance of Lumian 1 Inoculated with AMF for Verticillium Wilt Resistance

Over two years of independent field trials, we discovered that the application of AMF delayed the symptoms of *Verticillium* wilt disease. With the early outbreak of the disease, mycorrhizal plots exhibited fewer chlorotic and necrotic spots than the control, and typically, this was the only place that healthy plants were visible (Figure 5A); conversely, the control plots commonly displayed susceptibility in their leaves, and some plants eventually died (Figure 5B). These results indicated that AMF colonization is likely to effectively inhibit the spread of *V. dahliae* in cotton at its early occurrence phase. In 2015, the disease index (DI) of three mycorrhiza-treated plots showed reductions of 23.10%, 26.36%, and 38.18% compared to each control plot, respectively (Figure 5D). In 2016, the DI decreased by 22.98%, 28.29%, and 28.41% in three mycorrhiza-treated plots compared to the control, respectively (Figure 5E). Interestingly, we noticed three disease centers that were distributed among the three control plots in the field (Figure S2B). Consistent with the reduction of DI by applying the AMF, the *V. dahliae* biomass relative to the cotton leaves was significantly lower in the mycorrhiza-treated plots than in the control plots (Figure 5F). In addition to changes in symptoms in the leaves, we found less vascular discoloration in the mycorrhiza-treated plots than in the control plots (Figure 5C). Statistical analyses revealed that the AMF treatment significantly reduced vascular discoloration by 13.12% ($p = 0.05$) (Figure 5G). Therefore, we concluded that AMF treatment can efficiently enhance the resistance of Lumian 1 to *Verticillium* wilt under field conditions.

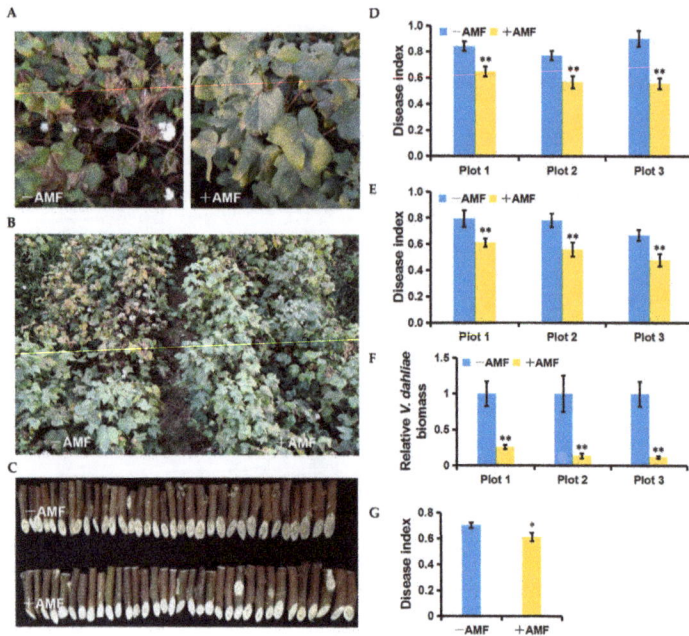

Figure 5. Effects of AMF on the *Verticillium* wilt resistance of Lumian 1 under field conditions. +AMF: mycorrhizal; −AMF: nonmycorrhizal. (**A**) The phenotypes of Lumian 1 at the early occurrence phase of *Verticillium* wilt; (**B**) The resistance phenotypes of Lumian 1 at 120 dpi in 2016; (**C**) The vascular discoloration phenotypes of Lumian 1 at 150 dpi in 2016; (**D,E**) The disease index of leaves at 120 dpi in 2015 (**D**) and 2016 (**E**), respectively. At least 40 plants were used for each experiment; (**F**) Quantitative detection of the *V. dahliae* biomass relative to cotton leaves at 120 dpi in 2016. The average fungal biomass was determined using at least 10 mycorrhiza-treated and 10 control cottons of each plot; (**G**) The disease index of the vascular bundle at 150 dpi in 2016. Error bars represent ± SD. * $p < 0.05$; ** $p < 0.01$.

2.5. Effect of Mycorrhizal Colonization on Cotton Resistance-Related Genes

SAR is a common resistance reaction directed against many types of phytopathogens. Some biochemical changes appear in the plant cells during SAR, such as the biosynthesis of pathogenesis-related (PR) proteins that are induced by stresses and play important roles in plant defense. *GhHSR203J* and *GhHIN1*, which are considered as marker genes for the hypersensitive response [46,47], were up-regulated by mycorrhizal colonization compared with the control (Figure 6A). *GhPR1* is a biomarker gene of SA signaling pathways [48]. *GhPR3* and *GhPR4* degrade chitin. *GhPR5* is thought to be involved in the synthesis of enzymes that metabolize and inhibit plant pathogens. *GhPR9* is a peroxidase that can reinforce the plant cell wall by catalyzing the synthesis of lignin to defend against the pathogen infection [48]. *GhPR1, GhPR3, GhPR4, GhPR5, and GhPR9* also exhibited increases of 3.8-, 2.0-, 2.0-, 3.3-and 28.8-fold in mycorrhiza-treated cotton plants, respectively, compared to the controls (Figure 6A). *GhPR10*, which may play a negative regulation role in SAR [49], was significantly reduced to 0.04-fold by applying AMF. These data suggest that the intensity of SAR can be further enhanced by mycorrhizal colonization.

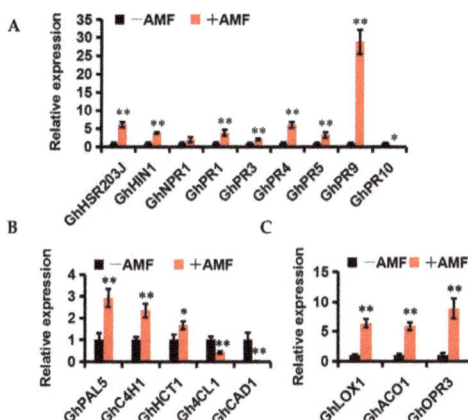

Figure 6. Expression patterns of cotton resistance-related genes in mycorrhizal (+AMF) and nonmycorrhizal (−AMF) Lumian 1. (**A**) Expression of *PR* genes; (**B**) Expression of JA synthesis-related genes; (**C**) Expression of lignin synthesis-related genes. The test was performed using reverse transcription quantitative PCR analysis of relative gene expression. Transcript abundance of genes was normalized to that of the reference gene *UBQ7* (GenBank Accession Number: DQ116441). Three biological replicates were used for each reaction with three technical replicates each. Mean values and standard errors were calculated from three biological replicates. * $p < 0.05$; ** $p < 0.01$.

Lignin synthesis is important in the resistance of cotton to *V. dahliae* [50,51]. To investigate its function during mycorrhizal colonization, the expression of lignin synthesis-related genes was examined, including *GhHCT1, GhPAL5, GhC4H1, Gh4CL1*, and *GhCAD1*. Compared with the controls, higher transcription levels of *GhHCT1, GhPAL5*, and *GhC4H1* and lower transcription levels of *Gh4CL1* and *GhCAD1* were detected in mycorrhiza-treated cotton (Figure 6B), indicating that AMF are capable of resisting *V. dahliae* by mediating lignin synthesis.

Jasmonic acid (JA) can accumulate during mycorrhizal symbiosis and contribute to MIR [22,52]. The expression of three biomarker genes of JA signaling pathways including *GhLOX1, GhACO1* and *GhOPR3* [53] was significantly up-regulated by 6.34-, 5.87- and 8.87-fold, respectively, after mycorrhizal colonization (Figure 6C), indicating that JA biosynthesis may be influenced by AMF colonization.

2.6. In Vitro Antifungal Activity Assay of AMF Symbionts

The ability of AMF hyphae, spores and endosymbionts to inhibit the growth of *V. dahliae* in in vitro systems was tested first. Surprisingly, we found the AMF symbionts, rather than AMF hyphae or spores, caused an obvious inhibitory effect on the mycelial growth of *V. dahliae* (Figure 7). As shown in Figure 7, the mycelia of *V. dahliae* could quickly spread to the M+ compartment when the mycorrhizal roots were absent (Figure 7A,B,E,F), but their growth was limited in the M− compartment (M medium without sugar) when the mycorrhizal roots were planted in the M+ compartment (Figure 7C,D). Intriguingly, we subsequently demonstrated that AMF symbionts could also remarkably inhibit the growth of *Fusarium oxysporum*, *Fusarium graminearum*, and *Rhizoctonia solani* at 7 dpi (Figure 8A). Even at 50 dpi, there was almost no mycelial expansion of *F. graminearum* and no sclerotial production of *R. solani* when AMF symbionts were present (Figure 8B). Therefore, it seems as if mycorrhizal colonization can not only enhance the resistance of host plants to *V. dahliae* via inducing the expression of resistance-related genes but may also have a direct adverse effect on fungi by its symbionts which may generate certain volatile compounds.

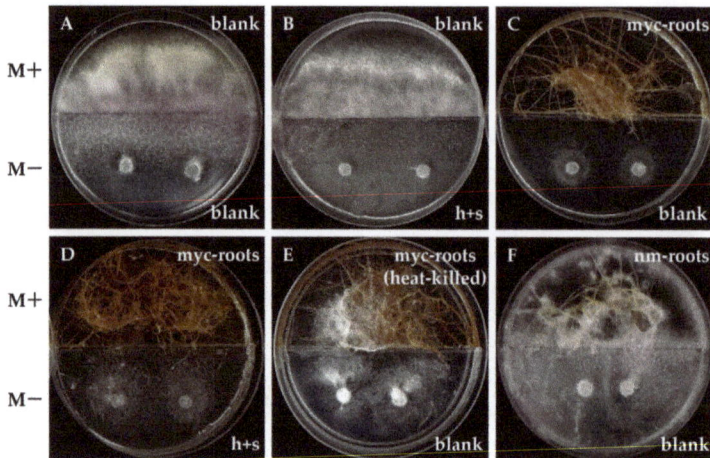

Figure 7. In vitro antimicrobial activities of AMF against *Verticillium dahliae*. Divided Petri dishes prevent nonvolatile solutes from diffusing between two compartments. The upper compartment of divided Petri dishes contained a complete growth medium (M+) used for the growth of AMF symbionts, and the lower compartment contained the same medium lacking sugar (M−), thus permitting the development of AMF hyphae and spores. Two discs covered with *V. dahliae* of 5 mm diameter were transferred to the M− compartment and incubated in the dark at 25 °C for 1 week. Before the inoculation of *V. dahliae* discs, fresh blank M+ and blank M− medium were decanted into divided Petri dishes. The height of M− was equal to the middle baffle while slightly higher than M+. The heights of M+ and M− in each plate were comparable to each other. Treatments were divided as follows: (**A**) empty M− and empty M+; (**B**) M− with AMF hyphae and spores (h + s) and empty M+; (**C**) empty M− and M+ with mycorrhizal roots (myc-roots); (**D**) M− with AMF h + s and M+ with myc-roots; (**E**) empty M− and M+ with dead myc-roots heated at 65 °C for 30 min; (**F**) empty M− and M+ with non-mycorrhizal roots (nm-roots). Experiments were repeated three times with similar results.

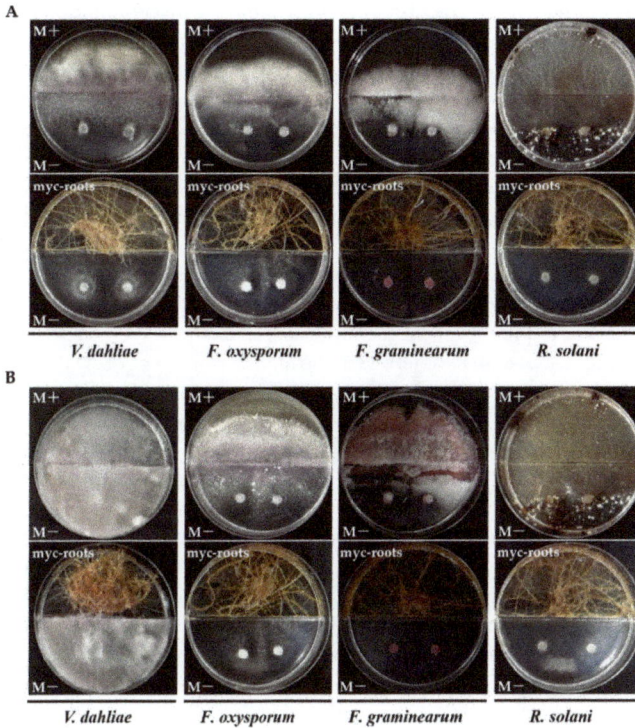

Figure 8. Antifungal activities of AMF symbionts against several soil-borne fungi at 7 dpi (**A**) and 50 dpi (**B**), respectively. The culture conditions were as described for Figure 7. The upper compartment contained M+ medium with or without AMF symbionts (myc-roots), and the downward compartment contained M− medium inoculated with two discs of *Verticillium dahliae*, *Fusarium oxysporum*, *Fusarium graminearum* and *Rhizoctonia solani*. Experiments were repeated three times with similar results.

3. Discussion

Numerous studies have shown that AMF can contribute to the growth and disease resistance of host plants. Our study presents interesting results about the effects of symbiotic efficiency of AMF (SEA) on the cotton growth improvement. We corroborated that high-SEA cotton Lumian 1 could perform well on both growth promotion and *Verticillium* wilt resistance following mycorrhizal colonization directly under field conditions. In addition to its roles in inducing plant disease resistance, we revealed that AMF symbionts could release certain volatile compounds, which may have broad-spectrum and long-term efficacy in terms of fungistasis.

AMF are the most widespread endomycorrhizal fungi, which play important roles in nutrient cycling processes and plant stress resistance by establishing symbiotic associations with plant roots. AMF can increase plant nutrient uptake of P [54,55], N [56,57], and K [58]. In addition, AMF are beneficial to the stabilization of soil aggregates [59], and improve resistance to water stress [36] and defense against pathogens [18]. However, a successful symbiotic relationship between AMF and its host involves the premise that the AMF will have an effect on improving plant growth and disease tolerance. The viability of AMF application in agricultural soils depends on many factors, including species compatibility, habitat niche availability, and competition with indigenous fungi [60]. To date, significant genetic variation of AMF species in their effects on host plants has been reported [61–63]. However, few studies have focused on the impact of diverse genotypes on both the plant response to the AMF as well as its mycorrhizal colonization level. Cotton is a mycotrophic plant in which growth and nutrient uptake is usually promoted by mycorrhizal colonization [64]. Here, we used SEA to appraise the level of mycorrhizal colonization in 17 cotton varieties under greenhouse conditions. Remarkably, the total SEA varied widely ranged from 3% (Jiaxing 1) to 94% (Lumian 1) among the cotton varieties tested (Figure 1). Eaton et al. inoculated 43 near-isogenic lines of *Trifolium repens* with AMF *Glomus mosseae* and observed a high degree of variation among individual lines in their mycorrhizal root infection rates [65]. Therefore, although AMF can colonize the roots of more than 90% of plant species, to a typical species, the ratio of mycorrhizal colonization is subject to the genotypes that imply its control by genetic characteristics.

We speculate that an effective SEA may be elementary for host plants to benefit from mycorrhizal colonization. Consistent with this hypothesis, AMF significantly improved plant growth of the high-SEA subgroup but did not do so on the low-SEA subgroup in the greenhouse (Figure 2A–D). Consistent with a role as a potential biofertilizer, the high performance of AMF-mediated PGP with Lumian 1 was also observed under field conditions (Figure 3A–E). This could be partially explained by helping the host plant to take up components such as phosphorus (P), which is a well-known contribution from mycorrhizal symbiosis [66]. Our results showed that mycorrhizal colonization significantly enhanced the inorganic phosphate (Pi) content in both the roots and leaves relative to the control (Figure 4A). Since *Pht 1–5* plays vital roles in Pi translocation and remobilization [67], its co-expression with the accumulation of Pi, the expression of certain cotton homologs of *AtPht 1–5*, including *Gh_A02G0202*, *Gh_A02G0203*, *Gh_D02G0263* and *Gh_D10G1372*, were induced in all of the mycorrhiza-treated plots (Figure 4B–D). In combination, an effective SEA is required for AMF-mediated PGP, including but not limited to providing predominantly Pi in exchange for the plant carbon source. However, SEA is highly variable in its host genotype. This genotypic variation is worth harnessing in genetic breeding to fully exploit the potentials of AMF, especially for regions of the world with Pi deficiency.

The other benefits of symbiotic systems include the promotion of the host resistance. In two-year field trials to control *Verticillium* wilt, we demonstrated a significant reduction of disease severity by AMF application with Lumian 1 (Figure 5A–G). Generally, the mechanisms of plant resistance can be divided into two broad categories: constitutive resistance and induced resistance. Lignin synthesis, a major constitutive resistance mechanism that involves the formation of new cell walls as a response to pathogens, has been shown to play a central role in the resistance of cotton to *V. dahliae*. The content of lignin in island cotton is higher than that in upland cotton, which is one of the core reasons that the former is more resistant to *V. dahliae* [50,51]. A reduction in the lignin content by silencing *GhHCT1* with VIGS in cotton had resulted in compromised resistance to *V. dahliae* [51]. This was accompanied by the reduced transcription level of upstream lignin synthesis-related genes such as *GhPAL5* and *GhC4H1* and the increased expression of genes involved in downstream lignin synthesis including *Gh4CL1* and *GhCAD1* [51]. Consistent with the higher level of resistance to *Verticillium* wilt, higher transcription levels of *GhHCT1*, *GhPAL5*, and *GhC4H1* and lower transcription levels of *Gh4CL1* and *GhCAD1* were detected in mycorrhiza-treated cotton, as compared to the control (Figure 6B), indicating that the AMF can resist *V. dahliae* by mediating lignin synthesis. MIR can be considered to be the other key reason to promote host resistance. The expression of many *PR* genes, including *GhHSR203J*, *GhHIN1*, *GhPR1*, *GhPR3*, *GhPR4*, *GhPR5*, and *GhPR9*, was significantly increased by mycorrhizal colonization. However, the expression of *GhPR10* was significantly reduced to 0.04-fold after AMF application (Figure 6A). Some PR10 homologs were shown to display antifungal activities in different plant species [68,69]. In contrast, the overexpression of *STH-2*, a member of the *PR10* family, in potato failed to enhance the resistance of potato to *Phytophthora infestans* and potato virus X [70]. Similar results were also identified in other studies [49,71]. In cotton, it appeared that a high resistance to *V. dahliae*, as well as more lignin, accumulated in the stems when *GhPR10* was silenced by VIGS, suggesting that *GhPR10* probably negatively regulates the resistance of cotton to *V. dahliae* [72,73]. AMF can affect the transcription of the *PR* genes and further enhance the intensity of SAR. In addition, the up-regulation of *GhLOX1*, *GhACO1* and *GhOPR3*, which encode key enzymes involved in JA biosynthesis pathway [53], was observed in mycorrhiza-treated cotton (Figure 6C), consistent with the concept that JA can accumulate during mycorrhizal symbiosis and contribute to MIR [22,52]. We conclude that the significant resistance to *V. dahliae* resulting from mycorrhizal colonization in high-SEA cotton Lumian 1 is associated with MIR and cell wall secondary metabolism.

In addition to MIR and cell wall defense, an antifungal activity of root exudates induced by AMF symbiosis was observed to promote resistance. *Glomus versiforme* could alter the exudation pattern of cotton roots and contribute to the bioactive effects on *V. dahliae* conidial germination [74]. *Pisolithus tinctorius* strain SMF, an ectomycorrhizal fungus, strongly inhibited the growth of *V. dahliae* [75]. Ericoid mycorrhizal fungi could protect the host plants from infections by pathogens such as *Phytophthora cinnamomi* and *Pythium* when they were sufficiently present in or on the roots [76]. Hage-Ahmed et al. demonstrated that a direct antibiotic activity of root exudates from tomatoes towards *F. oxysporum* f.sp. *lycopersici* could be induced by the interactions of the plant–AMF–pathogen [77]. Using HPLC-UV analyses, the antifungal substances were identified to be nonvolatile citrate and chlorogenic acid [77]. Intriguingly, our results showed that the mycorrhizal roots of carrots, rather than AMF hyphae, AMF spores, or non-mycorrhizal roots, suppressed the mycelial growth of *V. dahliae* (Figure 7). Carrot roots also strongly inhibit the growth of other soil-borne fungi such as *F. oxysporum*, *F. graminearum* and *R. solani*. In particular, this fungistasis could persist as long as 50 dpi, while in the Petri dish, the mycelia of *F. graminearum* could rarely expand, in addition to the effects on *R. solani* that included a lack of sclerotia production (Figure 8A,B). Given the blocking effect of the divided Petri dishes on the diffusion of nonvolatile solutes from root exudates, we propose that AMF symbionts can release certain volatile compounds that directly inhibit the growth and extension of fungal phytopathogens, which is to the best of our knowledge a novel biocontrol mechanism of mycorrhizal association. However, there are still many questions that need to be investigated further, such as identifying the volatile compounds, and whether they are generated in the *R. irregularis* CD1-Lumain 1 interaction system, and whether they are involved in promoting the resistance to *Verticillium* wilt.

4. Materials and Methods

4.1. AMF Inoculum Preparation

R. irregularis CD1 was used in this study and was maintained on Petri dishes as described by St-Arnaud [78]. Briefly, *R. irregularis* CD1 was co-cultivated with genetically transformed carrot roots in a two-compartment in vitro system. Half Petri dishes containing a complete growth medium (M+) were used for the growth of endosymbionts, while the other half containing the same medium excluding sugar (M−), thus permitting the development of AMF mycelia and spores. The height of M− was equal to the middle baffle, while M+ was slightly lower than that. The Petri dishes were incubated in the dark at 18 °C. After 2–3 months, a greater number of spores could be easily seen on the M− compartment. The medium containing spores was then blended with distilled water in a juice blender. The mixture of mycelia and spores was agitated by a magnetic stirrer, and then the spores were counted using a hemocytometer and stored temporarily in a 4 °C refrigerator.

4.2. Evaluation of SEA among 17 Cotton Cultivars

The seeds of 17 cotton cultivars (Figure 1) were provided by the collections from the Cotton Research Institute, Chinese Academy of Agricultural Sciences. All seeds were villus-shed and surface-sterilized by using 98% H_2SO_4 (100 °C, 1 min) and then soaked in water at 28 °C until they germinated.

Barren soil was collected at the fields of Shandong Agricultural University (Taian, China). The germinated seeds were planted in 15-cm diameter plastic pots (4 L) containing the soil-vermiculite-perlite mixture (3:2:1 ratio, v/v) sterilized by steaming (121 °C for 30 min). To ensure the direct contact of the seedling roots with *R. irregularis* CD1, we added a suspension of approximately 2000 spores into the center of each pot, 3 cm under the surface of the mixture. Each pot was planted with five seeds of the same cotton variety, and three seedlings of similar development were reserved for further study. The experiment was carried out in a greenhouse at 60% relative humidity under a 16/8 h and 26/20 °C (light/dark) photoperiod. Plants were harvested, and roots from each pot were collected separately at 40 days post-inoculation (dpi). The experiment was repeated twice. Each pot was repeated three times.

To assess the mycorrhizal colonization of each cotton variety by *R. irregularis* CD1, we used Trypan blue staining as described by Phillips [79]. Briefly, root samples were thoroughly washed and cut into approximately 1 cm sections, then steeped in 10% KOH (w/v) for 45 min at 60 °C and stained with 0.05% v/v Trypan blue in lactic acid. Separately, 100 1-cm-long fine roots were randomly selected from each line. Symbiotic efficiency was assessed using a Nikon Eclipse 90i light microscope. An estimate of Total AMF (%) was given as the ratio between root fragments colonized by an AMF structure. The parameter Vesicles (%) was an estimate of vesicle richness in the whole analyzed root system. The parameter Hyphae only (%) was the proportion of the root cortex infected by hyphae that lacked vesicles relative to the whole root system analyzed.

4.3. Experimental Design for Testing the PGP and Wilt Disease Resistance

A pot experiment to test the PGP with eight cotton varieties was described as above. The matched controls were given the equivalent volume of blank M− suspension. Four high-SEA subgroups (SEA ≥ 66%) were selected as Lumian 1, Binbei, Hai 3-79 and Yuzao 1. The other four low-SEA subgroups (SEA ≤ 26%) included Jiaxing 1, Lumianyan 22, Aizimian Sj-1 and Xinmian 33B. At 60 dpi, plant height and shoot and root fresh weights of the whole plants were measured. This experiment was repeated three times. Each pot planted three cottons and was repeated twice.

Field performance was investigated twice in the plain area of Taian city during 2015 and 2016 where *Verticillium* wilt occurred. The cotton variety Lumian 1 of the highest SEA was selected in this experiment. The experimental design was a complete randomized block (three mycorrhiza-treated plots and three control plots). Each plot was 8 m in length and 1 m in width (with a 1 m interval between blocks), and 50 plants were planted in two rows in each plot. Plot 1, Plot 2 and Plot 3 were used to refer to three repetitions (Figure S2A). We inoculated each germinated seed with a suspension of 800 spores or an equal volume of blank M− suspension. At 40, 55 and 72 dpi, the growth parameters were recorded for ten randomly selected plants from each experimental plot. At 60 dpi, six plants were randomly collected from mycorrhiza-treated and control plots respectively to measure the Pi content and the expression of several phosphate transport genes in both the root and leaves.

Subsequently, we evaluated the biological control of *Verticillium* wilt using AMF application during two years of field trials. Disease severity was recorded for each plant at 120 dpi using a disease index (DI) ranging from 0 to 4 rating scale based on to the percentage of foliage affected by acropetal chlorosis, necrosis, wilt, and/or defoliation as follows: 0 = no visible disease symptoms, 1 = 1–25%, 2 = 26–50%, 3 = 51–75%, and 4 = 76–100% or a dead plant. We randomly gathered a mixture of leaf and root samples from ten mycorrhiza-treated and control plants in each plot, which were used to extract DNA to measure the *V. dahliae* biomass and RNA to detect the expression of cotton defense-related genes, respectively. To better assess the severity of the disease, we also estimated the vascular discoloration of the cotton stems at 150 dpi. Twelve stems were randomly collected from each mycorrhiza-treated and control plot to be rated for vascular discoloration. We established a new rating scale according to the proportion of stem length that occurred vascular browning in longitudinal section as follows: 0, no vascular discoloration, 1 = 1–25%, 2 = 26–50%, 3 = 51–75%, 4 = 76–100%.

4.4. RNA Extraction and Reverse Transcription Quantitative PCR Analysis

Total RNA was isolated from 50 mg plant tissue by a E.Z.N.A.™ Plant RNA Kit (OMEGA-biotek, R6827-01, Doraville, GA, USA). One-half microgram RNA was used for first-strand cDNA synthesis using the *EasyScript* ®One-Step gDNA Removal and cDNA Synthesis SuperMix (TRANs, Beijing,

China). Quantitative PCR was performed with UltraSYBR Mixture (Comwin Biotech Co., Ltd., Beijing, China) on a QuantStudio™ 6 Flex Real-Time PCR System (Thermo Fisher, Waltham, MA, USA). The PCR program was as follows: 95 °C for 10 min, followed by 40 cycles of 95 °C for 15 s, 60 °C for 1 min. The specificity of the amplified PCR products was determined by melting curve analysis (95 °C for 15 s, 60 °C for 1 min, 95 °C for 15 s and 60 °C for 15 s). *UBQ7* of *G. hirsutum* (GenBank Accession Number: DQ116441) was used as internal control to standardize the results [80]. For each gene, quantitative PCR assays were repeated at least twice with triplicate runs. Relative expression levels were measured using the $2^{-\Delta\Delta Ct}$ analysis method. The sequences of gene-specific primers used in the assay are listed in Table S2.

4.5. Quantitative Detection of V. dahliae Biomass in Cotton

At 120 dpi, we randomly gathered the first true leaves of plants from six plots and extracted DNA by CTAB method [81]. To detect the *V. dahliae* biomass, quantitative PCR was performed with the same agents and amplification conditions as described above. The internal transcribed spacer region of the ribosomal DNA was targeted to generate a 200 bp amplicon using the fungus-specific primer ITS1-F, and the *V. dahliae*-specific reverse primer ST-VE1-R [82]. *UBQ7* was used for equilibration of the different DNA samples. The average fungal biomass was determined using ten mycorrhiza-treated and ten control cotton plants for each plot. This experiment was repeated three times.

4.6. Antimicrobial Activities of AMF against Phytopathogenic Fungi In Vitro Systems

To test whether AMF could contribute antifungal activity, the symbiotic carrot roots were cultured as described above. Phytopathogenic fungi were preincubated on PDA medium at 25 °C for four days. On the M− compartments, we inoculated two 5-mm diameter discs of cultures of *V. dahliae*, *F. oxysporum*, *F. graminearum* and *R. solani*, respectively. The Petri dishes were incubated in the dark at 25 °C and then observed at 7 or 50 dpi. This experiment was repeated three times.

4.7. Measurement of Pi Content

Approximately 0.5 g frozen samples were assayed using an improved method described by Nanamori et al. [83]. Briefly, the frozen sample was ground with liquid nitrogen and homogenized in 1 mL 10% (*w/v*) perchloric acid (PCA). The homogenate was diluted 10 times with 5% (*w/v*) PCA and then incubated on ice for 30 min. After centrifugation at $10,000\times g$ for 10 min at 4 °C, 500 μL supernatant was transferred and blended with 4.5 mL H_2O and 5 mL chromogenic agent. This mixture was incubated in a water bath at 45 °C for 25 min. After being cooled at 4 °C, the absorbance was measured at 820-nm wavelength. The Pi content was calculated using the normalization of fresh weight. The chromogenic agent contents were as follows: ddH$_2$O:3 M H_2SO_4:2.5% (*w/v*) Hexaammonium heptamolybdate tetrahydrate (Sinopharm Chemical Reagent Co., Ltd., Shanghai, China):5% (*w/v*) L-Ascorbic acid (Sinopharm Chemical Reagent Co., Ltd.) = 2:1:1:1 (volume ratio).

4.8. Data Analysis

The software DPS (Data Processing System, version 7.05, Hangzhou, China) was used to perform the statistical analyses. All data were subjected to analysis of variance by one-way analysis of variance (ANOVA). A post hoc analysis test LSD (least significant difference) was implemented to examine the significance of different treatment means against a standard control ($p = 0.05$ or 0.01).

5. Conclusions

In this study, we introduced the concept of SEA and ascertained a broad phenotypic variation of SEA between the interaction of *R. irregularis* CD1 and seventeen cotton cultivars. Intriguingly, we demonstrated that high-SEA is positively associated with the AMF-mediated PGP effects. In this study, we identified the cotton variety Lumian 1, which possessed the highest SEA and could

Int. J. Mol. Sci. **2018**, *19*, 241

perform well on both growth promotion and disease resistance following mycorrhizal colonization. These results suggest high-SEA is a key phenotype to accelerate the utilization of AMF that will contribute to research on the genetic improvement of SEA. The genotypic variation conferring high-SEA in Lumian 1 merits additional research to characterize this variation. In addition, we confirmed that the mycorrhizal roots of carrots could release unknown volatile compounds to directly inhibit the growth of several fungal phytopathogens, which to our knowledge, is a novel AMF-mediated biocontrol mechanism. The specific antifungal substances derived from the mycorrhizal symbionts merit additional research to identify these compounds.

Supplementary Materials: Supplementary materials can be found at www.mdpi.com/1422-0067/19/1/241/s1.

Acknowledgments: We thank the Cotton Research Institute, Chinese Academy of Agricultural Sciences provide the seeds of cotton. This work was supported by National Key R&D Program of China (2016YFD0100306), the Major Application Technology Innovation Project of Shandong Province (2016), the National Natural Science Foundation of China (31401428), the Funds of Shandong "Double Tops" Program (2017). FZ and XD were funded by the Taishan Scholar Project (20150621, 20161018).

Author Contributions: Zhaohui Chu and Fanchang Zeng designed and supervised the experiments. Qiang Zhang, Xinpeng Gao and Yanyun Ren performed the field investigation of cotton growth and wilt disease. Qiang Zhang and Jiajia Qiu performed the Pi content quantification. Qiang Zhang, Xinhua Ding and Ning Li performed the qRT-PCR and data analysis. Qiang Zhang and Zhaohui Chu wrote the manuscript. All authors read and approved the final manuscript.

Conflicts of Interest: The authors declare no conflict of interest.

Abbreviations

AMF	Arbuscular mycorrhizal fungi
PGP	Plant growth promotion
SEA	Symbiotic efficiency of AMF
MIR	Mycorrhiza-induced resistance
SAR	Systemic acquired resistance
ISR	Induced systemic resistance
MAMP	Microbe-associated molecular patterns
PTI	MAMP-triggered immunity
SA	Salicylic acid
ABA	Abscisic acid
DI	Disease index
dpi	Days post inoculation
JA	Jasmonic acid
PR	Pathogenesis-related proteins

References

1. Zhao, Y.; Wang, H.; Chen, W.; Zhao, P.; Gong, H.; Sang, X.; Cui, Y. Regional association analysis-based fine mapping of three clustered QTL for *Verticillium* wilt resistance in cotton (*G. hirsutum* L.). *BMC Genom.* **2017**, *18*, 661. [CrossRef] [PubMed]

2. Klosterman, S.J.; Atallah, Z.K.; Vallad, G.E.; Subbarao, K.V. Diversity, pathogenicity, and management of *Verticillium* species. *Annu. Rev. Phytopathol.* **2009**, *47*, 39–62. [CrossRef] [PubMed]

3. Meschke, H.; Walter, S.; Schrempf, H. Characterization and localization of prodiginines from *Streptomyces lividans* suppressing *Verticillium dahliae* in the absence or presence of *Arabidopsis thaliana*. *Environ. Microbiol.* **2012**, *14*, 940–952. [CrossRef] [PubMed]

4. Wang, Y.; Liang, C.; Wu, S.; Zhang, X.; Tang, J.; Jian, G.; Jiao, G.; Li, F.; Cui, C. Significant improvement of cotton *Verticillium* wilt resistance by manipulating the expression of *Gastrodia* antifungal proteins. *Mol. Plant* **2016**, *9*, 1436–1439. [CrossRef] [PubMed]

5. Wei, F.; Fan, R.; Dong, H.; Shang, W.; Xu, X.; Zhu, H.; Yang, J.; Hu, X. Threshold microsclerotial inoculum for cotton *Verticillium* wilt determined through wet-sieving and real-time quantitative PCR. *Phytopathology* **2015**, *105*, 220–229. [CrossRef] [PubMed]

6. Yadeta, K.A.; Hanemian, M.; Smit, P.; Hiemstra, J.A.; Pereira, A.; Marco, Y.; Thomma, B.P. The *Arabidopsis thaliana* DNA-binding protein AHL19 mediates *Verticillium* wilt resistance. *Mol. Plant Microbe Interact.* **2011**, *24*, 1582–1591. [CrossRef] [PubMed]

7. Zhou, H.; Fang, H.; Sanogo, S.; Hughs, S.E.; Jones, D.C.; Zhang, J. Evaluation of *Verticillium* wilt resistance in commercial cultivars and advanced breeding lines of cotton. *Euphytica* **2014**, *196*, 437–448. [CrossRef]

8. Miao, W.; Wang, X.; Li, M.; Song, C.; Wang, Y.; Hu, D.; Wang, J. Genetic transformation of cotton with a harpin-encoding gene hpa_{Xoo} confers an enhanced defense response against different pathogens through a priming mechanism. *BMC Plant Biol.* **2010**, *10*, 67. [CrossRef] [PubMed]

9. Zhang, Y.; Wang, X.F.; Ding, Z.G.; Ma, Q.; Zhang, G.R.; Zhang, S.L.; Li, Z.K.; Wu, L.Q.; Zhang, G.Y.; Ma, Z.Y. Transcriptome profiling of *Gossypium barbadense* inoculated with *Verticillium dahliae* provides a resource for cotton improvement. *BMC Genom.* **2013**, *14*, 637. [CrossRef] [PubMed]

10. Zhang, Y.; Wang, X.F.; Rong, W.; Yang, J.; Li, Z.K.; Wu, L.Q.; Zhang, G.Y.; Ma, Z.Y. Histochemical analyses reveal that stronger intrinsic defenses in *Gossypium barbadense* than in *G. hirsutum* are associated with resistance to *Verticillium dahliae*. *Mol. Plant Microbe Interact.* **2017**, *30*, 984–996. [CrossRef] [PubMed]

11. Zhang, W.W.; Zhang, H.C.; Qi, F.J.; Jian, G.L. Generation of transcriptome profiling and gene functional analysis in *Gossypium hirsutum* upon *Verticillium dahliae* infection. *Biochem. Biophys. Res. Commun.* **2016**, *473*, 879–885. [CrossRef] [PubMed]

12. Wang, W.N.; Yuan, Y.L.; Yang, C.; Geng, S.P.; Sun, Q.; Long, L.; Cai, C.W.; Chu, Z.Y.; Liu, X.; Wang, G.H.; et al. Characterization, expression, and functional analysis of a novel NAC gene associated with resistance to *Verticillium* Wilt and abiotic stress in Cotton. *G3* **2016**, *6*, 3951–3961. [CrossRef] [PubMed]

13. Yang, L.; Mu, X.Y.; Liu, C.; Cai, J.H.; Shi, K.; Zhu, W.J.; Yang, Q. Overexpression of potato miR482e enhanced plant sensitivity to *Verticillium dahliae* infection. *J. Integr. Plant Biol.* **2015**, *57*, 1078–1088. [CrossRef] [PubMed]

14. Yang, Y.W.; Ling, X.T.; Chen, T.Z.; Cai, L.W.; Liu, T.L.; Wang, J.Y.; Fan, X.H.; Ren, Y.Z.; Yuan, H.B.; Zhu, W.; et al. A cotton *Gbvdr5* gene encoding a leucine-rich-repeat receptor-like protein confers resistance to *Verticillium dahliae* in transgenic *Arabidopsis* and upland cotton. *Plant Mol. Biol. Report.* **2015**, *33*, 987–1001. [CrossRef]

15. Chen, J.Y.; Xiao, H.L.; Gui, Y.J.; Zhang, D.D.; Li, L.; Bao, Y.M.; Dai, X.F. Characterization of the *Verticillium dahliae* exoproteome involves in pathogenicity from cotton-containing medium. *Front. Microbiol.* **2016**, *7*, 1709. [CrossRef] [PubMed]

16. Chen, J.Y.; Li, N.Y.; Ma, X.F.; Gupta, V.K.; Zhang, D.D.; Li, T.G.; Dai, X.F. The ectopic overexpression of the cotton *Ve1* and *Ve2*-homolog sequences leads to resistance response to *Verticillium* wilt in *Arabidopsis*. *Front. Plant Sci.* **2017**, *8*, 844. [CrossRef] [PubMed]

17. Zhang, T.; Jin, Y.; Zhao, J.H.; Gao, F.; Zhou, B.J.; Fang, Y.Y.; Guo, H.S. Host-induced gene silencing of the target gene in fungal cells confers effective resistance to the cotton wilt disease pathogen *Verticillium dahliae*. *Mol. Plant* **2016**, *9*, 939–942. [CrossRef] [PubMed]

18. Jung, S.C.; Martinez-Medina, A.; Lopez-Raez, J.A.; Pozo, M.J. Mycorrhiza-induced resistance and priming of plant defenses. *J. Chem. Ecol.* **2012**, *38*, 651–664. [CrossRef] [PubMed]

19. Song, Y.Y.; Ye, M.; Li, C.Y.; Wang, R.L.; Wei, X.C.; Luo, S.M.; Zeng, R.S. Priming of anti-herbivore defense in tomato by arbuscular mycorrhizal fungus and involvement of the jasmonate pathway. *J. Chem. Ecol.* **2013**, *39*, 1036–1044. [CrossRef] [PubMed]

20. Mauch-Mani, B.; Baccelli, I.; Luna, E.; Flors, V. Defense priming: An adaptive part of induced resistance. *Annu. Rev. Plant Biol.* **2017**, *68*, 485–512. [CrossRef] [PubMed]

21. Song, Y.Y.; Chen, D.M.; Lu, K.; Sun, Z.X.; Zeng, R.S. Enhanced tomato disease resistance primed by arbuscular mycorrhizal fungus. *Front. Plant Sci.* **2015**, *6*, 786. [CrossRef] [PubMed]

22. Cameron, D.D.; Neal, A.L.; van Wees, S.C.; Ton, J. Mycorrhiza-induced resistance: More than the sum of its parts? *Trends Plant Sci.* **2013**, *18*, 539–545. [CrossRef] [PubMed]

23. Oldroyd, G.E. Speak, friend, and enter: Signalling systems that promote beneficial symbiotic associations in plants. *Nat. Rev. Microbiol.* **2013**, *11*, 252–263. [CrossRef] [PubMed]

24. Nadal, M.; Sawers, R.; Naseem, S.; Bassin, B.; Kulicke, C.; Sharman, A.; An, G.; An, K.; Ahern, K.R.; Romag, A.; et al. An *N*-acetylglucosamine transporter required for arbuscular mycorrhizal symbioses in rice and maize. *Nat. Plants* **2017**, *3*, 17073. [CrossRef] [PubMed]

25. Genre, A.; Chabaud, M.; Balzergue, C.; Puech-Pagès, V.; Novero, M.; Rey, T.; Fournier, J.; Rochange, S.; Bécard, G.; Bonfante, P.; et al. Short-chain chitin oligomers from arbuscular mycorrhizal fungi trigger nuclear Ca^{2+} spiking in *Medicago truncatula* roots and their production is enhanced by strigolactone. *New Phytol.* **2013**, *198*, 190–202. [CrossRef] [PubMed]

26. Zhang, J.; Zhou, J.M. Plant immunity triggered by microbial molecular signatures. *Mol. Plant* **2010**, *3*, 783–793. [CrossRef] [PubMed]

27. Luna, E.; Bruce, T.J.; Roberts, M.R.; Flors, V.; Ton, J. Next-generation systemic acquired resistance. *Plant Physiol.* **2012**, *158*, 844–853. [CrossRef] [PubMed]

28. Blilou, I.; Ocampo, J.A.; García-Garrido, J.M. Resistance of pea roots to endomycorrhizal fungus or *Rhizobium* correlates with enhanced levels of endogenous salicylic acid. *J. Exp. Bot.* **1999**, *50*, 1663–1668. [CrossRef]

29. De Jonge, R.; Bolton, M.D.; Thomma, B.P. How filamentous pathogens co-opt plants: The ins and outs of fungal effectors. *Curr. Opin. Plant Biol.* **2011**, *14*, 400–406. [CrossRef] [PubMed]

30. Ton, J.; Flors, V.; Mauch-Mani, B. The multifaceted role of ABA in disease resistance. *Trends Plant Sci.* **2009**, *14*, 310–317. [CrossRef] [PubMed]

31. Van Wees, S.C.; Van der Ent, S.; Pieterse, C.M. Plant immune responses triggered by beneficial microbes. *Curr. Opin. Plant Biol.* **2008**, *11*, 443–448. [CrossRef] [PubMed]

32. Cotton, T.A.; Dumbrell, A.J.; Helgason, T. What goes in must come out: Testing for biases in molecular analysis of arbuscular mycorrhizal fungal communities. *PLoS ONE* **2014**, *9*, e109234. [CrossRef] [PubMed]

33. Jiang, Y.N.; Wang, W.X.; Xie, Q.J.; Liu, N.; Liu, L.X.; Wang, D.P.; Zhang, X.W.; Yang, C.; Chen, X.Y.; Tang, D.Z.; et al. Plants transfer lipids to sustain colonization by mutualistic mycorrhizal and parasitic fungi. *Science* **2017**, *356*, 1172–1175. [CrossRef] [PubMed]

34. Luginbuehl, L.H.; Menard, G.N.; Kurup, S.; Van Erp, H.; Radhakrishnan, G.V.; Breakspear, A.; Oldroyd, G.E.D.; Eastmond, P.J. Fatty acids in arbuscular mycorrhizal fungi are synthesized by the host plant. *Science* **2017**, *356*, 1175–1178. [CrossRef] [PubMed]

35. Parniske, M. Arbuscular mycorrhiza: The mother of plant root endosymbioses. *Nat. Rev. Microbiol.* **2008**, *6*, 763–775. [CrossRef] [PubMed]

36. Garg, N.; Chandel, S. Arbuscular mycorrhizal networks: Process and functions. A review. *Agron. Sustain. Dev.* **2010**, *30*, 581–599. [CrossRef]

37. Smith, S.E.; Smith, F.A. Roles of arbuscular mycorrhizas in plant nutrition and growth: New paradigms from cellular to ecosystem scales. *Annu. Rev. Plant Biol.* **2011**, *62*, 227–250. [CrossRef] [PubMed]

38. Abdel-Fattah, G.M.; El-Haddad, S.A.; Hafez, E.E.; Rashad, Y.M. Induction of defense responses in common bean plants by arbuscular mycorrhizal fungi. *Microbiol. Res.* **2011**, *166*, 268–281. [CrossRef] [PubMed]

39. Liu, R.J. Effect of vesicular-arbuscular mycorrhizal fungi on *Verticillium* wilt of cotton. *Mycorrhiza* **1995**, *5*, 293–297. [CrossRef]

40. Eke, P.; Chatue, G.C.; Wakam, L.N.; Kouipou, R.M.T.; Fokou, P.V.T.; Boyom, F.F. Mycorrhiza consortia suppress the fusarium root rot (*Fusarium solani* f. sp. *Phaseoli*) in common bean (*Phaseolus vulgaris* L.). *Biol. Control* **2016**, *103*, 240–250. [CrossRef]

41. Sukhada, M.; Manjula, R.; Rawal, R.D. Evaluation of arbuscular mycorrhiza and other biocontrol agents against *Phytophthora parasitica* var. *nicotianae* infecting papaya (*Carica papaya* cv. Surya) and enumeration of pathogen population using immunotechniques. *Biol. Control* **2011**, *58*, 22–29. [CrossRef]

42. Spagnoletti, F.N.; Balestrasse, K.; Lavado, R.S.; Giacometti, R. Arbuscular mycorrhiza detoxifying response against arsenic and pathogenic fungus in soybean. *Ecotoxicol. Environ. Saf.* **2016**, *133*, 47–56. [CrossRef] [PubMed]

43. Zhang, H.; Franken, P. Comparison of systemic and local interactions between the arbuscular mycorrhizal fungus *Funneliformis mosseae* and the root pathogen *Aphanomyces euteiches* in *Medicago truncatula*. *Mycorrhiza* **2014**, *24*, 419–430. [CrossRef] [PubMed]

44. Cely, M.V.; de Oliveira, A.G.; de Freitas, V.F.; de Luca, M.B.; Barazetti, A.R.; Dos Santos, I.M.; Gionco, B.; Garcia, G.V.; Prete, C.E.; Andrade, G. Inoculant of arbuscular mycorrhizal fungi (*Rhizophagus clarus*) increase yield of soybean and cotton under field conditions. *Front. Microbiol.* **2016**, *7*, 720. [CrossRef] [PubMed]

45. Wang, C.X.; Wang, D.B.; Zhou, Q. Colonization and persistence of a plant growth-promoting bacterium *Pseudomonas fluorescens* strain CS85, on roots of cotton seedlings. *Can. J. Microbiol.* **2004**, *50*, 475–481. [CrossRef] [PubMed]

46. Takahashi, Y.; Uehara, Y.; Berberich, T.; Ito, A.; Saitoh, H.; Miyazaki, A.; Terauchi, R.; Kusano, T. A subset of hypersensitive response marker genes, including *HSR203J*, is the downstream target of a spermine signal transduction pathway in tobacco. *Plant J.* **2004**, *40*, 586–595. [CrossRef] [PubMed]

47. Lee, J.; Klessig, D.F.; Nürnberger, T. A harpin binding site in tobacco plasma membranes mediates activation of the pathogenesis-related gene HIN1 independent of extracellular calcium but dependent on mitogen-activated protein kinase activity. *Plant Cell* **2001**, *13*, 1079–1093. [CrossRef] [PubMed]

48. Lee, H.J.; Park, Y.J.; Seo, P.J.; Kim, J.H.; Sim, H.J.; Kim, S.G.; Park, C.M. Systemic immunity requires SnRK2.8-mediated nuclear import of NPR1 in Arabidopsis. *Plant Cell* **2015**, *27*, 3425–3438. [CrossRef] [PubMed]

49. Colditz, F.; Niehaus, K.; Krajinski, F. Silencing of PR-10-like proteins in *Medicago truncatula* results in an antagonistic induction of other PR proteins and in an increased tolerance upon infection with the oomycete *Aphanomyces euteiches*. *Planta* **2007**, *226*, 57–71. [CrossRef] [PubMed]

50. Xu, L.; Zhu, L.F.; Tu, L.L.; Liu, L.L.; Yuan, D.J.; Jin, L.; Long, L.; Zhang, X.L. Lignin metabolism has a central role in the resistance of cotton to the wilt fungus *Verticillium dahliae* as revealed by RNA-Seq-dependent transcriptional analysis and histochemistry. *J. Exp. Bot.* **2011**, *62*, 5607–5621. [CrossRef] [PubMed]

51. Guo, W.F.; Jin, L.; Miao, Y.H.; He, X.; Hu, Q.; Guo, K.; Zhu, L.F.; Zhang, X.L. An ethylene response-related factor, *GbERF1-like*, from *Gossypium barbadense* improves resistance to *Verticillium dahliae* via activating lignin synthesis. *Plant Mol. Biol.* **2016**, *91*, 305–318. [CrossRef] [PubMed]

52. López-Ráez, J.A.; Verhage, A.; Fernández, I.; García, J.M.; Azcón-Aguilar, C.; Flors, V.; Pozo, M.J. Hormonal and transcriptional profiles highlight common and differential host responses to arbuscular mycorrhizal fungi and the regulation of the oxylipin pathway. *J. Exp. Bot.* **2010**, *61*, 2589–2601. [CrossRef] [PubMed]

53. Fu, W.; Shen, Y.; Hao, J.; Wu, J.; Ke, L.; Wu, C.; Huang, K.; Luo, B.; Xu, M.; Cheng, X.; et al. Acyl-CoA N-acyltransferase influences fertility by regulating lipid metabolism and jasmonic acid biogenesis in cotton. *Sci. Rep.* **2015**, *5*, 11790. [CrossRef] [PubMed]

54. Volpe, V.; Giovannetti, M.; Sun, X.G.; Fiorilli, V.; Bonfante, P. The phosphate transporters LjPT4 and MtPT4 mediate early root responses to phosphate status in non mycorrhizal roots. *Plant Cell Environ.* **2016**, *39*, 660–671. [CrossRef] [PubMed]

55. Xie, X.; Lin, H.; Peng, X.; Xu, C.; Sun, Z.; Jiang, K.; Huang, A.; Wu, X.; Tang, N.; Salvioli, A.; et al. Arbuscular mycorrhizal symbiosis requires a phosphate transceptor in the *Gigaspora margarita* fungal symbiont. *Mol. Plant* **2016**, *9*, 1583–1608. [CrossRef] [PubMed]

56. Bender, S.F.; Conen, F.; Van der Heijden, M.G. Mycorrhizal effects on nutrient cycling, nutrient leaching and N$_2$O production in experimental grassland. *Soil Biol. Biochem.* **2015**, *80*, 283–292. [CrossRef]

57. Calabrese, S.; Pérez-Tienda, J.; Ellerbeck, M.; Arnould, C.; Chatagnier, O.; Boller, T.; Schüßler, A.; Brachmann, A.; Wipf, D.; Ferrol, N.; et al. GintAMT3-a low-Affinity Ammonium Transporter of the Arbuscular Mycorrhizal *Rhizophagus irregularis*. *Front. Plant Sci.* **2016**, *7*, 679. [CrossRef] [PubMed]

58. Garcia, K.; Chasman, D.; Roy, S.; Ané, J. Physiological responses and gene co-expression network of mycorrhizal roots under K$^+$ deprivation. *Plant Physiol.* **2017**, *173*, 1811–1823. [CrossRef] [PubMed]

59. Zhu, Y.G.; Miller, R.M. Carbon cycling by arbuscular mycorrhizal fungi in soil–plant systems. *Trends Plant Sci.* **2003**, *8*, 407–409. [CrossRef]

60. Verbruggen, E.; Heijden, M.G.; Rillig, M.C.; Kiers, E.T. Mycorrhizal fungal establishment in agricultural soils: Factors determining inoculation success. *New Phytol.* **2013**, *197*, 1104–1109. [CrossRef] [PubMed]

61. Koch, A.M.; Croll, D.; Sanders, I.R. Genetic variability in a population of arbuscular mycorrhizal fungi causes variation in plant growth. *Ecol. Lett.* **2006**, *9*, 103–110. [CrossRef] [PubMed]

62. Munkvold, L.; Kjøller, R.; Vestberg, M.; Rosendahl, S.; Jakobsen, I. High functional diversity within species of arbuscular mycorrhizal fungi. *New Phytol.* **2004**, *164*, 357–364. [CrossRef]

63. Colard, A.; Angelard, C.; Sanders, I.R. Genetic exchange in an arbuscular mycorrhizal fungus results in increased rice growth and altered mycorrhiza-specific gene transcription. *Appl. Environ. Microbiol.* **2011**, *77*, 6510–6515. [CrossRef] [PubMed]

64. Ibrahim, M. Effect of indigenous arbuscular mycorrhizal fungi combined with manure on the change in concentration some mineral elements in cotton (*Gossypium hirsutum* L.). *J. Plant Nutr.* **2017**, *40*, 2862–2871. [CrossRef]

65. Eason, W.R.; Webb, K.J.; Michaelson-Yeates, T.P.T.; Abberton, M.T.; Griffith, G.W.; Culshaw, C.M.; Hooker, J.E.; Dhanoa, M.S. Effect of genotype of *Trifolium repens* on mycorrhizal symbiosis with *Glomus mosseae*. *J. Agric. Sci.* **2001**, *137*, 27–36. [CrossRef]

66. MacLean, A.M.; Bravo, A.; Harrison, M.J. Plant signaling and metabolic pathways enabling arbuscular mycorrhizal symbiosis. *Plant Cell* **2017**, *29*, 2319–2335. [CrossRef] [PubMed]

67. Nagarajan, V.K.; Jain, A.; Poling, M.D.; Lewis, A.J.; Raghothama, K.G.; Smith, A.P. Arabidopsis Pht1; 5 mobilizes phosphate between source and sink organs and influences the interaction between phosphate homeostasis and ethylene signaling. *Plant Physiol.* **2011**, *156*, 1149–1163. [CrossRef] [PubMed]

68. Fernandes, H.; Michalska, K.; Sikorski, M.; Jaskolski, M.; Raghothama, K.G.; Smith, A.P. Structural and functional aspects of PR-10 proteins. *FEBS J.* **2013**, *280*, 1169–1199. [CrossRef] [PubMed]

69. Xie, Y.R.; Chen, Z.Y.; Brown, R.L.; Bhatnagar, D. Expression and functional characterization of two pathogenesis-related protein 10 genes from *Zea mays*. *J. Plant Physiol.* **2010**, *167*, 121–130. [CrossRef] [PubMed]

70. Constabel, C.P.; Bertrand, C.; Brisson, N. Transgenic potato plants overexpressing the pathogenesis-related *STH-2* gene show unaltered susceptibility to *Phytophthora infestans* and potato virus X. *Plant Mol. Biol.* **1993**, *22*, 775–782. [CrossRef] [PubMed]

71. Wang, C.S.; Huang, J.C.; Hu, J.H. Characterization of two subclasses of PR-10 transcripts in lily anthers and induction of their genes through separate signal transduction pathways. *Plant Mol. Biol.* **1999**, *40*, 807–814. [CrossRef] [PubMed]

72. Zhang, G.R. Analysis of PR Protein Family and Functional Study of PR10 and PR17 in Cotton. Master's Thesis, Agricultural University of Hebei, Baoding, China, 2015.

73. Liang, S. Isolation and Characterization of Cotton *PR10-Like* Gene Responsive to Infection by *Verticillium dahliae*. Master's Thesis, Hebei University, Baoding, China, 2009.

74. Zhang, G.; Raza, W.; Wang, X.; Ran, W.; Shen, Q. Systemic modification of cotton root exudates induced by arbuscular mycorrhizal fungi and *Bacillus vallismortis* HJ-5 and their effects on *Verticillium* wilt disease. *Appl. Soil Ecol.* **2012**, *61*, 85–91. [CrossRef]

75. Suh, H.W.; Crawford, D.L.; Korus, R.A.; Shetty, K. Production of antifungal metabolites by the ectomycorrhizal fungus *Pisolithus tinctorius* strain SMF. *J. Ind. Microbiol.* **1991**, *8*, 29–35. [CrossRef]

76. Grunewaldt-Stöcker, G.; von den Berg, C.; Knopp, J.; von Alten, H. Interactions of ericoid mycorrhizal fungi and root pathogens in Rhododendron: In vitro tests with plantlets in sterile liquid culture. *Plant Root* **2013**, *7*, 33–48. [CrossRef]

77. Hage-Ahmed, K.; Moyses, A.; Voglgruber, A.; Hadacek, F.; Steinkellner, S. Alterations in root exudation of intercropped tomato mediated by the arbuscular mycorrhizal fungus *Glomus mosseae* and the soilborne pathogen *Fusarium oxysporum* f.sp. *lycopersici*. *J. Phytopathol.* **2013**, *161*, 763–773. [CrossRef]

78. St-Arnaud, M.; Hamel, C.; Vimard, B.; Caron, M.; Fortin, J.A. Enhanced hyphal growth and spore production of the arbuscular mycorrhizal fungus *Glomus intraradices* in an in vitro system in the absence of host roots. *Mycol. Res.* **1996**, *100*, 328–332. [CrossRef]

79. Phillips, J.M.; Hayman, D.S. Improved procedures for clearing roots and staining parasitic and vesicular-arbuscular mycorrhizal fungi for rapid assessment of infection. *Trans. Br. Mycol. Soc.* **1970**, *55*, 158–161. [CrossRef]

80. Mo, H.J.; Sun, Y.X.; Zhu, X.L.; Wang, X.F.; Zhang, Y.; Yang, J.; Yan, G.J.; Ma, Z.Y. Cotton S-adenosylmethionine decarboxylase-mediated spermine biosynthesis is required for salicylic acid- and leucine-correlated signaling in the defense response to *Verticillium dahliae*. *Planta* **2016**, *243*, 1023–1039. [CrossRef] [PubMed]

81. Porebski, S.; Bailey, L.G.; Baum, B.R. Modification of a CTAB DNA extraction protocol for plants containing high polysaccharide and polyphenol components. *Plant Mol. Biol. Report.* **1997**, *15*, 8–15. [CrossRef]

82. Gkizi, D.; Lehmann, S.; L'Haridon, F.; Serrano, M.; Paplomatas, E.J.; Métraux, J.P.; Tjamos, S.E. The innate immune signaling system as a regulator of disease resistance and induced systemic resistance activity against *Verticillium dahliae*. *Mol. Plant Microbe Interact.* **2016**, *29*, 313–323. [CrossRef] [PubMed]

83. Nanamori, M.; Shinano, T.; Wasaki, J.; Yamamura, T.; Rao, I.M.; Osaki, M. Low phosphorus tolerance mechanisms: Phosphorus recycling and photosynthate partitioning in the tropical forage grass, *Brachiaria* hybrid cultivar Mulato compared with rice. *Plant Cell Physiol.* **2004**, *45*, 460–469. [CrossRef] [PubMed]

International Journal of
Molecular Sciences

MDPI

Review

Comparative Methods for Molecular Determination of Host-Specificity Factors in Plant-Pathogenic Fungi

Nilam Borah, Emad Albarouki and Jan Schirawski *

Institute of Applied Microbiology, RWTH Aachen University, Microbial Genetics, Worringerweg 1,
52074 Aachen, Germany; nilam.borah@rwth-aachen.de (N.B.); emad.albarouki@rwth-aachen.de (E.A.)
* Correspondence: jan.schirawski@rwth-aachen.de; Tel.: +49-241-802-6616

Received: 20 February 2018; Accepted: 14 March 2018; Published: 15 March 2018

Abstract: Many plant-pathogenic fungi are highly host-specific. In most cases, host-specific interactions evolved at the time of speciation of the respective host plants. However, host jumps have occurred quite frequently, and still today the greatest threat for the emergence of new fungal diseases is the acquisition of infection capability of a new host by an existing plant pathogen. Understanding the mechanisms underlying host-switching events requires knowledge of the factors determining host-specificity. In this review, we highlight molecular methods that use a comparative approach for the identification of host-specificity factors. These cover a wide range of experimental set-ups, such as characterization of the pathosystem, genotyping of host-specific strains, comparative genomics, transcriptomics and proteomics, as well as gene prediction and functional gene validation. The methods are described and evaluated in view of their success in the identification of host-specificity factors and the understanding of their functional mechanisms. In addition, potential methods for the future identification of host-specificity factors are discussed.

Keywords: host specificity; plant pathogen; fungi; effector; sequencing; genotyping

1. Introduction

The concept of host-specificity of plant-pathogenic fungi has always intrigued plant pathologists. Why can some fungal plant pathogens cause disease only in one specific host and not in others; or how can a fungal pathogen of a particular plant adapt and switch to a new host and thereby become a new pathogen for the new host, are questions that have shaped and will shape research in plant-fungus interaction. The term host-specificity refers to the capability of some fungal species or some members of one fungal species to cause disease only on particular plant species or only on some members of a specific plant species. Molecular models have been developed to explain the basis of host-specificity, like the gene-for-gene hypothesis developed by Henry H. Flor following his careful observations of the interactions of flax with flax rust [1]. According to this hypothesis, incompatible interactions result from the presence of resistance (R) proteins in the particular plants that recognize specific avirulence (AVR) proteins of the fungus, which results in successful plant defense against the pathogen. All interactions of plants carrying a particular R gene with pathogens carrying the corresponding *AVR* gene would be incompatible. This model has been refined by the guard and later the decoy models that predict that the AVR-R interaction is indirect, with the R protein guarding a target of the AVR protein, or guarding a decoy, a functionally inactive version of the target [1–3]. These models have led to a tremendous increase in the understanding of plant-pathogen interactions. The second great advancement in this understanding is the discovery of effectors—small, secreted, variable, genome-encoded proteins of the pathogens—that are essential to modulate the plant response [3–5]. The realization that effectors can have virulence and/or avirulence (AVR) functions elegantly links the two concepts and explains the multitude of observed species-specific effector genes on the pathogen side as well as the multitude of receptor genes with the role of R proteins on the plant side, by an evolutionary arms race [5,6].

Is there a difference between host specificity factors and virulence factors? The answer is: yes and no. It is yes, because some virulence factors are essential for basic and conserved virulence functions that are necessary for the processes of plant infection. These basic virulence factors, although essential for virulence of host-specific pathogens on their preferred host plants, do not contribute to incompatibility with the related plants and are, therefore, not involved in determining host specificity. However, the answer is also no, because all host-specificity factors should modulate virulence of the pathogen by either contributing to avirulence on the non-preferred or to virulence on the preferred host, or both. This means that all host-specificity factors should be virulence factors, but not all virulence factors are necessarily host-specificity factors.

Are all host-specificity factors effectors? Here the answer is no. While some host-specificity factors are indeed effectors, like for example the PWT3 and PWT4 effectors of *Magnaporthe oryzae* [7], the classical and best-known host-specificity toxins of *Alternaria alternata* are secondary metabolites generated by polyketide synthases (PKSs) [8–10]. The *PKS* genes reside on a conditionally dispensable chromosome, and transfer of the chromosome containing the PKS necessary for virulence of *A. alternata* tomato pathotype to a strawberry pathotype leads to strains able to produce both toxins needed for infection of tomato and strawberry [11].

While we have a fairly good understanding of host-adaptation in some systems, in others the mechanism is still unknown. Even if the mechanism can be inferred, identification of the participating factors is still quite challenging. For example, after knowledge that the *Ustilago hordei*-barley interaction is governed by AVR-R gene interactions, it took 32 years to clone an 80-kb region containing *AVR1*, and an additional 10 years to identify the gene encoding AVR1 [12–14]. Therefore, powerful methods are needed for the identification of host-specificity factors to unravel the molecular basis of host adaptation of phytopathogenic fungi to their plant hosts.

In the following, we review molecular methods that have been employed for the determination of host-specificity factors. We focus in this review on comparative analyses of two or more different pathovars, *formae speciales*, isolates, races or species of plant-pathogenic fungi. Where appropriate, we also included examples from oomycetes. We excluded from our review studies that investigate the contribution of individual factors to virulence on one host, even if these factors may be host-specificity factors. We also largely excluded literature on host-specificity of symbiotic mycorrhiza in spite of their importance, economic relevance and recent success in the identification of effectors necessary for plant infection [15]. We present the methods employed in chronological order, starting with methods for definition and characterization of the pathosystem, methods for genotyping, comparative -omics techniques like genomics, transcriptomics and proteomics, gene prediction and functional validation (Figure 1).

These methods are explained and evaluated in view of their success in the identification of host-specificity factors and the understanding of their functional mechanisms. For a review of known host-specificity mechanisms and factors, we would like to draw the reader's attention to previous reviews [16,17].

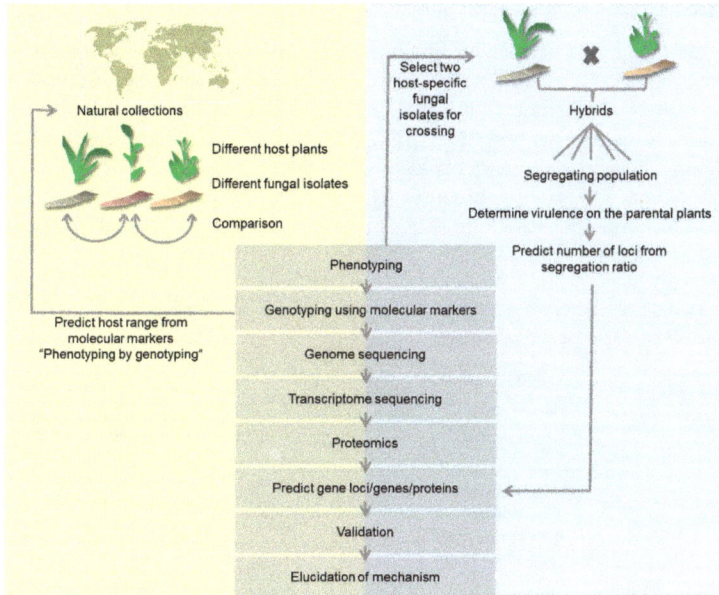

Figure 1. Graphical representation of comparative methods used to determine molecular host-specificity determinants. Methods starting with a collection of natural fungal isolates (yellow background) are contrasted to methods starting with segregating populations of defined strains (gray background). Both starting materials use the same set of analyzing methods.

2. Defining and Characterizing the Pathosystem

Any comparative molecular analysis is dependent on a well-characterized pathosystem. It is, therefore, not astounding that many studies collected and phenotypically characterized the available biodiversity of natural fungal isolates. Fungal isolates are collected from different parts of the world e.g., from fields of host plants that are infected by the fungus [18,19], or procured from various laboratories and culture collections [20]. Depending on the necessity and type of infection, fungal specimens can be isolated from any part of the plant, including infected leaves, stems or inflorescences [21]. After collection, the natural isolates are phenotypically characterized either by determining infection capability on different host plants coupled with microscopic analysis and analysis of the disease phenotype [22–34], or by the characterization of physiological traits, like their ability to produce or degrade various chemicals [29,31,35,36]. This way, various pathosystems were defined, like the *formae speciales* of *Fusarium oxysporum*, of *Puccinia*, and of *Alternaria*.

A well-defined pathosystem is the basis of any follow-up comparative analysis. From this pool of data, two different approaches are followed. In one approach, two isolates with different host-specificity are crossed or hybridized, generating a segregating population with differential capabilities to infect one or the other host plant. This segregating population is then analyzed to define genomic loci associated with differential infection capabilities. In the other approach, the complete pool of natural isolates is used for phenotypic and molecular analyses to link genomic markers to phenotypic traits (Figure 1). The approach using segregating populations generated by hybridization of host-specific isolates will be treated in a separate section below. In most examples, however, the approach using natural populations was followed.

3. Using Molecular Markers for Genotyping

To link genomic loci or markers to the host infection-specificity phenotype, many different methods have been developed that allow a molecular comparative characterization of the different isolates. These methods include analysis of restriction fragment-length polymorphism (RFLP), amplified fragment-length polymorphism (AFLP), random amplified polymorphic DNA (RAPD), micro- and minisatellites, mitochondrial haplotype, internal transcribed spacer (ITS) regions, as well as of complete proteomes and genome sequences. The advantages and disadvantages of these methods are summarized in Table 1. Examples of successful use in the genomic characterization of different fungal isolates are presented below.

Table 1. Method-specific advantages and disadvantages of molecular methods typically used for comparison of host-specific fungal isolates.

Method [1]	Advantages	Disadvantages
Restriction fragment-length polymorphism(RFLP)	Can detect allelic variants	Large DNA quantity needed, typically only 1–3 loci detected, usually radioactive labeling is used
Random amplified polymorphic DNA (RAPD)	Faster than RFLP, less DNA is needed, can detect 1–10 variant loci, suitable for detection of broad scale genetic structural differences	Cannot detect allelic variants (heterozygous alleles or homologous alleles normally give the same result), less reliable, polymerase chain reaction (PCR)-dependent assay
Simple sequence repeats; microsatellites (SSR)	More accurate than RAPD, suitable for discriminating different subpopulations	Microsatellite markers may not be evenly distributed in the genome, SSR are located in non-coding regions, false alleles or null alleles may be detected due to technical artifacts, blurry bands may occur
Amplified fragment-length polymorphism (AFLP)	Combines benefits of RAPD and RFLP	Difficult to develop locus-specific marker (fragment) proprietary technology to score heterozygous and homozygous
Analysis of mitochondrial DNA (mtDNA analysis)	Powerful tool for studying inheritance of mitochondrial genomes, for phylogenetic and population genetic analysis, for species identification and barcoding	In uniparental-mtDNA inheritances, no information about other parent: should be coupled with genomic-DNA analyses. In case of mtDNA recombination (bi-parental inheritance) many analysis not doable
Sequencing of internal transcribed spacer regions (ITS sequencing)	ITS1 and ITS2 regions are species-specific and have large copy numbers, ITS sequencing can be used in metagenomics studies (meta-barcode), can be coupled with NGS technique	Limited to discriminate intra- and intergeneric species
Analysis of protein abundance of all proteins (proteomics)	Many different techniques available, e.g., two-dimensional electrophoresis coupled to mass spectrometric protein identification, can analyze vast array of proteins at once, can do high throughput, high sensitivity possible, relative as well as absolute protein abundance quantification possible	Each technique has its own limitation, not all proteins can be identified by one single method. Results may be tissue- and environmental condition-dependent
Sequencing using next-generation sequencing techniques (NGS sequencing)	Identify millions of single nucleotide polymorphisms (SNPs) as well as insertions and deletions (INDELs) at once	PCR-born false variants, data analysis needs bioinformatic know-how and computing power

[1] See text for a description of the respective method.

A restriction fragment-length polymorphism (RFLP) is existent if a base substitution in the genomic DNA occurs within a restriction enzyme-recognition sequence leading to differently sized restriction fragments. Because a digest of genomic DNA with a restriction enzyme leads to too many restriction fragments to visualize individual length polymorphisms, the use of labeled probes hybridizing to specific regions of the genomic DNA is necessary. The probes may cover mating type genes [37], host-species specific repetitive DNA sequences of the pathogen [38], or fungal isolate-specific genomic regions [39]. The use of telomere-linked RFLP revealed a specific pattern for almost every strain of *Botrytis cinerea* isolated from different host plants from various regions, which showed the presence of extended polymorphism near telomeres in *B. cinerea* [40] and was, therefore, too specific to link a telomere-RFLP to the host infection phenotype. Identical sequence at the internal transcribed spacer (ITS) regions of the genes encoding ribosomal RNAs is used to verify close genetic relationship between isolates [41,42] but can also be used to differentiate host-specific fungal isolates. An ITS-based RFLP–polymerase chain reaction (PCR) method was successful in distinguishing different barley- and rye-infecting isolates of *Rhynchosporium* [43]. RFLP analysis using the 18–28S ribosomal transcriptional unit from *Aspergillus nidulans* as the probe revealed similar RFLP patterns among all pink and gray *Colletotrichum* isolates regardless of host origin. However, RFLP patterns of pink isolates were distinct from those of gray isolates [24]. This method was thus shown to be useful to differentiate two isolates based on initial morphological characterization, later supported by molecular marker polymorphism. Mitochondrial DNA (mtDNA) RFLP by *Hae*III-digestion of total genomic DNA was used to efficiently differentiate two *formae speciales* of the *Fusarium oxysporum* species complex, f. sp. *lycopersici* (Fol) and f. sp. *radicis-lycopersici* (Forl). These two *formae speciales*—although morphologically identical and capable of infecting the same tomato cultivar—cause distinct disease phenotypes [44]. PCR amplification of rDNA ITS regions followed by enzymatic digestion to perform RFLP has also been used to study host preferences of oomycete species within ecologically contrasting sites [45]. RFLP fingerprinting of oomycete isolates was used effectively in *Phytophthora infestans* to differentiate different clonal lineages [46,47]. From these examples it becomes clear that while RFLP can be used to differentiate specific isolates, the use of probes drastically limits the investigated genomic area so that the identification of an RFLP pattern that correlates with host-specificity depends on serendipity.

A genome-wide and less biased comparative approach is the use of amplified fragment length polymorphism (AFLP) [48]. In this technique, the genomic DNA is first digested with two different restriction enzymes, a frequent cutter and a rare cutter. Then specific adaptors are ligated that allow the PCR-amplification of restriction fragments. The number of amplified restriction fragments is reduced by extending the primers by 1 or 2 nt into the restriction fragment. To further reduce the number of amplified restriction fragments, a second PCR with labeled primers follows that extend by 3 to 4 nt into the restriction fragment. In this PCR, hybrid fragments with one rare-cutter adaptor and one frequent-cutter adaptor are preferentially amplified because two frequent-cutter adaptor primers form secondary structures that hamper product amplification. Primers with different extensions are tested to obtain about 50 to 120 amplified fragments, whose template-dependent lengths can then be compared by gel electrophoresis and label detection. AFLP has been successfully used to differentiate race composition, race variation and host-specific isolates, such as *P. infestans* isolates collected from cultivated potatoes and the native wild *Solanum* spp. *Solanum demissum* and *Solanum xendinense* in the Toluca Valley of central Mexico, and *Magnaporthe grisea* isolates associated primarily from perennial ryegrass and kikuyugrass in golf courses in California [21,49,50]. Host-specific groups could also be defined according to the AFLP pattern of various endophytic fungi following initial morphological identification [51]. *Blumeria graminis* has been classified into eight *formae speciales* with different host specificity. rDNA ITS region and β-tubulin gene-based phylogenetic analysis showed grouping of isolates according to their principal host genus, which was further supported by AFLP analysis [52]. AFLP data can also be used for the generation of phylogenetic analysis, revealing the origin of host-specific taxa [20].

A special form of AFLP is the investigation of the mitochondrial DNA (mtDNA) haplotype. In this technique, certain regions in mitochondrial genomes are first PCR-amplified and then subjected to restriction-enzyme digestion. mtDNA haplotype analysis has been successfully used to characterize different isolates of *P. infestans* [47,50]. The result can reveal whether the population structure is of monomorphic or polymorphic nature. The mtDNA haplotype has also been investigated in various geographical isolates of *Colletotrichum orbiculare* having a common host and in various other species of *Colletotrichum* [53]. The data helped to link host specificity with the haplotype pattern, thereby suggesting a way to recognize different host-specific isolates.

Random amplified polymorphic DNA (RAPD) is a PCR-based method that does not rely on restriction enzymes. A 10 nt primer of arbitrary sequence is used to amplify random segments of genomic DNA. This method has been widely used to identify and isolate molecular markers specific to a particular fungus, thereby helping in diagnosis of fungal infection in symptomless plants [54,55]. RAPD was used to characterize the genetic differentiation and correlation with host specificity among *Alternaria* spp. that cause brown spots on different *Citrus* spp. [26,56]. In *F. oxysporum*, a robust RAPD protocol was developed to identify economically important strains infecting specific hosts [57]. The RAPD technique was also used to discriminate between isolates of different host-specialized *Rynchosporium* species [58]. In *U. hordei*, RAPD was used in combination with microsatellites (see below) to assess genetic variation among different isolates in Tibetan areas of China [59]. In addition, RAPD has also been used as a tool to differentiate *formae speciales* of *Microbotryum violaceum* and to characterize the genetic diversity of the Italian population of *Ceratocystis fimbriata* f. sp. *platani* [60,61]. Although RAPD is a more unbiased technique covering a larger part of the fungal genome, the method has issues with reproducibility. Therefore, it is usually only used in combination with other markers, like micro- or mini-satellites.

Satellite DNA was first observed as a less-dense band of DNA clearly separated from the bulk of chromosomal DNA during density-gradient centrifugation. Microsatellites or simple sequence repeats (SSR), and their longer cousins, the minisatellites, consist of AT-rich repetitive DNA that seem to have a higher mutation rate resulting in a larger genetic diversity than other genomic regions. This genetic diversity can be used to discriminate different fungal isolates by PCR-amplifying the repetitive DNA [62]. Microsatellites have been found to be more informative for genotyping isolates from different hosts of the necrotrophic fungus *Botrytis cinerea* than RFLP patterns of the ADP-ATP translocase and nitrate reductase genes or MSB2 minisatellite sequence data [63]. In a study of the anther smut fungus *M. violaceum*, genetic diversity in sympatric, parapatric and allopatric populations of two host species was found using four polymorphic microsatellite regions [64]. In *Pyrenophora semeniperda*, that was shown to lack host specialization, weak yet significant population genetic structures as a function of host species could be observed by the use of seven polymorphic microsatellite loci [34]. Thus, the investigation of the diversity in microsatellite loci is a very sensitive tool to discriminate between different fungal isolates and to visualize even weak associations. However, microsatellite diversity has not yet been shown to be causally related to host specificity.

Other repetitive elements like retrotransposons or transposable elements have also been used for the characterization of host-specific fungal isolates. When using the reverse transcriptase gene of the LTR-retrotransposon *CfT-1* from *Cladosporium fulvum* as a probe, different *formae speciales* of *F. oxysporum* could clearly be differentiated [65]. PCR-detection of two transposable elements revealed different population structures of *B. cinerea* on a variety of different host plants [66].

It seems that a large number of different techniques have been developed that allow for molecular comparison of different organisms. Of the described techniques, AFLP and microsatellites seem to be the most sensitive methods to molecularly discriminate fungal isolates for which no or only very little sequence information is available. However, most studies using these methods did not come up with a clear genomic link to host-specificity of the investigated fungal strains. This means that other, even more precise techniques are needed to decipher the molecular basis of host-specificity. Here, the field has profited tremendously from the development of the next-generation sequencing (NGS) techniques that now allow cost-effective genome sequencing and data assembly of large fungal genomes. Since they have become available, the use of -omics (genomics, transcriptomics, and proteomics) approaches highly outnumber the use of the classical comparative approaches described above (Figure 2).

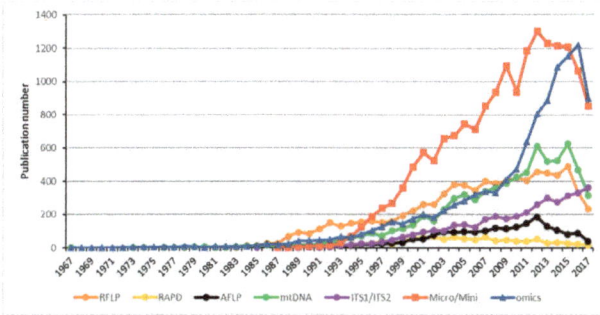

Figure 2. Hits of the search terms in the PubMed data base sorted by year of publication between 1967 and 2017. The -omics techniques, despite their relatively recent appearance and relatively high cost in comparison to other techniques, are increasingly popular. Analysis of micro- and minisatellites is still popular. The search terms were as follows: "Restriction fragment length polymorphism, RFLP analysis AND population genetic" for RFLP, "RAPD analysis AND population genetic" for RAPD, "Amplified fragment length polymorphism; AFLP analysis AND population genetic" for AFLP, "mitochondrial DNA; mtDNA; mitochondrial DNA Analysis AND population genetic" for mtDNA, "ITS1/ITS2; ITS1 OR ITS2 analysis" for ITS1/ITS2, "Micro/Minisatellite repeats OR micro/minisatellite analysis AND population genetic" for Micro/Mini, and "High throughput nucleotide sequencing, OR next generation sequencing analysis AND population genetic" for omics.

4. Comparing Whole Genome Sequences

One of the first eukaryotic genomes sequenced using next-generation sequencing techniques was that of *Sporisorium reilianum* f. sp. *zeae*, a close relative of *Ustilago maydis*. Both cause smut disease of maize but induce quite distinct symptoms. Genome comparison revealed that both genomes were highly syntenic but contained so called "divergence clusters" containing genes with below-average sequence conservation between the two organisms [67,68]. Within these divergence clusters, a high percentage of genes encoded proteins containing predicted secretion signal peptides and were thought to be involved in interaction with the plant, thereby explaining their increased evolution rate. Deletion analysis of complete cluster regions in *U. maydis* confirmed the role in virulence for four of six randomly selected clusters [67,68]. Genome sequencing of the related barley smut pathogen *U. hordei* allowed comparison with an organism showing a similar infection strategy as *S. reilianum* and being virulent on a different host plant. The three-way genome comparison revealed among others that most of the weakly conserved genes present in the *S. reilianum/U. maydis* divergence clusters have weakly conserved homologs in *U. hordei* [67,68], which supports the proposition that these proteins could play a function in adaptation to different hosts or lifestyles. Sequencing of the *Sporisorium scitamineum* genome allowed a four-genome comparison of effector genes [69]. This comparison revealed that evolution of effector-encoding clusters is driven by tandem gene duplication and the activity of transposable elements, and supports the conclusions drawn from analysis of the *U. hordei* genome [67–69]. Thus, genome comparison of smut fungi so far has resulted in lists of genes potentially encoding host-specificity factors.

Interesting gene candidates with a suspected role in host-specificity were also obtained when comparing the genomes of the closely related species *Colletotrichum graminicola* and *Colletotrichum sublineola* [70]. The main differences of the otherwise very similar genomes were found in genes for biosynthesis of specialized secondary metabolites and of small secreted protein effectors. However, whether these genes indeed contribute to host selection has yet to be tested. Key enzymes of fungal secondary metabolism and effector proteins were identified as potential host-specificity factors by genome comparison of *Rhynchosporium* species [71], and of members of the *Fusarium fujikuroi* species complex, where species-specific and isolate-specific differentiation in secondary metabolite-producing genes both in composition and expression were detected [72]. The comparison of whole genome sequences of four different strains of the *C. acutatum* species complex showed that changes in gene content were related to changes in host range with lineage-specific gene losses and gene-family expansions [73]. Gene loss was also suggested as a cause of fungal adaptation to a new dicot host after a host jump from a monocot plant by the smut fungus *Melanopsichium pennsylvanicum*. When compared to the genomes of three other smut fungi, *M. pennsylvanicum* was found to lack putative effector genes [74]. In addition to putative effectors, comparative whole genome and transcriptome analyses of *Lasiodiplodia theobromae* and five other *Botryosphaeriaceae* pathogens causing opportunistic infections in woody plants identified two in-planta expressed lignocellulose genes, whose overexpression increased virulence of the pathogens [75].

In a genome comparison study of the three species *Fusarium graminearum*, *Fusarium verticillioides* and *F. oxysporum* f. sp. *lycopersici* (Fol) lineage-specific (LS) genomic regions and LS chromosomes were discovered [76]. The authors of the study could prove experimentally that the presence of Fol LS chromosome 14 provides specificity for *Fusarium* adaptation towards the tomato [76], limiting the search for host-specificity factors to a single chromosome. This information was used in a recent study where the genomes of three legume-infecting *formae speciales* of *F. oxysporum* were compared to the genomes of the tomato-infecting Fol and the pea-infecting *Fusarium solani*. Combining comparative genome analysis with predicted LS gene content and *in-planta* transcription analysis revealed four candidate effectors conserved among legume-infecting *formae speciales* [77]. Clustering of presence/absence patterns of candidate effector genes following whole genome sequencing of five different *formae speciales* of *F. oxysporum* showed clustering of members of the same *forma specialis* [78] suggesting that effectors contribute to host specificity. This confirms and extends earlier reports of an association of the three *secreted in xylem* (SIX) effectors SIX1, SIX2 and SIX3 with tomato-infecting isolates [79], and of the suitability of the presence/absence determination of *SIX1* to *SIX5* as a robust method to differentiate different *formae speciales* [80].

For the identification of effectors, sequencing of RNA or cloning of expressed sequence tags (ESTs) can be very helpful. Sequencing of about 2000 cloned ESTs from *C. lentis*-infected lentil leaf tissues enabled annotation of 15 candidate effectors. Infection stage-specific gene expression was observed for the candidate effectors. One candidate effector, CICE6, was found to carry a single nucleotide polymorphism (SNP) that could be used efficiently to differentiate between two pathogenic races of *C. lentis* [81]. Comparison between transcriptional profiles of two races of *F. oxysporum* f. sp. *cubense* revealed a remarkably different gene-expression profile in response to a host cell wall [82] suggesting that differences in gene expression could contribute to host-specificity.

Without being exhaustive, this enumeration already shows that the field has profited tremendously from the development of next-generation sequencing techniques. While the classical methods described above using molecular markers have been mostly used for the characterization of host-specific isolates, no precise gene candidates have resulted. In contrast, genome comparison that is sometimes coupled to a comparison of the transcriptional profile, has often resulted in the identification of genomic regions, and sometimes even of genes that are associated with adaptation to a specific host. These genomic regions and these gene lists, therefore, contain strong candidates for the determination of host-specificity. It seems that the methods depending solely on molecular markers are not precise enough to result in the prediction of genes associated with a certain host-infection phenotype. Novel methods, like diversity array technology sequencing (DArTseq) have been developed that combine mapping technology with next-generation sequencing. In DArTseq, different restriction enzymes are used to cut the genomic DNA prior to sequencing, which separates the low-copy number DNA from repetitive sequences that are cut to small pieces. The combination of precise mapping of the avirulence phenotype on a certain wheat cultivar with gene-expression data in infected plant tissue resulted in the successful identification of AvrStb6 as responsible for avirulence of *Z. tritici* IPO323 on wheat cultivar Shafir [83]. Therefore, when aiming at identifying the responsible genes for host-specific infection, it is much more promising to resort to genome sequencing, to do genome comparison and to couple this data to the analysis of gene-expression profiles.

5. Comparing Complete Proteomes

An alternative approach to genome comparison and transcriptome analysis for the identification of proteins involved in host-specificity determination is the direct comparative analysis of the fungal proteomes. Several studies have compared the proteomes of different isolates or species with the idea of identifying proteins crucial for host-specificity. Tandem mass spectrometry was used as a tool to compare the proteomes of hyphae and germinating cysts of two closely related oomycetes, *Phytophthora pisi* and *Phytophthora sojae*, that cause disease on pea and soybean, respectively [84]. A global proteomic comparison of mycelium and germinating cysts was done in two other oomycete plant pathogens, *Phytophthora ramorum* and *P. sojae* [85]. A proteome comparison was also done with uredospores from two different populations of the rust fungus *Puccinia psidii* isolated from eucalyptus leaves and guava fruits [86]. Mycelial proteins from isolates of the brown rot fungus *Monilinia laxa* were obtained from apples and apricots, and were separated by 2-D gel electrophoresis, followed by LC-MS/MS of identified differentially expressed proteins [87]. In all of these studies, clear differences in the fungal proteomes were observed, and detected species- or population-specific proteins were suspected to have a role in host adaptation. However, the lists were long, and a causal connection of identified proteins to host specificity has not yet been shown.

One problem of comparing proteomes of infected plant material is that a high number of plant proteins may be differently expressed or differently modified because the two investigated fungal strains behave differently on the plant, rather than the different plant protein profiles being the reason for the different behavior of the fungal strains. Therefore, identifying fungal determinants of host-specific proliferation will be a difficult task to solve by proteome comparison alone.

Global comparative approaches have brought the scientific community much closer to the goal of identifying host-specificity factors by providing lists of genes or proteins that have a high probability of being involved in host adaptation. The follow-up of these approaches would now consist of testing individual high-probability candidates for their specific contribution to host selection. This can be challenging, especially if genetic tools for the investigated system are not yet available, or if lists contain several dozens of putative candidates. Success of the follow-up experiments will depend a great deal on the predictive quality of these lists, which will increase with closer relatedness of the compared pathogen genomes.

6. Investigating Segregating Populations

One way to overcome the problem of having to test candidates from long lists of putative host-specificity factors that have a certain possibility of being involved in host adaptation, is to investigate a well-defined population that segregates for the host-infection phenotype. The easiest method to obtain such a well-defined population is to generate it by targeted hybridization experiments or genetic crosses of closely related fungal isolates that differ in their infection phenotype.

Hybrids have been investigated before the knowledge of genome sequences. Mating compatible isolates of *Colletotrichum gloeosporioides* with different host specificities were crossed, and ascospore offspring were analyzed for virulence and RFLP patterns. However, no correlation between pathogenicity to the parental hosts and presence/absence of any RFLP marker could be found [88]. In a comparative study of various host-specific *Ascochyta* species, interspecific crosses could be obtained, and offspring analysis showed that AFLP markers segregated freely. While not being able to assign specific AFLP markers to host infection capacity, it was speculated that host-specificity may contribute to speciation [28]. F1 progeny of a cross of two *Leptosphaeria maculans* isolates that differed in their capacity to cause disease on *Brassica juncea*, were used to create a genetic map of different AFLP and RAPD markers as well as the mating type and a host-specificity locus that could be placed on the end of chromosome 9 [89]. Hybrids of *M. oryzae* isolates from rice and wheat were generated and the resultant F1 population was tested on wheat for pathogenicity [90]. The segregation ratio of avirulent to virulent offspring provides information of the number of loci involved. In this case, a ratio of 7:1 led to the prediction of three loci being involved in avirulence of the *M. oryzae* rice isolate on wheat. Allelism tests could associate two known loci, Pwt2 to papilla formation and Pwt1 for hypersensitive reaction. The third locus did not correlate with any known loci and was, therefore, named Pwt5 [90]. In a F1 population of a cross between wheat- and foxtail millet-pathogenic *M. grisea*, virulent isolates segregated in a 1:1 ratio on foxtail millet cultivars Beni-awa and Oke-awa but not on cultivar Kariwano-zairai, suggesting that specificity of *M. grisea* toward foxtail millet is governed by cultivar-dependent genetic mechanisms like gene-for-gene interactions [91].

Since these studies were all conducted without the knowledge of the genome sequences, it is conceivable that the combination of genome sequencing and the investigation of defined segregating populations will either allow much faster genome map-based cloning or, through the sequencing of complete populations, allow high-resolution single nucleotide polymorphism (SNP) mapping of host-specificity traits. Creating genetic linkage maps in fungi is an underdeveloped but potentially important field that is expected to gain future attention in the light of whole genome sequencing [92]. In this light, we have started to analyze hybrids of two compatible *formae speciales* of the head smut fungus *S. reilianum*, *S. reilianum* f. sp. *zeae* (SRZ) and *S. reilianum* f. sp. *reilianum* (SRS) that either infect maize (SRZ) or sorghum (SRS). On each other's hosts, both fungi can colonize but do not cause smut disease [93]. A population of meiotic progeny (SRSZ) of a mating event between SRS and SRZ was generated and for each individual strain the virulence potential on sorghum was tested and varied greatly between individuals (unpublished). To associate the virulence phenotype on sorghum to particular genomic-regions, about 190 strains were selected for genome sequencing. Mapping of sequencing reads to the parental genomes confirmed the presence of mosaic genomes in the offspring (unpublished). A detailed genome analysis will show whether specific parental genomic regions can be associated with the virulence phenotype on sorghum (Figure 3). As soon as such an associated region is known, knowledge of the genome sequence will allow direct prediction of candidate genes for validation of their role in determining host-specificity.

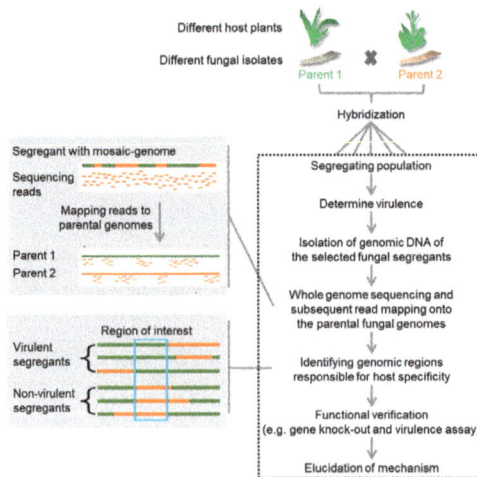

Figure 3. Strategy for identification of host-specificity factors using segregating populations of a hybridization event between two host-specific individuals. Virulence capacity of each offspring on one or both hosts is individually determined. Genome sequencing of fully virulent and avirulent offspring reveals parental origin of the mosaic genomes. Associating the parental origin of specific genomic loci to the virulence phenotype should lead to identification of genomic regions linked to host-specific virulence.

7. Validating Functional Contribution to Host-Specificity

Generating lists of the best potential candidates for involvement in host-specificity determination is already a great step forward. However, the goal is to show that a given candidate has indeed a role in host-specificity. Validation of a functional involvement of a specific candidate gene could be done by the generation of gene deletion or overexpression strains and monitoring of the mutant's host preference. As mentioned above, this can be a challenging task. In a few cases, functional validation was done and showed a clear relationship of the identified genes with host-specificity.

In many cases, genes involved in host-specificity were identified as a part of classical AVR-R gene interactions. For example, using crosses of *M. grisea* strains infecting different grass species, single genes that determine specificity towards the host weeping lovegrass (*Eragrostis curvula*) have been identified [94] and were shown to be part of a gene family [95]. The inability to infect weeping lovegrass could be associated with one member of the gene family, *PWL2*, where frequently occurring loss-of-function mutations of *PWL2* led to spontaneous pathogenic mutants [96]. Thus, *PWL2* has all the characteristics of a classical avirulence gene. *AVR1-CO39* of *M. oryzae* is another avirulence gene involved in host-specificity. The transfer of *AVR1-CO39* to a rice-pathogenic isolate resulted in transformants unable to cause disease on the rice cultivar CO39, while the rice cultivar 51583 that lacks the resistance gene *Pi-CO39(t)* could still be infected [97].

In other cases, lacking complementation of deleted virulence genes by orthologous genes from related species suggested a role in host-specific virulence. For example, the *SIX1* gene of *F. oxysporum* f. sp. *lycopersici* (Fol) was found to be necessary for full virulence of Fol on susceptible tomato but was also recognized by the I-3 resistance gene, leading to avirulence on I-3 tomato lines [98,99]. A homolog of *SIX1* was found to be responsible for virulence of *F. oxysporum* f. sp. *conglutinans* (Foc) on cabbage [100]. Interestingly, virulence of the Foc-*SIX1* deletion mutant on cabbage could be restored by reintroduction of *SIX1* of Foc but not of Fol. This suggested a host-specific virulence role for Fol-*SIX1* [100]. Similarly, virulence of mutants of the wheat pathogen *Zymoseptoria tritici* carrying deletions in the *Zt80707* or the *Zt89160* virulence genes, could be complemented with the respective genes from *Z. tritici*, but not with the respective orthologs from *Zymoseptoria pseudotritici* or *Zymoseptoria ardabiliae* [101] suggesting that the identified genes are involved in host-specific disease development.

In fungi where host specificity is governed by host-selective toxins, validation was done by heterologous expression of the toxin biosynthesis gene in a related species. For example, in the strawberry pathotype of *A. alternata*, transformation-mediated loss of the 1.05-Mb conditionally dispensable chromosome encoding all known toxin biosynthesis genes led to non-pathogenicity [9], which validated the role of the dispensable chromosome in host-specific pathogenicity. In *Alternaria citri*, mutation of a gene encoding an endo-polygalacturonase led to a reduction in virulence, while virulence was unchanged in mutants of *A. alternata* rough lemon pathotype lacking the same gene [102]. This showed that a cell-wall degrading enzyme contributes to virulence of one but not of the other pathotype. In contrast, PbToxB of the bromegrass pathogen *Pyrenophora bromi*, a homolog of the known host selective toxin PtrToxB from the wheat pathogen *Pyrenophora tritici-repentis*, was unexpectedly shown not to be toxic on bromegrass but on wheat, indicating that *P. bromi* has the potential to become a wheat pathogen [103].

These few examples and the surprises they contained illustrate the importance of validation experiments for genes suspected to have a role in host-specificity. Without knowledge of their contribution to host-specific virulence, it is impossible to generate meaningful hypotheses about potential mechanisms.

8. Deciphering Functional Mechanisms of Host-Specificity

Knowing the mechanisms that govern host-specific virulence capacities is the ultimate goal in plant-pathogen interaction research. The understanding gained allows the generation of novel plant-protection strategies and ever more reliable prediction of the danger of a particular fungus being involved in future disease outbreaks in novel hosts. In a few cases, more than just the virulence/avirulence-causing gene of the pathogen is known, which allows a better understanding of the basic mechanisms involved in host-specific interactions.

In *P. infestans*, virulence towards potato was shown to depend on the function of the RXLR effector AVR3a in inhibiting enzyme activity of the host ubiquitin proteasome system (UPS) [104]. When studying host adaptation of *P. infestans* and its *Mirabilis jalapa*-infecting sister species *P. mirabilis*, a single amino-acid polymorphism was identified in the host protease and a corresponding single amino-acid change in the pathogen effector as being responsible for virulence of their respective host plants, explaining ecological diversification [105]. In the case of *A. alternata*, the contribution of toxins to host-specific virulence success has been thoroughly proven. Unfortunately, this does not yet explain why a particular toxin would allow virulence only on a particular host plant. For the ACR toxin necessary for virulence of the *A. alternata* rough lemon pathotype, part of the functional mechanism was unraveled. Toxicity on rough citrus (*Citrus jambhiri* Lush.) was found to depend on the differential post-transcriptional processing of transcripts of the mitochondrial *ACRS* (ACR-toxin sensitivity) gene, which is present in both toxin-sensitive and toxin-insensitive citrus but processed to shorter transcripts in mitochondria of insensitive plants [106]. This work showed that host specificity of the rough lemon pathotype of *A. alternata* towards its host is due to altered mitochondrial RNA processing. In *Alternaria brassicicola*, that causes black spot on *Brassica* plants, the essential contribution of the AB-toxin to virulence is well known. It was shown that a host-derived factor, an oligosaccharide of 1.3 kDa, secreted from the plant just after *A. brassicicola* spore germination, was necessary to induce AB-toxin production [107] showing the involvement of a host-derived factor in the production of host-selective toxins.

9. Conclusions

A plethora of methods has been developed to help in the identification of host-specificity factors. While most molecular methods are excellent for describing molecular differences between host-selective strains, few are suited for rapid comparative identification of genes involved in host-specificity. With the development of next-generation sequencing technologies, at least the relatively rapid prediction of potential target genes now becomes possible. However, prediction alone is not enough. Without validation of functional involvement in host-specificity, the underlying functional mechanisms cannot be resolved. Resolving functional mechanisms is still a slow and challenging task, involving rigorous scientific work and extensive experimentation.

In most cases, in spite of great progress in the description of molecular differences of host-specific strains and in the prediction of genes possibly involved in host selection, we are still far from understanding how host-specific infection is achieved. What is the reason for this apparent lack of progress? Are the available methods insufficient? Do researchers just not search hard enough? Rapid identification of host-specificity factors could be hampered by the specific characteristics of the investigated system. So far, host-specificity factors have been identified only in systems where host-specificity depended on a single gene. However, this might not represent the majority of the cases, and adaptation might depend on more than just the presence of one gene or one altered amino acid in the other systems. In a lot of cases, secreted fungal proteins were suspected to modulate interaction with the plant and be the reason for being able to spread only on a specific host plant. However, this might not be a general solution. Most genes identified as being under positive selection pressure in the two *formae speciales* of *S. reilianum* encoded proteins that are internal to *S. reilianum* [108], suggesting that the capacity to proliferate successfully on a specific host might depend on adapted metabolic capacities. On the other hand, it could be that host-specific differences are explained not by the presence of particular genetic markers in the host-specific strains but are dependent on host-specific gene expression. In this case, identification of host-specificity factors will likely not be successful if considering only genome sequences for comparison, and gene-expression changes may need to be considered. Host-specific changes of fungal gene expression might not even be encoded in the pathogen genomes. As was shown in the *B. cinerea-Arabidopsis thaliana* interaction, small RNAs of the plant may migrate into the pathogen to regulate fungal gene expression [109]. If this turns out to be a general principle, most of the molecular comparative studies aimed at identifying host-specific molecular differences in the fungal pathogens are doomed to fail.

In spite of significant progress, the battle for understanding the principles of host selection in fungal plant–pathogen interaction is not yet won. A variety of comparative methods is available to help in the identification of host-specificity factors. An intelligent combination of classical genetics and next-generation sequencing in the consideration of gene-expression changes of both the pathogen and host may be needed in order to unravel the mechanistic basis of host-specificity of plant-pathogenic fungi.

Acknowledgments: This work was supported by grants from the German Research Foundation (DFG) to Jan Schirawski.

Author Contributions: Jan Schirawski conceived the paper, Nilam Borah conducted the literature research, Nilam Borah and Jan Schirawski wrote the paper, and Emad Albarouki prepared Figure 2, provided Table 1 and reviewed the text.

Conflicts of Interest: The authors declare no conflict of interest. The founding sponsors had no role in the design of the study; in the collection, analyses, or interpretation of data; in the writing of the manuscript; and in the decision to publish the results.

References

1. Flor, H.H. Current status of the gene-for-gene concept. *Annu. Rev. Phytopathol.* **1971**, *9*, 275–296. [CrossRef]
2. Van Der Biezen, E.A.; Jones, J.D.G. Plant disease-resistance proteins and the gene-for-gene concept. *Trends Biochem. Sci.* **1998**, *23*, 454–456. [CrossRef]
3. Van der Hoorn, R.A.L.; Kamoun, S. From guard to decoy: A new model for perception of plant pathogen effectors. *Plant Cell* **2008**, *20*, 2009–2017. [CrossRef] [PubMed]
4. Chisholm, S.T.; Coaker, G.; Day, B.; Staskawicz, B.J. Host-microbe interactions: Shaping the evolution of the plant immune response. *Cell* **2006**, *124*, 803–814. [CrossRef] [PubMed]
5. De Jonge, R.; Bolton, M.D.; Thomma, B.P.H.J. How filamentous pathogens co-opt plants: The ins and outs of fungal effectors. *Curr. Opin. Plant Biol.* **2011**, *14*, 400–406. [CrossRef] [PubMed]
6. Petit-Houdenot, Y.; Fudal, I. Complex interactions between fungal avirulence genes and their corresponding plant resistance genes and consequences for disease resistance management. *Front. Plant Sci.* **2017**, *8*, 1072. [CrossRef] [PubMed]
7. Inoue, Y.; Vy, T.T.P.; Yoshida, K.; Asano, H.; Mitsuoka, C.; Asuke, S.; Anh, V.L.; Cumagun, C.J.R.; Chuma, I.; Terauchi, R.; et al. Evolution of the wheat blast fungus through functional losses in a host specificity determinant. *Science* **2017**, *357*, 80–83. [CrossRef] [PubMed]
8. Meena, M.; Gupta, S.K.; Swapnil, P.; Zehra, A.; Dubey, M.K.; Upadhyay, R.S. Alternaria toxins: Potential virulence factors and genes related to pathogenesis. *Front. Microbiol.* **2017**, *8*, 1451. [CrossRef] [PubMed]
9. Hatta, R.; Ito, K.; Hosaki, Y.; Tanaka, T.; Tanaka, A.; Yamamoto, M.; Akimitsu, K.; Tsuge, T. A conditionally dispensable chromosome controls host-specific pathogenicity in the fungal plant pathogen *Alternaria alternata*. *Genetics* **2002**, *161*, 59–70. [PubMed]
10. Tsuge, T.; Harimoto, Y.; Hanada, K.; Akagi, Y.; Kodama, M.; Akimitsu, K.; Yamamoto, M. Evolution of pathogenicity controlled by small, dispensable chromosomes in *Alternaria alternata* pathogens. *Physiol. Mol. Plant Pathol.* **2016**, *95*, 27–31. [CrossRef]
11. Akagi, Y.; Akamatsu, H.; Otani, H.; Kodama, M. Horizontal chromosome transfer, a mechanism for the evolution and differentiation of a plant-pathogenic fungus. *Eukaryot. Cell* **2009**, *8*, 1732–1738. [CrossRef] [PubMed]
12. Sidhu, G.; Person, C. Genetic control of virulence in *Ustilago hordei* III. Identification of genes for host resistance and demonstration of gene-for-gene relations. *Can. J. Genet. Cytol.* **1972**, *14*, 209–213. [CrossRef]
13. Linning, R.; Lin, D.; Lee, N.; Abdennadher, M.; Gaudet, D.; Thomas, P.; Mills, D.; Kronstad, J.W.; Bakkeren, G. Marker-based cloning of the region containing the *UhavrI* avirulence gene from the basidiomycete barley pathogen *Ustilago hordei*. *Genetics* **2004**, *166*, 99–111. [CrossRef] [PubMed]

14. Ali, S.; Laurie, J.D.; Linning, R.; Cervantes-Chavez, J.A.; Gaudet, D.; Bakkeren, G. An immunity-triggering effector from the barley smut fungus *Ustilago hordei* resides in an Ustilaginaceae-specific cluster bearing signs of transposable element-assisted evolution. *PLoS Pathog.* **2014**, *10*, e1004223. [CrossRef] [PubMed]

15. Plett, J.M.; Kemppainen, M.; Kale, S.D.; Kohler, A.; Legue, V.; Brun, A.; Tyler, B.M.; Pardo, A.G.; Martin, F. A secreted effector protein of *Laccaria bicolor* is required for symbiosis development. *Curr. Biol.* **2011**, *21*, 1197–1203. [CrossRef] [PubMed]

16. Van der Does, H.C.; Rep, M. Virulence genes and the evolution of host specificity in plant-pathogenic fungi. *Mol. Plant Microbe Interact.* **2007**, *20*, 1175–1182. [CrossRef] [PubMed]

17. Mehrabi, R.; Bahkali, A.H.; Abd-Elsalam, K.A.; Moslem, M.; M'barek, S.B.; Gohari, A.M.; Jashni, M.K.; Stergiopoulos, I.; Kema, G.H.; de Wit, P.J. Horizontal gene and chromosome transfer in plant pathogenic fungi affecting host range. *FEMS Microbiol. Rev.* **2011**, *35*, 542–554. [CrossRef] [PubMed]

18. Andrie, R.M.; Pandelova, I.; Ciuffetti, L.M. A combination of phenotypic and genotypic characterization strengthens *Pyrenophora tritici-repentis* race identification. *Phytopathology* **2007**, *97*, 694–701. [CrossRef] [PubMed]

19. Chiapello, H.; Mallet, L.; Guerin, C.; Aguileta, G.; Amselem, J.; Kroj, T.; Ortega-Abboud, E.; Lebrun, M.H.; Henrissat, B.; Gendrault, A.; et al. Deciphering genome content and evolutionary relationships of isolates from the fungus *Magnaporthe oryzae* attacking different host plants. *Genome Biol. Evol.* **2015**, *7*, 2896–2912. [CrossRef] [PubMed]

20. Baayen, R.P.; O'Donnell, K.; Bonants, P.J.M.; Cigelnik, E.; Kroon, L.P.N.M.; Roebroeck, E.J.A.; Waalwijk, C. Gene genealogies and AFLP analyses in the *Fusarium oxysporum* complex identify monophyletic and nonmonophyletic formae speciales causing wilt and rot disease. *Phytopathology* **2000**, *90*, 891–900. [CrossRef] [PubMed]

21. Chen, C.-H.; Sheu, Z.-M.; Wang, T.-C. Host specificity and tomato-related race composition of *Phytophthora infestans* isolates in Taiwan during 2004 and 2005. *Plant Dis.* **2008**, *92*, 751–755. [CrossRef]

22. Kovacikova, E.; Kratka, J. Different manifestations of the pathogenity of some strains of *Fusarium oxysporum* f. sp. *pisi*. *Zentralbl. Bakteriol. Naturwiss.* **1979**, *134*, 159–166. [PubMed]

23. Heath, M.C. Host species specificity of the goldenrod rust fungus and the existence of rust resistance within some Goldenrod species. *Can. J. Bot.* **1992**, *70*, 2461–2466. [CrossRef]

24. Bernstein, B.; Zehr, E.I.; Dean, R.A.; Shabi, E. Characteristics of colletotrichum from peach, apple, pecan, and other hosts. *Plant Dis.* **1995**, *79*, 478–482. [CrossRef]

25. Orlicz-Luthardt, A.; Rieckmann, U.; Dercks, W. Pathogenicity of some species of *Fusarium* and formae speciales to china aster and carnation. *Gartenbauwissenschaft* **2000**, *65*, 137–143.

26. Peever, T.L.; Olsen, L.; Ibanez, A.; Timmer, L.W. Genetic differentiation and host specificity among populations of *Alternaria* spp. causing brown spot of grapefruit and tangerine × grapefruit hybrids in Florida. *Phytopathology* **2000**, *90*, 407–414. [CrossRef] [PubMed]

27. Yehuda, P.B.; Eilam, T.; Manisterski, J.; Shimoni, A.; Anikster, Y. Leaf rust on *Aegilops speltoides* caused by a new forma specialis of *Puccinia triticina*. *Phytopathology* **2004**, *94*, 94–101. [CrossRef] [PubMed]

28. Hernandez-Bello, M.A.; Chilvers, M.I.; Akamatsu, H.; Peever, T.L. Host specificity of *Ascochyta* spp. infecting legumes of the Viciae and Cicerae tribes and pathogenicity of an interspecific hybrid. *Phytopathology* **2006**, *96*, 1148–1156. [CrossRef] [PubMed]

29. Llorens, A.; Hinojo, M.J.; Mateo, R.; Gonzalez-Jaen, M.T.; Valle-Algarra, F.M.; Logrieco, A.; Jimenez, M. Characterization of *Fusarium* spp. isolates by PCR-RFLP analysis of the intergenic spacer region of the rRNA gene (rDNA). *Int. J. Food Microbiol.* **2006**, *106*, 297–306. [CrossRef] [PubMed]

30. Hedh, J.; Johansson, T.; Tunlid, A. Variation in host specificity and gene content in strains from genetically isolated lineages of the ectomycorrhizal fungus *Paxillus involutus* s. lat. *Mycorrhiza* **2009**, *19*, 549–558. [CrossRef] [PubMed]

31. Greco, M.; Patriarca, A.; Terminiello, L.; Fernandez Pinto, V.; Pose, G. Toxigenic *Alternaria* species from Argentinean blueberries. *Int. J. Food Microbiol.* **2012**, *154*, 187–191. [CrossRef] [PubMed]

32. Wolfe, B.E.; Pringle, A. Geographically structured host specificity is caused by the range expansions and host shifts of a symbiotic fungus. *ISME J.* **2012**, *6*, 745–755. [CrossRef] [PubMed]

33. Zuther, K.; Kahnt, J.; Utermark, J.; Imkampe, J.; Uhse, S.; Schirawski, J. Host specificity of *Sporisorium reilianum* is tightly linked to generation of the phytoalexin luteolinidin by *Sorghum bicolor*. *Mol. Plant Microbe Interact.* **2012**, *25*, 1230–1237. [CrossRef] [PubMed]

34. Beckstead, J.; Meyer, S.E.; Ishizuka, T.S.; McEvoy, K.M.; Coleman, C.E. Lack of host specialization on winter annual grasses in the fungal seed bank pathogen *Pyrenophora semeniperda*. *PLoS ONE* **2016**, *11*, e0151058. [CrossRef] [PubMed]

35. Yasuda, K.; Kojima, M. The role of stress metabolites in establishing host-parasite specificity between sweet potato and *Ceratocystis fimbriata*, black rot fungus. *Agric. Biol. Chem.* **1986**, *50*, 1839–1846. [CrossRef]

36. Masunaka, A.; Ohtani, K.; Peever, T.L.; Timmer, L.W.; Tsuge, T.; Yamamoto, M.; Yamamoto, H.; Akimitsu, K. An isolate of *Alternaria alternata* that is pathogenic to both tangerines and rough lemon and produces two host-selective toxins, ACT- and ACR-toxins. *Phytopathology* **2005**, *95*, 241–247. [CrossRef] [PubMed]

37. Zhan, J.; Kema, G.H.J.; Waalwijk, C.; McDonald, B.A. Distribution of mating type alleles in the wheat pathogen *Mycosphaerella graminicola* over spatial scales from lesions to continents. *Fungal Genet. Biol.* **2002**, *36*, 128–136. [CrossRef]

38. Sone, T.; Suto, M.; Tomita, F. Host species-specific repetitive DNA sequence in the genome of *Magnaporthe grisea*, the rice blast fungus. *Biosci. Biotechnol. Biochem.* **1993**, *57*, 1228–1230. [CrossRef] [PubMed]

39. Maharaj, A.; Rampersad, S.N. Genetic differentiation of *Colletotrichum gloeosporioides* and *C. truncatum* associated with anthracnose disease of papaya (*Carica papaya* L.) and bell pepper (*Capsium annuum* L.) based on ITS PCR-RFLP fingerprinting. *Mol. Biotechnol.* **2012**, *50*, 237–249. [CrossRef] [PubMed]

40. Levis, C.; Giraud, T.; Dutertre, M.; Fortini, D.; Brygoo, Y. Telomeric DNA of *Botrytis cinerea*: A useful tool for strain identification. *FEMS Microbiol. Lett.* **1997**, *157*, 267–272. [CrossRef] [PubMed]

41. Viji, G.; Wu, B.; Kang, S.; Uddin, W.; Huff, D.R. *Pyricularia grisea* causing gray leaf spot of perennial ryegrass turf: Population structure and host specificity. *Plant Dis.* **2001**, *85*, 817–826. [CrossRef]

42. Quecine, M.C.; Bini, A.P.; Romagnoli, E.R.; Andreote, F.D.; Moon, D.H.; Labate, C.A. Genetic variability in *Puccinia psidii* populations as revealed by PCR-DGGE and T-RFLP markers. *Plant Dis.* **2014**, *98*, 16–23. [CrossRef]

43. Zaffarano, P.L.; McDonald, B.A.; Linde, C.C. Two new species of Rhynchosporium. *Mycologia* **2011**, *103*, 195–202. [CrossRef] [PubMed]

44. Attitalla, I.H.; Fatehi, J.; Levenfors, J.; Brishammar, S. A rapid molecular method for differentiating two special forms (*lycopersici* and *radicis-lycopersici*) of *Fusarium oxysporum*. *Mycol. Res.* **2004**, *108*, 787–794. [CrossRef] [PubMed]

45. Nechwatal, J.; Wielgoss, A.; Mendgen, K. Diversity, host, and habitat specificity of oomycete communities in declining reed stands (*Phragmites australis*) of a large freshwater lake. *Mycol. Res.* **2008**, *112*, 689–696. [CrossRef] [PubMed]

46. Oyarzun, P.J.; Pozo, A.; Ordonez, M.E.; Doucett, K.; Forbes, G.A. Host Specificity of *Phytophthora infestans* on tomato and potato in Ecuador. *Phytopathology* **1998**, *88*, 265–271. [CrossRef] [PubMed]

47. Garry, G.; Forbes, G.A.; Salas, A.; Santa Cruz, M.; Perez, W.G.; Nelson, R.J. Genetic diversity and host differentiation among isolates of *Phytophthora infestans* from cultivated potato and wild solanaceous hosts in Peru. *Plant Pathol.* **2005**, *54*, 740–748. [CrossRef]

48. Vos, P.; Hogers, R.; Bleeker, M.; Reijans, M.; van de Lee, T.; Hornes, M.; Frijters, A.; Pot, J.; Peleman, J.; Kuiper, M.; et al. AFLP: A new technique for DNA fingerprinting. *Nucleic Acids Res.* **1995**, *23*, 4407–4414. [CrossRef] [PubMed]

49. Douhan, G.W.; de la Cerda, K.A.; Huryn, K.L.; Greer, C.A.; Wong, F.P. Contrasting genetic structure between *Magnaporthe grisea* populations associated with the golf course turfgrasses *Lolium perenne* (perennial ryegrass) and *Pennisetum clandestinum* (kikuyugrass). *Phytopathology* **2011**, *101*, 85–91. [CrossRef] [PubMed]

50. Flier, W.G.; Grünwald, N.J.; Kroon, L.P.N.M.; Sturbaum, A.K.; van den Bosch, T.B.M.; Garay-Serrano, E.; Lozoya-Saldaña, H.; Fry, W.E.; Turkensteen, L.J. The population structure of *Phytophthora infestans* from the toluca valley of central Mexico suggests genetic differentiation between populations from cultivated potato and wild *Solanum* spp. *Phytopathology* **2003**, *93*, 382–390. [CrossRef] [PubMed]

51. Karimi, S.; Mirlohi, A.; Sabzalian, M.R.; Sayed Tabatabaei, B.E.; Sharifnabi, B. Molecular evidence for Neotyphodium fungal endophyte variation and specificity within host grass species. *Mycologia* **2012**, *104*, 1281–1290. [CrossRef] [PubMed]

52. Wyand, R.A.; Brown, J.K. Genetic and forma specialis diversity in *Blumeria graminis* of cereals and its implications for host-pathogen co-evolution. *Mol. Plant Pathol.* **2003**, *4*, 187–198. [CrossRef] [PubMed]

53. Liu, B.; Wasilwa, L.A.; Morelock, T.E.; O'Neill, N.R.; Correll, J.C. Comparison of *Colletotrichum orbiculare* and several allied *Colletotrichum* spp. for mtDNA RFLPs, intron RFLP and sequence variation, vegetative compatibility, and host specificity. *Phytopathology* **2007**, *97*, 1305–1314. [CrossRef] [PubMed]

54. Rigotti, S.; Gindro, K.; Richter, H.; Viret, O. Characterization of molecular markers for specific and sensitive detection of *Botrytis cinerea* Pers.: Fr. in strawberry (*Fragariaxananassa duch.*) using PCR. *FEMS Microbiol. Lett.* **2002**, *209*, 169–174. [CrossRef]

55. Mahmodi, F.; Kadir, J.B.; Puteh, A.; Pourdad, S.S.; Nasehi, A.; Soleimani, N. Genetic diversity and differentiation of *Colletotrichum* spp. isolates associated with Leguminosae using multigene loci, RAPD and ISSR. *Plant Pathol.* **2014**, *30*, 10–24. [CrossRef] [PubMed]

56. Peever, T.L.; Canihos, Y.; Olsen, L.; Ibanez, A.; Liu, Y.C.; Timmer, L.W. Population genetic structure and host specificity of *Alternaria* spp. causing brown spot of minneola tangelo and rough lemon in Florida. *Phytopathology* **1999**, *89*, 851–860. [CrossRef] [PubMed]

57. Lievens, B.; Claes, L.; Vakalounakis, D.J.; Vanachter, A.C.; Thomma, B.P. A robust identification and detection assay to discriminate the cucumber pathogens *Fusarium oxysporum* f. sp. *cucumerinum* and f. sp. *radicis-cucumerinum. Environ. Microbiol.* **2007**, *9*, 2145–2161.

58. King, K.M.; West, J.S.; Brunner, P.C.; Dyer, P.S.; Fitt, B.D. Evolutionary relationships between *Rhynchosporium lolii* sp. nov. and other Rhynchosporium species on grasses. *PLoS ONE* **2013**, *8*, e72536. [CrossRef] [PubMed]

59. Zhou, Y.; Chao, G.M.; Liu, J.J.; Zhu, M.Q.; Wang, Y.; Feng, B.L. Genetic diversity of *Ustilago hordei* in Tibetan areas as revealed by RAPD and SSR. *J. Integr. Agric.* **2016**, *15*, 2299–2308. [CrossRef]

60. Perlin, M.H.; Hughes, C.; Welch, J.; Akkaraju, S.; Steinecker, D.; Kumar, A.; Smith, B.; Garr, S.S.; Brown, S.A.; Andom, T. Molecular approaches to differentiate subpopulations or formae speciales of the fungal phytopathogen *Microbotryum violaceum. Int. J. Plant Sci.* **1997**, *158*, 568–574. [CrossRef]

61. Santini, A.; Capretti, P. Analysis of the Italian population of *Ceratocystis fimbriata* f.sp. *platani* using RAPD and minisatellite markers. *Plant Pathol.* **2000**, *49*, 461–467.

62. Freeman, S.; Katan, T.; Shabi, E. Characterization of *Colletotrichum gloeosporioides* isolates from avocado and almond fruits with molecular and pathogenicity tests. *Appl. Environ. Microbiol.* **1996**, *62*, 1014–1020. [PubMed]

63. Asadollahi, M.; Fekete, E.; Karaffa, L.; Flipphi, M.; Arnyasi, M.; Esmaeili, M.; Vaczy, K.Z.; Sandor, E. Comparison of *Botrytis cinerea* populations isolated from two open-field cultivated host plants. *Microbiol. Res.* **2013**, *168*, 379–388. [CrossRef] [PubMed]

64. Van Putten, W.F.; Biere, A.; Van Damme, J.M.M. Host-related genetic differentiation in the anther smut fungus *Microbotryum violaceum* in sympatric, parapatric and allopatric populations of two host species *Silene latifolia* and *S. dioica. J. Evol. Biol.* **2005**, *18*, 203–212. [CrossRef] [PubMed]

65. Dipietro, A.; Anaya, N.; Roncero, M.I.G. Occurrence of a retrotransposon-like sequence among different formae speciales and races of *Fusarium oxysporum. Mycol. Res.* **1994**, *98*, 993–996. [CrossRef]

66. Samuel, S.; Veloukas, T.; Papavasileiou, A.; Karaoglanidis, G.S. Differences in frequency of transposable elements presence in *Botrytis cinerea* populations from several hosts in Greece. *Plant Dis.* **2012**, *96*, 1286–1290. [CrossRef]

67. Laurie, J.D.; Ali, S.; Linning, R.; Mannhaupt, G.; Wong, P.; Guldener, U.; Munsterkotter, M.; Moore, R.; Kahmann, R.; Bakkeren, G.; et al. Genome comparison of barley and maize smut fungi reveals targeted loss of RNA silencing components and species-specific presence of transposable elements. *Plant Cell* **2012**, *24*, 1733–1745. [CrossRef] [PubMed]

68. Schirawski, J.; Mannhaupt, G.; Munch, K.; Brefort, T.; Schipper, K.; Doehlemann, G.; Di Stasio, M.; Rossel, N.; Mendoza-Mendoza, A.; Pester, D.; et al. Pathogenicity determinants in smut fungi revealed by genome comparison. *Science* **2010**, *330*, 1546–1548. [CrossRef] [PubMed]

69. Dutheil, J.Y.; Mannhaupt, G.; Schweizer, G.; Sieber, C.M.K.; Munsterkotter, M.; Guldener, U.; Schirawski, J.; Kahmann, R. A tale of genome compartmentalization: The evolution of virulence lusters in smut fungi. *Genome Biol. Evol.* **2016**, *8*, 681–704. [CrossRef] [PubMed]

70. Buiate, E.A.; Xavier, K.V.; Moore, N.; Torres, M.F.; Farman, M.L.; Schardl, C.L.; Vaillancourt, L.J. A comparative genomic analysis of putative pathogenicity genes in the host-specific sibling species *Colletotrichum graminicola* and *Colletotrichum sublineola. BMC Genom.* **2017**, *18*, 67. [CrossRef] [PubMed]

71. Penselin, D.; Munsterkotter, M.; Kirsten, S.; Felder, M.; Taudien, S.; Platzer, M.; Ashelford, K.; Paskiewicz, K.H.; Harrison, R.J.; Hughes, D.J.; et al. Comparative genomics to explore phylogenetic relationship, cryptic sexual potential and host specificity of Rhynchosporium species on grasses. *BMC Genom.* **2016**, *17*, 953. [CrossRef] [PubMed]

72. Niehaus, E.M.; Munsterkotter, M.; Proctor, R.H.; Brown, D.W.; Sharon, A.; Idan, Y.; Oren-Young, L.; Sieber, C.M.; Novak, O.; Pencik, A.; et al. Comparative "Omics" of the *Fusarium fujikuroi* species complex highlights differences in genetic potential and metabolite synthesis. *Genome Biol. Evol.* **2016**, *8*, 3574–3599. [CrossRef] [PubMed]

73. Baroncelli, R.; Amby, D.B.; Zapparata, A.; Sarrocco, S.; Vannacci, G.; Le Floch, G.; Harrison, R.J.; Holub, E.; Sukno, S.A.; Sreenivasaprasad, S.; et al. Gene family expansions and contractions are associated with host range in plant pathogens of the genus Colletotrichum. *BMC Genom.* **2016**, *17*, 555. [CrossRef] [PubMed]

74. Sharma, R.; Mishra, B.; Runge, F.; Thines, M. Gene loss rather than gene gain is associated with a host jump from monocots to dicots in the smut fungus *Melanopsichium pennsylvanicum*. *Genome Biol. Evol.* **2014**, *6*, 2034–2049. [CrossRef] [PubMed]

75. Yan, J.Y.; Zhao, W.S.; Chen, Z.; Xing, Q.K.; Zhang, W.; Chethana, K.W.T.; Xue, M.F.; Xu, J.P.; Phillips, A.J.L.; Wang, Y.; et al. Comparative genome and transcriptome analyses reveal adaptations to opportunistic infections in woody plant degrading pathogens of Botryosphaeriaceae. *DNA Res.* **2018**, *25*, 87–102. [CrossRef] [PubMed]

76. Ma, L.-J.; van der Does, H.C.; Borkovich, K.A.; Coleman, J.J.; Daboussi, M.-J.; Di Pietro, A.; Dufresne, M.; Freitag, M.; Grabherr, M.; Henrissat, B.; et al. Comparative genomics reveals mobile pathogenicity chromosomes in Fusarium. *Nature* **2010**, *464*, 367–373. [CrossRef] [PubMed]

77. Williams, A.H.; Sharma, M.; Thatcher, L.F.; Azam, S.; Hane, J.K.; Sperschneider, J.; Kidd, B.N.; Anderson, J.P.; Ghosh, R.; Garg, G.; et al. Comparative genomics and prediction of conditionally dispensable sequences in legume-infecting *Fusarium oxysporum* formae speciales facilitates identification of candidate effectors. *BMC Genom.* **2016**, *17*, 191. [CrossRef] [PubMed]

78. Van Dam, P.; Fokkens, L.; Schmidt, S.M.; Linmans, J.H.; Kistler, H.C.; Ma, L.J.; Rep, M. Effector profiles distinguish formae speciales of *Fusarium oxysporum*. *Environ. Microbiol.* **2016**, *18*, 4087–4102. [CrossRef] [PubMed]

79. Van der Does, H.C.; Lievens, B.; Claes, L.; Houterman, P.M.; Cornelissen, B.J.; Rep, M. The presence of a virulence locus discriminates Fusarium oxysporum isolates causing tomato wilt from other isolates. *Environ. Microbiol.* **2008**, *10*, 1475–1485. [CrossRef] [PubMed]

80. Lievens, B.; Houterman, P.M.; Rep, M. Effector gene screening allows unambiguous identification of *Fusarium oxysporum* f. sp. *lycopersici* races and discrimination from other formae speciales. *FEMS Microbiol. Lett.* **2009**, *300*, 201–215. [CrossRef] [PubMed]

81. Bhadauria, V.; MacLachlan, R.; Pozniak, C.; Banniza, S. Candidate effectors contribute to race differentiation and virulence of the lentil anthracnose pathogen *Colletotrichum lentis*. *BMC Genom.* **2015**, *16*, 628. [CrossRef] [PubMed]

82. Qin, S.; Ji, C.; Li, Y.; Wang, Z. Comparative transcriptomic analysis of race 1 and race 4 of *Fusarium oxysporum* f. sp. *cubense* induced with different carbon sources. *G3 (Bethesda)* **2017**, *7*, 2125–2138. [CrossRef] [PubMed]

83. Kema, G.H.J.; Gohari, A.M.; Aouini, L.; Gibriel, H.A.Y.; Ware, S.B.; van den Bosch, F.; Manning-Smith, R.; Alonso-Chavez, V.; Helps, J.; Ben, M’ Barek, S.; et al. Stress and sexual reproduction affect the dynamics of the wheat pathogen effector AvrStb6 and strobilurin resistance. *Nat. Genet.* **2018**. [CrossRef] [PubMed]

84. Hosseini, S.; Resjo, S.; Liu, Y.; Durling, M.; Heyman, F.; Levander, F.; Liu, Y.; Elfstrand, M.; Funck Jensen, D.; Andreasson, E.; et al. Comparative proteomic analysis of hyphae and germinating cysts of *Phytophthora pisi* and *Phytophthora sojae*. *J. Proteom.* **2015**, *117*, 24–40. [CrossRef] [PubMed]

85. Savidor, A.; Donahoo, R.S.; Hurtado-Gonzales, O.; Land, M.L.; Shah, M.B.; Lamour, K.H.; McDonald, W.H. Cross-species global proteomics reveals conserved and unique processes in *Phytophthora sojae* and *Phytophthora ramorum*. *Mol. Cell. Proteom.* **2008**, *7*, 1501–1516. [CrossRef] [PubMed]

86. Quecine, M.C.; Leite, T.F.; Bini, A.P.; Regiani, T.; Franceschini, L.M.; Budzinski, I.G.F.; Marques, F.G.; Labate, M.T.V.; Guidetti-Gonzalez, S.; Moon, D.H.; et al. Label-free quantitative proteomic analysis of *Puccinia psidii* uredospores reveals differences of fungal populations infecting eucalyptus and guava. *PLoS ONE* **2016**, *11*, e0145343. [CrossRef] [PubMed]

87. Bregar, O.; Mandelc, S.; Celar, F.; Javornik, B. Proteome analysis of the plant pathogenic fungus *Monilinia laxa* showing host specificity. *Food Technol. Biotechnol.* **2012**, *50*, 326–333.

88. Cisar, C.R.; Spiegel, F.W.; TeBeest, D.O.; Trout, C. Evidence for mating between isolates of *Colletotrichum gloeosporioides* with different host specificities. *Curr. Genet.* **1994**, *25*, 330–335. [CrossRef] [PubMed]

89. Cozijnsen, A.J.; Popa, K.M.; Purwantara, A.; Rolls, B.D.; Howlett, B.J. Genome analysis of the plant pathogenic ascomycete *Leptosphaeria maculans*; mapping mating type and host specificity loci. *Mol. Plant Pathol.* **2000**, *1*, 293–302. [CrossRef] [PubMed]

90. Tosa, Y.; Tamba, H.; Tanaka, K.; Mayama, S. Genetic analysis of host species specificity of *Magnaporthe oryzae* isolates from rice and wheat. *Phytopathology* **2006**, *96*, 480–484. [CrossRef] [PubMed]

91. Murakami, J.; Tomita, R.; Kataoka, T.; Nakayashiki, H.; Tosa, Y.; Mayama, S. Analysis of host species specificity of *Magnaporthe grisea* toward foxtail millet using a genetic cross between isolates from wheat and foxtail millet. *Phytopathology* **2003**, *93*, 42–45. [CrossRef] [PubMed]

92. Foulongne-Oriol, M. Genetic linkage mapping in fungi: Current state, applications, and future trends. *Appl. Microbiol. Biotechnol.* **2012**, *95*, 891–904. [CrossRef] [PubMed]

93. Poloni, A.; Schirawski, J. Host specificity in *Sporisorium reilianum* is determined by distinct mechanisms in maize and sorghum. *Mol. Plant Pathol.* **2016**, *17*, 741–754. [CrossRef] [PubMed]

94. Valent, B.; Farrall, L.; Chumley, F.G. *Magnaporthe grisea* genes for pathogenicity and virulence identified through a series of backcrosses. *Genetics* **1991**, *127*, 87–101. [PubMed]

95. Kang, S.; Sweigard, J.A.; Valent, B. The PWL host specificity gene family in the blast fungus *Magnaporthe grisea*. *Mol. Plant Microbe Interact.* **1995**, *8*, 939–948. [CrossRef] [PubMed]

96. Sweigard, J.A.; Carroll, A.M.; Kang, S.; Farrall, L.; Chumley, F.G.; Valent, B. Identification, cloning, and characterization of PWL2, a gene for host species specificity in the rice blast fungus. *Plant Cell* **1995**, *7*, 1221–1233. [CrossRef] [PubMed]

97. Peyyala, R.; Farman, M.L. *Magnaporthe oryzae* isolates causing gray leaf spot of perennial ryegrass possess a functional copy of the AVR1-CO39 avirulence gene. *Mol. Plant Pathol.* **2006**, *7*, 157–165. [CrossRef] [PubMed]

98. Rep, M.; Meijer, M.; Houterman, P.M.; van der Does, H.C.; Cornelissen, B.J. *Fusarium oxysporum* evades I-3-mediated resistance without altering the matching avirulence gene. *Mol. Plant Microbe Interact.* **2005**, *18*, 15–23. [CrossRef] [PubMed]

99. Rep, M.; van der Does, H.C.; Meijer, M.; van Wijk, R.; Houterman, P.M.; Dekker, H.L.; de Koster, C.G.; Cornelissen, B.J. A small, cysteine-rich protein secreted by *Fusarium oxysporum* during colonization of xylem vessels is required for I-3-mediated resistance in tomato. *Mol. Microbiol.* **2004**, *53*, 1373–1383. [CrossRef] [PubMed]

100. Li, E.; Wang, G.; Xiao, J.; Ling, J.; Yang, Y.; Xie, B. A SIX1 homolog in *Fusarium oxysporum* f. sp. *conglutinans* is required for full virulence on cabbage. *PLoS ONE* **2016**, *11*, e0152273.

101. Poppe, S.; Dorsheimer, L.; Happel, P.; Stukenbrock, E.H. Rapidly evolving genes are key players in host specialization and virulence of the fungal wheat pathogen *Zymoseptoria tritici* (*Mycosphaerella graminicola*). *PLOS Pathog.* **2015**, *11*, e1005055. [CrossRef] [PubMed]

102. Isshiki, A.; Akimitsu, K.; Yamamoto, M.; Yamamoto, H. Endopolygalacturonase is essential for citrus black rot caused by *Alternaria citri* but not brown spot caused by *Alternaria alternata*. *Mol. Plant Microbe Interact.* **2001**, *14*, 749–757. [CrossRef] [PubMed]

103. Andrie, R.M.; Ciuffetti, L.M. *Pyrenophora bromi*, causal agent of brownspot of bromegrass, expresses a gene encoding a protein with homology and similar activity to Ptr ToxB, a host-selective toxin of wheat. *Mol. Plant Microbe Interact.* **2011**, *24*, 359–367. [CrossRef] [PubMed]

104. Birch, P.R.; Armstrong, M.; Bos, J.; Boevink, P.; Gilroy, E.M.; Taylor, R.M.; Wawra, S.; Pritchard, L.; Conti, L.; Ewan, R.; et al. Towards understanding the virulence functions of RXLR effectors of the oomycete plant pathogen *Phytophthora infestans*. *J. Exp. Bot.* **2009**, *60*, 1133–1140. [CrossRef] [PubMed]

105. Dong, S.M.; Stam, R.; Cano, L.M.; Song, J.; Sklenar, J.; Yoshida, K.; Bozkurt, T.O.; Oliva, R.; Liu, Z.Y.; Tian, M.Y.; et al. Effector specialization in a lineage of the Irish potato famine pathogen. *Science* **2014**, *343*, 552–555. [CrossRef] [PubMed]

106. Ohtani, K.; Yamamoto, H.; Akimitsu, K. Sensitivity to *Alternaria alternata* toxin in citrus because of altered mitochondrial RNA processing. *Proc. Natl. Acad. Sci. USA* **2002**, *99*, 2439–2444. [CrossRef] [PubMed]

107. Oka, K.; Akamatsu, H.; Kodama, M.; Nakajima, H.; Kawada, T.; Otani, H. Host-specific AB-toxin production by germinating spores of *Alternaria brassicicola* is induced by a host-derived oligosaccharide. *Physiol. Mol. Plant Pathol.* **2005**, *66*, 12–19. [CrossRef]

108. Schweizer, G.; Münch, K.; Mannhaupt, G.; Schirawski, J.; Kahmann, R.; Dutheil, J.Y. Positively selected effector genes and their contribution to virulence in the smut fungus *Sporisorium reilianum*. *Genome Biol. Evol.* **2018**, *10*, 629–645. [CrossRef] [PubMed]

109. Wang, M.; Weiberg, A.; Dellota, E.; Yamane, D.; Jin, H. Botrytis small RNA Bc-siR37 suppresses plant defense genes by cross-kingdom RNAi. *RNA Biol.* **2017**, *14*, 421–428. [CrossRef] [PubMed]

International Journal of
Molecular Sciences

MDPI

Article

Identification and Initial Characterization of the Effectors of an Anther Smut Fungus and Potential Host Target Proteins

Venkata S. Kuppireddy [1], Vladimir N. Uversky [2,3], Su San Toh [1,†], Ming-Chang Tsai [1], William C. Beckerson [1], Catarina Cahill [1], Brittany Carman [1] and Michael H. Perlin [1,*]

[1] Department of Biology, Program on Disease Evolution, University of Louisville, Louisville, KY 40208, USA; mail2swathi.k@gmail.com (V.S.K.); cloudysusan@gmail.com (S.S.T.); m0tsai02@louisville.edu (M.-C.T.); william.beckerson@louisville.edu (W.C.B.); catarina.cahill@louisville.edu (C.C.); brittany.carman@louisville.edu (B.C.)

[2] Department of Molecular Biology and University of South Florida Health Byrd Alzheimer's Research Institute, Morsani College of Medicine, University of South Florida, Tampa, FL 33612, USA; vuversky@health.usf.edu

[3] Laboratory of New Methods in Biology, Institute for Biological Instrumentation, Russian Academy of Sciences, Institutskaya Str., 7, Pushchino, Moscow Region 142290, Russia

*   Correspondence: michael.perlin@louisville.edu; Tel.: +1-502-852-5944

†   Current Address: Defence Medical and Environmental Research Institute, DSO National Laboratories, Singapore, Singapore.

Received: 24 October 2017; Accepted: 16 November 2017; Published: 22 November 2017

Abstract: (1) Background: Plant pathogenic fungi often display high levels of host specificity and biotrophic fungi; in particular, they must manipulate their hosts to avoid detection and to complete their obligate pathogenic lifecycles. One important strategy of such fungi is the secretion of small proteins that serve as effectors in this process. *Microbotryum violaceum* is a species complex whose members infect members of the Caryophyllaceae; *M. lychnidis-dioicae*, a parasite on *Silene latifolia*, is one of the best studied interactions. We are interested in identifying and characterizing effectors of the fungus and possible corresponding host targets; (2) Methods: In silico analysis of the *M. lychnidis-dioicae* genome and transcriptomes allowed us to predict a pool of small secreted proteins (SSPs) with the hallmarks of effectors, including a lack of conserved protein family (PFAM) domains and also localized regions of disorder. Putative SSPs were tested for secretion using a yeast secretion trap method. We then used yeast two-hybrid analyses for candidate-secreted effectors to probe a cDNA library from a range of growth conditions of the fungus, including infected plants; (3) Results: Roughly 50 SSPs were identified by in silico analysis. Of these, 4 were studied further and shown to be secreted, as well as examined for potential host interactors. One of the putative effectors, MVLG_01732, was found to interact with *Arabidopsis thaliana* calcium-dependent lipid binding protein (AtCLB) and with cellulose synthase interactive protein 1 orthologues; and (4) Conclusions: The identification of a pool of putative effectors provides a resource for functional characterization of fungal proteins that mediate the delicate interaction between pathogen and host. The candidate targets of effectors, e.g., AtCLB, involved in pollen germination suggest tantalizing insights that could drive future studies.

Keywords: biotrophic pathogen; anther smut; fungal effectors; *Microbotryum violaceum*

1. Introduction

During fungal infection of plants, a number of fungi secrete small proteins that serve to manipulate host responses and downstream events in host development during infection. Often such proteins allow biotrophic fungi to evade host defenses, but they can also redirect development so as to specifically benefit the fungus. Such proteins have been termed "effectors", and many share common characteristics among different fungi [1,2]. For instance, for oomycete pathogens, such as *Phytophthera* species, or rust species (e.g., *Melampsora lini*), small secreted proteins (SSPs) are secreted from specialized structures called haustoria that penetrate host plant cells to draw nutrients from their hosts. Such fungal effectors are SSPs that bear an N-terminal signal peptide; the effectors are usually unique to the pathogen. Most effectors are cysteine-rich and share no sequence similarity with other known proteins, thus revealing the specialized arsenal that each pathogen possesses and, most likely, their association with the specificity of the pathogen for its host. Some effectors are translocated directly from the infection structures, i.e., haustoria or appressoria, into plant cells, while others interact with host cell receptors and get internalized into the cell [1]. Some studies suggest that secreted proteins can act as structural effectors that could accumulate at the host/pathogen interface and stabilize the fungal filaments [2]. However, the mechanism of how these effectors work in the entry into the plant cell or in the proliferation of the fungus inside the host has yet to be fully elucidated. *Microbotryum lychnidis-dioicae* is an obligate biotrophic basidiomycete smut fungus and is a member of the *Microbotryum violaceum* species complex that infects members of the Caryophyllaceae family. *M. lychnidis-dioicae* infects the dioecious host plant, *Silene latifolia*. The fungal life cycle begins when the fungal spores are disseminated by wind or pollinator species and land on a suitable host. The diploid teliospores then undergo meiosis to produce yeast-like haploid sporidia that reproduce by budding. Conjugation takes place between sporidia of opposite mating type, under suitable conditions, such as low nutrients and cool temperatures. Conjugation results in the formation of an infectious dikaryotic hypha that is stabilized by host cues, allowing the fungus to produce an appressorium and penetrate the host tissue. The fungus overwinters in the meristematic tissue; infection becomes systemic in the following year, producing diseased flowers, in which the pollen has been replaced with fungal spores, thus rendering the male plants sterile. It is thus commonly referred to as the "anther smut" [3]. Karyogamy occurs in the dikaryotic hyphae resulting in the formation of diploid spores, thus completing the life cycle. The fungal life cycle thus exhibits both a saprobic haploid phase and a parasitic dikaryotic/diploid phase. The disease also aborts the development of female organs in female host plants. Moreover, the female plants develop immature male reproductive anthers, making this one of the most interesting cases of parasitic modification of host floral organs. Linnaeus was the first to notice the smut-induced anthers in the female host plants [4]. Since pollination drives disease transmission, anther smut is considered as a plant sexually transmitted disease (STD) [5].

Recently, the genome sequence and transcriptomes of *M. lychnidis-dioicae* and its interaction with the host *S. latifolia* have been produced [6]. However, there have been no experimental data provided to explain how this fungus can divert the host resources for its own propagation and survival. Here, we provide the first study to examine the function of the candidate proteins, i.e., the putative effectors that might be involved in the pathogenicity of this group of fungi.

2. Results

2.1. In Silico Analyses to Identify Potential Effectors

To provide a conservative estimate of proteins secreted by *Microbotryum lychnidis-dioicae*, several bioinformatic tools were employed, and only those proteins that passed all measures used were retained in the list of predicted secreted proteins (Table S1). Out of 7364 proteins, 279 were identified to have a signal peptide; from this group, 71 predicted proteins were smaller than 250 amino acids (hereafter referred to as small secreted proteins, SSPs). Of these, 46 appeared to be unique to *M. lychnidis-dioicae* or to the *Microbotryum* complex, and 60 lacked identifiable PFAM domains. Among the SSPs, 19 were also significantly upregulated during plant infection, suggesting that these may play a role during those stages of the fungal lifecycle and in pathogenicity [7].

2.2. Intrinsic Disorder in Predicted Small Secreted Proteins (SSPs)

Intrinsic disorder is known to play an important role in protein-protein interactions [8–14]; intrinsically disordered proteins (IDPs), hybrid proteins containing ordered domains, and intrinsically disordered protein regions (IDPRs) are common among pathogenic microbes [15], and play a number of roles in pathogen-host interactions [16,17]. Accordingly, we analyzed the overall intrinsic disorder predisposition of the 49 predicted secreted proteins from *M. lychnidis-dioicae* upregulated during infection, using a set of established disorder predictors from the PONDR family (PONDR® VSL2 [18], PONDR® VLXT [19], PONDR® VL3 [20], and PONDR® FIT [21]). We also used the ANCHOR algorithm [22,23] to evaluate the presence of the disorder-based protein-protein interaction sites, molecular recognition features (MoRFs), i.e., regions that might undergo the binding-induced disorder-to-order transition. Results of these analyses are summarized in Supplementary Materials, Table S2. These results draw a picture of an impressive prevalence of intrinsic disorder in the *M. lychnidis-dioicae* SSPs. In fact, all putative effectors have regions of intrinsic disorder, and many of the effectors are very disordered. In particular, 11 effectors (22.4%) can be classified as mostly disordered, since they have >50% disordered residues; 19 effectors (38.8%) are highly disordered, possessing between 30 and 50% of disordered residues; 18 effectors (36.7%) are moderately disordered, since they have between 10% and 30% of disordered residues; and just one protein (2.1%) has less than 10% disordered residues and therefore is mostly ordered. These values for disorder content are very high even for a eukaryotic organism and are rather atypical for groups of proteins that are not specifically selected for disorder. Furthermore, many effectors have disorder-based binding sites or MoRFs (i.e., sites that are disordered in the unbound state and undergo disorder-to-order transition at interaction with the binding partners). Finally, several effectors have more than one MoRF, suggesting that they can be engaged in interaction with multiple partners or, being engaged in interaction with one partner, utilize multivalent "wrapping around"-type binding mode. It is likely that the exceptionally high disorder levels and the presence of MoRFs can simplify interactions of these pathogenic effectors with host proteins or play some other role in regulation of the SSP functionality.

In line with the hypothesis that intrinsic disorder can be of functional importance for the SSPs from *M. lychnidis-dioicae*, Figure 1 represents in-depth analysis of the intrinsic disorder predisposition of four putative effector proteins that were up-regulated during infection and were shown to have important functions (Sections 2.3 and 2.4 for the detailed functional characterization of these proteins).

The corresponding disorder profiles were by the overlay of the outputs of six commonly used disorder predictors, PONDR® VSL2 [18], PONDR® VLXT [19], PONDR® VL3 [20], PONDR® FIT [21], as well as IUPred_short and IUPred_long [24]. Furthermore, for each of these four proteins, mean per-residue disorder probability was calculated by averaging disorder profiles generated by the individual predictors. The use of consensus for evaluation of intrinsic disorder is motivated by empirical observations that this approach usually increases the predictive performance compared to using a single predictor [25–27]. Figure 1 clearly shows that these four proteins are characterized by high levels of predicted disorder that range (as per the outputs of PONDR® VSL2 analysis) from 22.4% in MVLG_01732 to 61.0% in MVLG_06175, to 64.3% in MVLG_05720, and to 79.4% in MVLG_04106. According to the PONDR® VSL2-based analysis, there are four IDPRs in MVLG_01732 (residues 1–2, 40–50, 128–142 and 150–156) and three IDPRs in MVLG_04106 (residues 1–3, 25–37, and 39–107), whereas MVLG_06175 and MVLG_05720 have two IDPRs each (residues 1–4 and 51–118 and residues 1–3 and 50–129, respectively). Furthermore, according to the ANCHOR analysis, each of these four SSPs might have at least one MoRF (residues 145–153 in MVLG_01732, residues 113–118 in MVLG_06175, residues 7–12 in MVLG_05720, and residues 3–12 in MVLG_04106). The presence of MoRFs in these proteins was also analyzed by MoRF$_{CHiBi}$, which is a new computational approach for fast and accurate prediction of MoRFs in protein sequences. This analysis showed that although there is no MoRF$_{CHiBi}$-identified MoRF in MVLG_01732, this protein has two regions with some potential to act as MoRFs (residues 29–39 and 139–156). Similarly, there are no MoRF$_{CHiBi}$-identified MoRFs in MVLG_05720, which, however, have four regions with some potential to act as MoRFs (residues 1–14,

66–76, 97–104, and 120–129). On the other hand, MVLG_06175 has two MoRFs (residues 97–111 and 113–118), and almost the entire chain of MVLG_04106 can act as disorder-based binding region, since this protein has two MoRFs, residues 1–70 and 87–104, that cover almost 83% of its sequence.

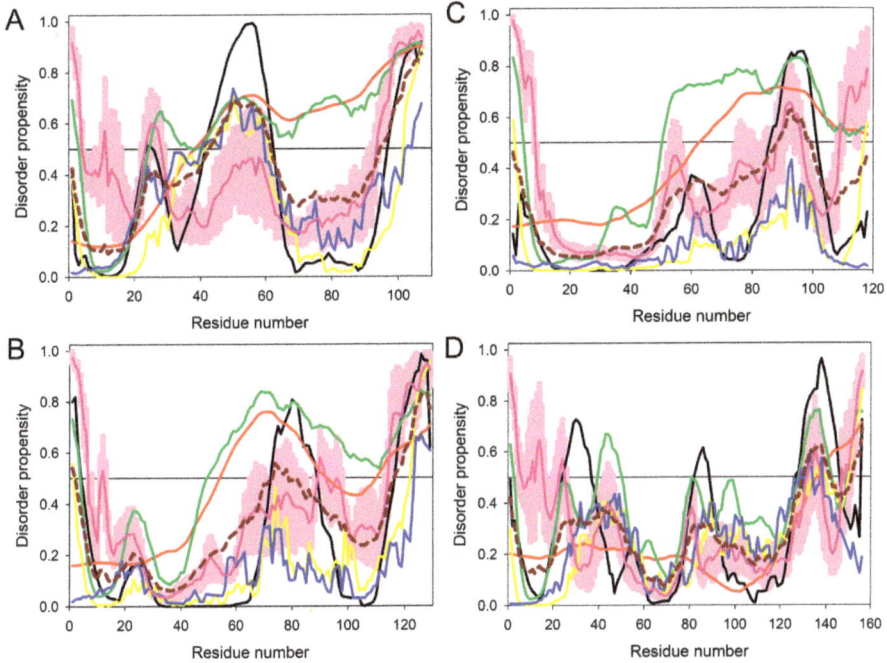

Figure 1. Evaluating intrinsic disorder propensity of protein effectors (**A**) MVLG_04106; (**B**) MVLG_05720; (**C**) MVLG_06175 and (**D**) MVLG_01732 by a series of per-residue disorder predictors. Disorder profiles generated by PONDR® VLXT, PONDR® VL3, PONDR® VSL2, IUPred_short, IUPred_long, and PONDR® FIT, are shown by black, red, green, yellow, blue, and pink lines, respectively. Dark red dashed line shows the mean disorder propensity calculated by averaging disorder profiles of individual predictors. Light pink shadow around the PONDR® FIT shows error distribution. In these analyses, the predicted intrinsic disorder scores above 0.5 are considered to correspond to the disordered residues/regions, whereas regions with the disorder scores between 0.2 and 0.5 are considered flexible.

2.3. Yeast Secretion Trap to Verify the Secretory Nature of Predicted Effectors

We used Yeast Secretion Trap (YST) [28], a molecular genetic approach, to confirm the secretory nature of a small subset of the SSP putative effector proteins (MVLG_01732, MVLG_04106, MVLG_05720, and MVLG_06175; Table 1), each of which was also up-regulated during infection. Three of these proteins were also Cys-rich (MVLG_04106, MVLG_05720, and MVLG_06175), another hallmark of effectors in a number of fungal species [29]. YST employs a mutant strain of yeast, SEY 6210, that has a deletion in the *SUC2* locus encoding the enzyme, invertase. Invertase catalyzes hydrolysis of the disaccharide, sucrose, to glucose and fructose, so that the yeast cell can then take up glucose and metabolize this sugar. Thus, the SEY 6210 mutant yeast strain is normally unable to grow on media where sucrose is the sole carbon source. The method uses a vector, pYSTO-0, bearing the coding region of Suc2 invertase without its signal peptide and its start codon. The protein of interest can be cloned as a translational fusion protein with the invertase driven by a constitutive promoter from *ADH1*. If the protein of interest is secreted, this will result in the reconstituted functional activity of the invertase and enable the yeast cells to grow on sucrose medium. All four predicted effectors from *M. lychnidis-dioicae* examined experimentally with the yeast secretion trap assay indeed appeared to be secreted, since the signal peptide of each allowed Suc2p to be secreted and thus provide for growth of the yeast SEY 6210 mutant on sucrose medium (Figure 2). In contrast, SEY 6210 cells transformed with the vector only were unable to grow on such media.

Table 1. Candidate SSPs chosen for further analyses.

Predicted Protein	Expression [a]	Size (Amino Acids)	No. of Cys	Function
MVLG_01732	144 rsem vs. 0	156	1	Candidate effector
MVLG_04106	86 rsem vs. 0	107	6	Candidate effector
MVLG_05720	1164 rsem vs. 0	129	12	Candidate effector
MVLG_06175	127 rsem vs. 0	118	10	Candidate effector

[a] rsem normalized counts for infected male *S. latifolia* vs. expression in YPD or nutrient-limited agar [7].

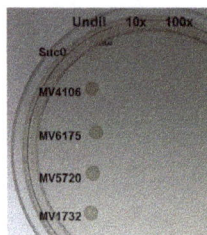

Figure 2. Results of secretion trap experiment with four *M. lychnidis-dioicae* predicted small secreted proteins (SSP) effectors. Suc0, yeast cells transformed with vector alone on sucrose, leu drop-out medium. Undil, undiluted; 10× and 100× dilutions.

2.4. Yeast Two-Hybrid Experiment

Our goal was to determine the function of these fungal proteins that are predicted, and now confirmed, to be secreted, as well as being highly expressed, during infection. We employed yeast two-hybrid genetic screening to identify the possible host interactors for these fungal proteins. As mentioned above, we chose a small subset of the SSPs that were also found to be induced in expression *in planta*.

2.4.1. MVLG_04106 Autoactivates the Reporter Genes in Yeast Two-Hybrid Assay

We expressed MVLG_04106 lacking its signal peptide as a fusion protein to Gal4BD in the bait vector (pGBKT7-MVLG_04106ΔSP) and tested its activity in expressing the reporter genes. It was found that the yeast strain transformed with this construct activated all three of the reporter genes-*HIS3*, *ADE2*, and *MEL1*, when mated with the opposite mating strain containing only the control prey vector (Figure 3). This indicates the cells' ability to grow on media lacking the essential nutrients histidine and adenine because of the activation of the enzymes aminoimidazole ribonucleotide carboxylase 2 (*ADE2*) and imidazole glycerol phosphate dehydratase 3 (*HIS3*). Moreover, the cells were also able to express α-galactosidase, the gene product of the Melibiase 1 (*MEL1*) reporter gene that enables the yeast cells to turn blue-green in the presence of the chromogenic substrate X-α-gal. This was unexpected, so we generated the reciprocal set of constructs to further investigate possible transcriptional activation by MVLG_04106. In this case, a fusion protein was generated with MVLG_04106 and Gal4AD in the prey vector to test if the reporter genes could again be activated. Surprisingly, in this case the reporter genes were not activated. This suggests that MVLG_04106 could activate the transcription of the reporter genes only when attached to the corresponding DNA binding domain for those genes (i.e., Gal4BD). One possibility is that this fungal protein acts as a transcription factor in modulating the host gene expression during infection. In line with the known fact that transcription factors are typically characterized by high levels of intrinsic disorder [30–32], MVLG_04106 was predicted to possess 79.4% disordered residues (see Figure 1A) and is shown to contain long disorder-based interaction regions. The predicted protein contains 106 amino acid residues and is cysteine rich, with approximately 5% Cys residues. Further domain analysis using PROSITE did not yield any information, but prediction of post translational modification sites indicated proteolytic cleavage at residue D35, which could allow the mature protein to function as a transcriptional regulator [33] (Supplemental Table S2). Structural modelling using Swiss-Model yielded chorismite mutase for residues 30–65, for which there was 22.22% similarity in the 3-dimensional structure. When we compared the amino acid sequence of MVLG_04106 with predicted proteins of *M. silenes-dioicae* [34], there was 99.07% identity with the corresponding orthologue, whereas that for the *M. violaceum sensu lato* species [35], only had 63.04% identity.

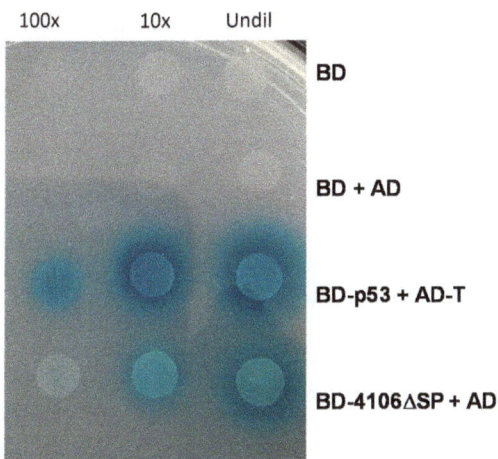

Figure 3. Autoactivation of three reporter genes by MVLG_04106 on QDO/X-α-gal + 3-AT (5 mM) plates. Undil, undiluted; 10× and 100× dilutions. QDO (Quadruple drop out media), 3-AT (3-Amino-1,2,4-triazole), BD (DNA binding domain in pGBKT7 vector), AD (Activation domain in pGADT7 vector), BD-p53 (pGBKT7-53 positive control plasmid), AD-T (pGADT7-T positive control plasmid), and BD-4106ΔSP (MVLG_04106 lacking signal peptide).

2.4.2. MVLG_05720 Fungal Protein Interacts with Fungal Proteins

Yeast two-hybrid screening with MVLG_05720 yielded 614 colonies after the initial stringent selection on QDO medium with 5 mM 3AT, along with screening for α-galactosidase expression on X-α-gal (as blue-green colonies). Further selection on 50 mM 3AT to reduce leaky *HIS* selection yielded 129 colonies for examination via sequence analysis. Of the 129 sequenced clones, we recovered only fungal interactors: 99 of the clones represented MVLG_07305, 27 of the clones were found to be MVLG_04206, and 3 of the clones matched MVLG_04267. Figure 1B illustrates that there are 64.3% disordered residues in MVLG_05720, and this protein has several MoRFs. It contains 129 amino acids and is highly cysteine rich with roughly 9% Cys residues. When the amino acid sequence of MVLG_05720 was compared to the genomes of *M. silenes-dioicae* and *M. violaceum sensu lato*, the corresponding orthologues showed 96.9% identity and 85.93% identity, respectively.

2.4.3. MVLG_06175 Interacts with a Host Protein and a Fungal Protein

Yeast two-hybrid screening with MVLG_06175 initially yielded 1000 colonies after the stringent selection on QDO/X-α-gal + 3AT (5 mM) medium. Further selection to reduce leaky *HIS* selection yielded 201 colonies for examination via sequence analysis. Of the 39 sequenced clones that we recovered, 4 of them were full length clones that encode CASPL2C1, 1 was a fungal protein encoded by MVLG_06379, and the rest of the sequenced clones were for the fungal protein encoded by MVLG_07305 mentioned above. The *S. latifolia* genomic region matching the sequence for CASPL2C1 is found on contig m.88187 (GenBank: FMHP01040264.1) in NCBI for the *Silene latifolia* genome assembly (taxid:37657). It also corresponded to c93454_g1 RNA detected in RNA-Seq experiments [6,7]. According to Figure 1C, 61.0% of residues in MVLG_06175 are predicted to be intrinsically disordered and this protein can be engaged in disorder-based protein-protein interactions. It contains 118 amino acids and is Cys-rich (roughly 8% Cys residues). When the amino acid sequence of MVLG_6175 was compared to the genomes of *M. silenes-dioicae* and *M. violaceum sensu lato*, there was 95.76%, but only 59.83% identity, respectively, with the corresponding orthologues.

2.4.4. MVLG_01732 Interacts with Host Proteins

Yeast two-hybrid screening with MVLG_01732 yielded 401 colonies after the initial stringent selection on QDO/X-α-gal + 3AT (5 mM) medium. Further selection to reduce leaky *HIS* selection yielded 65 colonies for examination via sequence analysis. From the 65 sequenced clones, yeast two-hybrid screening of MVLG_01732 revealed interesting host plant interactors. One of the interactors, represented by 52 clones, was found from blastp searches of the *Arabidopsis thaliana* genome (TAIR; https://www.arabidopsis.org/) as an orthologue of the AT3G61050.2 gene, which encodes a calcium-dependent lipid binding protein (AtCLB). AtCLB has both coiled coil regions and C2 domain similar to synaptotagmins, and synaptotagmins were also identified as hits in blastp searches of the ncbi database. Synaptotagmins are class of proteins with an N terminal transmembrane and two cytoplasmic C_2 domains (Figure S1). The *S. latifolia* genomic region matching the AtCLB sequence is found on contig m.108787 (GenBank: FMHP01019528.1) in the *S. latifolia* genome assembly. It also corresponded to c85332_g3 RNA detected in RNA-Seq experiments [6,7].

The other interactor identified by yeast two-hybrid was cellulose synthase Interactive protein 1 (CSI1; Figure S2), represented by 13 clones. The *S. latifolia* genomic region matching this sequence is found on contig m.23209 (GenBank: FMHP01009449.1) in the *S. latifolia* genome assembly. It also corresponded to c93789_g3 RNA detected in RNA-Seq experiments [6,7]. Figure 1D shows that with 22.4% disordered residues, MVLG_01732 is the least disordered protein analyzed in this study. However, despite relatively low disorder content, MVLG_01732 contains MoRFs and, therefore, is expected to use intrinsic disorder for protein-protein interactions. The protein is 156 amino acids long and is not rich in Cys residues. When the amino acid sequence of MVLG_1732 was compared to the genome of *M. silenes-dioicae*, a 94.23% identity match was found in the corresponding orthologue, whereas only a 48.99% identity match was observed for the orthologue from *M. violaceum sensu lato*.

3. Discussion

In this study, we were able to predict from in silico analyses a conservative estimate of the secretome of *M. lychnidis-dioicae*. Furthermore, among this group, we identified candidate effectors as SSPs that were also highly expressed during plant infection. For the four putative effectors examined in greater detail in this study, amino acid sequence comparisons between *M. lychnidis-dioicae* and *M. silenes-dioicae* [34] revealed that these two organisms share close similarity in their predicted SSPs. In contrast, comparisons of most of the orthologues identified in *M. violaceum sensu lato* [35], the species that infects *Silene paradoxa*, had significantly lower amino acid similarities to those of the other species. These findings suggest that the latter organism has diverged substantially from the other two species, a finding supported by the phylogenetic relationships of the three respective fungal species [36] and the lack of cross-infectivity for *M. violaceum sensu lato* on either *S. latifolia* or *S. dioicae*; similarly, neither *M. lychnidis-dioicae* nor *M. silenes-dioicae* have been found to infect *S. paradoxa*. Of note, all SSPs were shown to contain IDPRs, and the vast majority of these proteins (>61.0%) were classified as mostly or highly disordered. Many SSPs were also predicted to have at least one MoRF, with some of the putative effectors possessing multiple MoRFs that can be utilized in promiscuous interactions with the fungal and host proteins. To test some of these predictions, a subset of the SSPs predicted in silico were confirmed to be secreted by YST experiment. We conducted yeast two-hybrid analysis for these SSPs to identify their host interactors and hence to understand their role in the mechanism of the infection (Figure 4).

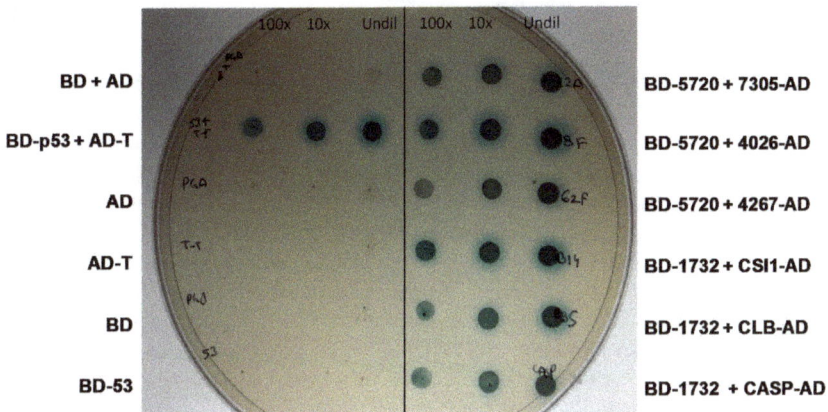

Figure 4. Yeast two-hybrid spot test results for all four proteins and positive and negative controls on quadruple drop-out medium (QDO)/X-α-gal + 3AT (5 mM) plates. For the spot test, each strain bearing the plasmid was grown in 3 ml of appropriate drop out liquid media at 30 °C and shaken at 280 rpm for 2 days. To reconfirm the interaction, 10 ML of each culture (washed and resuspended in 0.9% w/v NaCl) was mixed and spotted on QDO plates containing X-α-gal at the indicated dilutions and incubated at 30 °C for 3–5 days. Undil, undiluted; 10× and 100× dilutions.

Studies suggest that the proteins that undergo post translational modifications (PTMs) are considered to interact more with other proteins by engaging in more physical contacts and are known to be in the central network pathways more than non-PTM proteins. For example, all but four of the 49 secreted proteins we examined in detail were predicted to be targets for amidation. C-terminal amidation has been shown to be involved in membrane interactions for some proteins. In one case, an antimicrobial peptide, maximin H5, was able to penetrate and lyse erythrocyte membranes when amidated, but the ability to penetrate lipid membranes was severely reduced with deamidated peptide [37]. For the 4 SSPs, we examined in detail by yeast two-hybrid analysis, MVLG_04106 and MVLG_05720, which were predicted to have amidation targets (but, not at the C-terminus), while MVLG_06175 and MVLG_01732 each have a predicted target closer to the C terminus. If amidation plays a similar role for these SSPs as it does for maximin H5, this could indicate that these effectors penetrate host cells as part of their normal function.

3.1. MVLG_04106 Could Serve as a Transcriptional Regulator

The finding that MVLG_04106 was able to autoactivate all the reporter genes—*HIS3*, *ADE2*, and *MEL1*—in the yeast two-hybrid screen suggests its role as a transcriptional regulator. However, analysis by structural modelling reveals that a portion of this protein is similar to chorismate mutase, a vital enzyme that catalyzes the conversion of chorismate to prephenate in the shikimate pathway, leading to the production of aromatic amino acids, phenylalanine, and tyrosine, and regulating their balance. Chorismate also serves as a substrate for the production of salicylic acid (SA), which is a major signaling defense molecule in plants. This fungal protein chorismate mutase could deviate the flow of available chorismate for the production of prephenate and hence channel down its availability for SA production. In fact, studies show that *Ustilago maydis*, an obligate biotrophic pathogen that causes corn smut, also secretes an effector called Cmu1, a chorismate mutase taken up by plant cells and spread to adjacent cells causing metabolic priming in the infected cells [38]. However, if MVLG_04106 is a chorismate mutase, this still begs the question of how it autoactivates the reporter genes in yeast two-hybrid assay. Transcriptome analysis revealed that it is highly expressed during infection but not under in vitro conditions, all of which suggests its role in pathogenicity. Thus, further investigation is requited to better define its true role during infection.

3.2. MVLG_05720 Possibly Regulated by Additional Fungal Proteins

Three fungal proteins were identified as interactors with MVLG_05720: MVLG_07305, MVLG_04026, and MVLG_04267. None of the three were predicted via bioinformatic tools to be secreted. The differential expression data [6,7] indicated that MVLG_07305 was downregulated in late infection stages *in planta* and upregulated in mated conditions in vitro (while MVLG_05720 was downregulated during mating) [6]. Thus, MVLG_07305 may play some role in mating or in the transition to dikaryotic filaments. The gene is located on the mating-type chromosome, but its expression is similar in both a1 and a2 mating-type strains on either rich or nutrient-limited media [7]. Blastp predicted its function as a putative fimbrial outer membrane usher protein, containing a mannose binding domain. Of note, fimbrial appendages were first observed serendipitously on the haploid cells of an anther smut fungus [39]. They are involved in cell-to-cell communication and adhesion during mating before pathogenesis, as enzymatic and mechanical removal of these structures were shown to delay mating until the regeneration of fimbriae occurred [40,41].

The second fungal interactor, MVLG_04026, followed the same expression pattern as that of MVLG_07305; its predicted function was as a Fibrillin-like protein. Fibrillins are secreted proteins that constitute the backbone of extracellular macromolecular microfibrils [42]. The C terminus of fibrillins can undergo multimerization as a consequence of intermolecular disulfide bonding with itself or other proteins soon after secretion [43]. However, MVLG_04026 was predicted via bioinformatic tools not to be secreted. Its transcription was also upregulated during mating and downregulated during infection.

MVLG_04267 was not found to be differentially expressed under any of the conditions examined. It belongs to the DUF1212 superfamily, a class of membrane proteins with unknown function. Perhaps this protein plays a role in transport of MVLG_05720. If the MVLG_07305 and 4026 proteins are translated during mating and persist during infection, they may interact with MVLG_05720, to sequester it until it is needed for manipulation of the host. However, the mechanism of their action remains a mystery and requires further investigation.

3.3. MVLG_06175 Role in Host Entry During the Infection and in Reproduction

From yeast two-hybrid screening, the protein product of fungal gene *MVLG_06175* interacts with a host CASP-like protein 2C1 orthologue of *Spinacia oleracea* (LOC110788005), transcript mRNA (XM_021992637.1); it also matched CASP-like protein 2C1, AT4G25830.1 of *A. thaliana*. The corresponding transcript from *Silene* expression data (Toh et al. submitted) similarly matched the same *S. oleracea* protein. CASP-like proteins (CASPLs) are homologues of Casparian strip membrane domain proteins (CASPs). With respect to the functions of CASPLs, previous research showed CASPLs might function as protein barriers on the cell membrane of the endodermis and form protein scaffolds for the synthesis of the Casparian strip. Some CASPLs were shown to be expressed in the root endodermis, peripheral root cap, root meristem zone, trichomes, lateral root primordia, young leaves, and the floral organ abscission zone in *Arabidopsis thaliana* [44]. This last role, in floral organs, would be an appropriate target for a fungal effector from an anther smut. Alternatively, since CASPLs are orthologous with MARVEL domain proteins associated with the function of epithelial tight junctions [45], CASPLs might be related to tight junction functions in plant cells as well. Thus, the interaction between MVLG_06175 and CASPL2C1 of *Silene* could indicate that *Microbotryum* alters the functions of tight junctions to enter the host tissues during infection.

MVLG_06379 was also found as a fungal interactor of MVLG_06175. MVLG_06379 contains a PFAM domain (PF03328.7) for ATP citrate lyase (ACL) β subunit. The enzyme converts cytosolic citrate into acetyl-CoA for further fatty acid synthesis, but in the parasitic fungi the transcription and translation of *ACL* appears to be associated with infection and reproduction. *Cryptococcus neoformans* increased transcriptional level of *ACL1* within macrophages. Additionally, mutants lacking *ACL1* showed higher susceptibility to antifungal drugs, a lower survival rate within macrophages, and defects in expression of virulence factors [46].

3.4. MVLG_01732 Role in Altering the Vesicular Traffic in the Host and Male Sterility

In blastp analyses searching the *Arabidopsis* genome we found this interactor as the orthologue of the AT3G61050.2 gene. This encodes a calcium-dependent lipid binding protein (AtCLB) that has both coiled coil regions and C2 domain (see Figure S1) similar to synaptotagmins. Many coiled coil proteins are involved in regulating gene expression as transcription factors. The motif is present in the nucleotide binding site leucine rich repeat (NBS-LRR) proteins of R genes. *Arabidopsis* encodes 150 NBS-LRR-type proteins and they are either the coiled coil (CC) type or the TIR type [47]. C2 domains in animal cells are involved in signal transduction and vesicle trafficking, but in plant cells they are not well characterized. They could be involved in plant stress signal transduction as positive or negative regulators of stress signaling cascades (Figure S3 for predicted interaction partners of AtCLB). AtCLB expression is highly detected in rosette leaves and flowers and low in roots, stems, and cauline leaves. Transcriptome analyses studies on pollen germination and tube growth shows its expression in mature pollen, hydrated pollen, and pollen tube growth, which suggest its role in the development of the male gametophyte [48]. Studies show that AtCLB acts as a DNA-binding protein, and binds specifically to the promoter sequence of Thalional synthase 1 (*THAS1*), a key enzyme in the synthesis of the triterpenoid, thalionol. AtCLB negatively regulates *THAS1* transcription, as part of a response involved in drought and stress tolerance. AtCLB becomes localized on the nuclear membrane and can bind to ceramides, a glycolipid present in cellular membranes that acts as a second messenger in cell signaling, cell differentiation, and apoptosis [45]. Since it is a membrane protein, its activation by membrane lipid ceramide could result in a proteolytic cleavage and translocate it to nucleus to activate transcription of a different set of genes [49]. Interestingly, analysis of the amino acid sequence showed that there are two proteolytic cleavages at positions 28 (after the transmembrane region (1–22)) and 383 (close to where the C2 domain (264–361) ends and the coiled coil region (390–417) begins). Although this is purely speculative at this point, if the MVLG_01732 effector were to become intracellular, upon Ca^{2+} triggering due to conformational changes, the AtCLB protein could interact with the effector at the coiled coil region to mediate AtCLB activation by the membrane lipid ceramide resulting in proteolytic cleavage after the TM to yield mature protein and translocation to the nucleus to regulate transcription of target genes.

In blastp analyses against the NCBI database, the same host interactor for MVLG_01732 matched a portion of the C_2 domain and mostly the C-terminal coiled coil region of synatotagmin-5 of *Beta vulgaris* subsp. *vulgaris* (LOC104905441), transcript variant X2, mRNA (XM_010693990.2); the corresponding transcript from *Silene* expression data (Toh et al. submitted) similarly matched the same *B. vulgaris* protein. Synaptotagmins are a family of membrane proteins concentrated on secreted vesicles, including synaptic vesicles. They are composed of a short uncleaved N-terminal signal peptide that overlaps a transmembrane (TM) domain, a synaptotagmin-like mitochondrial and lipid-binding protein (SMP) domain, and two tandem cytosolic calcium binding domains (C_2A and C_2B) at the C-terminus required to bind to phospholipids or different ligands in response to calcium signals [50]. Ca^{2+} plays an important role as a second messenger in response to variety of stimuli like cold, drought, salt, oxidative, and biotic stress. Ca^{2+} binding confers two roles in membrane targeting process. One is to provide a bridge between C_2 domain and anionic phospholipids, and the second is to induce intra or inter domain conformational changes, which further triggers membrane protein interactions.

The second host interactor for MVLG_01732 matched the C_2 domain of Cellulose synthase interactive protein 1 (CSI1) of *Spinach oleracea* (Accession number: XP_021846375) and *Beta vulgaris* (Accession no: XP_010680591), orthologues of *Arabidopsis* AT2G22125.1 gene. Again, the C_2 domain (Figures S1 and S2) is a Ca^{2+} binding motif originally identified in Protein kinase C [51]. However, not all C_2 domains are regulated by Ca^{2+}, with some functioning in a Ca^{2+}-independent manner and others having mainly a structural role. C_2 domains interact with cellular membranes and mediate key intracellular processes like insulin secretion and neurotransmitter release in eukaryotic cells. It binds to a multitude of different ligands and substrates that include Ca^{2+}, inositol polyphosphates, intracellular proteins, and phospholipids.

Mutant analyses found that CSI physically interacts with microtubules and plays a crucial role in anther dehiscence. This is interesting because the events leading to anther dehiscence are coordinated with pollen differentiation, flower development, and opening for successful pollination. CSI 1 disruption mutants exhibited complete sterility and defective anther dehiscence, with crumpled pollen and defective pollen release from the anther. Moreover, such mutants had morphological changes in the epidermal and endothecial cell length and width necessary for anther maturation, indicating the reason for defective dehiscence may be due to unstable microtubules. CSI mutants also exhibited altered sensitivity to exogenous Ca^{2+} levels, which indicates that there is Ca^{2+}-mediated regulation in microtubule stability and anther dehiscence [52]. Of note, CSI 1 mutants also exhibited decreased number of ovules per gynoecium but were viable, indicating an additional effect of CSI in early gynoecial development.

We hypothesize that the fungal effector MVLG_01732 modulates the function of CSI1 by interacting with the C_2 domain (suggested by our yeast two-hybrid results), thereby altering the stability of microtubules, resulting in delayed anther development and dehiscence. This could provide the fungus an opportunity to hijack anther development, replacing the pollen grains with its teliospores. Studies show that calcium binding proteins and calcium dependent signaling are involved in both the development of embryo sacs and during the development of pollen [53]. In both the host interactors we identified, the MVLG_01732 effector binding could modulate C_2 domains and their interaction with Ca^{2+}, triggering several signaling pathways for the benefit of the fungus.

In sum, in silico analyses predicted a number of fungal small secreted proteins that could serve as effectors to modulate the plant host. For a subset of these, we identified the host interactors that are candidates for targets of these effectors. Recognizing that, even with appropriate controls, yeast two-hybrid analysis can give false positive results, we are planning to further verify the predicted interactions in future experiments using co-immunoprecipitation from infected plants. There is also a need to characterize the function of these interactors and their roles in the plant. Future experiments to express the fungal effectors in transgenic plants might recapitulate phenotypes observed during infection. Although the natural host, *S. latifolia*, currently lacks a transformation system, heterologous plant systems like *A. thaliana* are amenable for such experiments and should help in the characterization of these fungal proteins. Moreover, expressing the fungal proteins with a fluorescent marker like GFP or mCherry would help determine the localization of these fungal proteins inside the host. Pull down assays can also be conducted to identify additional host proteins, if any, that may have been missed by yeast two-hybrid analysis. This experimental model could then be expanded in the future to provide mechanistic insights into the interplay of this biotrophic pathogen with its host.

4. Materials and Methods

4.1. Plant and Fungal Growth

Silene latifolia seeds that were used in this study were originally collected from a field population in Clover Hollow near Mountain Lake Biological Station, Virginia. Sterilized seeds were plated on sterile 0.3% phytagar (Life Technologies/Thermo Fisher, Waltham, MA, USA), one half strength Murashige and Skoog salts (Sigma Aldrich, St. Louis, MO, USA), and 0.05% MES (2-(*N*-morpholino) ethanesulphonic acid) buffer (Sigma-Aldrich). Seeds were kept at 4 °C for 5 days to encourage germination and then were transferred to a 20 °C growth chamber with 13 h of fluorescent light. Humidity was kept high initially by using dome covers and flood trays, and was gradually decreased to lower levels. Seedlings were transplanted to bigger pots for the emerging new roots to provide hydration requirement when the volume of soil was not sufficient. Plants were grown in Sunshine MVP professional growing mix (Sun gro Horticulture Canada Ltd, cat no: 02392868, Agawam, MA, USA) and were watered every other day with 100-ppm fertilizer (Peters Professional 15-16-17 Peat-Lite Special, Formula no: S12893, JR Peters, Inc. Allentown, PA, USA) [7].

Fungal strains of *M. lychnidis dioiceae*, p1A1 and p1A2, were axenically grown separately on nutrient rich media (yeast peptone dextrose media (YPD); 1% yeast extract, 10% dextrose, 2% peptone, and 2% agar) at 28 °C for 5 days and nutrient-free water agar media for 2 days (2% water agar).

Plant infection employed haploid *M. lychnidis-dioicae* p1A1 and p1A2 cells that were grown on nutrient rich media (YPD; 1% yeast extract, 10% dextrose, 2% peptone, and 2% agar) at 28 °C; these were harvested and adjusted to a concentration of 1×10^9 cells/mL in equal proportion before being spotted onto nutrient free media (2% agar). Once conjugation tubes were identified microscopically, the cells were resuspended to a concentration of 1×10^6 in distilled water. Then, 5 μL of this was dropped onto the floral meristem of 11–12 day old *S. latifolia* seedlings [7].

4.2. In Silico Analyses

4.2.1. Prediction of Small Secreted Proteins (SSPs)

Prediction of the secretome used a pipeline of software packages (TargetP1.1, SignalP3.0, SignalP4.0 (http://www.cbs.dtu.dk/services/SignalP/), TMHMM2.0, PredGPI, Phobius, NucPred, Prosite, and WoLF PSORT) to provide a stringent determination of likely secretion [7] (Table S1 and Figure 5).

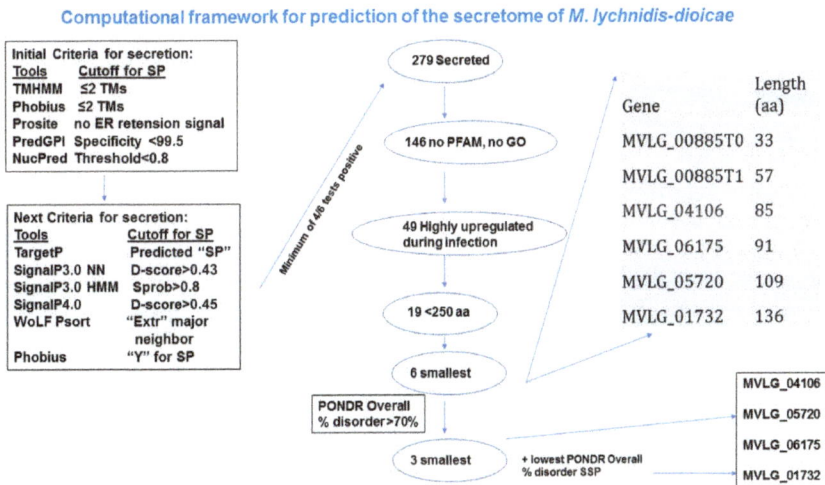

Figure 5. Computational framework for prediction of secretome for *M. lychnidis-dioicae* and selection of candidate effectors for further analyses. Detailed description of tools and cut-off criteria for secretome prediction and prediction of disorder are provided in Supplementary Methods. Numbers tally for proteins at each stage of secretome prediction are provided in tab 3 of Table S1. TM, trans-membrane domain; ER, endoplasmic reticulum' SP, secreted protein; PFAM, protein family; aa, amino acid; GO, gene ontology; MVLG designations refer to specific *Microbotryum lychnidis-dioicae* proteins.

4.2.2. Prediction of Intrinsic Disorder

In order to analyse the residue level of disorder propensity of 49 putative effector proteins, four intrinsic disorder predictors were used: PONDR® VSL2 [18], PONDR® VLXT [19], PONDR® VL3 [20], and PONDR® FIT [21]. While evaluating the intrinsic disorder predisposition of four SSPs targeted for functional analysis (MVLG_01732, MVLG_04106, MVLG_05720, and MVLG_06175), in addition to the members of the PONDR family, IUPred_short and IUPred_long were used [24].

Molecular recognition features (MoRFs) are short segments with increased order propensity located within longer disordered regions. MoRFs bind to globular protein domains and undergo disorder-to-order transition. These disorder-based binding sites are categorized into three types: α-MoRFs (form α-helices upon binding), β-MoRFs (form β-strands), and ι-MoRFs (form irregular structures). For all 49 predicted secreted proteins whose transcription was upregulated during infection, the ANCHOR algorithm (http://anchor.enzim.hu/) was used to predict such protein binding regions that are disordered in isolation but can undergo disorder-to-order transition upon binding [22]. This computational tool finds segments within disorder regions that cannot form stable intra-chain interactions to fold on their own, but are likely to gain stabilizing energy by interacting with a globular protein partner [22]. Furthermore, the presence of MoRFs in MVLG_01732, MVLG_04106, MVLG_05720, and MVLG_06175 was further evaluated by another computational tool, MoRF$_{chibi}$ [54].

4.2.3. Additional Bioinformatic Analyses

Alignment of nucleotide and/or amino acid sequences to find regions of similarity between such biological sequences employed Basic Local Alignment Search Tool (BLAST; https://blast.ncbi.nlm.nih.gov/Blast.cgi). Further domain analysis for prediction of post translational modification sites used ModPred [33]. Structural modelling of predicted proteins utilized Swiss-Model [55,56]. Results of these analyses are found in Table S2. Additional analysis methods are provided in Supplementary Materials and associated references [57–60].

4.3. Yeast Secretion Trap (YST) Experiment

For each candidate effector, validation of secretion employed a yeast-based secretion trap method [28]. Putative secretion signals for each fungal gene were cloned into the pYSTO-0 vector. In such analyses, if the putative signal peptide from a protein provides for secretion of the Suc2p invertase, *S. cerevisiae* cells will be able to grow on sucrose as a sole carbon source; inability to promote growth would indicate that the fungal protein of interest is not normally secreted.

The signal peptide sequence of each fungal protein was determined by Signal P software and amplified by PCR. Standard PCR cycle was used with initial denaturation set at 94 °C for 4 min and 35 cycles of 94 °C for 30 s, 60 °C for 30 s, and 72 °C for 30 s, with a final extension time of 5 min at 72 °C. The product was held at 4 °C at the end of the cycle.

The PCR products were separated by gel electrophoresis through 1.8% agarose (Agarose LE; USB Corp., Cleveland, OH, USA). The fragments were excised from the gel and purified using the Zymo Gel DNA recovery kit (Orange, CA, USA). The purified fragments were subjected to restriction digestion with *Eco*RI and *Not*I enzymes. The digested fragment was purified and cloned into the pYST-0 vector to obtain a translational fusion with the invertase expressed from the *ADH1* promoter and transformed into *Escherichia coli* DH5 α cells. Cells were plated on LB plates with ampicillin (100 mg·L^{-1}) and incubated at 37 °C overnight. *E. coli* strain, DH5 α (Bethesda research Laboratories, Bethesda, MD, USA), was utilized for all cloning purposes. *E. coli* strains were grown at 37 °C in Circle Grow media (MP Biomedicals, LLC, Solon, OH, USA) and plasmid DNA was isolated from potential clones using the alkaline lysis procedure [61]. The presence of each signal peptide encoded in-frame with the *SUC2* coding region was confirmed by DNA sequencing at the Nucleic Acids Core Facility (Center for Genetics and Molecular Medicine, University of Louisville, Louisville, KY, USA).

Invertase-deficient (suc2⁻) *S. cerevisiae* strain (SEY 6210 (MATαleu2-3, 112 ura3-52 his-Δ200 trp1-Δ901 lys2-801 suc2⁻ Δ9 GAL)) [62] cells were transformed with the constructs using the lithium acetate/single-stranded carrier DNA/PEG method [63]. Selection was on Synthetic Dropout medium, with SD/-Leu (Clontech, Mountain View, CA, USA) selection plates containing glucose as the sole carbon source. The dropout medium contained glucose (20 g·L⁻¹), yeast nitrogen base (6.7 g·L⁻¹), dropout mix minus leucine (2 g·L⁻¹), agar (15 g·L⁻¹), and water. The plates were incubated at 30 °C for 6–10 days. The colonies were restreaked for purification onto SD/-Leu drop out selection plates with sucrose as the sole carbon source to select the positive clones that were able to utilize sucrose by secreting invertase enzyme. Such strains were grown overnight in 3 mL of SD/-Leu broth with sucrose, and 10-fold dilutions were spotted onto SD/-Leu with glucose or sucrose as the carbon source and incubated for 5 days at 30 °C. Clones harboring functional signal peptides with reconstituted invertase activity were able to grow on sucrose as the sole carbon source. Untransformed mutant yeast strain SEY 6210 and the same strain, transformed with empty pYST-0 vector, were used as negative controls. Plasmid DNA was extracted from the positive clones and used to retransform *E. coli*. The constructs were again checked for the presence of signal peptide sequence by DNA sequencing.

4.4. RNA Extraction and cDNA Library Construction

RNA for generating the cDNA library was obtained from the axenically grown cultures of p1A1 and p1A2 haploid strains [7] on nutrient rich media for 5 days (YPD) at 28 °C, nutrient-free water agar media (2% water agar) for 2 days, and the fungal infected *Silene latifolia* tissue [7]. This latter set of RNAs was extracted from floral stem (pedicle, and remaining cluster and sepals), floral buds (male and female) at different stages (male: 2–6 mm buds, 8 mm to fully opened smutted flowers; female: 3–6 mm, 7–14 mm, and 15–24 mm). The quality of the RNA was checked by Agilent Bioanalyzer and all the samples indicated highly intact RNA with the RNA integrity scores of at least 7.8. The total samples were pooled equally based on Bioanalyzer quantification to generate a normalized cDNA library. The cDNA library was constructed in a Gal4 based prey vector, pGADT7 (Clontech), by CD Genomics (Shirley, NY, USA) for yeast two-hybrid screening.

4.5. Yeast Two Hybrid Screen

The yeast two-hybrid system allows for an initial screening of possible protein-protein interactions [64,65]. A "bait" protein of interest is expressed from a yeast (*Saccharomyces cerevisiae*) expression vector as a fusion with the Gal4 DNA binding domain (BD). Interactors with bait are identified by screening "prey" expressed from a yeast vector where the fusion is with the Gal4 transcriptional activation domain (AD). pGBKT7 was used as a "bait" vector with the GAL4 DNA-binding domain and pGADT7 was used as a "prey" vector with the GAL4 DNA activation domain. While neither the BD, nor AD alone, can activate transcription of the reporter genes used in this system, if two proteins physically interact (i.e., if prey Y interacts with bait X), then the BD and AD are brought together and reporter genes will be expressed. In our studies, the prey proteins were all expressed from normalized cDNA libraries of the different stages of *M. lychnidis-dioicae*, including in association with its host, *S. latifolia*. Initial selection of interactors involves ability to grow on increasingly more stringent auxotrophic media, since the yeast strains have auxotrophic mutations that require them to either be provided with the missing nutrients or to have a functional interaction that activates transcription of reporter genes whose read-out is complementation of the growth defect. Additionally, an α-galactosidase gene serves as a reporter, whereby color change to blue-green occurs via cleavage of 5-bromo-4-chloro-3-indolyl α-D-galactopyranoside (X-α-gal) in the medium. In order to avoid false positives, a number of controls were employed, including comparisons using (1) vectors alone/without bait or prey (i.e., pGBKT7 or pGADT7, respectively); (2) bait in BD vector alone; (3) prey in AD vector alone; (4) re-transformation of yeast strains with identified interactors and bait; (5) repetition of the experiment with vectors, in which the bait has been fused to AD and the identified prey interactor has been fused to the BD, so as to avoid artifacts associated with the particular fusion used originally. The interaction of pGBKT7-53 (containing p53 coding sequence) and pGADT7-T (containing T antigen coding sequence) was used as a positive control. Only those candidate interactors that passed these stringent tests were considered worthy of further investigation.

The coding sequences of each of the effector candidates, lacking signal peptides and stop codon, were PCR amplified using cDNA as template, generated from fungal infected *S. latifolia* floral buds, using the primer pairs described in Table S3. The effector candidates tested in this study were MVLG_004106, MVLG_005720, MVLG_06175, and MVLG_001732, for which sequences are available in the JGI Fungal Genome database [66]. The PCR products were cloned into the pCR 2.1 TOPO entry vector (Invitrogen/Thermo Fisher, Waltham, MA, USA). *Escherichia coli* strains, DH5 α (Bethesda research Laboratories, Bethesda, MD, USA), were utilized for all cloning purposes. Plasmid DNA was isolated and the inserts were digested out of this vector with *Eco*RI and *Bam*HI. Purified fragments were subsequently cloned into a pGBKT7 destination vector (Clontech) where transcription of the cloned gene would be driven by an *ADH1* promoter, producing fusion proteins at their N termini with the DNA binding domain of the Gal 4 transcription factor.

S. cerevisiae strain Y187 (Library host strain) (MATα, ura3-52, his3-200, ade2-101, trp1-901, leu2-3, 112, gal4Δ, met–, gal80Δ, URA3 : : GAL1$_{UAS}$-GAL1$_{TATA}$-lacZ) [67], containing the *MEL1/lacz* reporter gene, was transformed with the prey vector containing the cDNA library using the Frozen-EZ Yeast Transformation II kit (Zymo Research) and selected on SD drop out medium lacking Leucine (SD/-Leu). MELIBIASE1 (*MEL1*) reporter gene encodes α-galactosidase and enables yeast cells to turn blue-green in the presence of the chromogenic substrate, 5-bromo-4-chloro-3-indolyl α-D-galactopyranoside (X-α-gal). Cell density of the library was calculated by tittering 10^{-4}, 10^{-5}, 10^{-6}, and 10^{-7} dilutions on SD/-Leu plates.

The AH109 yeast strain (Mating partner) (MATa, trp1-901, leu2-3, 112, ura3-52, his3-200, gal4Δ, gal80Δ, LYS2::GAL1$_{UAS}$-GAL1$_{TATA}$-HIS3, GAL2$_{UAS}$-GAL2$_{TATA}$-ADE2, URA3::MEL1$_{UAS}$-MEL1$_{TATA}$-lacZ) [68], containing *HIS3*, *ADE2*, and *MEL1/lacz*) reporter genes, was used as the host for the bait constructs. The *HIS3*, *ADE2* reporter gene products enable the cells to biosynthesize required nutrients to grow on plates lacking histidine and adenine. The three reporter genes are under the control of distinct *GAL4* upstream sequences and promoter elements *GAL1*, *GAL2*, and *MEL1*, respectively, yielding strong and specific responses. In AH109, the entire *HIS3* promoter (including both TATA boxes) was replaced by the entire *GAL1* promoter, leading to tight regulation of the *HIS3* reporter gene in this strain. The bait constructs were transformed into AH109 by the lithium acetate/single-stranded carrier DNA/PEG method [63] and selected on SD drop out medium lacking Trp.

The yeast two-hybrid screening was conducted following the Matchmaker Library Construction and Screening Kits User manual (Clontech). Initial screening was conducted on high stringent quadruple drop out media (QDO) SD/-Ade/-His/-Leu/-Trp plates with X-α-gal and 5 mM 3AT. Subsequently, colonies were restreaked onto QDO/X-α-gal plates with 5 mM 3AT initially, and then on to QDO/x-α-gal plates containing 50 mM 3AT to select strong interactors. 3AT was used to inhibit the leaky expression that reduces the effectiveness of histidine selection, and to inhibit X-α-gal to allow detection of the *MEL1/lacz* reporter. 3-AT is a competitive inhibitor of the yeast HIS3 protein (His3p), blocking low levels of His3p expression, and thus suppressing background growth on SD medium lacking His. Only the positive blue-green clones (indicating α-galactosidase activity) that survived on the highest 3AT levels were used for further screening. To estimate the mating efficiency and to calculate the total number of screened colonies dilution serials were prepared and 100 µL of each dilution was spread on SD/-Trp, SD/-leu, and DDO plates. Plasmids were isolated from the surviving colonies and were individually used to transform *E. coli*. The prey plasmids were isolated from *E. coli*, and sequenced and analyzed by BLAST screens against the NCBI database [69]. The dropout medium contained Glucose (20 g·L^{-1}), yeast nitrogen base (6.7 g·L^{-1}), appropriate dropout mix (2 g·L^{-1}), agar (15 g·L^{-1}), and water.

5. Conclusions

In this paper we identified, for the first time, the interactors of the putative effectors of *M. lychnidis-dioicae*. We believe that the protein product of MVLG_04106 codes for a transcriptional regulator/activator of host responses to allow successful infection. The fungal interactors of MVLG_05720 protein product, MVLG_07305 and MVLG_04026, could potentially sequester this effector until it is required during infection. MVLG_06175 appears to interact with a CASP-like homologue and may be involved in cell-cell junctions. The identification of two host interactors of MVLG_01732-AtCLB and CSI I, which play roles in anther/pollen development and dehiscence, provides exciting targets for future studies, as we hypothesize this effector may be crucial in redirecting anther and pollen development in such a way as to benefit the reproductive program of the fungus. Plant infection studies with knockouts or over expression of these effector genes will further our understanding in characterizing the function of these key players in the infection. This strongly suggests the need to also characterize the remaining candidate effector proteins for a more complete understanding of the mechanisms of infection and development of this fascinating plant parasite.

Supplementary Materials: Supplementary materials can be found at www.mdpi.com/1422-0067/18/11/2489/s1.

Acknowledgments: The authors are indebted to Jocelyn Rose for the components of the yeast secretion trap system and to Jen Goeckler-Fried and Jeff Brodsky for yeast strain SEY6210. We also thank two anonymous reviewers whose constructive criticisms led to an improved final article. This project was supported by National Science Foundation (NSF) award #0947963 to MHP and by National Institutes of Health (NIH) sub-award#OGMB131493C1 to MHP from grant P20GM103436 (Nigel Cooper, PI). The contents of this work are solely the responsibility of the authors and do not represent the official views of the NIH. No funds were received for covering the costs to publish in open access.

Author Contributions: Venkata S. Kuppireddy and Michael H. Perlin conceived and designed the experiments; Venkata S. Kuppireddy, Su San Toh, Catarina Cahill, Brittany Carman, William C. Beckerson, and Ming-Chang Tsai performed the experiments; Venkata S. Kuppireddy, Vladimir N. Uversky, Ming-Chang Tsai, William C. Beckerson, and Michael H. Perlin analyzed the data; Venkata S. Kuppireddy, Vladimir N. Uversky, and Michael H. Perlin wrote the paper.

Conflicts of Interest: The authors declare no conflict of interest.

References

1. Ellis, J.; Catanzariti, A.-M.; Dodds, P. The problem of how fungal and oomycete avirulence proteins enter plant cells. *Trends Plant Sci.* **2006**, *11*, 61–63. [CrossRef] [PubMed]

2. Kemen, E.; Kemen, A.; Ehlers, A.; Voegele, R.; Mendgen, K. A novel structural effector from rust fungi is capable of fibril formation. *Plant J. Cell Mol. Biol.* **2013**, *75*, 767–780. [CrossRef] [PubMed]

3. Giraud, T.; Yockteng, R.; Lopez-Villavicencio, M.; Refregier, G.; Hood, M.E. Mating system of the anther smut fungus Microbotryum violaceum: Selfing under heterothallism. *Eukaryot. Cell* **2008**, *7*, 765–775. [CrossRef] [PubMed]

4. Uchida, W.; Matsunaga, S.; Sugiyama, R.; Kazama, Y.; Kawano, S. Morphological development of anthers induced by the dimorphic smut fungus Microbotryum violaceum in female flowers of the dioecious plant Silene latifolia. *Planta* **2003**, *218*, 240–248. [CrossRef] [PubMed]

5. Sloan, D.B.; Giraud, T.; Hood, M.E. Maximized virulence in a sterilizing pathogen: The anther-smut fungus and its co-evolved hosts. *J. Evol. Biol.* **2008**, *21*, 1544–1554. [CrossRef] [PubMed]

6. Toh, S.S.; Chen, Z.; Schultz, D.J.; Cuomo, C.A.; Perlin, M.H. Transcriptional analysis of mating and pre-infection stages of the anther smut, Microbotryum lychnidis-dioicae. *Microbiology* **2017**. [CrossRef] [PubMed]

7. Perlin, M.H.; Amselem, J.; Fontanillas, E.; Toh, S.S.; Chen, Z.; Goldberg, J.; Duplessis, S.; Henrissat, B.; Young, S.; Zeng, Q.; et al. Sex and parasites: Genomic and transcriptomic analysis of Microbotryum lychnidis-dioicae, the biotrophic and plant-castrating anther smut fungus. *BMC Genom.* **2015**, *16*, 461. [CrossRef] [PubMed]

8. Dunker, A.K.; Cortese, M.S.; Romero, P.; Iakoucheva, L.M.; Uversky, V.N. Flexible nets. The roles of intrinsic disorder in protein interaction networks. *FEBS J.* **2005**, *272*, 5129–5148. [CrossRef] [PubMed]

9.   Patil, A.; Nakamura, H. Disordered domains and high surface charge confer hubs with the ability to interact with multiple proteins in interaction networks. *FEBS Lett.* **2006**, *580*, 2041–2045. [CrossRef] [PubMed]
10. Ekman, D.; Light, S.; Bjorklund, A.K.; Elofsson, A. What properties characterize the hub proteins of the protein-protein interaction network of Saccharomyces cerevisiae? *Genome Biol.* **2006**, *7*, R45. [CrossRef] [PubMed]
11. Haynes, C.; Oldfield, C.J.; Fei, J.; Klitgord, N.; Cusick, M.E.; Radivojac, P.; Uversky, V.N.; Vidal, M.; Iakoucheva, L.M. Intrinsic Disorder Is a Common Feature of Hub Proteins from Four Eukaryotic Interactomes. *PLoS Comput. Biol.* **2006**, *2*, e100. [CrossRef] [PubMed]
12. Dosztányi, Z.; Chen, J.; Dunker, A.K.; Simon, I.; Tompa, P. Disorder and Sequence Repeats in Hub Proteins and Their Implications for Network Evolution. *J. Proteome Res.* **2006**, *5*, 2985–2995. [CrossRef] [PubMed]
13. Singh, G.P.; Dash, D. Intrinsic disorder in yeast transcriptional regulatory network. *Proteins* **2007**, *68*, 602–605. [CrossRef] [PubMed]
14. Singh, G.P.; Ganapathi, M.; Dash, D. Role of intrinsic disorder in transient interactions of hub proteins. *Proteins* **2007**, *66*, 761–765. [CrossRef] [PubMed]
15. Mohan, A.; Sullivan, W.J., Jr.; Radivojac, P.; Dunker, A.K.; Uversky, V.N. Intrinsic disorder in pathogenic and non-pathogenic microbes: Discovering and analyzing the unfoldomes of early-branching eukaryotes. *Mol. Biosyst.* **2008**, *4*, 328–340. [CrossRef] [PubMed]
16. Dolan, P.T.; Roth, A.P.; Xue, B.; Sun, R.; Dunker, A.K.; Uversky, V.N.; LaCount, D.J. Intrinsic disorder mediates hepatitis C virus core-host cell protein interactions. *Protein Sci.* **2015**, *24*, 221–235. [CrossRef] [PubMed]
17. Schneider, D.R.; Saraiva, A.M.; Azzoni, A.R.; Miranda, H.R.; de Toledo, M.A.; Pelloso, A.C.; Souza, A.P. Overexpression and purification of PWL2D, a mutant of the effector protein PWL2 from Magnaporthe grisea. *Protein Expr. Purif.* **2010**, *74*, 24–31. [CrossRef] [PubMed]
18. Peng, K.; Radivojac, P.; Vucetic, S.; Dunker, A.K.; Obradovic, Z. Length-dependent prediction of protein intrinsic disorder. *BMC Bioinform.* **2006**, *7*, 208. [CrossRef] [PubMed]
19. Romero, P.; Obradovic, Z.; Li, X.H.; Garner, E.C.; Brown, C.J. Sequence complexity of disordered protein. *Proteins* **2001**, *42*, 38–48. [CrossRef]
20. Peng, K.; Vucetic, S.; Radivojac, P.; Brown, C.J.; Dunker, A.K.; Obradovic, Z. Optimizing long intrinsic disorder predictors with protein evolutionary information. *J. Bioinform. Comput. Biol.* **2005**, *3*, 35–60. [CrossRef] [PubMed]
21. Xue, B.; Dunbrack, R.L.; Williams, R.W.; Dunker, A.K.; Uversky, V.N. PONDR-FIT: A meta-predictor of intrinsically disordered amino acids. *Biochim. Biophys. Acta* **2010**, *1804*, 996–1010. [CrossRef] [PubMed]
22. Dosztányi, Z.; Meszaros, B.; Simon, I. ANCHOR: Web server for predicting protein binding regions in disordered proteins. *Bioinformatics* **2009**, *25*, 2745–2746. [CrossRef] [PubMed]
23. Mészáros, B.; Simon, I.; Dosztanyi, Z. Prediction of protein binding regions in disordered proteins. *PLoS Comput. Biol.* **2009**, *5*, e1000376. [CrossRef] [PubMed]
24. Dosztányi, Z.; Csizmok, V.; Tompa, P.; Simon, I. The pairwise energy content estimated from amino acid composition discriminates between folded and intrinsically unstructured proteins. *J. Mol. Biol.* **2005**, *347*, 827–839. [CrossRef] [PubMed]
25. Peng, Z.; Kurgan, L. On the complementarity of the consensus-based disorder prediction. *Pac. Symp. Biocomput.* **2012**, 176–187.
26. Fan, X.; Kurgan, L. Accurate prediction of disorder in protein chains with a comprehensive and empirically designed consensus. *J. Biomol. Struct. Dyn.* **2014**, *32*, 448–464. [CrossRef] [PubMed]
27. Walsh, I.; Giollo, M.; Di Domenico, T.; Ferrari, C.; Zimmermann, O. Comprehensive large-scale assessment of intrinsic protein disorder. *Bioinformatics* **2015**, *31*, 201–208. [CrossRef] [PubMed]
28. Lee, S.J.; Rose, J.K. A yeast secretion trap assay for identification of secreted proteins from eukaryotic phytopathogens and their plant hosts. *Methods Mol. Biol.* **2012**, *835*, 519–530. [PubMed]
29. Lu, S.; Edwards, M.C. Genome-Wide analysis of small secreted cysteine-rich proteins identifies candidate effector proteins potentially involved in fusarium graminearum-wheat interactions. *Phytopathology* **2016**, *106*, 166–176. [CrossRef] [PubMed]
30. Liu, J.; Perumal, N.B.; Oldfield, C.J.; Su, E.W.; Uversky, V.N. Intrinsic disorder in transcription factors. *Biochemistry* **2006**, *45*, 6873–6888. [CrossRef] [PubMed]
31. Bhalla, J.; Storchan, G.B.; MacCarthy, C.M.; Uversky, V.N.; Tcherkasskaya, O. Local flexibility in molecular function paradigm. *Mol. Cell. Proteom.* **2006**, *5*, 1212–1223. [CrossRef] [PubMed]

32. Minezaki, Y.; Homma, K.; Kinjo, A.R.; Nishikawa, K. Human transcription factors contain a high fraction of intrinsically disordered regions essential for transcriptional regulation. *J. Mol. Biol.* **2006**, *359*, 1137–1149. [CrossRef] [PubMed]

33. Pejaver, V.; Hsu, W.-L.; Xin, F.; Dunker, A.K.; Uversky, V.N.; Radivojac, P. The structural and functional signatures of proteins that undergo multiple events of post-translational modification. *Protein Sci.* **2014**, *23*, 1077–1093. [CrossRef] [PubMed]

34. Branco, S.; Badouin, H.; Rodríguez de la Vega, R.C.; Gouzy, J.; Carpentier, F.; Aguileta, G.; Siguenza, S.; Brandenburg, J.-T.; Coelho, M.A.; Hood, M.E.; et al. Evolutionary strata on young mating-type chromosomes despite the lack of sexual antagonism. *Proc. Natl. Acad. Sci. USA* **2017**, *114*, 7067–7072. [CrossRef] [PubMed]

35. Branco, S.; Carpentier, F.; de la Vega, R.C.R.; Badouin, H.; Snirc, A.; le Prieur, S.; Coelho, M.A.; Bergerow, D.; Hood, M.E.; Giraud, T. Multiple convergent events of supergene evolution in mating-type chromosomes. *Submitt. Nat. Commun.* Unpublished work. **2017**.

36. Fontanillas, E.; Hood, M.E.; Badouin, H.; Petit, E.; Barbe, V.; Gouzy, J.; de Vienne, D.M.; Aguileta, G.; Poulain, J.; Wincker, P.; et al. Degeneration of the nonrecombining regions in the mating-type chromosomes of the anther-smut fungi. *Mol. Biol. Evol.* **2015**, *32*, 928–943. [CrossRef] [PubMed]

37. Dennison, S.R.; Mura, M.; Harris, F.; Morton, L.H.; Zvelindovsky, A.; Phoenix, D.A. The role of C-terminal amidation in the membrane interactions of the anionic antimicrobial peptide, maximin H5. *Biochim. Biophys. Acta* **2015**, *1848*, 1111–1118. [CrossRef] [PubMed]

38. Djamei, A.; Schipper, K.; Rabe, F.; Ghosh, A.; Vincon, V.; Kahnt, J.; Osorio, S.; Tohge, T.; Fernie, A.R.; Feussner, I.; et al. Metabolic priming by a secreted fungal effector. *Nature* **2011**, *478*, 395–398. [CrossRef] [PubMed]

39. Poon, H.; Day, A.W. 'Fimbriae' in the fungus Ustilago violacea. *Nature* **1974**, *250*, 648–649. [CrossRef] [PubMed]

40. Poon, N.H.N.; Day, A.W. Fungal fimbriae. I. Structure, origin, and synthesis. *Can. J. Microbiol.* **1975**, *21*, 537–546. [CrossRef] [PubMed]

41. Celerin, M.; Ray, J.M.; Schisler, N.J.; Day, A.W.; Stetler-Stevenson, W.G.; Laudenbach, D.E. Fungal fimbriae are composed of collagen. *EMBO J.* **1996**, *15*, 4445–4453. [PubMed]

42. Hubmacher, D.; Bergeron, E.; Fagotto-Kaufmann, C.; Sakai, L.Y.; Reinhardt, D.P. Early Fibrillin-1 assembly monitored through a modifiable recombinant cell approach. *Biomacromolecules* **2014**, *15*, 1456–1468. [CrossRef] [PubMed]

43. Reinhardt, D.P.; Gambee, J.E.; Ono, R.N.; Bächinger, H.P.; Sakai, L.Y. Initial Steps in Assembly of Microfibrils: Formation of disulfide-cross-linked multimers containing fibrillin-1. *J. Biol. Chem.* **2000**, *275*, 2205–2210. [CrossRef] [PubMed]

44. Roppolo, D.; Boeckmann, B.; Pfister, A.; Boutet, E.; Rubio, M.C.; Dénervaud-Tendon, V.; Vermeer, J.E.M.; Gheyselinck, J.; Xenarios, I.; Geldner, N. Functional and evolutionary analysis of the casparian strip membrane domain protein family. *Plant Physiol.* **2014**, *165*, 1709–1722. [CrossRef] [PubMed]

45. De Silva, K.; Laska, B.; Brown, C.; Sederoff, H.W.; Khodakovskaya, M. Arabidopsis thaliana calcium-dependent lipid-binding protein (AtCLB): A novel repressor of abiotic stress response. *J. Exp. Bot.* **2011**, *62*, 2679–2689. [CrossRef] [PubMed]

46. Griffiths, E.J.; Hu, G.; Fries, B.; Caza, M.; Wang, J. A defect in ATP-citrate lyase links acetyl-CoA production, virulence factor elaboration and virulence in Cryptococcus neoformans. *Mol. Microbiol.* **2012**, *86*, 1404–1423. [CrossRef] [PubMed]

47. Meyers, B.C.; Kozik, A.; Griego, A.; Kuang, H.; Michelmore, R.W. Genome-wide analysis of NBS-LRR-encoding genes in arabidopsis. *Plant Cell* **2003**, *15*, 809–834. [CrossRef] [PubMed]

48. Wang, Y.; Zhang, W.-Z.; Song, L.-F.; Zou, J.-J.; Su, Z.; Wu, W.-H. Transcriptome analyses show changes in gene expression to accompany pollen germination and tube growth in arabidopsis. *Plant Physiol.* **2008**, *148*, 1201–1211. [CrossRef] [PubMed]

49. Wang, X.; Sato, R.; Brown, M.S.; Hua, X.; Goldstein, J.L. SREBP-1, a membrane-bound transcription factor released by sterol-regulated proteolysis. *Cell* **1994**, *77*, 53–62. [CrossRef]

50. Saheki, Y.; De Camilli, P. The Extended-Synaptotagmins. *Biochim. Biophys. Acta* **2017**, *1864*, 1490–1493. [CrossRef] [PubMed]

51. Coussens, L.; Parker, P.J.; Rhee, L.; Yang-Feng, T.L.; Chen, E.; Waterfield, M.D.; Francke, U.; Ullrich, A. Multiple, distinct forms of bovine and human protein kinase C suggest diversity in cellular signaling pathways. *Science* **1986**, *233*, 859–866. [CrossRef] [PubMed]

52. Mei, Y.; Gao, H.B.; Yuan, M.; Xue, H.W. The Arabidopsis ARCP protein, CSI1, which is required for microtubule stability, is necessary for root and anther development. *Plant Cell* **2012**, *24*, 1066–1080. [CrossRef] [PubMed]

53. Lalanne, E.; Michaelidis, C.; Moore, J.M.; Gagliano, W.; Johnson, A.; Patel, R.; Howden, R.; Vielle-Calzada, J.P.; Grossniklaus, U.; Twell, D. Analysis of transposon insertion mutants highlights the diversity of mechanisms underlying male progamic development in Arabidopsis. *Genetics* **2004**, *167*, 1975–1986. [CrossRef] [PubMed]

54. Malhis, N.; Jacobson, M.; Gsponer, J. MoRFchibi SYSTEM: Software tools for the identification of MoRFs in protein sequences. *Nucleic Acids Res.* **2016**, *44*, 488–493. [CrossRef] [PubMed]

55. Biasini, M.; Bienert, S.; Waterhouse, A.; Arnold, K.; Studer, G.; Schmidt, T.; Kiefer, F.; Gallo Cassarino, T.; Bertoni, M.; Bordoli, L.; et al. SWISS-MODEL: Modelling protein tertiary and quaternary structure using evolutionary information. *Nucleic Acids Res.* **2014**, *42*, W252–W258. [CrossRef] [PubMed]

56. Arnold, K.; Bordoli, L.; Kopp, J.; Schwede, T. The SWISS-MODEL workspace: A web-based environment for protein structure homology modelling. *Bioinformatics* **2006**, *22*, 195–201. [CrossRef] [PubMed]

57. Petersen, T.N.; Brunak, S.; von Heijne, G.; Nielsen, H. Signalp 4.0: Discriminating signal peptides from transmembrane regions. *Nat. Methods* **2011**, *8*, 785–786. [CrossRef] [PubMed]

58. Sigrist, C.J.; de Castro, E.; Cerutti, L.; Cuche, B.A.; Hulo, N.; Bridge, A.; Bougueleret, L.; Xenarios, I. New and continuing developments at prosite. *Nucleic Acids Res.* **2013**, *41*, D344–D347. [CrossRef] [PubMed]

59. De Castro, E.; Sigrist, C.J.; Gattiker, A.; Bulliard, V.; Langendijk-Genevaux, P.S.; Gasteiger, E.; Bairoch, A.; Hulo, N. Scanprosite: Detection of prosite signature matches and prorule-associated functional and structural residues in proteins. *Nucleic Acids Res.* **2006**, *34*, W362–W365. [CrossRef] [PubMed]

60. Berardini, T.Z.; Reiser, L.; Li, D.; Mezheritsky, Y.; Muller, R.; Strait, E.; Huala, E. The Arabidopsis information resource: Making and mining the "gold standard" annotated reference plant genome. *Genesis* **2015**, *53*, 474–485. [CrossRef] [PubMed]

61. Sambrook, J.R.D. *Molecular Cloning: A Laboratory Manual*; Cold Spring Harbor Laboratory Press: Cold Spring Harbor, NY, USA, 2001.

62. Robinson, J.S.; Klionsky, D.J.; Banta, L.M.; Emr, S.D. Protein sorting in Saccharomyces cerevisiae: Isolation of mutants defective in the delivery and processing of multiple vacuolar hydrolases. *Mol. Cell. Biol.* **1988**, *8*, 4936–4948. [CrossRef] [PubMed]

63. Gietz, R.D.; Schiestl, R.H. High-efficiency yeast transformation using the LiAc/SS carrier DNA/PEG method. *Nat. Protoc.* **2007**, *2*, 31–34. [CrossRef] [PubMed]

64. McAlister-Henn, L.; Gibson, N.; Panisko, E. Applications of the Yeast Two-Hybrid System. *Methods* **1999**, *19*, 330–337. [CrossRef] [PubMed]

65. Greener, M.; Perkel, J.M. The yeast two-hybrid assay. *Scientist* **2005**, *19*, 32.

66. Nordberg, H.; Cantor, M.; Dusheyko, S.; Hua, S.; Poliakov, A.; Shabalov, I.; Smirnova, T.; Grigoriev, I.V.; Dubchak, I. The genome portal of the Department of Energy Joint Genome Institute: 2014 updates. *Nucleic Acids Res.* **2014**, *42*, D26–D31. [CrossRef] [PubMed]

67. Wade Harper, J.; Adami, G.R.; Wei, N.; Keyomarsi, K.; Elledge, S.J. The p21 Cdk-interacting protein Cip1 is a potent inhibitor of G1 cyclin-dependent kinases. *Cell* **1993**, *75*, 805–816. [CrossRef]

68. James, P.; Halladay, J.; Craig, E.A. Genomic Libraries and a Host Strain Designed for Highly Efficient Two-Hybrid Selection in Yeast. *Genetics* **1996**, *144*, 1425–1436. [PubMed]

69. Altschul, S.F.; Gish, W.; Miller, W.; Myers, E.W.; Lipman, D.J. Basic local alignment search tool. *J. Mol. Biol.* **1990**, *215*, 403–410. [CrossRef]

International Journal of
Molecular Sciences

MDPI

Article

Transcriptome Analysis of Tomato Leaf Spot Pathogen *Fusarium proliferatum*: *De novo* Assembly, Expression Profiling, and Identification of Candidate Effectors

Meiling Gao [1,2], Siyu Yao [3], Yang Liu [2], Haining Yu [1], Pinsan Xu [1], Wenhui Sun [1], Zhongji Pu [1], Hongman Hou [2,*] and Yongming Bao [1,4,*]

[1] School of Life Science and Biotechnology, Dalian University of Technology, Dalian 116024, China;
    gaomeiling@mail.dlut.edu.cn (M.G.); yuhaining@dlut.edu.cn (H.Y.); xupinsan@dlut.edu.cn (P.X.);
    sunwenhui123@mail.dlut.edu.cn (W.S.); puzhongji@mail.dlut.edu.cn (Z.P.)
[2] School of Food Science and Technology, Dalian Polytechnic University, Dalian 116034, China;
    liuyang_vsw@126.com
[3] College of Food Science and Engineering, Northwest A&F University, Yangling 712100, China;
    yaosiyu@nwafu.edu.cn
[4] School of Food and Environmental Science and Technology, Dalian University of Technology,
    Panjin 124221, China
*   Correspondence: houhongman@dlpu.edu.cn (H.H.); biosci@dlut.edu.cn (Y.B.);
    Tel.: +86-411-8232-2020 (H.H.); +86-427-263-1777 (Y.B.)

Received: 10 October 2017; Accepted: 17 December 2017; Published: 22 December 2017

Abstract: Leaf spot disease caused by the fungus *Fusarium proliferatum* (Matsushima) Nirenberg is a destructive disease of tomato plants in China. Typical symptoms of infected tomato plants are softened and wilted stems and leaves, leading to the eventual death of the entire plant. In this study, we resorted to transcriptional profile analysis to gain insight into the repertoire of effectors involved in *F. proliferatum*–tomato interactions. A total of 61,544,598 clean reads were *de novo* assembled to provide a *F. proliferatum* reference transcriptome. From these, 75,044 unigenes were obtained, with 19.46% of the unigenes being assigned to 276 Kyoto Encyclopedia of Genes and Genomes (KEGG) pathways, with 22.3% having a homology with genes from *F. fujikuroi*. A total of 18,075 differentially expressed genes (DEGs) were identified, 720 of which were found to code for secreted proteins. Of these, 184 were identified as candidate effectors, while 79.89% had an upregulated expression. Moreover, 17 genes that were differentially expressed in RNA-seq studies were randomly selected for validation by quantitative real-time polymerase chain reaction (qRT–PCR). The study demonstrates that transcriptome analysis could be an effective method for identifying the repertoire of candidate effectors and may provide an invaluable resource for future functional analyses of *F. proliferatum* pathogenicity in *F. proliferatum* and tomato plant–host interactions.

Keywords: *Fusarium proliferatum*; tomato plants; effector; pathogenicity; DEGs

1. Introduction

The tomato (*Solanum lycopersicum*) is one of the most important crop plants and one that also serves as a model system for fruit development [1]. According to the Food and Agriculture Organization Statistical Database (http://faostat3.fao.org), China was the largest tomato producer in 2013, with a total output of 50 million tons [2]. However, Chinese tomato production is threatened by diseases. Among these is tomato leaf spot caused by *Fusarium proliferatum* (Matsushima) Nirenberg, one of the most destructive fungal diseases for tomatoes [3]. The typical symptoms of tomato leaf spot on infected leaves and stems are necrotic spots that have a dark brown appearance and may continue to grow, causing the stems to soften and wilt, eventually leading to the death of the entire plant [3].

Fusarium proliferatum, a broadly distributed saprophytic pathogen, can cause destructive diseases to an extremely wide range of hosts that span several plant families, including maize [4], tomato [3], garlic [5] and soybean [6]. Additionally, *F. proliferatum* has been reported to produce a number of mycotoxins (including fusaric acid, fumonisin, fusaproliferin, beauvericin and moniliformin) that pose a serious threat to global food security and human health [7–10]. *Fusarium proliferatum* is an endophyte that dwells on the plant and produces a large number of conidia that can survive for many years in the soil [11]. As the weather becomes wetter and warmer, the conidia germinate and spread via atmospheric dust and rainwater movement. As a result of this, they end up infecting seeds, soil, as well as other plant materials. The germinated conidia enter the tomato plants through the stomata, and they form hyphae that grow along the vascular tissues and extend into the leaves and stem. Subsequently, the mycelia colonize the tomato plants, and this results in the appearance of leaf spots. As of now, the *F. oxysporum* infection process in tomatoes has been well documented [12–14], but *F. proliferatum's* infection process on tomatoes has not been as thoroughly described. This is therefore an important problem that needs urgent solving.

At present, fungicides are the main management strategy for controlling fungal tomato diseases, but there is a lack of research on the response of *F. proliferatum* to different fungicides. Additionally, tomato cultivars resistant to tomato leaf spot are currently unavailable in the conventional market. Even if these management strategies were available, they would face a huge challenge due to the genetic variability of *F. proliferatum* rapidly emerging in the population [15].

Numerous genes of the filamentous plant pathogens have been shown to undergo diversified selection during the host–pathogen interaction [16]. More recently, it has been recognized that evolution has equipped *Fusarium* plant pathogens with a diverse range of infection strategies. These include the production and secretion of proteins and other molecules, collectively known as effectors, that successfully facilitate the infection process by reprogramming the host metabolism and by manipulating the immune responses of host cells to enable parasitic colonization [16,17]. Effectors are active outside the fungal cell and alter the host-cell structure and its function in order to generally facilitate the fungal lifestyle inside the plant and enhance access to nutrients [18]. Fungal effectors that trigger resistance or susceptibility in specific host plants have been identified in a number of ascomycetes that include *Fusarium* species. Most recent molecular studies of *Fusarium* pathogens have focused on investigating the secreted effectors produced by the pathogens during infection. Here, the use of transcriptomic analysis has helped identify secreted effectors in various *Fusarium* pathogens, such as the *F. oxysporum* species complex, *F. graminearum*, *F. verticillioides* and *F. virguliforme* [13,19–21]. For example, several effector genes secreted in xylem (*SIX*) were found during the *F. oxysporum*, f. sp. *Lycopersici*/tomato interaction, some of which were shown to be essential for pathogenicity [12–14]. Using transcriptional analysis, Lu and Edwards [19] have identified a list of potential small, secreted cysteine-rich protein-derived effectors produced by *F. graminearum* in the course of *F. graminearum*–wheat interaction. Brown et al. [20] indicated that secreted gene expression (*SGE1*) was required for pathogenicity and can affect synthesis of multiple secondary metabolites. *SGE1* has a role in the global regulation of transcription in *F. verticillioides*. Unraveling the secreted effectors of *F. proliferatum* produced during pathogenesis is therefore important to improve the control strategies for this pathogen.

The *F. proliferatum* genome has still not been sequenced, but genome sequences of several *Fusarium* species (*F. oxysporum*, *F. fujiluroi* and *F. graminearum*) have been deposited in the National Center of Biotechnology Information (NCBI) database. These can provide a rich trove of reference information from which to build an effective strategy for gaining more insight into effectors secreted in the *F. proliferatum*–tomato interaction [22–24]. Next-generation sequencing technologies have advanced rapidly, with ribonucleic acid sequencing (RNA-seq) becoming an instrumental assay for the analysis of fungal transcriptomes. By combing these with bioinformatics tools, putative secreted proteins can be predicted; of these, some that are relatively small (fewer than 200 amino acids) and contain a high percentage of cysteine residues (usually 2% to 20%) have been considered as effector molecules [18].

As of now, *F. proliferatum*'s infection process on tomatoes has not yet been as thoroughly described. In this work, we characterized the infection process of *F. proliferatum* in tomatoes in order to better understand the molecular basis of the *F. proliferatum*–tomato interaction. We then performed a *de novo* transcriptome analysis to predict putative effectors that may be contributing to the pathogenicity of *F. proliferatum* in tomatoes.

2. Results and Discussion

2.1. Characterization of Time Course of F. proliferatum-Infected Tomato Leaves

In order to obtain an overview of the *F. proliferatum* transcriptome and effector gene activity during the different phases of infection, we used a scanning electron microscope to observe six samples of *F. proliferatum*-infected tomato leaves at different infection time points (0, 12, 24, 48, 72 and 96 h after *F. proliferatum* inoculation) (Figure 1). Irregular epidermal cells were found on these leaves and the extent of this irregularity grew with a longer infection time. By 48 h post-inoculation (hpi), the stomata were invaded by the spores (Figure 1), and by 72 hpi, the number of spores on the leaf surface increased, accompanied by a corresponding increase in the number of invading spores on the stomata. By 96 hpi, the stomata and epidermal cells of the tomato leaves were covered with spores, and the leaves's epidermal cells became more irregular and wrinkled. At the same time, the hyphae showed obvious growth on the stomata (Figure S1). Nguyen et al. [11] also showed that infection of maize leaf tissues by *F. proliferatum* occurred via the stomata, and that the microconidia of *F. proliferatum* that formed inside the leaf tissues sporulated through the stomata, which may provide nutrients to the pathogen. Based on this data, mycelia collected from 7-day-old potato dextrose agar (PDA) cultures (KC) and samples of infected leaf tissues taken 96 hpi (KS_1) were subjected to an RNA-seq analysis to reveal more candidate effector genes involved in *F. proliferatum*–tomato interactions.

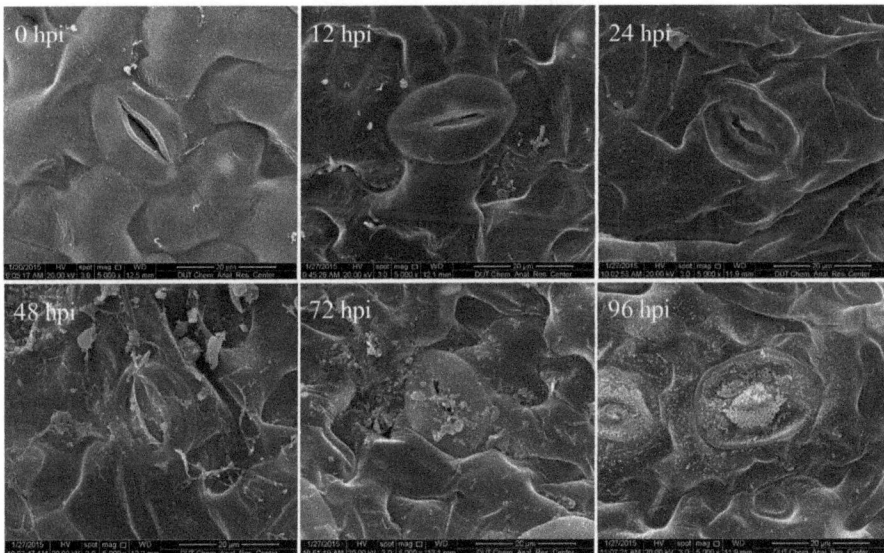

Figure 1. Preparation of *Fusarium proliferatum*-infected tomato leaves (*Solanum lycopersicum*) for transcriptome analysis. Leaves were obtained 0, 12, 24, 48, 72 and 96 h after inoculation (hpi) with a conidial suspension of *F. proliferatum*.

2.2. De novo Assembly of F. proliferatum Transcriptome

By using RNA-seq technology, two transcriptomic datasets were generated from KC and KS_1 in order to better understand the pathogenicity of *F. proliferatum*. We obtained a total of 14.16 Gb of sequencing data, including 115,988,950 raw reads and 98,705,420 clean reads with a base average error rate below 0.03%. An overview of the transcriptome assembly statistics is shown in Table 1. After removing low-quality and adapter sequences, 37,091,012 and 61,614,408 clean reads were obtained for KC and KS_1 samples, respectively. The Q20 percentage and Q30 percentages were more than 96% and 92%, respectively. The average GC percentages in the KC and KS_1 samples were 51.58% and 53.15%, respectively.

Table 1. Summary of the RNA-Seq data.

cDNA Library	Raw Reads	Clean Reads	Clean Bases (Gb)	Error (%)	Q20 (%) [1]	Q30 (%) [2]	GC (%)
KC	52,509,182	37,091,012	6.46	0.03	96.28	92.16	51.58
KS_1	63,479,768	61,614,408	7.70	0.03	96.13	92.29	53.15

[1] Q20: percentage of bases with a Phred value >20; [2] Q30: percentage of bases with a Phred value >30.

To establish the *F. proliferatum* transcriptome in the absence of a reference genome, the clean reads of KS_1 were mapped against the tomato genome [25]. A total of 24,453,586 clean reads did not map to the tomato genome. The clean reads of KC were consequently further pooled, yielding a total number of 61,544,598 reads, which were then used to perform the Trinity program for the *de novo* assembly of the *F. proliferatum* reference transcriptome. This analysis yielded 89,716 transcripts expressed from 75,044 unigenes (Table 2). The length of the unigenes ranged from 201 to 17,632 bp, with an N50 length of 1283 bp, a mean length of 767 bp and a median length of 419 bp. The transcript and unigene length distribution is shown in Figure 2. Around 26.80% of the transcripts were longer than 1 kb. To assess the quality of the sequencing and *de novo* assembly, all the assembled clean reads were mapped onto the *F. proliferatum* reference transcriptome. Mapping ratios of 93.31% and 74.65% were obtained for KC and KS_1, respectively.

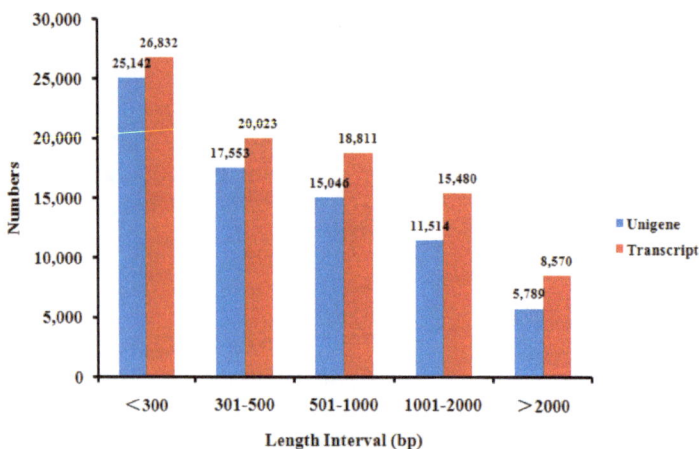

Figure 2. Distribution of transcript and unigene length in the assembled *F. proliferatum* reference transcriptome.

Table 2. Sequence summary of transcriptome assembly statistics.

Category	Total Number	Min Length (bp)	Mean Length (bp)	Median Length (bp)	Max Length (bp)	N50
Transcripts	89,716	201	853	471	17,632	1443
Unigenes	75,044	201	767	419	17,632	1283

2.3. F. proliferatum Reference Transcriptome Annotation

In a search against the NCBI non-redundant (Nr) protein database, the annotation of the *de novo* assembled gene annotation revealed 46,292 unigenes (61.68%) with significant homology hits (*e*-value = 1×10^{-5}) (Table 3). Half of the unigene sequences were more than 95% identical to the mapped sequences in the Nr database, while 70.2% of the unigenes had significant homology hits (*e*-value < 1×10^{-30}) (Figure 3A, B). A total of 39,854 unigenes (53.10%) were matched to the SwissPort database. The mapping rates of the unigenes against the NCBI nucleotide (Nt), Kyoto Encyclopedia of Genes and Genomes Orthology (KO), Protein family (Pfam) and euKaryotic Ortholog Groups (KOG) databases were 83.30%, 28.21%, 50.09% and 32.43%, respectively. As shown in Table 3, a total of 71,878 unigenes (95.78%) were annotated in at least one database. According to a species classification analysis, only 16,734 unigenes (22.3%) had a high homology with the *Fusarium fujikuroi* genes, followed by the *Botrytis cinerea* (4352 unigenes, 5.8%) and *Penicillium oxalicum* (3226 unigenes, 4.3%) genes, while 46,752 unigenes (62.3%) had a high homology with sequences from other organisms (Figure 3C). According to BLASTx results, half of the *F. proliferatum* unigenes did not have any annotation, even though some of them were highly expressed. These unigenes may code for new proteins, which would account for the fact that no homologous genes from other *Fusarium* species could be found in the databases used this study.

Table 3. Annotationresults of the assembled unigenes.

Database	Number of Unigenes	Percentage (%)
Annotated in Nr	46,292	61.68
Annotated in Nt	62,514	83.3
Annotated in KO	21,172	28.21
Annotated in SwissPort	39,854	53.10
Annotated in Pfam	37,595	50.09
Annotated in GO	38,947	51.89
Annotated in KOG	24,338	32.43
Annotated in all databases	12,073	16.08
Annotated in at least one database	71,787	95.78
Total unigenes	75,044	100

KO: Kyoto Encyclopedia of Genes and Genomes Orthology; GO: Gene ontology; KGO: euKaryotic Ortholog Groups.

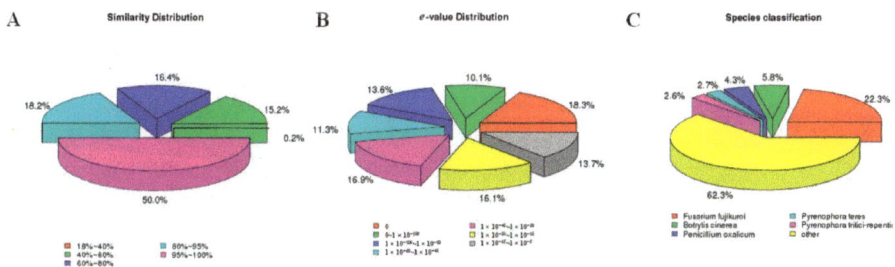

Figure 3. Species distribution of *F. Proliferatum* unigenes. (**A**) Similarity distribution of top Basic Local Alignment Search Tool (BLAST) hits for each unigene; (**B**) *e*-value distribution of BLAST hits with a cut off *e*-value of 1×10^{-5}; (**C**) Species distribution for top BLAST hits in the Nr database.

The gene annotation showed that a total of 38,947 unigenes (51.89%) had at least one annotation characterized by gene ontology (GO) terms. According to the three Blast2GO categories, the GO terms of *F. proliferatum* unigenes could be grouped into the following categories: biological process, molecular function and cellular component (Figure S2). As shown in Table S1, the unigenes were assigned with one or more GO terms. In the biological process category, unigenes annotated to cellular process (18.14%) and single-organism process (14.02%) terms were the most dominant, followed by metabolic process (13.92%) and regulation of biological process (10.02%) terms. In the cellular component category, unigenes assigned to cell part (22.13%), membrane (16.12%) and membrane part (12.02%) terms were highly represented. Some of the unigenes were assigned to cell part terms, as a result of being of a similar category to the recently reported transcriptomes of *Fusarium oxysporum* f. sp. *Ciceris*, the fungus that causes vascular wilt in chickpeas [26]. For the molecular function category, unigenes related to catalytic activity (37.65%) and binding (32.71%) were found to be most abundant.

After the GO analysis, the unigenes were assigned to the KOG database for a functional prediction and classification. A total of 24,338 unigenes were grouped into 26 KOG classifications (Figure 4). A high percentage of unigenes was assigned to the KOG's general function prediction (3674, 15.09%) category, followed by the following categories: posttranslational modification, protein turnover, chaperones (3206, 13.17%), translation, ribosomal structure and biogenesis (2604, 10.69%), energy production and conversion (2014, 8.27%), and signal transduction mechanisms (1924, 7.90%).

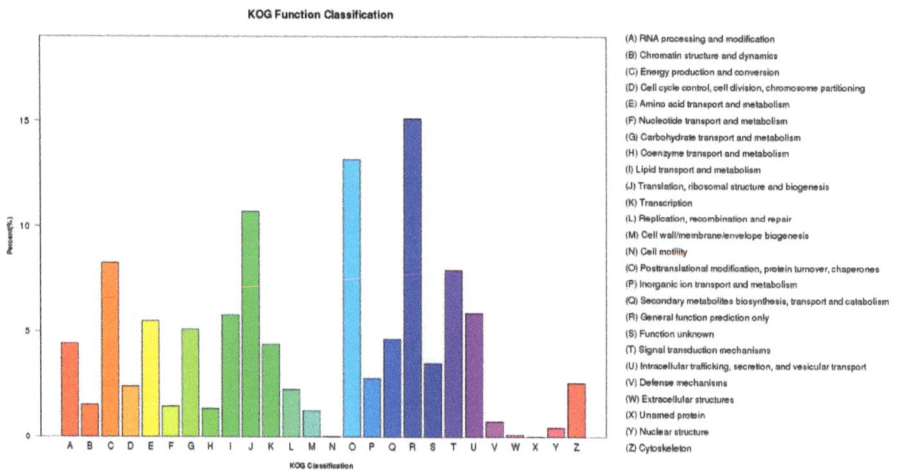

Figure 4. euKaryotic Ortholog Groups (KOG) functional categories for *F. proliferatum*.

Meanwhile, the Kyoto Encyclopedia of Genes and Genomes (KEGG) database was used to search for active biochemical pathways in all unigenes of *F. proliferatum*. A total of 14,607 unigenes were clustered into 276 KEGG pathways (Figure 5 and Table S2), and the most represented classification was based on metabolism categories. Carbohydrate metabolism (2694 unigenes), amino acid metabolism (1879 unigenes), energy metabolism (1651 unigenes) and lipid metabolism (1306 unigenes) were the main metabolic pathways. KEGG organismal-system categories included the endocrine system (1089 unigenes), nervous system (681 unigenes) and immune system (635 unigenes). In the genetic-information processing categories, the most significant enriched KEGG pathways were translation (2619 unigenes), followed by folding, sorting and degradation (1594 unigenes). Additionally, the most abundant subcategories in KEGG environmental-information processing and cellular processes categories were signal transduction (1941 unigenes) and transport and catabolism (1360 unigenes), respectively. The above functional annotations indicated that the clustered unigenes represented an extensive catalog encompassing a large proportion of the genes expressed in *F. proliferatum*.

KEGG Classification

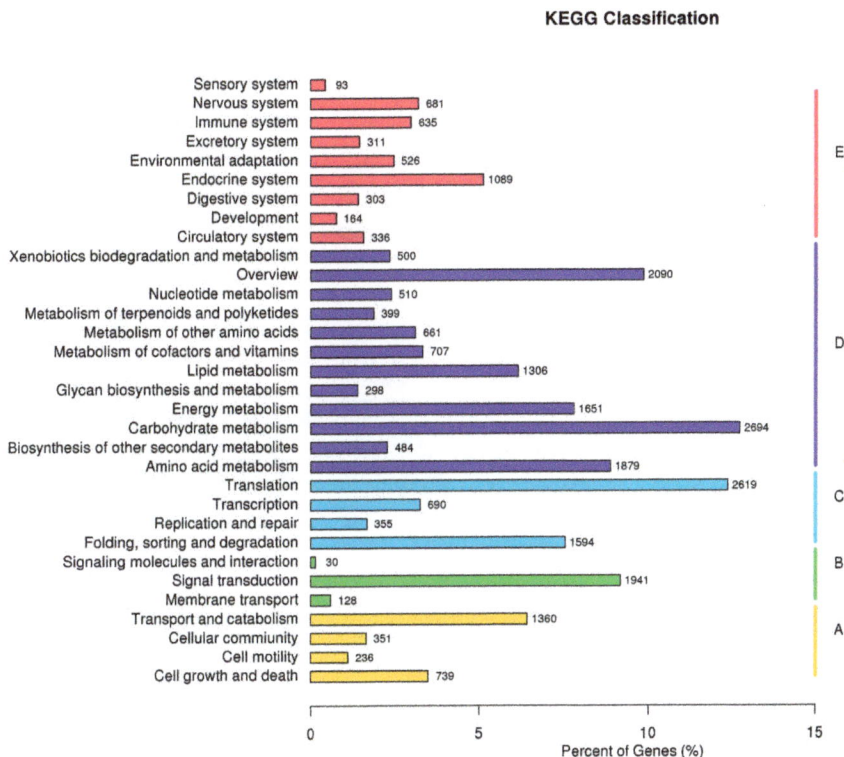

Figure 5. Pathway assignment based on the Kyoto Encyclopedia of Genes and Genomes (KEGG). (**A**) Classification based on cellular processes categories; (**B**) classification based on environmental-information processing categories; (**C**) classification based on genetic-information processing categories; (**D**) classification based on metabolism categories; and (**E**) classification based on organismal-system categories.

2.4. F. proliferatum Differential Gene Expression in KS_1

After mapping against the *de novo F. proliferatum* transcriptome, a total of 18,075 differentially expressed genes (DEGs) were found to display a significant differential expression in KS_1 compared to KC (q-value < 0.005 and $|\log_2 (\text{foldchange})| > 1$) (Figure S3). Overall, the majority of the DEGs (14,766, 81.69%) in KS_1 were upregulated genes, while only 3309 DEGs were downregulated genes, suggesting a strong interaction between *F. proliferatum* and the tomato leaves.

Our transcriptome analysis provides a further view on the expression of *F. proliferatum* genes at 96 hpi. However, it is worth pointing out that our RNA-seq analyses were generated from two treatments, one from KS_1 and the other from KC. As a consequence, the interpretation of the transcriptome data may be potentially biased. To confirm the RNA-seq profiles, quantitative real-time PCR (qRT–PCR) was used to examine 17 randomly selected genes using three independent biological replicates of KS_1 and KC (Figure 6). The fold change ($\log_2$ ratio) was used to validate the results when comparing qRT–PCR gene-expression levels with the RNA-seq gene FPKM (fragments per kilobase of exon per million fragments mapped) values (Table S3). The result indicated that the unigenes' expression profiles obtained from the two approaches were basically consistent.

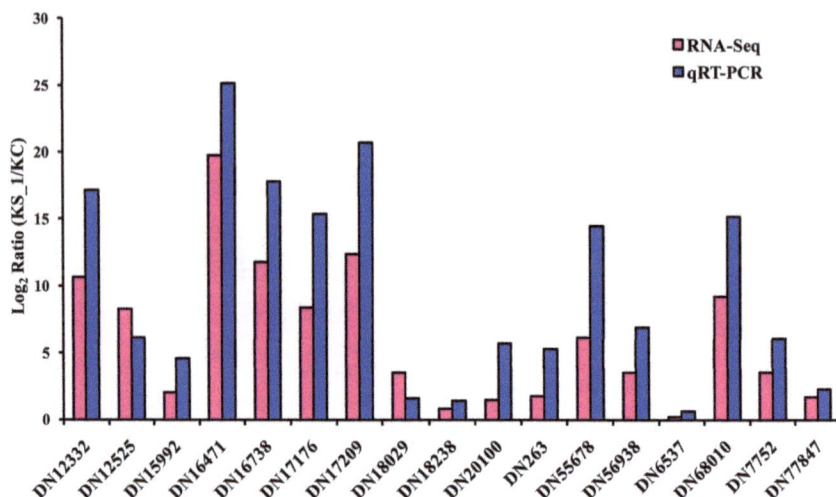

Figure 6. Validation of the expression of differentially expressed genes (DEGs) in *F. proliferatum* by Quantitative real-time PCR (qRT-PCR).

Using the *F. proliferatum* reference transcriptome, a GO enrichment analysis compared KS_1 and KC in order to understand the functional differences among DEGs. A corrected *p*-value below 0.05 indicated that the function was enriched. In the GO enrichment analysis, the biological process category was most abundant, followed by the molecular function category and the cellular component category (Figure S4). Highly enriched DEGs in KS_1were involved in the macromolecular complex, hydrolase activity, protein complex, ion transport and substrate-specific transporter activity. The macromolecular complex has been shown to be involved in the catabolic process of chitin, which is usually associated with the biosynthesis of fungal cell walls [26]. Hydrolase activity, released during the infection process, is key to the maintenance of wall plasticity associated with the fungal cell [27]. Our data indicated that the genes for ion transport were significantly enriched in the enriched DEGs in KS_1, suggesting that these DEGs may be essential for the pathogen to absorb nutrients from its host and for it to export fungal secondary metabolites and toxic compounds to the outside.

The KEGG enrichment analysis also compared KS_1 and KC in order to elucidate the significantly enriched biochemical pathway of DEGs in *F. proliferatum* (Figure S5). The most upregulated DEGs in KS_1 were those that were involved in the oxidative phosphorylation pathway, followed by the starch and sucrose metabolism (Figure S6A). Other upregulated DEGs in KS_1 included those associated with protein processing in the endoplasmic reticulum, ribosome, endocytosis, and amino sugar and nucleotide sugar metabolism. The major pathways triggered by *F. oxysporum*. f. sp. *ciceris* during conidial germination include the starch and sucrose metabolism, the amino sugar and nucleotide sugar metabolism, and the propanoate metabolism [26]. We also found these pathways in our study, and a large number of upregulated DEGs were associated with them. However, most of the DEGs in KS_1 involved in RNA transport, the cysteine and methionine metabolism, various types of *N*-glycan biosyntheses, lysine degradation, the purine metabolism and RNA degradation were upregulated (Figure S6B).

2.5. F. proliferatum Candidate Effectors

In plant–microbe interactions, some secreted proteins play an important role in promoting the fungal infection of host plants. Filamentous pathogens are known to secrete an arsenal of effector proteins that regulate innate immunity in plants and that facilitate the development of plant diseases [28]. Numerous effectors of *Fusarium* have been identified, such as *F. oxysporum*, *F. graminearum*, *F. verticillioides* and *F. virguliforme* [16,18–20]. However, the effectors in *F. proliferatum* have not been well studied on the molecular level. A total of 184 candidate effector genes were identified from the DEGs that were used to encode secreted proteins (Table S4), and most of these candidate effector genes (147,79.89%) were upregulated in KS_1. The cysteine content of 184 candidate-effectors encoded proteins ranged from 2.02% to 15.31%, with 24 candidate effectors having more than 10% of cysteine, while the majority (128, 69.65%) had less than 5% cysteine. This result is similar to the cysteine content previously identified in small secreted cysteine-rich proteins in *F. graminearum* [18].

As anticipated, 39.13% of the candidate effectors in this study lacked homology with known proteins and were annotated as hypothetical proteins in the Nr database. This is consistent with the identified effectors from filamentous fungi [29,30]. Only 37 candidate effectors (20.10%) had functional annotations in the Nr database (Table S4). Further analysis revealed that some of the secreted proteins reportedly associated with fungal pathogenicity were also found in our study: for example, the glycosylphosphatidylinol (GPI)-anchored cysteine-rich fungal effector motif (CFEM) domain protein, the cell-wall protein, and the hydrophobin and blastomyces yeast-phase-specific (BYS1) domain protein. The CFEM domain is conserved in ascomycetes and is an inadenylate cyclase (MAC1)-interacting (ACI1) protein first discovered and isolated from *Magnaporthe grisea* [31]. CFEM-containing proteins play important roles in the pathogenesis of fungi, acting as signal transducers, cell-surface receptors, or adhesion molecules in host–pathogen interactions [31]. After BLAST, three candidate effectors (DN78913, DN18169 and DN67233) were found to contain the CFEM domain against the domain conserved in the Nr database. The functional annotation indicated that DN18169 and DN67233 may belong to the GPI-anchored CFEM domain protein. The GPI-anchored CFEM domain protein can interact with a fungal adenylate cyclase controlling appressorium formation, which is a critical step in the development of rice blast disease [31] and fusarium head blight (a devastating disease in wheat) [18]. Hydrophobins are secreted proteins that are expressed during plant–fungus interaction. In addition, they are located on the outer surfaces of the cell walls of the mycelia and conidia, and they have been shown to mediate fungus–host interaction [32]. We found that two candidate effectors (DN18029 and DN2243) are homologous to hydrophobins. The BYS1-domain protein was originally purified from the pathogenic dimorphic fungus *Blastomyces dermatitidis* [33], which was found in two candidate effectors (DN33323 and DN69307), suggesting that the BYS1-domain protein may be involved in the pathogenicity of *F. proliferatum*-infected tomato plants. On the other hand, concanavalin A-like lectin/glucanase has been identified as an effector protein in *Pyrenophorateres* f. *teres* [34], which was also found in the candidate effector DN57353. Candidate effectors homologous to acid phosphatase, adhesin, clock-controlled protein-like protein, glycoside hydrolase protein and tyrosinase were also identified from the secreted proteins of *F. proliferatum*-infected tomato plants.

Finally, to investigate the expression profiles of the candidate effector genes at six different time points following infection during *F. proliferatum*–tomato interaction, ten of the 17 candidate effector genes confirmed above by qRT–PCR were subjected again to qRT–PCR using infected leaves from 0, 12, 24, 48, 72 and 96 hpi. The expression of DN18029 increased steadily from 0 to 96 hpi, but high expression levels of DN263, DN20100 and DN12525 were observed only at 96 hpi (Figure 7). However, DN7752 and DN18238 were expressed at relatively high levels at 96 hpi, whereas DN56938 and DN15992 were expressed at high levels at 48 hpi, and DN6537 and DN77847 at 72hpi.

3. Materials and Methods

3.1. Biological Material and Inoculation Assays

The *Fusarium proliferatum* (Matsushima) Nirenberg strain used in this study was originally isolated from the leaves and stems of infected tomato plants obtained from a commercial tomato greenhouse in Dalian, Liaoning Province, China in January 2014, and it was obtained as previously described [3]. Tomato plants (*S. lycopersicum* 'Zaofen No. 2') were grown individually in plastic pots (10 cm diameter × 8 cm height) and placed in a growth chamber set at 25 °C under a 16 h light/8 h dark photoperiod. Mycelia and spores were harvested from the fungus grown on PDA medium at 25 °C in the dark for 7 days. To simulate a *F. proliferatum* infection, three-week-old tomato plants were sprayed with a conidial suspension of *F. proliferatum* containing 1×10^8 spores/mL. After spraying, the plants were placed in a 25 °C dark growth chamber with 100% relative humidity for 24 h.

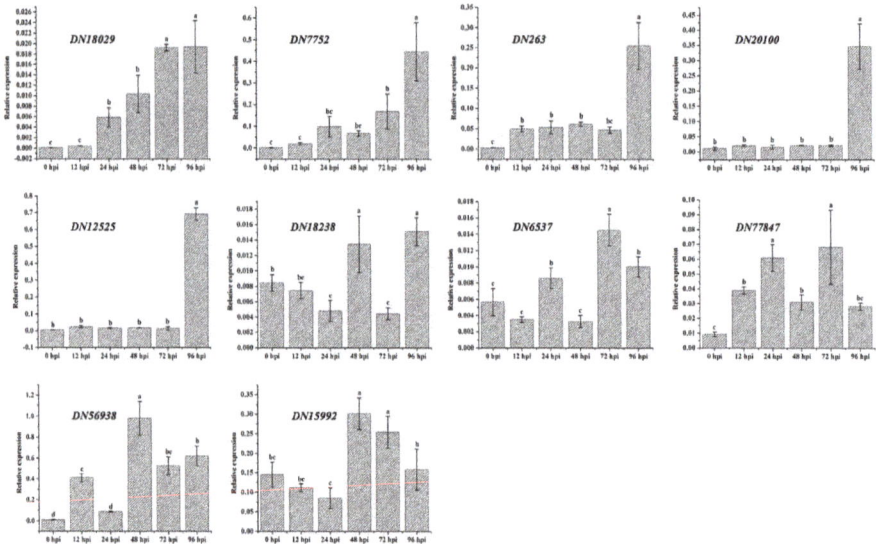

Figure 7. Expression levels of ten *F. proliferatum* candidate effector genes at six different time points (0, 12, 24, 48, 72 and 96 hpi) as determined by qRT–PCR analyses. The β-tubulin and ubiquitin genes were used as two internal control genes for normalization.

3.2. Evaluation of Time Course of F. proliferatum Infection

The inoculation assays of *F. proliferatum*-infected tomato plants were performed as described above. Infected leaf tissues were collected at 0, 12, 24, 48, 72 and 96 hpi after the *F. proliferatum* were sprayed with a conidial suspension containing 1×10^8 spores/mL. The mycelia and infected leaf materials were immediately frozen in liquid nitrogen and stored at −80 °C. After the RNA extraction, the samples were subjected to RNA-seq sequencing and qRT-PCR to verify the observed gene expression patterns. For microscopic observation, *F. proliferatum*-infected tomato leaves were manually cut into 2 to 3 mm pieces with a sterile scalpel and immediately placed in a pre-cooled 2.5% glutaraldehyde fixative for at least 2 h at 4 °C at different time points. The samples were observed using the Nova NanoSEM450 scanning electron microscope (FEI Corporation, San Francisco, CA, USA). Our observations were carried out for leaves taken at different hpis.

3.3. Transcriptome Profiling

The tomato plants were inoculated with *F. proliferatum* as described previously. The *Fusarium proliferatum* (Matsushima) Nirenberg strain was cultured in a PDA medium at 25 °C in the dark for 7 days. The mycelia was then gently scraped from the plates with a sterile lab spoon as a KC sample. For total RNA extraction, mycelia and infected leaf materials were ground to a powder in liquid nitrogen using a mortar and pestle using TRIzol® LS Reagent (Invitrogen, Carlsbad, CA, USA) following the manufacturer's instructions. For the RNA-seq analysis, RNA isolation was conducted using mycelia collected from 7-day-old PDA cultures (KC), and infected leaf tissues taken 96 hpi (KS_1) after

F. proliferatum's inoculation via a conidial suspension. For each specimen, RNA was extracted from a mixture of three independent biological replicates. Polyadenylated (Poly(A)) mRNA was isolated from the total RNA, and cDNA libraries were then constructed and sequenced at the 2×150 bp paired-end read mode with Illumina HiSeq®2500, performed at Novogene Corporation, Beijing, China, in accordance with the manufacturer's standard protocol.

After discarding the adapter-only reads, subsequent analyses first used low-quality reads with ambiguous bases and reads with more than 50% Qphred ≤ 20 bases from generated raw paired-reads, then used the clean reads from the filtered and trimmed reads. The quality of the two libraries was controlled with FastQC. As the KS_1 sample contained both host and fungal transcriptomes and the KC sample contained only the fungal transcriptome, the clean reads from the KS_1 library were aligned with the *S. lycopersicum* SL2.50 genome (https://www.ncbi.nlm.nih.gov/genome/?term=Solanum+lycopersicum+L), and the unmapped clean reads were generated by using TopHat version 2.0.6 [35]. Following this, the clean reads from the KC library and the unmapped clean reads from the KS_1 library were used as input to generate a preliminary assembly via the *de novo* assembly of the *F. proliferatum* transcriptome using Trinity software with a default *k*-mer length of 25 [36]. The assembled contigs were processed with CD–HIT–EST with an identity threshold of 95% in order to remove redundant transcripts [37]. Low-complexity sequences were masked using DustMasker, and sequences with fewer than 200 bp were discarded. After the assembly, a unigene dataset was produced by performing clustering on scaffolds using TGI Clustering tools [38]. Again, the unigenes were mapped to the *S. lycopersicum* genome to further exclude the contaminating sequences. Finally, the resulting transcript dataset was taken as a *F. proliferatum* reference transcriptome for further analysis.

3.4. Transcriptome Annotation and DEG

An open reading frames (ORFs) prediction was performed using the European molecular biology open software suite [39], and the longest ORF was extracted from each unigene. After this, all unigenes were annotated using BLASTx alignments by comparing their sequences with various protein databases, including the Nr (http://www.ncbi.nlm.nih.gov/), SwissPort (http://www.ebi.ac.uk/uniprot/), KOG (http://www.ncbi.nlm.nih.gov/KOG/) and KEGG (http://www.genome.jp/kegg/) databases, with an *e*-value cutoff of 1×10^{-5}. A functional annotation by GO term of all the assembled unigenes was performed with the Blast2GO program (https://www.blast2go.com/). Finally, the WEGO software (http://wego.genomics.org.cn/) was used to perform a GO function classification and reveal the distribution of gene functions in *F. proliferatum* at the macromolecular level. Gene expression levels were estimated by RNA-seq expression estimation for the two samples by expectation maximization (RSEM) [40], and FPKM was the most commonly used method to normalize gene expression levels [41].

Prior to the differential gene expression analysis, the edgeR program package was used to adjust the read counts in each sequenced library. To identify DEGs, a comparison between the two samples was performed using the DEGseq 1.12.0 R package (R Foundation for Statistical Computing, Vienna, Austria) [42]. The *p*-value was adjusted using the *q*-value [43], and the fold change ($\log_2$ratio) was estimated according to the normalized gene expression level in the two samples. In this paper, *q*-value < 0.005 and $|\log_2 (\text{foldchange})| > 1$ were set as the threshold for the DEGs.

3.5. Candidate Effector Gene Prediction

Several prediction algorithms were utilized to predict the putative secreted proteins. The program TargetP 1.1 (http://www.cbs.dtu.dk/services/TargetP/) was used to predict the cleavage sites for the predicted presequences. Signal peptide cleavage sites were identified using SignalP 4.1 (http://www.cbs.dtu.dk/services/SignalP/), after which transmembrane helices were detected with TMHMM 2.0 (http://www.cbs.dtu.dk/services/TMHMM-2.0/) [4]. Putative candidate effector proteins identified through the transcriptomic analysis of *F. proliferatum* were selected based on four conditions: (1) the presence of N-terminal signal peptide cleavage sites; (2) the effector proteins having fewer than

200 amino acids; (3) the percentage of cysteine content being greater or equal to 2%; and (4) no transmembrane helices in the mature proteins.

3.6. Quantitative RT-PCR Assay

The tomato plants were inoculated with *F. proliferatum* as described previously. Infected leaf tissues were collected at 0, 12, 24, 48, 72 and 96 hpi after the *F. proliferatum* was sprayed with a conidial suspension containing 1×10^8 spores/mL. Following this, the total RNA of each sample was extracted as described previously. Seventeen of the DEGs identified via the RNA-seq analysis were randomly selected for confirmation by qRT–PCR. The primers were designed from ten candidate-effector gene sequences using Primer-Premier 5 software (Premier Biosoft Interpairs, Palo Alto, CA, USA). The β-tubulin [44] and ubiquitin [23] genes were both used simultaneously as two internal control genes in order to obtain a more accurate quantitative result (Table S5). For the qRT–PCR, a first-strand cDNA synthesis was performed using RNA samples from KC, KS_1 and the tomato leaf tissues collected at six different time points (0, 12, 24, 48, 72 and 96 hpi) following *F. proliferatum*'s inoculation via a conidial suspension. This synthesis was performed using a PrimeScript™ RT reagent kit with a gDNA Eraser (TaKara, Dalian, China), in accordance with the manufacturer's instruction. The Mx3005p™ detection system (Agilent Stratagene, Santa Clara, CA, USA) was used to determine gene expression viaqRT–PCR. Quantitative RT–PCR was performed on cDNA samples (diluted 1:10) using SYBR®*Premix Ex Taq*™ II (TliRNaseH Plus) (Takara), in accordance with the manufacturer's instruction. The samples were first incubated at 95 °C for 30 s. This was followed by 40 amplification cycles at 95 °C for 10 s, 60 °C for 25 s, and then 72 °C for 25 s. Using the geNorm program manual, the threshold cycle (CT) values of two reference genes and ten candidate effector genes were quantified with the comparative CT Method. The normalization factor values of the two reference genes were then automatically calculated via geNorm. The standard deviation was calculated according to the mathematical formulae in the geNorm manual [45]. The data for KC, KS_1 and the tomato leaf tissues collected at six different time points were analyzed by one way-ANOVA contained in the SPSS 17.0 software. Differences between samples were considered to be statistically significant at the $p < 0.05$ level.

3.7. Accession Numbers

All RNA-seq reads generated in this study were deposited at the GenBank SRA database (http://www.ncbi.nlm.nih.gov/sra) under the BioProject ID PRJNA397359.

4. Conclusions

In conclusion, this study is the first to report the use of transcriptome analysis as a means of screening effector genes that are involved in *F. proliferatum*-infected tomato interactions. *De novo* sequencing of the *F. Proliferatum* transcriptome yielded new insights into the molecular pathogenicity of this important tomato plant fungus. Using bioinformatics and functional analysis, a total of 184 candidate effector genes were identified from a high number of DEGs. Most of the candidate effector genes were expressed as hypothetical proteins, so the functional verification of these candidate effector genes and their respective roles in *F. proliferatum* would need further investigation. There is no doubt that the result of such findings will accelerate the identification of effector genes that play a key role in the resistant or susceptible responses of the tomato plants.

Supplementary Materials: Supplementary materials can be found at www.mdpi.com/1422-0067/19/1/31/s1.

Acknowledgments: This work was supported by the Panjin Campus for Food Science and Technology Research Initiative, the Dalian University of Technology, the science and technology foundation of Liaoning province (201602049), and the National Key Technology Support Program (2015BAD17B00).

Author Contributions: Meiling Gao, Hongman Hou, and Yongming Bao conceived and designed the experiments; Meiling Gao, Siyu Yao, Yang Liu, Pinsan Xu, and Wenhui Sun performed the experiments; Meiling Gao, Haining Yu, and Zhongji Pu analyzed the data; Meiling Gao wrote the paper.

Conflicts of Interest: The authors declare no conflict of interest.

Abbreviations

DEGs	Differentially expressed genes
qRT–PCR	Quantitative real-time PCR
SIX	Secreted in xylem
SGE1	Secreted gene expression
NCBI	National Center of Biotechnology Information
RNA-seq	Ribonucleic acid sequencing
Hpi	Hourspost-inoculation
PDA	Potato dextrose agar
Nr	Non-redundant
Nt	Nucleotide
KEGG	Kyoto encyclopedia of genes and genomes
KO	Kyoto encyclopedia of genes and genomes Orthology
Pfam	Protein family
KOG	euKaryotic Ortholog Groups
BLAST	Basic Local Alignment Search Tool
GO	Gene ontology
FPKM	Fragments per kilobase of exon per million fragments mapped
GPI	Glycosylphosphatidylinol
CFEM	Cysteine-rich fungal effector motif
BYS1	Blastomyces yeast-phase-specific
Poly(A)	Polyadenylated (Poly(A))
ORFs	Open reading frames
RSEM	RNA-seq expression estimation by expectationmaximization
CT	Threshold cycle

References

1. Consortium, T.T.G. The tomato genome sequence provides insights into fleshy fruit evolution. *Nature* **2012**, *485*, 635–641. [CrossRef]
2. Suwannarach, N.; Kumla, J.; Nitiyon, S.; Limtong, S.; Lumyong, S. First report of sour rot on tomato caused by *Galactomyces reessii* in Thailand. *J. Gen. Plant Pathol.* **2016**, *82*, 228–231. [CrossRef]
3. Gao, M.L.; Luan, Y.S.; Yu, H.N.; Bao, Y.M. First report of tomato leaf spot caused by *Fusarium proliferatum* in China. *Can. J. Plant Pathol.* **2016**, *38*, 400–404. [CrossRef]
4. Peltomaa, R.; Vaghini, S.; Patiño, B.; Benito-Peña, E.; Moreno-Bondi, M.C. Species-specific optical genosensors for the detection of mycotoxigenic *Fusarium* fungi in food samples. *Anal. Chim. Acta* **2016**, *935*, 231–238. [CrossRef] [PubMed]
5. Seefelder, W.; Gossmann, M.; Humpf, H.U. Analysis of Fumonisin B1 in *Fusarium proliferatum*-Infected Asparagus Spears and Garlic Bulbs from Germany by Liquid Chromatography–Electrospray Ionization Mass Spectrometry. *J. Agric. Food Chem.* **2002**, *50*, 2778–2781. [CrossRef] [PubMed]
6. Chang, K.F.; Hwang, S.F.; Conner, R.L.; Ahmed, H.U.; Zhou, Q.; Turnbull, G.D.; Strelkov, S.E.; Mclaren, D.L.; Gossen, B.D. First report of *Fusarium proliferatum* causing root rot in soybean (*Glycine max* L.) in Canada. *Crop Prot.* **2015**, *67*, 52–58. [CrossRef]
7. Palmero, D.; De Cara, M.; Nosir, W.; Galvez, L.; Cruz, A.; Woodward, S.; Gonzalez-Jaen, M.T.; Tello, J.C. *Fusarium proliferatum* isolated from garlic in Spain: Identification, toxigenic potential and pathogenicity on related *Allium* species. *Phytopathol. Mediterr.* **2012**, *51*, 207–218.
8. Rheeder, J.P.; Marasas, W.F.O.; Vismer, H.F. Production of Fumonisin Analogs by *Fusarium* Species. *Appl. Environ. Microbiol.* **2002**, *68*, 2101–2105. [CrossRef] [PubMed]
9. Gil-Serna, J.; Gálvez, L.; París, M.; Palmero, D. *Fusarium proliferatum* from rainwater and rooted garlic show genetic and pathogenicity differences. *Eur. J. Plant Pathol.* **2016**, *146*, 199–206. [CrossRef]

10. Isack, Y.; Benichis, M.; Gillet, D.; Gamliel, A. A selective agar medium for isolation, enumeration and morphological identification of *Fusarium proliferatum*. *Phytoparasitica* **2014**, *42*, 541–547. [CrossRef]

11. Nguyen, T.T.; Dehne, H.W.; Steiner, U. Histopathological assessment of the infection of maize leaves by *Fusarium graminearum*, *F. proliferatum*, and *F. verticillioides*. *Fungal Biol.* **2016**, *120*, 1094–1104. [CrossRef] [PubMed]

12. Taylor, A.; Vágány, V.; Jackson, A.C.; Harrison, R.J.; Rainoni, A.; Clarkson, J.P. Identification of pathogenicity-related genes in *Fusarium oxysporum* f. sp. *cepae*. *Mol. Plant Pathol.* **2016**, *17*, 1032–1047. [CrossRef] [PubMed]

13. Rocha, L.O.; Laurence, M.H.; Ludowici, V.A.; Puno, V.I.; Lim, C.C.; Tesoriero, L.A.; Summerell, B.A.; Liew, E.C.Y. Putative effector genes detected in *Fusarium oxysporum* from natural ecosystems of Australia. *Plant Pathol.* **2016**, *65*, 914–929. [CrossRef]

14. Rep, M.; van der Does, H.C.; Meijer, M.; van Wijk, R.; Houterman, P.; Dekker, H.; de Koster, C.; Cornelissen, B. A small, cysteine-rich protein secreted by *Fusarium oxysporum* during colonization of xylem vessels is required for I-3-mediated resistance in tomato. *Mol. Microbiol.* **2010**, *53*, 1373–1383. [CrossRef] [PubMed]

15. Dong, Y.; Li, Y.; Zhao, M.; Jing, M.; Liu, X.; Liu, M.; Guo, X.; Zhang, X.; Chen, Y.; Liu, Y. Global Genome and Transcriptome Analyses of *Magnaporthe oryzae* Epidemic Isolate 98-06 Uncover Novel Effectors and Pathogenicity-Related Genes, Revealing Gene Gain and Lose Dynamics in Genome Evolution. *PLoS Pathog.* **2015**, *11*, e1004801. [CrossRef] [PubMed]

16. Sperschneider, J.; Gardiner, D.M.; Thatcher, L.F.; Lyons, R.; Singh, K.B.; Manners, J.M.; Taylor, J.M. Genome-Wide Analysis in Three *Fusarium* Pathogens Identifies Rapidly Evolving Chromosomes and Genes Associated with Pathogenicity. *Genome Biol. Evol.* **2015**, *7*, 1613–1627. [CrossRef] [PubMed]

17. Hogenhout, S.A.; van der Hoorn, R.A.; Terauchi, R.; Kamoun, S. Emerging concepts in effector biology of plant-associated organisms. *Mol. Plant Microbe Interact.* **2009**, *22*, 115–122. [CrossRef] [PubMed]

18. Win, J.; Chaparro-Garcia, A.; Belhaj, K.; Saunders, D.G.; Yoshida, K.; Dong, S.; Schornack, S.; Zipfel, C.; Robatzek, S.; Hogenhout, S.A.; et al. Effector biology of plant-associated organisms: Concepts and perspectives. *Cold Spring Harb. Symp. Quant. Biol.* **2012**, *77*, 235–247. [CrossRef] [PubMed]

19. Lu, S.; Edwards, M.C. Genome-Wide Analysis of Small Secreted Cysteine-Rich Proteins Identifies Candidate Effector Proteins Potentially Involved in *Fusarium graminearum*-Wheat Interactions. *Phytopathology* **2016**, *106*, 166–176. [CrossRef] [PubMed]

20. Brown, D.W.; Busman, M.; Proctor, R.H. *Fusarium verticillioides* SGE1 is required for full virulence and regulates expression of protein effector and secondary metabolite biosynthetic genes. *Mol. Plant Microbe Interact.* **2014**, *27*, 809–823. [CrossRef] [PubMed]

21. Chang, H.X.; Domier, L.; Radwan, O.; Yendrek, C.; Hudson, M.; Hartman, G.L. Identification of multiple phytotoxins produced by *Fusarium virguliforme* including a phytotoxic effector (FvNIS1) associated with sudden death syndrome foliar symptoms. *Mol. Plant Microbe Interact.* **2016**, *29*, 96–108. [CrossRef] [PubMed]

22. Ma, L.J.; Shea, T.; Young, S.; Zeng, Q.; Kistler, H.C. Genome Sequence of *Fusarium oxysporum* f. sp. *melonis* Strain NRRL 26406, a Fungus Causing Wilt Disease on Melon. *Genome Announc.* **2014**, *2*, e00730-14. [CrossRef] [PubMed]

23. Wiemann, P.; Sieber, C.M.K.; Bargen, K.W.V.; Studt, L.; Niehaus, E.M.; Espino, J.J.; Huß, K.; Michielse, C.B.; Albermann, S.; Wagner, D. Deciphering the Cryptic Genome: Genome-wide Analyses of the Rice Pathogen *Fusarium fujikuroi* Reveal Complex Regulation of Secondary Metabolism and Novel Metabolites. *PLoS Pathog.* **2013**, *9*, 371–376. [CrossRef] [PubMed]

24. Gardiner, D.M.; Stiller, J.; Kazan, K. Genome Sequence of *Fusarium graminearum* Isolate CS3005. *Genome Announc.* **2014**, *2*, e00227-14. [CrossRef] [PubMed]

25. Aoki, K.; Yano, K.; Suzuki, A.; Kawamura, S.; Sakurai, N.; Suda, K.; Kurabayashi, A.; Suzuki, T.; Tsugane, T.; Watanabe, M. Large-scale analysis of full-length cDNAs from the tomato (*Solanum lycopersicum*) cultivar Micro-Tom, a reference system for the Solanaceae genomics. *BMC Genom.* **2010**, *11*, 210. [CrossRef] [PubMed]

26. Sharma, M.; Sengupta, A.; Ghosh, R.; Agarwal, G.; Tarafdar, A.; Nagavardhini, A.; Pande, S.; Varshney, R.K. Genome wide transcriptome profiling of *Fusarium oxysporum* f sp. *ciceris* conidial germination reveals new insights into infection-related genes. *Sci. Rep.* **2016**, *6*, 37353. [CrossRef] [PubMed]

27. Adams, D.J. Fungal cell wall chitinases and glucanases. *Microbiology* **2004**, *150*, 2029–2035. [CrossRef] [PubMed]

28. Kamoun, S. Groovy times: Filamentous pathogen effectors revealed. *Curr. Opin. Plant Biol.* **2007**, *10*, 358–365. [CrossRef] [PubMed]

29. Ellis, J.G.; Rafiqi, M.; Gan, P.; Chakrabarti, A.; Dodds, P.N. Recent progress in discovery and functional analysis of effector proteins of fungal and oomycete plant pathogens. *Curr. Opin. Plant Biol.* **2009**, *12*, 399–405. [CrossRef] [PubMed]

30. Stergiopoulos, I.; de Wit, P.J. Fungal Effector Proteins. *Annu. Rev. Phytopathol.* **2009**, *47*, 233–263. [CrossRef] [PubMed]

31. Kulkarni, R.D.; Kelkar, H.S.; Dean, R.A. An eight-cysteine-containing CFEM domain unique to a group of fungal membrane proteins. *Trends Biochem. Sci.* **2003**, *28*, 118–121. [CrossRef]

32. Guzmán-Guzmán, P.; Alemán-Duarte, M.I.; Delaye, L.; Herrera-Estrella, A.; Olmedo-Monfil, V. Identification of effector-like proteins in *Trichoderma* spp. and role of a hydrophobin in the plant-fungus interaction and mycoparasitism. *BMC Genet.* **2017**, *18*, 16. [CrossRef] [PubMed]

33. Burg, E.F., III; Smith, L.H., Jr. Cloning and characterization of *bys1*, a temperature-dependent cDNA specific to the yeast phase of the pathogenic dimorphic fungus *Blastomyces dermatitidis*. *Infect. Immun.* **1994**, *62*, 2521–2528. [PubMed]

34. Ismail, I.A.; Able, A.J. Secretome analysis of virulent *Pyrenophora teres* f. *teres* isolates. *Proteomics* **2016**, *16*, 2625–2636. [CrossRef] [PubMed]

35. Trapnell, C.; Pachter, L.; Salzberg, S.L. TopHat: Discovering splice junctions with RNA-Seq. *Bioinformatics* **2009**, *25*, 1105–1111. [CrossRef] [PubMed]

36. Grabherr, M.G.; Haas, B.J.; Yassour, M.; Levin, J.Z.; Thompson, D.A.; Amit, I.; Adiconis, X.; Fan, L.; Raychowdhury, R.; Zeng, Q. Full-length transcriptome assembly from RNA-Seq data without a reference genome. *Nat. Biotechnol.* **2011**, *29*, 644–652. [CrossRef] [PubMed]

37. Li, W.; Godzik, A. Cd-hit: A fast program for clustering and comparing large sets of protein or nucleotide sequences. *Bioinformatics* **2006**, *22*, 1658–1659. [CrossRef] [PubMed]

38. Pertea, G.; Huang, X.; Liang, F.; Antonescu, V.; Sultana, R.; Karamycheva, S.; Lee, Y.; White, J.; Cheung, F.; Parvizi, B. TIGR Gene Indices clustering tools (TGICL): A software system for fast clustering of large EST datasets. *Bioinformatics* **2003**, *19*, 651–652. [CrossRef] [PubMed]

39. Rice, P.; Longden, I.; Bleasby, A. EMBOSS: The European Molecular Biology Open Software Suite. *Trends Genet.* **2000**, *16*, 276–277. [CrossRef]

40. Li, B.; Dewey, C.N. RSEM: Accurate transcript quantification from RNA-Seq data with or without a reference genome. *BMC Bioinform.* **2011**, *12*, 323. [CrossRef] [PubMed]

41. Trapnell, C.; Williams, B.A.; Pertea, G.; Mortazavi, A.; Kwan, G.; van Baren, M.J.; Salzberg, S.L.; Wold, B.J.; Pachter, L. Transcript assembly and quantification by RNA-Seq reveals unannotated transcripts and isoform switching during cell differentiation. *Nat. Biotechnol.* **2010**, *28*, 511–515. [CrossRef] [PubMed]

42. Wang, L.; Feng, Z.; Wang, X.; Wang, X.; Zhang, X. DEGseq: An R package for identifying differentially expressed genes from RNA-seq data. *Bioinformatics* **2010**, *26*, 136–138. [CrossRef] [PubMed]

43. Storey, J.D. The Positive False Discovery Rate: A Bayesian Interpretation and the *q*-Value. *Ann. Stat.* **2003**, *31*, 2013–2035. [CrossRef]

44. Thatcher, L.F.; Gardiner, D.M.; Kazan, K.; Manners, J.M. A highly conserved effector in *Fusarium oxysporum* is required for full virulence on *Arabidopsis*. *Mol. Plant Microbe Interact.* **2011**, *25*, 180–190. [CrossRef] [PubMed]

45. Vandesompele, J.; Preter, K.D.; Pattyn, F.; Poppe, B.; Roy, N.V.; Paepe, A.D.; Speleman, F. Accurate normalization of real-time quantitative RT-PCR data by geometric averaging of multiple internal control genes. *Genome Biol.* **2002**, *3*, research0034. [CrossRef] [PubMed]

International Journal of
Molecular Sciences

MDPI

Article

Transcriptome Analysis of Kiwifruit in Response to *Pseudomonas syringae* pv. *actinidiae* Infection

Tao Wang, Gang Wang, Zhan-Hui Jia, De-Lin Pan, Ji-Yu Zhang * and Zhong-Ren Guo

Institute of Botany, Jiangsu Province and Chinese Academy of Sciences, Nanjing 210014, China;
immmorer@163.com (T.W.); wg20092011@163.com (G.W.); 13915954315@163.com (Z.-H.J.);
PPxsperfect@163.com (D.-L.P.); zhongrenguo@cnbg.net (Z.-R.G.)
* Correspondence: maxzhangjy@cnbg.net; Tel.: +86-025-8434-7033

Received: 21 November 2017; Accepted: 23 January 2018; Published: 26 January 2018

Abstract: Kiwifruit bacterial canker caused by *Pseudomonas syringae* pv. *actinidiae* (Psa) has brought about a severe threat to the kiwifruit industry worldwide since its first outbreak in 2008. Studies on other pathovars of *P. syringae* are revealing the pathogenesis of these pathogens, but little about the mechanism of kiwifruit bacterial canker is known. In order to explore the species-specific interaction between Psa and kiwifruit, we analyzed the transcriptomic profile of kiwifruit infected by Psa. After 48 h, 8255 differentially expressed genes were identified, including those involved in metabolic process, secondary metabolites metabolism and plant response to stress. Genes related to biosynthesis of terpens were obviously regulated, indicating terpens may play roles in suppressing the growth of Psa. We identified 283 differentially expressed resistant genes, of which most U-box domain containing genes were obviously up regulated. Expression of genes involved in plant immunity was detected and some key genes showed differential expression. Our results suggest that Psa induced defense response of kiwifruit, including PAMP (pathogen/microbe-associated molecular patterns)-triggered immunity, effector-triggered immunity and hypersensitive response. Metabolic process was adjusted to adapt to these responses and production of secondary metabolites may be altered to suppress the growth of Psa.

Keywords: kiwifruit; bacterial canker; Psa; resistance

---

1. Introduction

Kiwifruit bacterial canker disease was first reported on *Actinidiae chinesis* var. *deliciosa* in Shizuoka, Japan in 1984 [1]. In 2010, the pathogen *Pseudomonas syringae* pv. *actinidiae* (Psa) was detected in New Zealand, and within two years it infected 37% of New Zealand orchards and continues to increase [2]. To date, Psa has been detected in the main kiwifruit producing countries, including China, Chile, and European countries [3,4]. Pathovars of the species *P. syringae* cause important diseases in a wide range of plant species. To look for the way to control these diseases, researchers worldwide are trying to find the pathogenesis of *P. syringae*. Plants hold a complete immune system which is composed of two lines of defense. The first is PAMP-triggered immunity (PTI), which recognizes molecular microbial determinants, termed pathogen/microbe-associated molecular patterns (PAMPs/MAMPs), via pattern recognition receptors (PPRs) [5]. The second line is termed effector-triggered immunity (ETI) which detects injected effector proteins in the cytoplasm by resistance proteins and elicits further immunity. PTI and ETI can combine to cause hypersensitive response (HR) at infection site, which involves programmed cell death.

Studies on the pathogenesis of Psa are limited, but works on other pathovars of *P. syringae* especially the model species *P. syringae* pv. tomato DC3000 give us the chance to view the interactivity between *P. syringae* and the host. A functional hypersensitive response and pathogenicity (*hrp* pathogenicity island [PAI]) type III secretion system (T3SS) that directs the delivery of effector proteins into host cells has been shown to be the key pathogenicity factor required for *P. syringae* to colonize and parasitize host plants [6]. Plant immune system is a major target of type III effectors. *P. syringae* suppresses plant immune system by translocating immune-suppressing effector proteins through T3SS into plant cell [7]. Although effector proteins suppress immunity in some plants, in other plants, they trigger ETI [8] upon their recognition by cognate resistance proteins which, in turn, activate a secondary defense reaction HR [9].

The genome of Psa has been analyzed by different groups and genes possibly involved in pathogenesis were identified. McCann and his colleagues [10] identified 51 known type III effectors from four different clades of Psa and only 17 were found in all Psa genomes. This raised the possibility that the capacity to cause disease in kiwifruit resided primarily in the core genome of Psa. Meanwhile Psa also displayed a set of genes involved in degradation of lignin derivatives and other phenolics [11]. In-depth studies on Psa genomes have shown that this pathovar can rapidly adapt to a new host and new environments through the acquisition and/or loss of mobile genetic elements and virulence factors, thereby resulting in a multi-faceted plant pathogen [12]. In this study we analyzed the transcriptomic profile of kiwifruit infected by Psa, hoping to explore the response of kiwifruit on the molecular level and to lay foundation for understanding the pathogenesis of kiwifruit canker disease.

2. Results

2.1. De Novo RNA-Seq Assembly and Annotation of Unigenes

The valid reads from all samples were merged for de novo assembly using trinity software. A total of 110,134 unigenes with a N50 of 1226 bp were obtained (Table 1). All unigenes were longer than 200 bp and the average length was 759 bp. Functional annotation of unigenes were performed by blasting against various databases. Of all the 110,134 unigenes, 50,305 (45.68%) matched to known sequences, with 49,897 (45.31%) matching to sequences in Nr (non-redundant protein sequences) database, 34,331 (31.17%) matching to Swissprot database, 30,430 (27.63%) matching KOG (Clusters of eukaryotic ortholog groups of proteins) database, 20,524 (18.64%) matching KEGG (Kyoto encyclopedia of genes and genomes) database.

Table 1. Functional annotation of the kiwifruit unigenes.

Database	Number of Unigenes	Percentage
Nr	49,897	45.31
Swissprot	34,331	31.17
KOG	30,430	27.63
KEGG	20,524	18.64
Annotation gene	50,305	45.68
Without annotation gene	59,829	54.32
Total unigenes	110,134	100.00

Nr: non-redundant protein sequence; KOG: Clusters of eukaryotic ortholog groups of proteins; KEGG: Kyoto encyclopedia of genes and genomes.

2.2. Functional Classification of Unigenes

To better understand functions of the unigenes, we did GO (Gene Ontology) analysis and categorized the 20,524 unigenes matching to KOG database into three GO trees (biological processes, cellular components, and molecular functions), which were further classified into 48 functional groups (Figure 1). The three groups with the most number of unigenes in the category of biological process were cellular process, metabolic process and single-organism process. Groups with the most unigenes in cellular component were cell, cell part and organelle. Binding and catalytic activity were the biggest groups in molecular function.

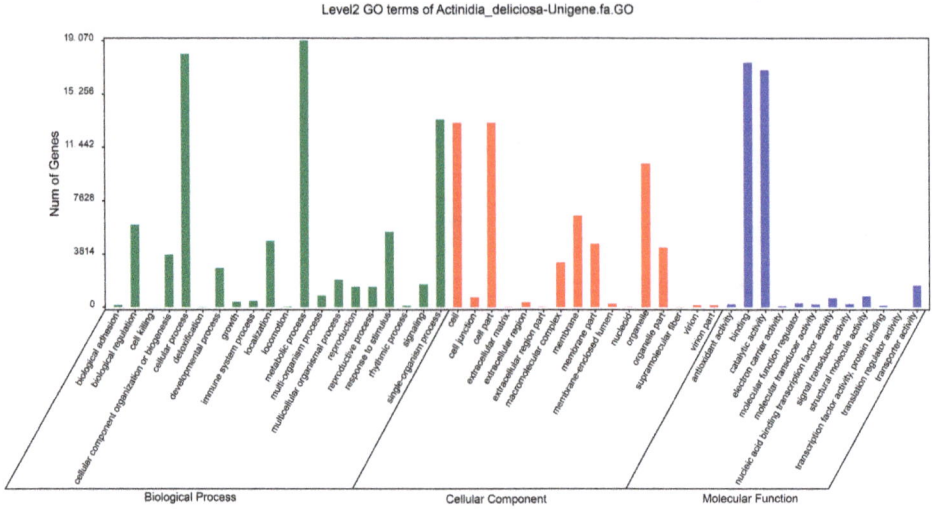

Figure 1. Gene Ontology classification of unigenes.

2.3. Analysis of Differentially Expressed Unigenes

Expression level of each unigene was calculated and differentially expressed genes (DEGs) between Psa-treated sample and control at 48 and 96 h time points were identified. Cluster analysis of unigenes of the five samples was done (Figure 2). Compared with the other four samples, expression pattern of CK was obviously different and presented the most number of DEGs. PY3, XJ3, PY4, and XJ4 gradually showed more expression differences.

Twelve unigenes were randomly selected for qRT-PCR analysis (Figure S1) to validate the results of the RNA sequencing data. All of the selected unigenes exhibited similar expression patterns to those from RNA sequencing data, so indicating that the results of RNA sequencing were credible.

2.4. Functional Classification of Differentially Expressed Genes

Enrichment and classification of the DEGs were performed by searching GO and KEGG database (Tables S1 and S2). In the category of biological process, GO terms of cellular process, metabolic process and single-organism process enriched the most DEGs (Figure 3). In the category of cellular component, cell, cell part and organelle part enriched the most DEGs. Binding and catalytic activity got the most DEGs in the category of molecular function. Most of GO terms in the category of biological process had more DEGs up regulated by Psa treatment than those down regulated, and it was the opposite case in the cellular component category.

We compared the expression patterns between Psa-treated samples and control at 48 and 96 h respectively. There were totally 8255 (7.50%) DEGs (Table S3) between PY3 and XJ3, of which 2733 DEGs were down regulated in XJ3 relative to PY3, and 5522 DEGs were up regulated. Only 4281 DEGs (Table S4) were identified between PY4 and XJ4, and the numbers of DEGs down regulated and up regulated were similar. Among the most differentially expressed genes between PY3 and XJ3, we identified DEGs participating in terpene synthesis, salicylic acid-binding, jasmonate, disease resistance, ethylene response and WRKY transcription factor which were all related to disease resistance of plant. There were also many DEGs participating in biosynthesis of secondary metabolites, environmental adaptation and carbohydrate metabolism.

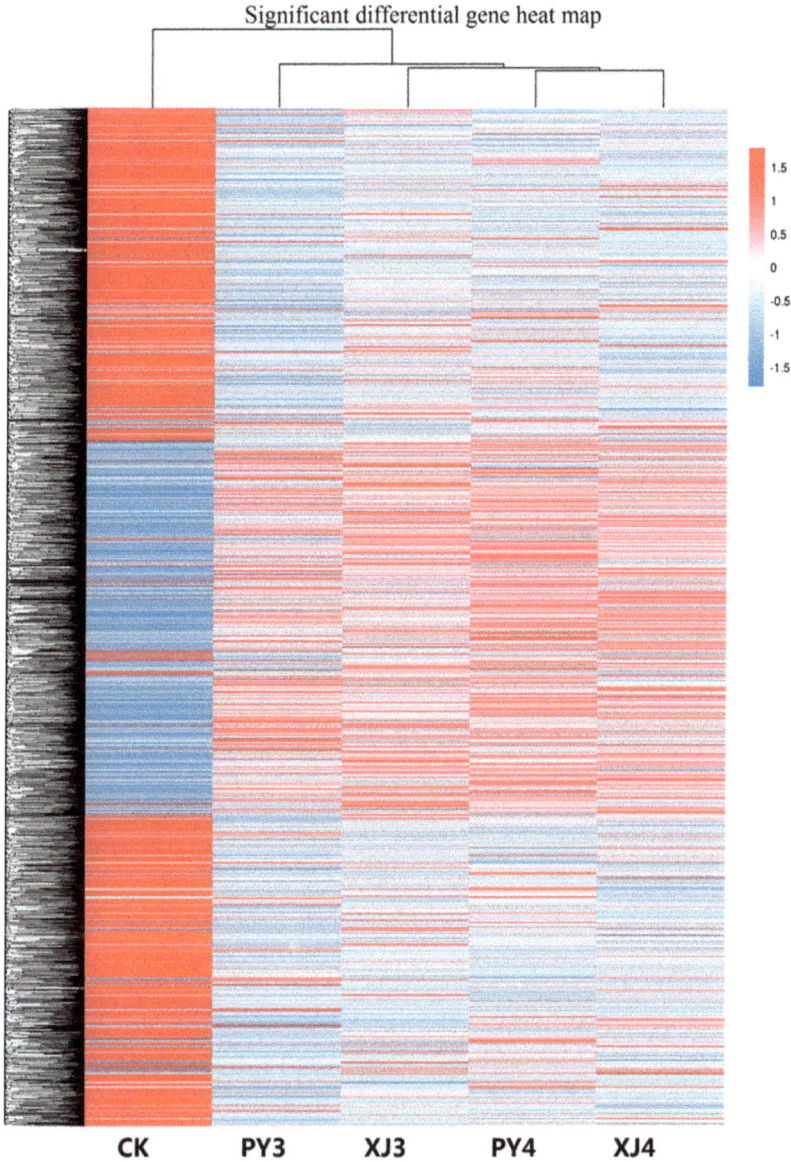

Figure 2. Cluster of significant differentially expressed genes of the five experimental samples. The RPKM (reads per kb per million reads) values of unigenes were used for hierarchical cluster analysis. Expression level was showed by different colors, the redder the higher expression and the bluer the lower. Five treatments were set: CK, only carved; PY3, inoculated with water and sampled 48 h after inoculation; PY4, inoculated with water and sampled at 96 h; XJ3, inoculated with Psa and sampled at 48 h; XJ4, inoculated with Psa and sampled at 96 h.

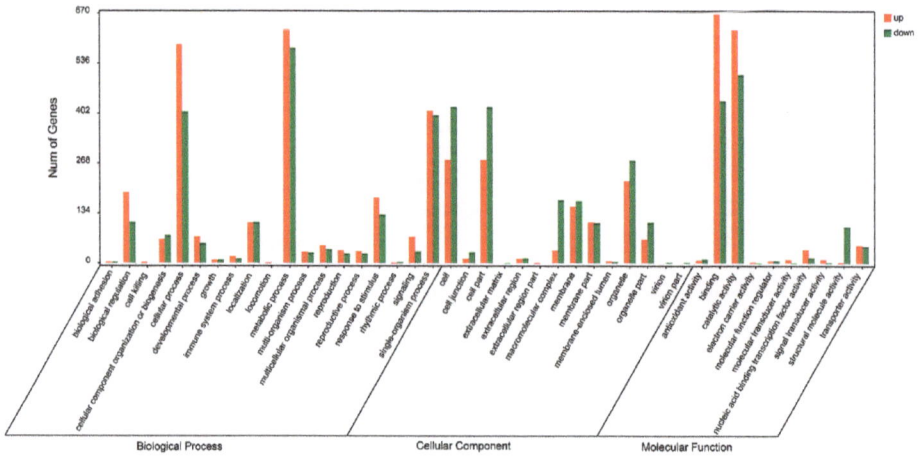

Figure 3. GO classification of differentially expressed genes.

The top 20 progresses influenced by Psa treatment were summarized in Table 2. The top one and top four progresses were both related to terpene metabolism. Metabolic process enriched the most abundant DEGs (1203) in the 20 progresses. Four of the 20 progresses: regulation of defense response, regulation of response to stress, response to bacterium, and regulation of multi-organism process were directly related to plant response to stress. KEGG analysis resulted that the pathway of translation was the mostly influenced pathway and enriched the most abundant DEGs in the top 20 pathways. Other pathways influenced by Psa treatment included metabolism of terpenoids, polyketides, and other secondary metabolites, and metabolism of carbohydrate, amino acid, and lipid.

Table 2. The enriched differential progress top 20.

GO ID	Description	DEGs Genes with Pathway Annotation (1521)	All Genes with Pathway Annotation (24,936)
1	Terpene biosynthetic process	7	7
2	Tricarboxylic acid metabolic process	20	120
3	Regulation of defense response	15	77
4	Terpene metabolic process	7	19
5	Citrate metabolic process	19	119
6	Regulation of response to stress	15	84
7	Acetate metabolic process	5	18
8	Positive regulation of microtubule Polymerization or depolymerization	2	2
9	Metabolic process	1203	19,066
10	Lignin metabolic process	5	21
11	Apoptotic process	2	3
12	Regulation of cell size	2	3
13	phenylpropanoid metabolic process	15	128
14	Cellular amino acid catabolic process	6	33
15	Response to bacterium	27	282
16	Energy coupled proton transmembrane Transport, against electrochemical gradient	11	88
17	Nucleoside diphosphate metabolic process	5	26
18	Cellular respiration	11	89
19	Regulation of multi-organism process	3	10
20	Arachidonic acid metabolic process	2	4

Multiple pathways involved in the metabolism of secondary metabolites were enriched, including those related to metabolism of the three main kinds of secondary metabolites: terpenes, phenols, and alkaloids (Table 3). Terpenes are the richest natural production and all of the monoterpenoid, diterpenoid, triterpenoid, and sesquiterpenoid (carotenoid) and terpenoid backbone biosynthesis pathways were regulated by Psa infection. Expression of as many as 64 (6.92%) genes in phenylpropanoid biosynthesis pathway which is the key progress in biosynthesis of phenols was regulated. Gene encoding the key enzyme phenylalanine ammonia-lyase (PAL) in phenol biosynthesis was found to be up regulated. Meanwhile, metabolism of the three main intermediate products tryptophan, tyrosine, and phenylalanine of phenol biosynthesis was also influenced. Two pathways in alkaloids biosynthesis were changed but the number of genes regulated only accounted a small part of all the genes annotated.

Table 3. Enriched pathways involved in secondary metabolism.

Metabolites	Pathway ID	Pathway	DEGs Genes with Pathway Annotation (925)	All Genes with Pathway Annotation (11,433)
Terpenes	ko00902	Monoterpenoid biosynthesis	8 (0.86%)	17 (0.15%)
	ko00903	Limonene and pinene degradation	8 (0.86%)	32 (0.28%)
	ko00904	Diterpenoid biosynthesis	8 (0.86%)	32 (0.28%)
	ko00909	Sesquiterpenoid and triterpenoid biosynthesis	4 (0.43%)	17 (0.15%)
	ko00906	Carotenoid biosynthesis	8 (0.86%)	51 (0.45%)
	ko00905	Brassinosteroid biosynthesis	2 (0.22%)	17 (0.15%)
	ko00900	Terpenoid backbone biosynthesis	9 (0.97%)	117 (1.02%)
Phenols	ko00940	Phenylpropanoid biosynthesis	64 (6.92%)	317 (2.77%)
	ko00380	Tryptophan metabolism	28 (3.03%)	105 (0.92%)
	ko00350	Tyrosine metabolism	14 (1.51%)	86 (0.75%)
	ko00360	Phenylalanine metabolism	9 (0.97%)	84 (0.73%)
	ko00941	Flavonoid biosynthesis	9 (0.97%)	91 (0.8%)
	ko00944	Flavone and flavonol biosynthesis	1 (0.11%)	11 (0.1%)
	ko00400	Phenylalanine, tyrosine and tryptophan biosynthesis	6 (0.65%)	89 (0.78%)
Alkaloids	ko00950	Isoquinoline alkaloid biosynthesis	7 (0.76%)	39 (0.34%)
	ko00960	Tropane, piperidine and pyridine alkaloid biosynthesis	4 (0.43%)	42 (0.37%)

Pathogenesis-related (PR) proteins are thought to participate in plant-defense mechanism. We searched the kiwifruit transcriptomic profile and identified genes encoding PR proteins belonging to 10 families of the all 17 PR protein families (Table 4 and Table S5). We identified the most unigenes in PR-9, PR-5 and PR-14 family with properties of peroxidase, thaumatin-like and lipid-transfer protein respectively. Unigenes belonging to PR-1, PR-4, PR-5, PR-9, PR-10, and PR-12 showed differentially expression in Psa-infected kiwifruit. Meanwhile we also identified genes encoding three kinds of antimicrobial peptides: hevein-like peptide, knottin-type peptide and snakin peptide, of which hevein-like peptide and snakin peptide were regulated by Psa infection.

Table 4. Differentially expressed genes encoding pathogenesis-related proteins.

Family	Properties	All Expressed Unigenes	Differentially Expressed Unigenes
PR-1	Unknown	6	2
PR-2	β-1,3-glucanase	1	0
PR-4	Chitinase type I, II	2	1
PR-5	Thaumatin-like	39	6
PR-6	Proteinase-inhibitor	9	0
PR-9	Peroxidase	96	8
PR-10	"Ribonuclease-like"	1	1
PR-12	Defensin	3	1
PR-13	Thionin	3	0
PR-14	Lipid-transfer protein	21	0

We screened out the resistant genes from the RNA sequencing data by searching the R-Gene database (PRGdb). In total, 4773 resistant genes were identified which could be grouped into 22 classes. The biggest class was RLP which acts as receptors to recognize avirulence genes. Of all the resistant genes identified, expression of 283 was changed by Psa treatment (Table 5). The class with the most differentially expressed resistant genes was NL which holds domains NBS (nucleotide-binding site) and LRR (leucine-rich repeat). Classes CN, CNL, N, and TNL also hold the NBS domain. Classes CNL, NL and TNL hold the LRR domain while RLK, RLK-GNK2, and RLP hold an extracellular leucine-rich repeat (eLRR). So, most of the differentially expressed resistant genes hold a NBS domain or a LRR domain. For most of the big classes, a certain number of genes exhibited regulated expression patterns in XJ3. Of the resistant genes, 211 (Table S6) hold the U-box domain and 25 were differentially expressed in Psa infected kiwifruit. Of the 25 differentially expressed U-box domain containing genes, 22 were up regulated in Psa infected kiwifruit and only three down regulated with small extent.

Table 5. The differentially expressed resistant genes.

Class	Number of All Identified Resistant Gene	Number of Differentially Expressed Genes
CN	123	6
CNL	411	25
L	19	0
Mlo-like	34	0
N	706	36
NL	828	69
Other	93	8
PTO	2	1
Pro-like	74	4
RLK	281	20
RLK-GNK2	246	13
RLK-Kinase	1	0
RLK-Malectina	1	0
RLK-Pro-like	1	0
RLP	1265	60
RLP-Malectin	5	0
RLP-Malectina	1	0
RPW8-NL	11	1
T	98	5
TNL	572	35
TNL-OT	1	0

2.5. Analysis of Genes in Plant-Pathogen Interaction

From KEGG analysis, we enriched genes participating in the pathway of plant-pathogen interaction. A total of 593 related genes were identified (Table S2) and they involve almost all of the processes of the plant-pathogen interaction pathway, including PTI, HR, stomatal closure, ETI, and programmed cell death. Of all the 593 plant-pathogen interaction related genes, 59 showed altered expression in Psa treatment (Figure S2). Two DEGs encoding a pattern recognition receptor (PRR) *CERK1* were identified, but it belonged to PRRs which recognize chitin PAMPs of fungus. Expression of a protein kinase encoding gene *CDPK* was also regulated. It plays important roles in regulating gene transcriptional changes and other cellular response. Two transcription factors *WRKY25* and *WRKY29* also displayed changed expression. They participate in the MAPK (mitogen-activated protein kinase) signal pathway and induce expression of defense-related genes. The other differentially expressed transcription factors were pathogenesis-related genes transcriptional activator *Pti1* and *Pti4*. The *PR1* gene which plays an important role in plant disease resistance takes part in multiple biological processes including MAPK signaling pathway, plant hormone signal transduction and plant-pathogen interaction. Its expression change in Psa treatment indicated its disease resistance function. We also identified two differentially expressed disease resistance genes *RPM1* and *RPS2*. They both function with another resistance gene *RIN4* and induce hypersensitive response. Expression regulation of the above resistance genes in the plant-pathogen interaction pathway indicates that Psa treatment induced plant immunity and led to functioning of the related resistance genes to protect the plant body.

3. Discussion

Plants are organisms which cannot move like animals and therefore they cannot escape potential threats from environment, including pathogens, arthropods, and abiotic stress. They survive depending on constitutive physical and chemical defense mechanisms such as waxy cuticles, cell walls, and phytoanticipins [13]. Besides these common defense mechanisms, plants also evolve a specific immune system, namely PTI and ETI to defend themselves against various pathogens around them [14]. To successfully infect a plant, pathogens have to penetrate the physical layer and make use of plant nutrition which would induce expression changes of various genes.

Secondary metabolites play important roles in regulation of plant growth and defense to pests and pathogens [15]. Plants deploy numerous secondary metabolites to facilitate interaction with biotic and abiotic environment. In our study, gene expression of the three main kinds of secondary metabolites including terpenes, phenols, and alkaloids were influenced by Psa infection. Terpenes are the largest class of natural products, many of which are toxic to insects [16], fungi [17], and bacteria [18]. Expression changes of genes involved in biosynthesis of monoterpenoid, diterpenoid, triterpenoid, sesquiterpenoid, and terpenoid backbone indicated that terpenes may play an important role in interaction between kiwifruit and Psa. Changes in diterpenoid and triterpenoid were particularly important, for the former is involved in gibberellin biosynthesis, and the latter involved in brassinosteroid biosynthesis. Both of the two plant hormones were found to function in plant innate immunity [19]. Expressions of eight genes related to biosynthesis of carotenoid which is tetraterpene were all increased after Psa infection. Carotenoids are important antioxidants to sweep reactive oxygen species produced by plant under stress [20].

Plant phenols are secondary metabolites with various structures. They work as signal compounds, pigments, internal physiological regulators or chemical messengers, and function in the resistance mechanism of plants against pathogens [21]. Most phenols biogenetically arise from the shikimate-phenylpropanoid-flavonoids pathway where phenylalanine ammonia-lyase (PAL) plays the key role in phenols production. Gene expression of *PAL* was up regulated in a sample infected by Psa meanwhile the phenylpropanoid biosynthesis pathway and phenylalanine metabolism pathway were also influenced. Metabolisms of tryptophan, tyrosine, and phenylalanine which are the mean precursors of phenols were altered as well. The results above indicate that the whole metabolism pathway of phenols was regulated and plant phenols may play an important role in kiwifruit resistance to Psa.

Pathogenesis-related (PR) proteins are induced under various biotic and abiotic stresses. They play an important role in plant-defense mechanism. We identified genes encoding PR proteins belonging to 10 families of the all 17 PR protein families characterized to date [22]. Unigenes belonging to PR-1, PR-4, PR-5, PR-9, PR-10, and PR-12 showed differentially expression in Psa-infected kiwifruit. In another study on kiwifruit by Beatrice [23], PR-1 and PR-5 expression was also induced by Psa. PR-1 proteins act as a molecular marker for systemic acquired resistance response. PR-5 acts as antifungal; glucanase and xylanase inhibitors; and α-amylase and trypsin inhibitors. Its down-regulation leads to susceptibility of resist *Piper colubrinum* to the oomycete pathogen *Phytophthora capsici* [24]. PR-4 proteins bind to chitin, and play an important role in enhancing the chitinase activity. The induction of PR4 transcripts in wheat coleoptils and roots is correlated with the expression of the corresponding proteins that are expressed only in the infected tissues [25]. PR-9 catalyzes cross-linking of macromolecules in plant cell wall and produces a free radical like H_2O_2 against a wide range of pathogens [26]. PR-12 proteins are small cysteine rich peptides providing protection against a broad range of organisms. They are known to inhibit protein synthesis, enzyme activity and ion channel function [27]. Among these PR proteins, PR-12 protein (defensin) and PR-13 protein (thionin) also act as antimicrobial peptides which are found as host defenses against pathogens and pests in diverse organisms [28]. Genes encoding two other antimicrobial peptides hevein-like peptide and snakin peptide were found to express differentially in Psa-infected kiwifruit. Snakin peptide is involved in plant-pathogen interactions [29] and snaking-Z derived from *Zizyphus jujube* fruits displayed antimicrobial activity against different bacterial and fungal [30].

Most disease resistance genes in plants encode nucleotide-binding site leucine-rich repeat (NBS-LRR) proteins. NBS-LRR proteins are involved in detection of diverse pathogens, including bacteria, viruses, fungi, nematodes, insects, and oomycetes. Expressions of many members of the two subfamilies CNL (CC-domain-containing) and TNL (TIR-domain-containing) of NBS-LRR family were detected in kiwifruit and 70 of them were found to be regulated in the Psa-infected kiwifruit. Meanwhile, most of the differentially expressed resistance genes hold an NBS domain or an LRR domain. The NBS domain is also called NB-ARC (nucleotide binding adaptor shared by NOD-LRR proteins, APAF-1, R proteins and CED4) domain. It is thought to result in conformational changes that regulate downstream signaling [31]. The LRR domain is involved in specific recognition of pathogen effector molecules [32] and it also functions as a regulatory domain [33].

Ubiquitination regulates diverse cellular processes, including floral transition, circadian rhythm, photomorphogenesis, and cell death [34,35]. In the study of Avr9/Cf-9 interaction, Gonzálezlamothe and his colleagues [36] found that two of the three *Avr9/Cf-9 Rapidly Elicited* (*ACRE*) genes essential for *Cf-9*- and *Cf-4*-dependent hypersensitive response encode putative E3 ubiquitin ligase components. Our results identified 211 U-box domain-containing protein encoding genes, of which 25 were differentially expressed in Psa-infected kiwifruit and 22 were up regulated. U-box is a derived version of RING-finger domain that lacks the hallmark metal-chelating residues of the latter but is likely to function similarly to the RING-finger in mediating ubiquitin-conjugation of protein substrates [37,38]. *ACRE74* which encodes a U-box E3 ligase homolog was induced in Cf9 tobacco and Cf9 tomato after Avr9 elicitation and its overexpression induced a stronger HR. This shows that the E3 ligase ACRE74 is essential for plant defense and disease resistance. PUB13 (plant U-box protein 13) is a well-studied example in plant disease resistance. Silencing of the PUB13 induced spontaneous cell death, elevated resistance to biotrophic pathogens but increased susceptibility to necrotrophic pathgenes [39]. Another study showed that PUB13 is also involved in regulating the FLS2-mediated PTI [40]. In our study, most of the differentially expressed PUB encoding genes were up regulated by Psa infection in kiwifruit, indicating they may play an important role in the interaction between kiwifruit and Psa.

During the long term of interaction between plants and pathogens, plants have evolved a complete defense system, namely PTI and ETI. This immune system will be triggered by recognition of PAMPs or effector secreted by invading pathogens and induces expression of resistance genes. Meanwhile, pathogens also can escape recognition by plants by lose or change of PAMPs and disturb ETI by new evolved effectors [41]. Approximately 50 pathovars of *P. syringae* have been recognized [42], and they cause economically important diseases in a wide range of plant species. Psa was first identified in Japan in 1984 [1] and it might evolved from ancestor of other hosts [43]. Transcriptomic analysis of kiwifruit infected by Psa contributes to explore the interaction between Psa and kiwifruit. Expressions of many genes involved in PTI and ETI were detected and several important genes showed differential expression in Psa-infected kiwifruit. *CDPK* and *Rboh* were PAMP induced genes which displayed increased expression in Psa-infected kiwifruit. These two genes regulate the production of reactive oxygen species [44,45] which induce HR. WRKY TFs are a large family involved in various plant processes but most notably in coping with diverse biotic and abiotic stresses [46,47]. In this study, four WRKY genes *WRKY22*, *WRKY25*, *WRKY29*, and *WRKY33* were all up regulated by Psa infection. *WRKY22*, *WRKY29* and *WRKY33* were also found to be up-regulated in Arabidopsis induced by chitin [48]. Overexpression of *WRKY25* resulted in increased disease symptoms to *P. syringae* infection, possibly by negatively regulating salicylic acid (SA)-mediated defense responses [49]. Two *Pti* genes were induced by Psa. *Pti1* is involved in a Pto-mediated signaling pathway, probably by acting as a component downstream of Pto in a phosphorylation cascade. Its expression in tobacco plants enhanced the hypersensitive response to a *P. syringae* pv. tabacoo strain carrying the avirulence gene *avrPto* [50]. *Pti4* confers resistance to *P. syringae* pv tomato that causes bacterial speck disease in tomato [51]. RIN4 in Arabidopsis is targeted by type III effectors AvrRpt2 and AvrRpm1 which inhibit PAMP-induced signaling and compromise the host's basal defense system. The R proteins, RPS2 and RPM1 whose encoding genes were regulated by Psa sense type III effectors-induced perturbation of RIN4 and guard the plant against pathogens [52]. One heat shock protein (HSP) encoding gene *HSP90* was also down regulated. HSP90 is required for functioning of RPS2 and its inhibition reduces the HR and abolishes resistance against *P. syringae* pv. tomato DC3000 [53].

4. Materials and Methods

4.1. Plant Materials and Treatments

The kiwifruit (*Actinidia chinensis* var. *deliciosa*) cultivar "Jinkui" kept in Institute of Botany, Jiangsu Province and Chinese Academy of Science, China, was used in this study. Shoots of good growth vigor were collected from kiwifruit trees and stuck in MS medium, and maintained in growth chambers. The condition was set with a temperature of 25 °C and 12 h/12 h (light/dark) cycles. After one week, seedlings were inoculated with the canker-causing bacteria *Pseudomonas syringae* pv. *actinidiae* (Psa). Bacterial cells were suspended in distilled water and adjusted to an $OD_{600} = 0.2$, and injected into the seedling stems which were carved with a knife. Five treatments were set: CK, only carved; PY3, inoculated with water and sampled 48 h after inoculation; PY4, inoculated with water and sampled at 96 h; XJ3, inoculated with Psa and sampled at 48 h; XJ4, inoculated with Psa and sampled at 96 h. Phloem of each sample was collected with three biological replicates.

4.2. RNA Extraction, Transcriptome Sequencing and De Novo Assembly

Total RNA was isolated from phloem samples according the method of Cai [54], and mRNA was enriched by Oligo (dT) bead. Then the enriched mRNA was fragmented into short fragments (approximately 200–700 nt) and reverse transcribed into cDNA with random primers, and then the second-strand cDNA were synthesized. Sequencing was done using Illomina HiSeq™4000 by Gene Denovo Biotechnology Co. (Guangzhou, China).

After filtering of low-quality raw reads, transcriptome de novo assembly was carried out with short reads assembling program Trinity [55]. The assembled transcript whose length was larger than 200 bp was kept. The longest transcript in each locus was taken as the unigene.

4.3. Functional Annotation of Unigenes

BLASTx program (http://www.ncbi.nlm.nih.gov/BLAST) was used to annotate the unigenes, with an *E*-value threshold of 10^{-5} to NCBI non-redundant protein (Nr) database (http://www.ncbi.nlm.nih.gov), the Swiss-Prot protein database (http://www.expasy.ch/sprot), and the COG/KOG database (http://www.ncbi.nlm.nih.gov/COG). The best alignment results were for protein functional annotations. GO annotation of unigenes was analyzed by Blast2GO software [56], and functional classification of unigenes was performed using WEGO software [57]. Kyoto Encyclopedia of Genes and Genomes (KEGG) annotations were obtained in http://www.genome.jp/kegg.

For R-Gene analysis, protein coding sequences of unigenes were aligned by BLASTp to R-Gene database PRGdb (http://prgdb.crg.eu/wiki/Main_Page).

4.4. Functional Analysis of Differentially Expressed Unigenes

The unigene expression was calculated and normalized to RPKM (Reads Per kb per Million reads) [58]. The edgeR package (http://www.r-project.org/) was used to identify differentially expressed genes (DEGs) between Psa treated sample and control. Genes with a fold change ≥ 2 and a false discovery rate (FDR) ≤ 0.05 in a comparison were defined as significant DEGs. All DEGs were mapped to GO terms in the Gene Ontology database (http://www.geneontology.org), and gene numbers were calculated for every term. Significantly enriched GO terms in DEGs comparing to the genome background were defined by hypergeometric test. The rigorous FDR correction method was for *q* value correction, and GO terms were defined as being significantly enriched when the *q* value was ≤0.05. KEGG pathway enrichment analysis was done to identify significantly enriched metabolic pathways or signal transduction pathways in DEGs comparing with the whole genome background. Pathways with *q* value ≤ 0.05 were defined as significantly enriched.

4.5. Quantitative RT-PCR Analysis

To test the expression results from transcriptome sequencing, we determined the expression levels of 12 randomly selected unigenes through the method of qRT-PCR. Primers were designed using Primer5 software (Table S7), and *AdActin* was used as internal control. We performed qRT-PCR using the SYBR® *Premix Ex Taq*™ (Perfect Real Time, Dalian, China) (TaKaRa Code: DRRO41A), with PCR conditions of 40 cycles of 95 °C for 20 s, 60 °C for 20 s, and 72 °C for 40 s. Relative gene expression was calculated according to the $2^{-\triangle\triangle Ct}$ method [59].

5. Conclusions

In order to explore the interaction between Psa and its host kiwifruit plants, we analyzed the transcriptome of kiwifruit infected by Psa. In total, 8255 differentially expressed genes were identified, including those involved in secondary metabolites metabolism, NBS-LRR protein encoding genes, and genes of plant immunity system PTI and ETI. Expression changes of genes involved in the secondary metabolism especially the biosynthesis of terpenes were evident, indicating the probable role of secondary metabolites in plant defense. Expressions of genes encoding NBS-LRR proteins which are usually products of resistance genes were also found to be regulated. Among these NBS-LRR protein genes, we noted that U-box domain containing genes were obviously differentially expressed. PUB proteins mediate ubiquitin-conjugation of protein substrates, and may function in HR. Expression of genes involved in PTI and ETI was detected and some key genes showed differential expression in Psa-infected kiwifruit. These genes play important roles in plant immunity system, such as PAMP and effector recognition, signal transduction, HR and defense related gene induction. We hope our results will facilitate the future study of interaction between Psa and kiwifruit.

Supplementary Materials: Supplementary materials can be found at www.mdpi.com/1422-0067/19/2/373/s1.

Acknowledgments: This study was funded by the National Natural Science Foundation of China (NSFC) (Grant No. 31401854), the Foundation of Jiangsu Key Laboratory for the Research and Utilization of Plant Resources (Institute of Botany, Jiangsu Province and Chinese Academy of Sciences) (Grant No. JSPKLB201602) and the Natural Science Foundation of Jiangsu Province (Grant No. BK20171328).

Author Contributions: Zhong-Ren Guo and Ji-Yu Zhang designed the experiments. Tao Wang, Ji-Yu Zhang and Gang Wang performed the experiments. Tao Wang, Zhan-Hui Jia, and De-Lin Pan analyzed the data. Tao Wang and Ji-Yu Zhang wrote the manuscript. All authors read and approved the final manuscript.

Conflicts of Interest: The authors declare no conflict of interest.

References

1. Serizawa, S.; Ichikawa, T.; Takikawa, Y.; Tsuyumu, S.; Goto, M. Occurrence of bacterial canker of kiwifruit in Japan, description of symptoms, isolation of the pathogen and screening of bacteriocides. *Jpn. J. Phytopathol.* **1989**, *55*, 427–436. [CrossRef]

2. Vanneste, J.L.; Yu, J.; Cornish, D.A.; Tanner, D.J.; Windner, R. Identification, virulence and distribution of two biovars of *Pseudomonas syringae* pv. *actinidiae* in New Zealand. *Plant Dis.* **2013**, *97*, 708–719. [CrossRef]

3. Chapman, J.R.; Taylor, R.K.; Weir, B.S.; Romberg, M.K.; Vanneste, J.L. Phylogenetic relationships among global populations of *Pseudomonas syringae* pv. *actinidiae*. *Phytopathology* **2012**, *102*, 1034–1044. [CrossRef] [PubMed]

4. Vanneste, J.L.; Poliakoff, F.; Audusseau, C.; Cornish, D.A.; Paillard, S. First report of *Pseudomonas syringae* pv. *actinidiae*, the causal agent of bacterial canker of kiwifruit in France. *Plant Dis.* **2011**, *95*, 1311–1312. [CrossRef]

5. Boller, T.; Felix, G. A renaissance of elicitors, perception of microbe-associated molecular patterns and danger signals by pattern-recognition receptors. *Annu. Rev. Plant Biol.* **2009**, *60*, 379–406. [CrossRef] [PubMed]

6.   Alfano, J.R.; Collmer, A. Type III secretion system effector proteins, double agents in bacterial disease and plant defense. *Annu. Rev. Phytopathol.* **2004**, *42*, 385–414. [CrossRef] [PubMed]

7.   Guo, M.; Tian, F.; Wamboldt, Y.; Alfano, J.R. The majority of the type iii effector inventory of *pseudomonas syringae* pv. *tomato* dc3000 can suppress plant immunity. *Mol. Plant. Microbe Interact.* **2009**, *22*, 1069–1080. [CrossRef] [PubMed]

8.   Chisholm, S.T.; Coaker, G.; Day, B.; Staskawicz, B.J. Host-microbe interactions, shaping the evolution of the plant immune response. *Cell* **2006**, *124*, 803–814. [CrossRef] [PubMed]

9.   Greenberg, J.T.; Yao, N. The role and regulation of programmed cell death in plant-pathogen interactions. *Cell. Microbiol.* **2004**, *6*, 201–211. [CrossRef] [PubMed]

10.   Mccann, H.C.; Rikkerink, E.H.A.; Bertels, F.; Fiers, M.; Lu, A.; Reesgeorge, J. Genomic analysis of the kiwifruit pathogen *pseudomonas syringae* pv. *actinidiae* provides insight into the origins of an emergent plant disease. *PLoS Pathog.* **2013**, *9*, e1003503. [CrossRef]

11.   Scortichini, M.; Marcelletti, S.; Ferrante, P.; Petriccione, M.; Firrao, G. *Pseudomonas syringae* pv. *actinidiae*: A re-emerging, multi-faceted, pandemic pathogen. *Mol. Plant Pathol.* **2012**, *13*, 631–640. [CrossRef] [PubMed]

12.   Marcelletti, S.; Ferrante, P.; Petriccione, M.; Firrao, G.; Scortichini, M. *Pseudomonas syringae* pv. *actinidiae* draft genome comparisons reveal strain-specific features involved in adaptation and virulence to *Actinidia* species. *PLoS ONE* **2011**, *6*, e27297. [CrossRef] [PubMed]

13.   Agrios, G.N. How plants defend themselves against pathogens. In *Plant Pathology*; Elsevier Inc.: Amsterdam, The Netherlands, 2005; pp. 105–160.

14.   Dodds, P.N.; Rathjen, J.P. Plant immunity, towards an integrated view of plant-pathogen interactions. *Nat. Rev. Genet.* **2010**, *11*, 539–548. [CrossRef] [PubMed]

15.   Bennett, R.N.; Wallsgrove, R.M. Tansley Review No. 72. Secondary metabolites in plant defence mechanisms. *New Phytol.* **1994**, *127*, 617–633. [CrossRef]

16.   Lee, S.; Peterson, C.J.; Coats, J.R. Fumigation toxicity of monoterpenoids to several stored product insects. *J. Stored Prod. Res.* **2003**, *39*, 77–85. [CrossRef]

17.   Hammer, K.A.; Carson, C.F.; Riley, T.V. Antifungal activity of the components of *Melaleuca alternifolia* (tea tree) oil. *J. Appl. Microbiol.* **2003**, *95*, 853–860. [CrossRef] [PubMed]

18.   Friedman, M.; Henika, P.R.; Mandrell, R.E. Bactericidal activities of plant essential oils and some of their isolated constituents against *Campylobacter jejuni*, *Escherichia coli*, *Listeria monocytogenes*, and *Salmonella enterica*. *J. Food Prot.* **2002**, *65*, 1545–1560. [CrossRef] [PubMed]

19.   Bruyne, L.D.; Hfte, M.; Vleesschauwer, D.D. Connecting growth and defense: The emerging roles of brassinosteroids and gibberellins in plant innate immunity. *Mol. Plant* **2014**, *7*, 943–959. [CrossRef] [PubMed]

20.   Canfield, L.M.; Forage, J.W.; Valenzuela, J.G. Carotenoids as cellular antioxidants. *Proc. Soc. Exp. Biol. Med.* **1992**, *200*, 260–265. [CrossRef] [PubMed]

21.   Slatnar, A.; Mikulič-Petkovšek, M.; Veberič, R.; Štampar, F. Research on the involment of phenoloics in the defence of horticultural plants. *Acta Agric. Slov.* **2016**, *107*, 183–189. [CrossRef]

22.   Christensen, A.B.; Cho, B.H.; Næsby, M.; Gregersen, P.L.; Brandt, J.; Madrizordeñana, K.; Collinge, D.B.; Thordal-Christensen, H. The molecular characterization of two barley proteins establishes the novel pr-17 family of pathogenesis-related proteins. *Mol. Plant Pathol.* **2002**, *3*, 135–144. [CrossRef] [PubMed]

23.   Beatrice, C.; Linthorst, J.M.H.; Cinzia, F.; Luca, R. Enhancement of PR1, and PR5, gene expressions by chitosan treatment in kiwifruit plants inoculated with *pseudomonas syringae* pv. *actinidiae*. *Eur. J. Plant Pathol.* **2017**, *148*, 163–179. [CrossRef]

24.   Anu, K.; Jessymol, K.K.; Chidambareswaren, M.; Gayathri, G.S.; Manjula, S. Down-regulation of osmotin (PR5) gene by virus-induced gene silencing (VIGS) leads to susceptibility of resistant *piper colubrinum* link. to the oomycete pathogen *phytophthora capsici* leonian. *Indian J. Exp. Biol.* **2015**, *53*, 329–334. [PubMed]

25.   Bertini, L.; Leonardi, L.; Caporale, C.; Tucci, M.; Cascone, N.; Berardino, I.D.; Buonocore, V.; Caruso, C. Pathogen-responsive wheat *PR4* genes are induced by activators of systemic acquired resistance and wounding. *Plant Sci.* **2003**, *164*, 1067–1078. [CrossRef]

26.   Hegde, Y.R.; Keshgond, R.S. Role of pathogenesis-related proteins in plant disease management—A review. *Agric. Rev.* **2013**, *34*, 145–151.

27.   Vriens, K.; Cammue, B.P.; Thevissen, K. Antifungal plant defensins: Mechanisms of action and production. *Molecules* **2014**, *19*, 12280–12303. [CrossRef] [PubMed]

28. Egorov, T.A.; Odintsova, T.I.; Pukhalsky, V.A.; Grishin, E.V. Diversity of wheat anti-microbial peptides. *Peptides* **2005**, *26*, 2064–2073. [CrossRef] [PubMed]

29. Bindschedler, L.V.; Whitelegge, J.P.; Millar, D.J.; Bolwell, G.P. A two component chitin-binding protein from french bean—Association of a proline-rich protein with a cysteine-rich polypeptide. *FEBS Lett.* **2006**, *580*, 1541–1546. [CrossRef] [PubMed]

30. Daneshmand, F.; Zare-Zardini, H.; Ebrahimi, L. Investigation of the antimicrobial activities of Snakin-Z, a new cationic peptide derived from *Zizyphus jujuba* fruits. *Nat. Prod. Res.* **2013**, *27*, 2292–2296. [CrossRef] [PubMed]

31. Tameling, W.I.; Elzinga, S.D.; Darmin, P.S.; Vossen, J.H.; Takken, F.L.; Haring, M.A.; Cornelissen, B.J. The tomato R gene products I-2 and MI-1 are functional ATP binding proteins with ATPase activity. *Plant Cell* **2002**, *14*, 2929–2939. [CrossRef] [PubMed]

32. Luck, J.E.; Lawrence, G.J.; Dodds, P.N.; Shepherd, K.W.; Ellis, J.G. Regions outside of the leucine-rich repeats of flax rust resistance proteins play a role in specificity determination. *Plant Cell* **2000**, *12*, 1367–1377. [CrossRef] [PubMed]

33. Tao, Y.; Yuan, F.; Leister, R.T.; Ausubel, F.M.; Katagiri, F. Mutational analysis of the *Arabidopsis* nucleotide binding site-leucine-rich repeat resistance gene *RPS2*. *Plant Cell* **2000**, *12*, 2541–2554. [PubMed]

34. Yee, D.; Goring, D.R. The diversity of plant U-box E3 ubiquitin ligases, from upstream activators to downstream target substrates. *J. Exp. Bot.* **2009**, *60*, 1109–1121. [CrossRef] [PubMed]

35. Craig, A.; Ewan, R.; Mesmar, J.; Gudipati, V.; Sadanandom, A. E3 ubiquitin ligases and plant innate immunity. *J. Exp. Bot.* **2009**, *60*, 1123–1132. [CrossRef] [PubMed]

36. Gonzálezlamothe, R.; Tsitsigiannis, D.I.; Ludwig, A.A.; Panicot, M.; Shirasu, K.; Jones, J.D. The U-box protein CMPG1 is required for efficient activation of defense mechanisms triggered by multiple resistance genes in tobacco and tomato. *Plant Cell* **2006**, *18*, 1067–1083. [CrossRef] [PubMed]

37. Aravind, L.; Koonin, E.V. The U box is a modified RING finger—A common domain in ubiquitination. *Curr. Biol.* **2000**, *10*, R132–R134. [CrossRef]

38. Lorick, K.L.; Jensen, J.P.; Fang, S.; Ong, A.M.; Hatakeyama, S.; Weissman, A.M. Ring fingers mediate ubiquitin-conjugating enzyme (E2)-dependent ubiquitination. *Proc. Natl. Acad. Sci. USA* **1999**, *96*, 11364–11369. [CrossRef] [PubMed]

39. Li, W.; Ahn, I.P.; Ning, Y.; Park, C.H.; Zeng, L.; Whitehill, J. The U-box/ARM E3 ligase PUB13 regulates cell death, defense and flowering time in Arabidopsis. *Plant Physiol.* **2012**, *159*, 239–250. [CrossRef] [PubMed]

40. Lu, D.; Lin, W.; Gao, X.; Wu, S.; Cheng, C.; Avila, J. Direct ubiquitination of pattern recognition receptor FLS2 attenuates plant innate immunity. *Science* **2011**, *332*, 1439–1442. [CrossRef] [PubMed]

41. Zhou, J.M.; Chai, J. Plant pathogenic bacterial type III effectors subdue host responses. *Curr. Opin. Microbiol.* **2008**, *11*, 179–185. [CrossRef] [PubMed]

42. Gardan, L.; Shafik, H.; Belouin, S.; Broch, R.; Grimont, F.; Grimont, P.A. DNA relatedness among the pathovars of *Pseudomonas syringae* and description of *Pseudomonas tremae* sp. nov. and *Pseudomonas cannabina* sp. nov. (ex Sutic and Dowson 1959). *Int. J. Syst. Bacteriol.* **1999**, *49*, 469–478. [CrossRef] [PubMed]

43. Monteil, C.L.; Cai, R.; Liu, H.; Llontop, M.E.; Leman, S.; Studholme, D.J. Nonagricultural reservoirs contribute to emergence and evolution of *pseudomonas syringae* crop pathogens. *New Phytol.* **2013**, *199*, 800–811. [CrossRef] [PubMed]

44. Kobayashi, M.; Ohura, I.; Kawakita, K.; Yokota, N.; Fujiwara, M.; Shimamoto, K. Calcium-dependent protein kinases regulate the production of reactive oxygen species by potato NADPH oxidase. *Plant Cell* **2007**, *19*, 1065–1080. [CrossRef] [PubMed]

45. Kurusu, T.; Kuchitsu, K.; Tada, Y. Plant signaling networks involving Ca^{2+} and Rboh/Nox-mediated ROS production under salinity stress. *Front. Plant Sci.* **2015**, *6*, 427. [CrossRef] [PubMed]

46. Eulgem, T. Regulation of the Arabidopsis defense transcriptome. *Trends Plant Sci.* **2005**, *10*, 71–78. [CrossRef] [PubMed]

47. Naoumkina, M.; He, X.; Dixon, R. Elicitor-induced transcription factors for metabolic reprogramming of secondary metabolism in Medicago truncatula. *BMC Plant Biol.* **2008**, *8*, 132. [CrossRef] [PubMed]

48. Wan, J.; Zhang, S.; Stacey, G. Activation of a mitogen-activated protein kinase pathway in Arabidopsis by chitin. *Mol. Plant Pathol.* **2004**, *5*, 125–135. [CrossRef] [PubMed]

49. Zheng, Z.; Mosher, S.; Fan, B.; Klessig, D.; Chen, Z. Functional analysis of Arabidopsis WRKY25 transcription factor in plant defense against *Pseudomonas syringae*. *BMC Plant Biol.* **2007**, *7*, 2. [CrossRef] [PubMed]

50. Zhou, J.; Loh, Y.T.; Bressan, R.A.; Martin, G.B. The tomato gene pti1encodes a serine/threonine kinase that is phosphorylated by pto and is involved in the hypersensitive response. *Cell* **1995**, *83*, 925–935. [CrossRef]
51. Wu, K. Functional analysis of tomato Pti4 in Arabidopsis. *Plant Physiol.* **2002**, *128*, 30. [CrossRef] [PubMed]
52. Kim, M.G.; Da, C.L.; Mcfall, A.J.; Belkhadir, Y.; Debroy, S.; Dangl, J.L. Two *pseudomonas syringae* type III effectors inhibit RIN4-regulated basal defense in Arabidopsis. *Cell* **2005**, *121*, 749–759. [CrossRef] [PubMed]
53. Takahashi, A.; Casais, C.; Ichimura, K.; Shirasu, K. HSP90 interacts with RAR1 and SGT1 and is essential for RPS2-mediated disease resistance in Arabidopsis. *Proc. Natl. Acad. Sci. USA* **2003**, *100*, 11777–11782. [CrossRef] [PubMed]
54. Cai, B.H.; Zhang, J.Y.; Gao, Z.H.; Qu, S.C.; Tong, Z.G.; Mi, L.; Qiao, Y.S.; Zhang, Z. An improved method for isolation of total RNA from the leaves of *Fragaria spp.. Jiangsu J. Agric. Sci.* **2008**, *24*, 875–877.
55. Grabherr, M.G.; Haas, B.J.; Yassour, M.; Levin, J.Z.; Thompson, D.A.; Amit, I. Full-length transcriptome assembly from RNA-Seq data without a reference genome. *Nat. Biotechnol.* **2011**, *29*, 644–652. [CrossRef] [PubMed]
56. Conesa, A.; Götz, S.; Garcíagómez, J.M.; Terol, J.; Talón, M.; Robles, M. Blast2GO, a universal tool for annotation, visualization and analysis in functional genomics research. *Bioinformatics* **2005**, *21*, 3674–3676. [CrossRef] [PubMed]
57. Ye, J.; Fang, L.; Zheng, H.; Zhang, Y.; Chen, J.; Zhang, Z. WEGO: A web tool for plotting Go annotations. *Nucleic Acids Res.* **2006**, *34*, W293. [CrossRef] [PubMed]
58. Mortazavi, A.; Williams, B.A.; Mccue, K.; Schaeffer, L.; Wold, B. Mapping and quantifying mammalian transcriptomes by RNA-Seq. *Nat. Methods* **2008**, *5*, 621–628. [CrossRef] [PubMed]
59. Livak, K.J.; Schmittgen, T.D. Analysis of relative gene expression data using real-time quantitative PCR and the 2(-Delta Delta C (T)) method. *Methods* **2001**, *25*, 402–408. [CrossRef] [PubMed]

International Journal of
Molecular Sciences

MDPI

Article

Protein Activity of the *Fusarium fujikuroi* Rhodopsins CarO and OpsA and Their Relation to Fungus–Plant Interaction

Alexander Adam [1], Stephan Deimel [1], Javier Pardo-Medina [2], Jorge García-Martínez [2], Tilen Konte [3], M. Carmen Limón [2], Javier Avalos [2] and Ulrich Terpitz [1,*]

[1] Department of Biotechnology and Biophysics, Biocenter, Julius Maximilian University of Würzburg, D-97074 Würzburg, Germany; alexander.adam@uni-wuerzburg.de (A.A.); stephan.deimel@stud-mail.uni-wuerzburg.de (S.D.)

[2] Department of Genetics, Faculty of Biology, University of Seville, E-41012 Seville, Spain; jpardo6@us.es (J.P.-M.); jorgegarmar@gmail.com (J.G.-M.); carmenlimon@us.es (M.C.L.); avalos@us.es (J.A.)

[3] Institute of Biochemistry, Faculty of Medicine, University of Ljubljana, Sl-1000 Ljubljana, Slovenia; tilen.konte@mf.uni-lj.si

* Correspondence: ulrich.terpitz@uni-wuerzburg.de; Tel.: +49-931-31-84226

Received: 27 November 2017; Accepted: 8 January 2018; Published: 11 January 2018

Abstract: Fungi possess diverse photosensory proteins that allow them to perceive different light wavelengths and to adapt to changing light conditions in their environment. The biological and physiological roles of the green light-sensing rhodopsins in fungi are not yet resolved. The rice plant pathogen *Fusarium fujikuroi* exhibits two different rhodopsins, CarO and OpsA. CarO was previously characterized as a light-driven proton pump. We further analyzed the pumping behavior of CarO by patch-clamp experiments. Our data show that CarO pumping activity is strongly augmented in the presence of the plant hormone indole-3-acetic acid and in sodium acetate, in a dose-dependent manner under slightly acidic conditions. By contrast, under these and other tested conditions, the *Neurospora* rhodopsin (NR)-like rhodopsin OpsA did not exhibit any pump activity. Basic local alignment search tool (BLAST) searches in the genomes of ascomycetes revealed the occurrence of rhodopsin-encoding genes mainly in phyto-associated or phytopathogenic fungi, suggesting a possible correlation of the presence of rhodopsins with fungal ecology. In accordance, rice plants infected with a CarO-deficient *F. fujikuroi* strain showed more severe bakanae symptoms than the reference strain, indicating a potential role of the CarO rhodopsin in the regulation of plant infection by this fungus.

Keywords: fungal rhodopsins; CarO; OpsA; *Fusarium fujikuroi*; *Oryza sativa*; rice–plant infection; green light perception; indole-3-acetic acid (IAA); bakanae; patch-clamp

1. Introduction

Fungi inhabit almost every ecological niche in all kind of ecosystems, where they have to face sudden changes in their growth conditions. Their survival capacity relies on the successful adaption of their physiology to diverse environmental scenarios, which requires efficient control of gene expression. A remarkable example is found in the colonization of host organisms, that in fungi implies differential regulation of many genes [1]. One of the most important physical parameters impacting on fungal life is sunlight [2]. Fungi possess various photosensory proteins that allow them to perceive different light wavelengths, and to adapt to changing light conditions [2,3]. While flavin-based proteins, such as those of the white collar, photolyase, vivid, and cryptochrome families perceive blue light, red light is sensed by phytochromes via biliverdin [4]. Many fungi are also equipped with membrane-embedded green light-sensing rhodopsins [5]. However, only a small number of these photoreceptors have been biophysically or biologically characterized [6–9].

Fungal rhodopsins belong to the super family of G-protein coupled receptors, and accordingly, consist of seven transmembrane helices (TMs). In ascomycetes, four clades of opsin-like proteins are distinguished [10], that are grouped around either HSP30 (heat shock protein 30 [11]), ORP-1 (opsin related protein 1 [12]), NOP-1 (*Neurospora* opsin protein 1 [13]), or CarO (from *Fusarium fujikuroi* CarO [14], also known as auxiliary ORP-like rhodopsins [15]). Only NOP-1-like and CarO-like opsins are microbial rhodopsins that exhibit the lysine residue required for covalent binding of the chromophore all-*trans*-retinal. In the rhodopsin protein, the retinal is located in an interior pocket surrounded by the TMs. Retinal isomerization from all-*trans* to 13-*cis* initiates a series of consecutive conformational alterations (photointermediates), which are correlated with protein function, known as photocycle. The NOP-1-related opsin clade can be subdivided into *Leptosphaeria* rhodopsin (LR)-like [9] and *Neurospora* rhodopsin (NR)-like [13] rhodopsins, due to their characteristic protein function [15]. NR-like fungal rhodopsins most likely provide sensory functions, and their photocycle is very slow [13,16]. By contrast, LR-like and CarO-like rhodopsins exhibit fast photocycles [8,14], and provide proton pumping activity [8,9,16]. A characteristic of CarO-like rhodopsins is the occurrence of a presumptive interaction site for a so far unknown transducer protein [7].

We investigate the rhodopsins of the fungus *F. fujikuroi*, an ascomycete that provokes bakanae disease in rice (*Oryza sativa*) plants [17]. The disease is accompanied by an increase in stem elongation, which is provoked by the plant hormone gibberellic acid (GA) produced by the fungus [18,19]. The infected plants become chlorotic (paler green tissues) and exhibit less leaves and internodes. In severe cases, even growth stagnation (stunted growth) can be observed [20], an effect provoked by the action of a cytotoxic secondary metabolite, fusaric acid, produced by the fungus [21,22]. *F. fujikuroi* possesses two genes coding for rhodopsins, *carO* and *opsA* [14,23]. The *carO* gene is linked to and co-regulated with genes coding for enzymes for retinal synthesis, whose expression is strongly induced by light, and the CarO protein is an effective proton pump highly expressed in light-exposed conidia, where it slows down their germination [8,24]. On the other hand, the *opsA* gene is only moderately upregulated by light, and the function of the OpsA rhodopsin is not known.

The prevalence of fungal rhodopsins in the genomes of phytopathogenic or phyto-associated fungi suggests a potential role in the host–fungus interaction [8]. Supporting this assumption, a recent analysis of the microbiome in the leaf environment of different plants revealed high abundance of rhodopsins in the phyllosphere [25]. Furthermore, the LR-like rhodopsin Sop1 from the fungus *Sclerotinia sclerotiorum* plays an essential role in the virulence of the fungus. Also, the two rhodopsins of the ascomycete *Alternaria brassicola* (gb | AB08921 and gb | AB06529) are upregulated in the plant host environment [26]. Similarly, the rhodopsin *ops3* (gb | Um04125) of the basidiomycete *Ustilago maydis* is upregulated during the infection process in corn plants [27]. In general, pH plays an important role on fungal virulence during the plant infection [28,29], which could be related with the involvement of proton pumping rhodopsins in fungal pathogenicity.

In the present study, we aim to improve our understanding of the biological role of the rhodopsins in *F. fujikuroi*. We expressed either CarO::YFP or OpsA::YFP in HEK293 cells, performed electrophysiology experiments, and observed that the presence of the auxin indole-3-acetic acid (IAA) and acetate increases the CarO pumping activity up to 10-fold, whereas OpsA did not show any net charge transfer under the tested conditions. We conducted a basic local alignment search tool (BLAST) search for the occurrence of fungal rhodopsins in 108 Ascomycota genomes, and observed that the majority of rhodopsin genes are found in phytopathogenic and phyto-associated fungi. Furthermore, we tested the ability of conidia from mutant *F. fujikuroi* strains to infect rice plants, and observed more severe bakanae symptoms in plants infected by null CarO mutants, compared to those infected by the reference strain, as indicated by their increased internodal length and reduced chlorophyll content. Based on RT-PCR experiments, we found no indication for any co-regulation of CarO and G proteins.

Taken together, our data provide evidence on the participation of fungal rhodopsins in fungi-plant interactions and phytopathogenesis.

2. Results

2.1. CarO Pumping Activity Is Enhanced by Indole-3-Acetic Acid (IAA) and Acetate Whereas OpsA Does Not Exhibit Pumping Activity

During its pathogenic growth in the plant apoplast, *F. fujikuroi* faces different organic compounds, including weak organic acids (WOAs), and plant hormones like auxin [30]. Recently, we found that at low pH, the proton pump activity of CarO increased in gluconate solution [8]. Further investigation revealed that this supporting effect was mainly provoked by low concentrations of acetate ions used as counter ion for Mg^{2+} and Ca^{2+} in the gluconate solution, rather than by gluconate itself. As the pumping signal of CarO expressed in yeast cells was very low (Supplementary Figure S1), to further investigate this WOA effect, CarO::YFP and OpsA::YFP were expressed in HEK293 human cells (Figure 1a,b), and investigated with patch-clamp technique in a whole cell configuration. The rhodopsins were activated by illumination with green laser light (532 nm; 2–3×10^{17} photons s^{-1} mm^{-1}). Under our experimental conditions, chloride channels are activated in HEK293 cells, interfering with the small pump signal [31]. Therefore, for a better signal-to-noise ratio, the WOA effect was measured in sodium gluconate-based (chloride-free) extracellular solution. Nevertheless, similar results were obtained in a NaCl-based solution.

Upon illumination under standard conditions (NaCl pH 7.4 bath solution) CarO exhibited the expected outward-directed signal [8], with a transient response that relaxed in the dark to the stationary level with a decay in a bi-exponential manner (Figure 1c). By contrast, in the presence of sodium acetate or the plant hormone IAA the activity of the proton pump CarO was substantially increased, and the ratio between peak and stationary current was clearly reduced (Figure 1c).

In the case of OpsA, no electrogenicity was found under all tested conditions, including those providing high pumping activity in CarO (Figure 1d), as expected from the assignation of OpsA to the group of NR-like rhodopsins [15]. This was further supported by sequence comparisons that revealed that OpsA is phylogenetically closer to NOP-1 than to CarO [23]. Like NOP-1 [32], these proteins are supposed to be non-pumping rhodopsins. Accordingly, NOP-1 was shown to exhibit a very slow photocycle, and did not provoke currents when expressed in neurons [16].

We further investigated the effect of WOA on the pump activity of CarO. When gluconate was gradually replaced by acetate, an increased pump activity of CarO was observed along the whole range of voltages, from -120 mV to $+40$ mV (Figure 2a). At the same time, the characteristic voltage dependency with higher activity at positive membrane potentials remained unaffected. A notable increase of pump activity became already observable in the two-digit micromolar range. Similar results were also obtained when gluconate was gradually replaced by IAA (Supplementary Figure S2). For better visualization, we plotted the dependence of pump activity at 0 mV membrane potential on the WOA concentration, and fit it with a Hill function (Figure 2b). The increase by IAA was in a similar range as by acetate, but it could not be analyzed in the full range of concentrations, due to its low solubility in water (Figure 2b). When acetate-activated CarO was faced with pH 9, a transient increment of the pump activity was observed, similar to the one recently reported in a gluconate solution ([8]; Supplementary Figure S3). The activity increased to a maximal 3.3-fold compared to that at 0.7 mM sodium acetate pH 5, and decreased within a few minutes to activities lower than those in NaCl pH 7.4. In bacteriorhodopsin, the pump activity did not increase in presence of WOAs, suggesting that this effect is specific for the CarO protein (Supplementary Figure S4).

In halorhodopsin and bacteriorhodopsin, the protonation of the Schiff base counter ion leads to a shift of the absorption spectrum [33,34]. In order to figure out if the presence of acetate or IAA might affect the protonation state, we recorded the action spectrum of CarO in sodium chloride pH 7.4, sodium gluconate pH 5, sodium gluconate supplemented with 1 mM IAA, or sodium gluconate supplemented with 10 mM sodium acetate. For all conditions, maximal pump activity was observed at 532 nm, and all data were in a similar range, suggesting unchanged protonation state of the counter ion of the Schiff base (Figure 2c).

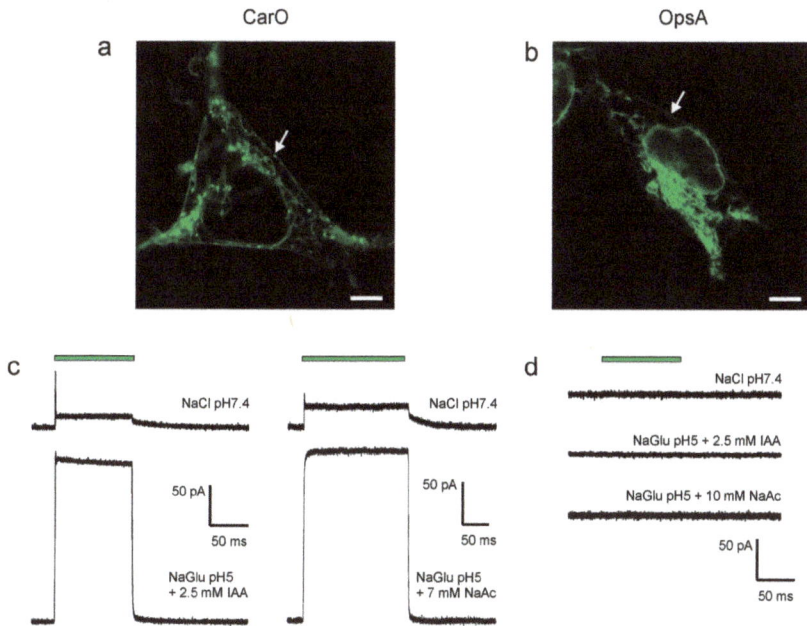

Figure 1. Microscopic and patch-clamp analysis of the fungal rhodopsins CarO and OpsA expressed in HEK293 cells. (**a**,**b**) Representative confocal laser scanning micrographs showing the localization of the fungal rhodopsins in the cells. Both rhodopsins are partly trafficked to the plasma membrane (indicated by **white arrows**), and thus, accessible to the patch-clamp pipette. Scale bars represent 5 µm. (**c**,**d**) Typical traces of either CarO (**c**) or OpsA (**d**) recorded in whole cell mode at 0 mV holding potential at intracellular pH 7.4 and diverse extracellular conditions, as indicated. Cells were illuminated with green light (532 nm DPSS laser), as illustrated by the green bar. The pump activity of CarO is clearly augmented by indole-3-acetic acid (IAA) and sodium acetate (NaAc) at pH 5, whereas for OpsA no electrogenic activity was detected under any of the tested conditions, suggesting a protein function unrelated with ion pumping.

2.2. Fungal Rhodopsins Are Predominant in Phyto-associated Fungi

In our previous study, we found that rhodopsin genes are recurrent in the genomes of phytopathogenic fungi [8]. In order to find out if the presence of rhodopsins is more related to the phylogeny of the respective species than to its ecology, we analyzed the occurrence of rhodopsins in different fungal species by BLAST searches of their genome sequences (Supplementary Table S1). For this investigation, we focused our attention on the set of species used in a recent evolutionary analysis of fungal effector proteins [35], with some additions, arranged according to a former 6-gene maximum-likelihood phylogeny of Ascomycota [36]. Our phylogenetic analysis (Figure 3a) revealed that these photoreceptors are not present in all taxa. Out of 108 analyzed species from 42 orders, rhodopsins were only found in 38 fungal genomes. Interestingly, 74% of the species containing rhodopsins are either phytopathogens or phyto-associated (28 species). In accordance, rhodopsins were found in 9 out of 13 orders that include mainly phytopathogenic or phyto-associated fungi, while only in 1 out of 15 orders that include mainly wood inhabiting or saprophytic fungi.

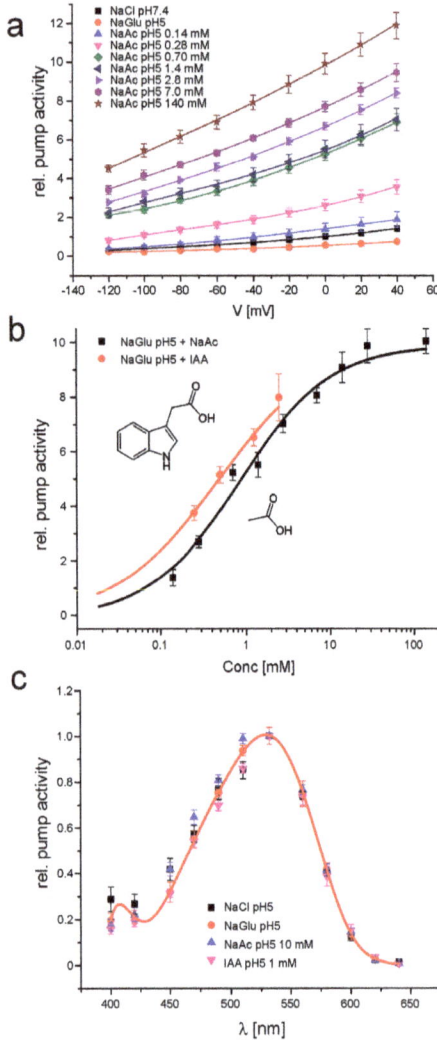

Figure 2. Patch-clamp analysis of the influence of sodium acetate and IAA in the pump activity of CarO. (**a**) Current–voltage relationship of the CarO pump activity, as indicated in sodium chloride pH 7.4, sodium gluconate pH 5, various concentration of sodium acetate in sodium gluconate pH 5, or only sodium acetate pH 5. Mean ± SEM of at least 5 cells are given. For better visualization, data were described by a cubic fit. (**b**) Dose–response relationship of CarO pump activity and IAA and sodium acetate dissolved in a bath solution of sodium gluconate pH 5. Data were described using a standard Hill equation. (**c**) Action spectrum of CarO as indicated in either NaCl pH 5, sodium gluconate (NaGlu) pH 5, sodium gluconate pH 5 + 10 mM sodium acetate, or sodium gluconate pH 5 + 1 mM IAA. For better visualization, the data points of NaCl pH 5 were fit using a polynomial fit. Note, that the spectrum is not influenced by the presence of the weak organic acids tested.

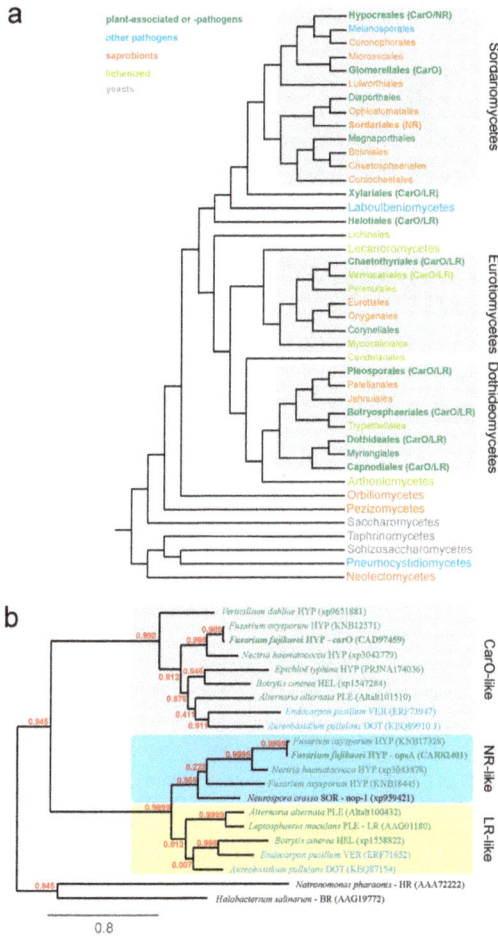

Figure 3. Phylogenetic analysis of ascomycetes and fungal rhodopsins. (**a**) Phylogenetic tree of the phylum Ascomycota. For each order, the typical ecological niche is indicated by the font color (green: phyto-associated or phytopathogenic fungi; blue: other pathogens; orange: saprobionts or wood inhabiting fungi; pale green: lichenized fungi; grey: yeasts). Orders that possess rhodopsins are printed in bold font, and the classification of the fungal rhodopsin is given in brackets. Three different classes of fungal rhodopsins are distinguished, NR-like (from *N. crassa*), LR-like (from *L. nodorum*, proton pumps), and CarO-like (auxiliary ORP-like rhodopsin, proton pumps with a putative capacity to interact with other proteins). NR-like and LR-like rhodopsins are merged in the clade of NOP-1 like opsins. Many phytopathogenic and phyto-associated fungi possess genes coding for rhodopsins, while most saprophytic fungi do not, indicating a potential role for these green light-sensing photoreceptors in fungus–plant interaction. A complete list with gene accession numbers is given in Supplementary Table S1. (**b**) Phylogenetic tree of fungal rhodopsins from selected species from Hypocreales (HYP), Helotiales (HEL), Pleosporales (PLE), Verrucariales (VER), Dothideales (DOT), and Sordariales (SOR), as indicated after the species name. Data in brackets represent gene database accession numbers. The tree that was rooted to the microbial halorhodopsin (HR) from *Natronomonas pharaonis*, and bacteriorhodopsin (BR) from *Halobacterium salinarum*, shows the phylogenetic relationships of fungal rhodopsins, including OpsA and CarO from *F. fujikuroi*. Branch support values are given in red, and branch lengths are as indicated by the bar. A complete tree is given in Supplementary Figure S4).

As indicated above, the rhodopsins have been classified in three types, NR-like rhodopsins, with slow photocycle and presumable photosensory functions, and CarO-like and LR-like rhodopsins, with fast photocycles and proton pumping activities [15]. Though functionally distinct, LR-like and NR-like rhodopsins are clearly distinguished from the CarO-like rhodopsins, phylogenetically very closely related, and merge in one clade that was recently denoted as NOP1-like opsins [10]. We compared the amino acid sequences of the rhodopsins from the 38 fungi through different bioinformatic tools (see Experimental Procedures), and we used the generated data to construct a phylogenetic tree (Figure 3b, Supplementary Figure S4). The distribution of the rhodopsins in the tree supports the above-mentioned classification. Strikingly, no fungus was found to possess all three types of rhodopsins. By contrast, some fungi possess several copies or variations of the same type of rhodopsins. For example, the plant pathogen *Fusarium oxysporum* (Hypocreales), exhibits two different NR-like rhodopsins, while the human-pathogenic fungus *Coniosporium epidermidis* (Chaetothyriales) possesses two different CarO-like rhodopsins.

2.3. Bakanae Symptoms of Rice Plants Are Affected by the Null CarO Mutation in the Infective Fungus

Among the investigated species, CarO-like rhodopsins are spread in different taxonomic groups, but they are only present in phytopathogenic, human pathogenic, or rock-inhabiting species. However, rhodopsins are not found in all phytopathogenic fungi, e.g., they are absent in the rice pathogen *Magnaporthe grisea*. NR-like rhodopsins seem to be unique of the Sordariomycetes, while the combination of CarO-like and LR-like rhodopsins are mainly found in Eurotiomycetes, Dothideomycetes, and Helotiales. In the Sordariomycetes, only the Hypocreales and Glomerellales, both encompassing phytopathogenic fungi, exhibit CarO-like rhodopsins (Figure 3a, Supplementary Table S1). Overall, the presence of rhodopsins in mainly phyto-associated orders, independently of their phylogeny, suggests a positive selection for rhodopsins in plant environments.

As rhodopsins are abundant in phyto-associated fungi, we aimed to find out if *F. fujikuroi* strains that are deficient in either OpsA [23] or CarO [14] would exhibit a distinct phenotype when infecting rice plants. In this experiment, previously established fungal strains were used, namely wild type (FKMC1995 *Fusarium fujikuroi* (*G. fujikuroi* mating population C)); ΔopsA (mutant FKMC1995 strain with deleted rhodopsin gene *opsA* [23]), CarO⁻ (mutant FKMC1995 strain with disrupted rhodopsin gene *caro* (shift in open reading) [14]), and CarO⁺ (isogenic reference of the CarO⁻ strain with intact rhodopsin gene *caro*). To analyze bakanae symptoms, we infected two-day old rice germlings (cultivar Sendra) with fresh conidia (50 conidia/rice seed) of the respective *F. fujikuroi* strain or water (control) under green light and grew the developing rice plants in vermiculite pots in light–dark cycles at 28 °C. All plants that were infected with fungi exhibited clear bakanae symptoms (Figure 4). Nevertheless, for each fungal strain, we noticed high variability in the morphology of the plant replicates. Thus, statistical analysis was required, and the symptoms of bakanae were assessed by differences of internodal length and changes in the plant chlorophyll content (Figure 5). While internodes of uninfected control plants exhibited after 10 days a mean length of 4.74 cm (SEM 0.15, $n = 42$), all infected plants exhibited longer internodes (CarO⁺ 5.38 cm, SEM 0.16, $n = 58$; CarO⁻ 6.08 cm, SEM 0.21, $n = 57$, Figure 5a; WT 5.53 cm, SEM 0.15, $n = 56$; ΔOpsA 5.81 cm, SEM 0.18, $n = 56$, Figure 5c). The CarO-deficient mutant produced a significant ($p < 0.05$; two tailed student *t*-test) increase in the internodal length of the rice plant (Figure 5a), compared to its reference strain CarO⁺, and the wild type, thus showing more severe symptoms of the disease. Also, the plants infected by the CarO-deficient mutant exhibited a paler pigmentation, and their content of chlorophylls/carotenoids was significantly reduced in comparison to control plants, or to those infected with the CarO⁺ reference (Figure 5b). Moreover, colonization of the rice seeds by fungal mycelium was faster with conidia of the CarO-deficient strain than with those of the CarO⁺ control (data not shown). By contrast, we did not detect any significant difference between the chlorophylls /carotenoids content of plants infected with wild type conidia, and that of plants infected with ΔopsA conidia, though both samples showed lowered content of these pigments when compared to non-infected plants. These data suggest that OpsA does not influence the bakanae symptoms (Figure 5c,d).

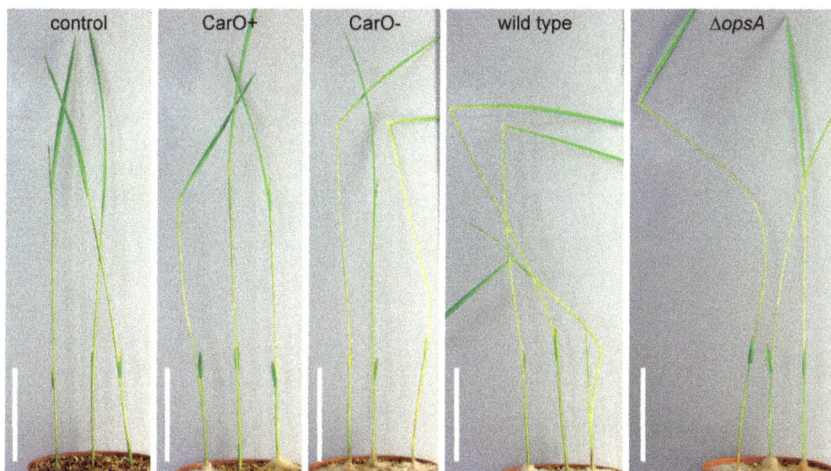

Figure 4. Rice plant infection with rhodopsin-deficient *F. fujikuroi* strains. Analysis of potential effects of the rhodopsins OpsA and CarO on the virulence of *F. fujikuroi* in rice plants (cultivar Sendra). Representative images of 13-day-old rice plants 10 days after infection with conidia from *F. fujikuroi*, strains or water (non-infected control), as indicated, are shown. Note that the plant shape varies even under the same conditions between the individuals. Therefore, to quantify the bakanae symptoms, the length of the first internode and the chlorophyll/carotenoid content were chosen to estimate disease severity (see Figure 5). White bars represent 5 cm.

Addition of gibberellin GA3 to the nutrient solution results in an elongation of the rice plants (data not shown). As the infection with the CarO⁻ strain induces stem elongation, we aimed to find out if gibberellic acid (GA) synthesis was enhanced in the absence of CarO. However, the *carO⁻* mutation did not affect the gibberellin production under laboratory conditions (Supplementary Figure S5). Similarly, OpsA does not affect gibberellin production [23]. These data suggest that the observed differences in bakanae symptoms might be due to the differential capacity of both strains to grow in the plant. In support to this conclusion, a higher number of mycelial colonies were able to grow from stems of infected rice plants when the infection was achieved with the CarO deficient mutant than with the CarO⁺ control strain (Supplementary Figure S6).

2.4. Light Exerts Minor Influence on mRNA Levels for Genes of G Proteins

The finding of a presumptive interaction site for a putative transducer protein in the CarO-like protein of *Phaeosphaeria nodorum* [7] suggests that CarO might interact with a regulatory partner. Because of the participation of heterotrimeric G complexes in signal transduction from different animal opsins, we postulate that CarO might interact with a specific Gα protein. Considering that CarO expression is strongly light dependent, we might expect some level of co-regulation with its putative transducing partner. To check this hypothesis, the genome of *F. fujikuroi* was screened for the presence of genes for putative G proteins, based on a previous scrutiny carried out in *Gibberella zeae* (*Fusarium graminearum*) [37]. We found in *F. fujikuroi* five genes for putative Gα subunits (FUJ_06643, 443 aa; FFUJ_04487, 355 aa; FFUJ_07379, 354 aa; FFUJ_08667, 353 aa; FFUJ_05248, 636 aa), and single genes for a Gβ subunit (FUJ_09550, 359 aa) and a Gγ subunit (FFUJ_03226, 93 aa). A phylogram of the Gα and Gβ sequences confirmed that the five α subunits are more closely related to each other than to the β subunit (Figure 6b).

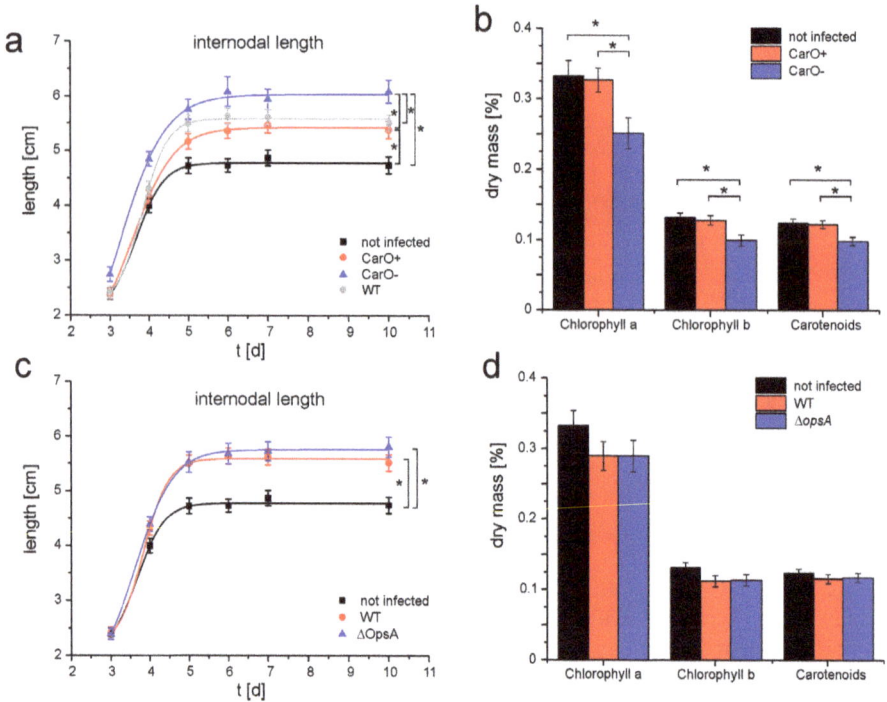

Figure 5. Influence of fungal infection on internodal length and chlorophyll content. (**a,c**) Mean length and standard error (data pooled from four independent experiments with 15 or 12 seedlings each for infected and control plants, respectively) of the first internode of rice plants. (**a**) Rice plants were either not infected (black, $n = 42$) or infected with the CarO$^+$ (red, $n = 58$), the CarO$^-$ (blue, $n = 57$) or wild type (gray, $n = 56$) strains. (**c**) Rice plants were either not infected (black, $n = 42$) or infected with wild type (red, $n = 56$) or the *opsA* deletion mutant (blue, $n = 56$) strains. (**b,d**) Chlorophyll and carotenoid extraction from rice plants 8 days after infection. Data are mean values and SEM of at least 12 trials out of 3 independent test series. (**b**) Rice plants were either not infected (black, $n = 12$) or infected with CarO$^+$ (red, $n = 15$) or CarO$^-$ (blue, $n = 14$) conidia. (**d**) Rice plants were either not infected (black, $n = 12$) or infected with wild type (red, $n = 14$) or *opsA* deletion (blue, $n = 15$) conidia. Differences in internodal length and chlorophyll/carotenoid content were tested for significance in a two-tailed student *t*-test. Significant distinctions are highlighted by an asterisk.

To check a possible co-regulation with *carO*, we investigated the effect of illumination on the transcript levels of the seven genes for G proteins in comparison to those of the gene *carO* and the structural genes of the carotenoid pathway *carRA* and *carB*, linked to *carO* in a co-regulated cluster [38]. As expected, the mRNA levels for these genes increased rapidly following light onset to reach 200–300-fold in the case of *carRA* and *carB*, and more than 1000-fold in the case of *carO* (Figure 6a). By contrast, the genes for G proteins were very moderately affected by light, with maximal variations of only 3-fold (Figure 6a–d). Such changes are very small compared to the strong photoinduction exhibited by *carO*, suggesting lack of regulatory connections.

Recently the gene expression profile of *F. fujikuroi* strain FKMC1995 in vitro and in planta was published [39]. We compared the expression levels of the above-mentioned proteins at 6 mM glutamine (22.7 mM in our experiments), with the in planta conditions (Supplementary Table S2). In accordance with the RT-PCR data we found the level of G proteins to be almost unchanged (ratios between 0.7 and

1.4), while *carO* (3.3) and *carB* (2.7) were upregulated in the plant environment. In contrast, *carRA* was almost unchanged (1.1) and *carX* (0.3) even downregulated in planta.

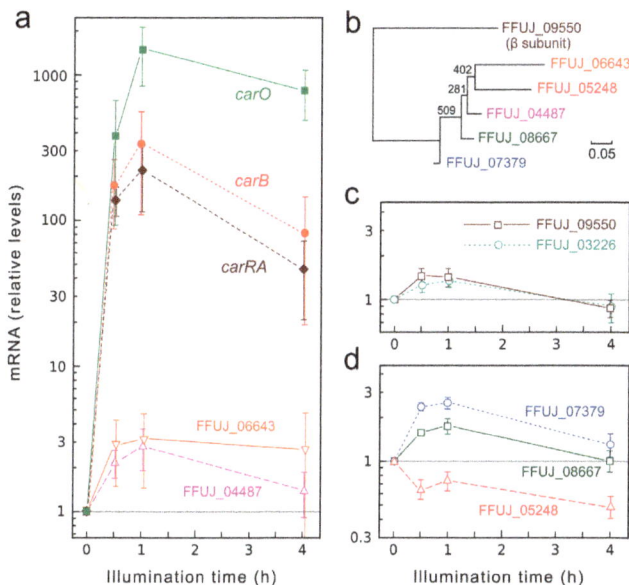

Figure 6. Effect of illumination on transcript levels for the genes *carRA*, *carB*, and *carO* and the seven genes for G proteins in the *F. fujikuroi* genome. (**a,c,d**) Real-time RT–PCR analyses of the indicated genes in total RNA samples from the wild type FKMC1995. The cultures were grown for three days in the dark before illumination for 30 min, 1 h, or 4 h. Relative levels are referred to the value of the wild type in the dark. Data show means and standard errors of the mean from three biological replicates. The data are represented in different graphs to avoid overlapping. (**b**) Phylogram of the sequences of the five Gα proteins (FUJ_06643, 04487, 07379, 08667 and 05248). The Gβ protein (FUJ_09550) was used as outgroup. The Gγ protein (FUJ_09550) was not included in the phylogram because of its small size (93 residues). The scale indicates the proportion of changes. Numbers in nodes indicate bootstrap values.

3. Discussion

Despite the efforts dedicated to the detailed characterization of several fungal rhodopsins [7–9,13,14,40], the biological role and impact of these proteins in filamentous fungi is still elusive. In previous investigations, the phenotypic differences of the rhodopsin-deficient mutants compared to the wild type were absent or hardly detectable [13,14,23,40]. Strikingly, in our recent study, CarO was found to slow down the germination of *F. fujikuroi* conidia in light [8]—the first phenotype associated with a fungal rhodopsin. A recent, detailed analysis of NOP-1 from *N. crassa* now revealed a clear phenotype of the fungal rhodopsin in the regulation of the sexual–asexual switch [10]. However, the analyses were performed using cultures grown in defined nutrient media under standard laboratory conditions, in the absence of a host plant.

To our knowledge, there is yet only one investigation dealing with the phenotype of rhodopsin-deficient mutants of phytopathogenic fungi in their host plant environment [41]. In this investigation, the LR-like rhodopsin Sop1 was shown to be essential for growth, sclerotial development, and full virulence of *S. sclerotiorum*. In accordance, our BLAST search for rhodopsins in a hundred sequenced fungal genomes (Figure 3; Supplementary Figure S5 and Supplementary Table S1) reveals that the majority of genes coding for fungal rhodopsins are found in phyto-associated or phytopathogenic fungi (fungi colonizing living plant tissue), whereas only a small percentage of saprotrophic fungi (fungi decaying organic matter) exhibit such genes in their genomes. The number of phyto-associated fungi might be even bigger, as the ecology of many fungi is not well investigated, and might conceal some surprises, as happened recently with the discovery of an unknown phyto-associated lifestyle of *N. crassa* [42].

To figure out a potential role of NR-like and CarO-like rhodopsins in colonization of plants, we tested the ability of rhodopsin-deficient *F. fujikuroi* strains to colonize rice plants and to provoke the characteristic symptoms of the bakanae disease [17,20,21]. Recent RNA-seq data have shown that both rhodopsins are expressed in the plant environment [39] (Supplementary Table S2). According to this study, transcription of *carO* is only slightly (ratio 3.2), and *opsA* highly (ratio 28.6) upregulated in plants. In our experimental analysis, the extension of the plant shoot, measured as the length of first internode, and the content of chlorophyll and carotenoids, were chosen as indicators for infection severity. Indeed, our data show, for the first time, that a fungal rhodopsin of a phytopathogenic fungus affects the severity of rice plant infection (Figures 4 and 5). When investigating the role of CarO and OpsA in the development of bakanae symptoms, the shoots between seed and first node are significantly enlarged in all infected plants, compared to uninfected control plants. Unexpectedly, the elongation of the shoot was significantly enhanced in plants infected with CarO$^-$ strain, in comparison to those infected with the CarO$^+$ or wild type strain. Moreover, the chlorophyll content in the CarO$^-$ strain infected plants was lower than in those infected with the CarO$^+$ strain, suggesting a more aggressive phytopathogenic phenotype in absence of the CarO rhodopsin. Similar symptoms could also be induced by treatment of plants with GA [20] and GA-producing *Fusarium* species are more efficient rice pathogens than those unable to produce GAs [43]. Moreover, comparison of pathogenic capacities of wild type *F. fujikuroi* and a GA-defective mutant on rice revealed an impaired ability of the mutant hyphae to penetrate the plant cells [44]. In our case, the amount of GAs produced in shake cultures of the CarO-deficient strain was similar to those produced by the control CarO$^+$ strain (Supplementary Figure S6). We cannot exclude that the GA-production of *F. fujikuroi* in the plant environment differs, and might be influenced by CarO, a task to be investigated in the future. However, based on our data, the longer internodes in the CarO$^-$-infected plants should be interpreted as a result of a higher amount of fungal biomass compared to those infected with the CarO$^+$ control strain. Accordingly, the CarO deficient strain was found to be more effective in colonizing the plants (Supplementary Figure S7). Thus, one might conclude that the increased severity of bakanae symptoms in absence of CarO is caused by an increased virulence.

Fungal virulence is directly related to the speed of germination [45]. The faster the fungus invades the host tissue, the lower the risk for the conidia to be displaced or outcompeted by other fungi [46]. In this respect, it must be noted that under laboratory conditions, the conidia of the CarO$^-$ mutant germinated earlier and faster than those of the CarO$^+$ strain [24], whereas in later developmental stages, both strains exhibit a similar colony growth, i.e., mycelial development [8]. Therefore, the faster germination may not be sufficient to explain the observed increase in virulence. Recent findings about the role of rhodopsins suggest negative regulatory role under light, like CarO in conidia germination and NOP-1 in protoperithecial production in the sexual cycle [8,10]. It can be speculated that rhodopsin activity might also negatively regulate the virulence of *F. fujikuroi*. The molecular mechanisms underlying the observed effects will require further investigation. Also, the potential importance of carotenoid production for rhodopsin function has to be taken into account. In *F. fujikuroi*, the *carO* gene is located and co-regulated in a cluster with genes coding for enzymes involved in carotenoid synthesis, namely *carX*, *carRA*, and *carB*. Interestingly, in 8 fungi out of the 11 fungal species shown in Figure 3b, the *carO* counterpart was clustered with orthologs for these *car* genes (Supplementary Table S3). Recent RNA-seq data show that *carB* is upregulated, *carRA* almost unchanged, and *carX* downregulated during plant colonization. Thus, *F. fujikuroi* with deleted *car* genes should be considered in future pathogenesis studies in rice plants.

To date, the role of light in phytopathogenesis by fungi has received very limited attention. Nevertheless, some investigations have revealed connections between fungal light perception and pathogenicity. In *Magnaporthe oryzae*, pre-illumination of conidia for one hour did not affect spore germination, but stimulated their chemically assayed superoxide production, resulting in spores more tolerant to toxic diffusate of rice leaf compared to those that were non-illuminated [47]. In the grey mold *Botrytis cinerea*, the proteins of the white collar photoreceptor complex are required for coping with excessive light or oxidative stress, and also to achieve full virulence [48], and in *Colletotrichum acutatum*, the light quality influences the virulence of the fungus [49]. Remarkably, the fungus *Diplodia mutila* leaves its endophytic niche to become a pathogen in response to high–intense light. This is most likely triggered by light-induced production of H_2O_2 by the fungus, which results in hypersensitivity, cell death, and tissue necrosis in the infected palm [50]. In this respect, it is of interest that the CarO protein of the closely related fungus *F. graminearum* was recently shown to be highly upregulated after initiation of the sexual cycle [10]. Besides that, recent investigations in *N. crassa* and *S. sclerotiorum* suggest a role of rhodopsins in oxidative and osmotic stress [10,41].

Another environmental factor that plays a critical role during infection processes in phytopathogenic fungi is pH [29]. Because of its effective proton pumping activity, CarO may contribute to local pH changes in infected tissues. Fungi, in general, use proton gradients for the signaling and transport of substances through membranes [51]. A green light-driven proton pump could thus play a role as an energy saving mechanism. In accordance, our microscopic data show that the CarO protein is expressed in light-exposed hyphae [8]. The *carO* gene is actively induced by light in older mycelia (four day old submerged cultures [14]), and although the physiological conditions in the plant may be very different than under ex planta laboratory conditions, the data suggest that the gene is also induced by light in the plant, a fact that could be favored by the higher efficiency of the rhodopsin under green light. By contrast, the OpsA$^-$ mutant did not exhibit altered bakanae symptoms in comparison to the wild type (Figure 5c,d). OpsA is a NR-like rhodopsin, only present in the Sordariamycetes, and similar to NOP-1 of *N. crassa*, a rhodopsin with no electrogenicity [16] and with a putative sensory function. In accordance, we could not observe any pumping activity of OpsA, though we tested it under several experimental conditions (Figure 1d). Nevertheless, we cannot exclude that OpsA might behave differently in its native environment, and the localization and biological role of OpsA will be matter of future investigations.

The strong increment in pump activity of the outward-directed proton pump CarO, in response to the plant hormone IAA and to acetate at moderate acid pH (Figure 2a–c), is of high interest, because of the predictable implications in the fungus–plant interaction. In principle, under such conditions, we would expect a reduction of pump activity due to the increased proton gradient. A number of different ion species, such as sodium, chloride, or imidazole, are known to be pumped or guided by microbial rhodopsins [52]. Therefore, we may speculate that the ion species transported by CarO might be not restricted to protons. Acetate is a small anion, and thus, theoretically it could be considered as a potentially pumped species. This argumentation is refuted by our comparative experiments with IAA, which is too big to pass through the ca 0.6 nm pore of a microbial rhodopsin [53,54], and shows similar results as with acetate (Figure 2b). IAA and acetate are capable of diffusing in the undissociated form through the membrane with permeability coefficients of 6.9×10^{-3} cm s^{-1} [55] and 2.3×10^{-4} cm s^{-1} [56], respectively. Intracellularly, the compounds can dissociate, as indicated by the capacity of acetate to symmetrically acidify the cytosol in frog oocytes [57,58]. Suchlike perfusion through the membranes is the method by which IAA is distributed in the growing plant. However, in our patch-clamp experiments, the cytosolic conditions are well defined, as the cytosol is rapidly replaced by the pipette solution [59]. We therefore assume that the increase of pumping activity is not due to a potential uncoupling mechanism of acetate or IAA. Instead, it is more likely that the acetate group interacts with the free hydrogen in the pump pore, and consequently enables faster release of protons, and therefore, speeds up the pumping activity. A similar behavior was formerly described for other microbial rhodopsins [60], but we could not observe a comparable response of bacteriorhodopsin to acetate or IAA at pH 5 (Supplementary Figure S4). However, we found a supporting effect of WOAs for other fungal rhodopsins (unpublished data), indicating that this effect is of biological relevance in fungal rhodopsins. This could especially be the case in the plant environment, where IAA and acetate and other WOAs are present in the plant sap [30]. IAA was recently shown to be produced by *Fusarium* species [61], and to play a role in the bakanae symptoms evoked in rice plants by *Fusarium proliferatum* [62].

It has to be taken into account that our electrophysiological investigation was performed in rhodopsins expressed in mammalian cells. Though this is a common procedure for characterization of microbial rhodopsins, it must be taken into account that the protein might behave differently in its native environment, e.g., due to the absence of potential interacting/transducing proteins. We considered the possible participation of a G protein as a CarO regulatory partner, an interaction solidly established in the case of the visual opsins in vertebrates, and some non-visual proteins in lower animals [63,64]. However, the apparent lack of strong regulation by light of any of the G protein genes of *F. fujikuroi* does not favor this hypothesis, but it cannot be discarded. More experiments are needed to confirm if CarO-like rhodopsins only have proton-pumping activity, or whether they also have a signal-transducing activity through the interaction with a G protein or with a membrane-embedded regulatory protein.

The functions of rhodopsins in fungi are puzzling and fascinating biological issues. Our data suggest that the two *F. fujikuroi* rhodopsins play very different roles, reflected by their distinct biochemical properties and regulatory patterns. In the case of CarO, we could show for the first time that a rhodopsin influences the behavior of a fungus during plant infection by reducing the severity of disease symptoms. Although its mechanism of action and its participation in other cellular processes remain to be elucidated, we may speculate that green light is interpreted by the fungus as a signal to attenuate the strength of the pathogenic colonization of the plant. The ways in which the photobiochemical activities of different fungal rhodopsins influence the physiology of these organisms, and their ability to survive in their natural environment, will be the aim of further studies.

4. Materials and Methods

4.1. Fungal Strains and Culture Conditions

All fungal strains used in this study are derivatives of *Fusarium fujikuroi* FKMC1995. Phenotypic analyses of CarO were performed with the strains SF100 (CarO$^-$; transformant 4 R3) and SF101 (CarO$^+$ control; transformant 4 R2), both derived from a former transformant with a *carO$^+$ carO$^-$* duplication generated by homologous integration with a *carO$^-$* plasmid [14]. SF100 contains a disrupted CarO allele, whereas SF101 is a revertant strain from the same transformant exhibiting an intact CarO allele. Phenotypic analysis of OpsA was done with strain SF223 (Δ*opsA*; [23]) that is lacking the OpsA gene, and FKMC1995 wild type strain was used as control. Fluorescent strains WF2 (*F. fujikuroi* CarO::YFP [8]) was described before. Fungi were cultured in Dextrose-*Gibberella* (DG) minimal medium [65] where NaNO$_3$ was replaced by 3 g L^{-1} glutamine (DG$_{gln}$), or in potato dextrose agar (PDA). Fungi were grown in a Peltier-controlled incubator (Memmert IPP110) equipped with an appropriate light module. In addition, a customized illumination module consisting of 10 green high-power LEDs (205 lm, LUXEON Rebel LXML PM01, Lumileds; San Jose, CA, USA) controlled by a customized power supply was used for green light illumination of fungal cultures. Dark-grown cultures were protected against light by storing them in an opaque box inside the incubator.

4.2. Cell Cultures

Stable cell lines of Flp-In™ 293 T-REx (Invitrogen, Carlsbad, CA, USA) expressing either CarO::YFP [8] or OpsA::YFP (this study; implemented according to manufactures manual) were cultured at 37 °C and 5% CO$_2$ in Dulbecco's Modified Eagle Medium (DMEM) supplemented with fetal calf serum (10%), L-glutamine (2 mM), penicillin (100 U mL^{-1}), streptomycin (100 µg mL^{-1}), and as selection, antibiotics hygromycin B (100 μg mL^{-1}) and blasticidin HCl (15 μg mL^{-1}). For experiments, the cells were grown in a 6-well plate with 2 mL medium/well. Protein expression was induced by tetracycline (3 μg mL^{-1}) for 12–16 h. Cells were treated with all-*trans*-retinal (1 μM) to ensure sufficient availability of chromophore for the tested rhodopsins.

4.3. Molecular Biology

For the construction of pcDNA5/FRT/TO-opsA::YFP, the cDNA of the *opsA* gene without stop codon and fused to *eyfp* with a 21-bp linker was synthesized (Gene Art, Life Technologies, Waltham, MA, USA) and delivered in the vector pMA-RQ. This 1.6 kb fragment was cut by *Xho*I and ligated to the pcDNA5/FRT/TO-backbone cut with *Xho*I, resulting in pcDNA/FRT/TO-opsA::YFP.

4.4. Patch-Clamp

Patch-clamp experiments were performed as described by García-Martínez et al. [8] with some modifications. For activation of the rhodopsins, a 532 nm DPSS laser (MGL-III-532, 150 mW, TTL 1 Hz–1 kHz; Changchun New Industries Optoelectronics, Changchun, China) was coupled into the fluorescence beam path of an inverse microscope (Axiovert 200, Zeiss, Jena, Germany). Pipettes (GB150F-8P, Scientific-Instruments, Hofheim, Germany) had a tip opening diameter of 1.5 mm and exhibited a resistance of 3–5 MΩ in standard bath solution. Whole cell currents were recorded with an Axopatch 200B amplifier coupled to a DigiData 1440 interface (Molecular Devices Corporation, Union City, NJ, USA), low-pass filtered at 5 kHz, and digitized at a sampling rate of 100 kHz with the software Clampex 10.3 (Molecular Devices Co.). Data analysis was performed using Clampfit 10.6 (Molecular Devices Co.) and Origin Pro 9.1G or 2016G (OriginLab Corporation, Northampton, MA, USA). The experiments were performed in a customized perfusion chamber offering convenient slow solution exchange with a flow rate of ca 2 mL min^{-1}. Standard bath solution contained 140 mM NaCl, 2 mM MgCl$_2$, 2 mM CaCl$_2$, and 10 mM 4-(2-hydroxyethyl)-1-piperazineethanesulfonic acid (HEPES) pH 7.4. For pH screening, 2-(N-morpholino)ethanesulfonic acid (TRIS) or 2-(N-morpholino)ethanesulfonic acid (MES) was used instead of HEPES, and the pH of the bath solution was adjusted to 5 or 9 with HCl or NaOH. The influence of WOAs on the rhodopsins was analyzed in solution buffered with either 10 mM HEPES, TRIS, or MES. The gluconate solution contained 140 mM sodium gluconate, 2 mM magnesium gluconate, and 2 mM CaOH, the respective pH was adjusted with either gluconic acid or NaOH. The acetate solution consisted of 140 mM sodium acetate, 2 mM magnesium acetate, and 2 mM calcium acetate, and the respective pH was adjusted with either acetic acid or NaOH. Acetate solution was mixed with either standard bath solution or gluconate solution to yield the final concentration. IAA was dissolved in ethanol to a final concentration of 250 mM, and added directly to the respective physiological solution.

4.5. Fluorescence Microscopy

Conidia of *F. fujikuroi* were inoculated in DG$_{gln}$-medium in 8-well Labtek II chambers which were treated 1 h with 0.5 M NaOH and coated with 0.01% poly-D-lysine (125 μL per well) for at least 2 h at RT. Conidia were fixed directly or germinated at 28 °C in presence of light (6 W m^{-2}) for 15–18 h. Samples were fixed for 10 min in 1% formaldehyde in phosphate buffered saline (PBS, pH 7.4) and stored in PBS at 4 °C. Fluorescence microscopy was performed at a confocal laser scanning microscope (SP700, Zeiss, Jena, Germany). Images were processed with ZEN software (ZEN 2012, Zeiss, Jena, Germany) or Fiji [66] with ImageJ 1.50f [67].

4.6. Rice Plant Infection Assays

Rice seeds belong to the Japonica variety J. Sendra, used in different rice fields in Spain [68], kindly provided by the Federación de Arroceros de Sevilla (Isla Mayor, Sevilla, Spain). Rice seeds were sterilized in a desiccator for 15 h with chlorine gas (33 mL 25% sodium hypochlorite mixed with

2.6 mL 25% HCl) immediately before their use. Sterilized seeds were transferred to 96-well plates, and germinated in 350 μL/well 0.1$\times$ yeast peptone dextrose (YPD) medium for 48 h at 28 °C in 12/12 light–dark cycle (12 h light: 1 h 3 W m^{-2}, 3 h 6 W m^{-2}, 4 h 10 W m^{-2}, 3 h 6 W m^{-2}, 1 h 3 W m^{-2}) and additionally illuminated with green light (see section about fungal strains and culture conditions). Only seedlings that had developed shoots/roots of 1–2 mm length were chosen for further experiments. Seedlings were washed with 1 mL H$_2$O, transferred into 24-well plates on sterilized filter paper, and exposed to 500 μL H$_2$O supplemented with 50 conidia freshly harvested from light-exposed cultures of the respective strain. Control plants were exposed only to H$_2$O. Infection of the seeds occurred within 24 h under continuous green light (2×10^{-15} photons s^{-1} mm^{-2}) and 12/12 light–dark cycles with white light (3×10^{-15} photons s^{-1} mm^{-2}). The green light was used to guarantee full activity of the rhodopsin during the infection process. Infected rice seedlings were transferred to pots filled with vermiculite (2–3 mm) and 35 mL H$_2$O. Four independent assays were performed with 15 and 12 seedlings/assay for infected (CarO$^+$, CarO$^-$, ΔOpsA, wild type) and control plants, respectively. To ensure sufficient nutrition of the rice plants, 20 mL 0.5 $\times$ modified Rigaud–Puppo (RP) medium [69] were added; RP medium composition (g L^{-1}): KNO$_3$ 0.5, KH$_2$PO$_4$ 0.2, K$_2$SO$_4$ 0.2, MgSO$_4$·7H$_2$O 0.2, CaCl$_2$·2H$_2$O 0.1, and 1 mL/L DG microelement solution [65]. Plants were grown at 28 °C in 12/12 light–dark cycles with additional green light. Three days after embedment, plants were fed with 20 mL 0.5$\times$ modified RP medium, and 4 days later with 35 mL H$_2$O. Plant growth was documented at days 3–7 and 10 after infection by photography. Plant length and internodal distances were determined with the image analysis software ImageJ, version 1.50f [67], and data from 4 assays were pooled for further analysis.

4.7. Determination of Chlorophyll and Carotenoid Plant Content

After the experiment, the overall chlorophyll content and the dry weight of the plants were determined. Plants were cut directly at the seed and freeze-dried for 1 day before their dry masses were determined. Plants were pulverized in a mortar and extracted in 80% acetone overnight. The plant debris was removed by centrifugation (1200$\times$ *g*, 5 min, 4 °C). The optical density was measured at 663.2 nm and 646.8 nm for chlorophylls, and 470 nm for carotenoids, and their contents were calculated as previously described [70].

4.8. Gibberellin Production

For gibberellin measurements, 10^6 spores of *Fusarium* strains were inoculated in 250 mL of low nitrogen ICI minimal medium [71] with 10% of NH$_4$NO$_3$ in 500 mL flasks. Cultures were incubated at 30 °C with 200 rpm rotary shaking. After 7 days of growth in the dark or under illumination (3 Wm^{-2}), cultures were filtered and centrifuged to get rid of mycelia and spores, respectively. Mycelia were dried at 80 °C for 24 h to measure dry weight, and culture supernatants were kept at -20 °C until measurement of gibberellins. The method used to determine gibberellin was an improvement of a protocol described by Candau et al. [72].

Fluorescent derivatives of gibberellins were obtained by shaking together 50 μL of culture supernatant, 50 μL of ethanol (96%, *v/v*), and 500 μL of a cooled mixture of equal volumes of sulfuric acid, and 96% ethanol, and incubating the reaction mixture at 48 °C in 1.5 mL reaction tube for 30 min. After that, 200 μL of each reaction was pipetted in a 96-well plate. Fluorescence was measured in a black microtiter box on a Synergy microplate reader (Biotek, Winooski, VT, USA) at excitation and emission wavelengths of 360 nm and 460 nm. GA3 (Sigma-Aldrich, St. Louis, MO, USA) was used as a standard. Measurements were done from two different experiments, each one with two culture replicates.

4.9. Gene Expression Analyses

Cultures consisted in 500 mL flasks with 100 mL of DG media, inoculated with 10^6 conidia of the wild type strain FKMC1995, and grown for 3 days in an orbital shaker at 150 rpm at 30 °C in the dark.

Afterwards, 25 mL of the cultures were transferred to Petri dishes under red safe light and incubated for 8 h in the dark at the same temperature for adaptation to the new conditions. Then, the Petri dishes were used for mycelia collection before or after exposure to white light (7 W m^{-2}) for 30, 60, and 240 min. Mycelia from each Petri dish was filtered, frozen in liquid nitrogen, and ground with a mortar and a pestle for RNA extraction with the TRIzol™ (Ambion, Life Technologies, Waltham, MA, USA) reagent. Total RNA concentrations were estimated with a Nanodrop ND-1000 spectrophotometer (Nanodrop Technologies, Wilmington, DE, USA). RNA (2.5 µg) was treated with DNAse (Affymetrix, Thermo Fisher Scientific, Waltham, MA, USA), and reverse transcribed to cDNA with Transcriptor first-strand cDNA synthesis kit (Roche, Mannheim, Germany). Final concentrations were set to 25 ng µL^{-1}. qRT-PCR analyses were performed in a LightCycler 480 real-time instrument (Roche) with the LightCycler 480 SYBR green I Master (Roche). Genes and primer sets (forward vs reverse in 5'->3' orientation) are described in Table 1. Transcript levels for each gene were normalized against the tubulin beta chain gene under the same conditions. Data of three biological replicates, each one with three technical replicates, were averaged.

Table 1. Primers used in the qRT-PCR assays.

Gene	Forward Primer	Reverse Primer
FFUJ_11802 (*carRA*)	CAGAAGCTGTTCCCGAAGACA	TGCGATGCCCATTTCTTGA
FFUJ_11803 (*carB*)	TCGGTGTCGAGTACCGTCTCT	TGCCTTGCCGGTTGCTT
FFUJ_11804 (*carO*)	TGGGCAACGCAGTGACAT	TGCGCAGACAGCCCAGTA
FFUJ_04487	CAACTACCGGCCAACTGTCT	TCTGCATGTGCCTTGTTCTC
FFUJ_05248	TTCGGAAGCTTGCAACAACG	TCGGTGGGTTGATTCGTGAG
FFUJ_06643	CAGCTATCCTGCAGAAGCGA	CATGCTCATCGCCGAAAAGG
FFUJ_07379	TAACCCCGACAACGAGAAAC	GTCTACCCACAGGGCTTTGA
FFUJ_08667	GATGTCCTCCGATCTCGTGT	CTTTCGCTCGGATCTTTGAC
FFUJ_09550 (Gβ type)	ATCACCTCGGTGGCTACATC	ATGTCCCAAACCTTGCACTC
FFUJ_03226 (Gγ type)	ACCGAGCTCAACAATCGTCT	TGCAGTAGGCAATGATGCTC
FFUJ_04397 (tubulin β)	CCGGTGCTGGAAACAACTG	CGAGGACCTGGTCGACAAGT

4.10. Sequence Analyses

Protein sequences from the rhodopsins genes identified in 38 fungal genomes were aligned with T-Coffee [73] and used for PhyML [74] analysis with a SH-like and Chi2-based approximate likelihood-ratio test (aLRT) using the LG substitution mode. Phylogenetic tree was generated with the TreeDyn analysis [75]. G proteins in the *F. fujikuroi* genome were identified by protein BLAST analyses through the NCBI server [76] with the sequences of formerly identified G proteins of *Gibberella zeae* [37]. The bootstrap NJ tree was obtained with the ClustalX 1.83 program [77], excluding gaps, and applying the correction for multiple substitutions. The tree was represented with the NJPlot program [78].

Supplementary Materials: Supplementary materials can be found at www.mdpi.com/1422-0067/19/1/215/s1.

Acknowledgments: We thank Javier Suárez, from the Federación de Arroceros de Sevilla, for providing the rice plant seeds, Wolfgang Dröge-Laser for help with rice seed sterilisation, and Michael Brunk, Kristina Frohn, and Sebastian Sputh for great assistance in the lab. We are grateful to the Bavarian Research Alliance for travel funding (BaylntAn_Uni_Würzburg_2012_39). The German Research Foundation (TE 832/4-1), the Spanish Government (Ministerio de Ciencia y Tecnología, project BIO2012-39716), and the Andalusian Government (project CTS-6638) funded this study. Spanish grants included support from the European Regional Development Fund (ERDF).

Author Contributions: Alexander Adam and Ulrich Terpitz carried out CLSM measurements, mammalian cell and fungal culture, and patch-clamp experiments. Stephan Deimel performed rice plant infection and Jorge García-Martínez the molecular biology work. Javier Pardo-Medina performed fungal cultures and RT-PCR experiments, M.Carmen Limón carried on fungal cultures and analyzed gibberellin content, and Tilen Konte and Ulrich Terpitz conceived phylogenetic analysis of fungal rhodopsins. Alexander Adam, Ulrich Terpitz, Stephan Deimel, Tilen Konte, M. Carmen Limón, and Javier Avalos analyzed data. All authors provided discussions and contributed to the manuscript. Ulrich Terpitz and Javier Avalos planned the experiments, Ulrich Terpitz wrote the manuscript.

Conflicts of Interest: The authors declare no conflict of interest.

Abbreviations

BLAST	Basic Local Alignment Search Tool
BR	Bacteriorhodopsin
DPSS	Diode Pumped Solid State
GA	Gibberellic Acid
IAA	Indole-3-Acetic Acid
LR	*Leptosphaeria* Rhodopsin
NaAc	Sodium Acetate
NaGlu	Sodium Gluconate
NR	*Neurospora* Rhodopsin
ORP	Opsin Related Protein
SEM	Standard Error of the Mean
TM	Transmembrane Helix
WOA	Weak Organic Acid

References

1. Turra, D.; Segorbe, D.; Di Pietro, A. Protein kinases in plant-pathogenic fungi: Conserved regulators of infection. *Annu. Rev. Phytopathol.* **2014**, *52*, 267–288. [CrossRef] [PubMed]
2. Fuller, K.K.; Loros, J.J.; Dunlap, J.C. Fungal photobiology: Visible light as a signal for stress, space and time. *Curr. Genet.* **2015**, *61*, 275–288. [CrossRef] [PubMed]
3. Corrochano, L.M.; Galland, P. Photomorphogenesis and Gravitropism in Fungi. In *Growth, Differentiation and Sexuality—The Mycota I*; Wendland, J., Ed.; Springer: Berlin, Germany, 2016; pp. 235–263.
4. Corrochano, L.M. Fungal photoreceptors: Sensory molecules for fungal development and behaviour. *Photochem. Photobiol. Sci.* **2007**, *6*, 725–736. [CrossRef] [PubMed]
5. Brown, L.S. Proton-pumping microbial rhodopsins—Ubiquitous structurally simple helpers of respiration and photosynthesis. *Struct. Basis Biol. Energy Gener.* **2014**, *39*, 1–20. [CrossRef]
6. Brown, L.S.; Dioumaev, A.K.; Lanyi, J.K.; Spudich, E.N.; Spudich, J.L. Photochemical reaction cycle and proton transfers in *Neurospora rhodopsin*. *J. Biol. Chem.* **2001**, *276*, 32495–32505. [CrossRef] [PubMed]
7. Fan, Y.; Solomon, P.; Oliver, R.P.; Brown, L.S. Photochemical characterization of a novel fungal rhodopsin from *Phaeosphaeria nodorum*. *Biochim. Biophys. Acta Bioenerg.* **2011**, *1807*, 1457–1466. [CrossRef] [PubMed]
8. García-Martínez, J.; Brunk, M.; Avalos, J.; Terpitz, U. The CarO rhodopsin of the fungus *Fusarium fujikuroi* is a light-driven proton pump that retards spore germination. *Sci. Rep.* **2015**, *5*, 7798. [CrossRef] [PubMed]
9. Waschuk, S.A.; Bezerra, A.G.; Shi, L.; Brown, L.S. *Leptosphaeria* rhodopsin: Bacteriorhodopsin-like proton pump from a eukaryote. *Proc. Natl. Acad. Sci. USA* **2005**, *102*, 6879–6883. [CrossRef] [PubMed]
10. Wang, Z.; Wang, J.; Li, N.; Li, J.; Trail, F.; Dunlap, J.C.; Townsend, J.P. Light sensing by opsins and fungal ecology: NOP-1 modulates entry into sexual reproduction in response to environmental cues. *Mol. Ecol.* **2017**. [CrossRef] [PubMed]
11. Piper, P.W.; Ortiz-Calderon, C.; Holyoak, C.; Coote, P.; Cole, M. Hsp30, the integral plasma membrane heat shock protein of *Saccharomyces cerevisiae*, is a stress-inducible regulator of plasma membrane H^+-ATPase. *Cell Stress Chaperones* **1997**, *2*, 12–24. [CrossRef]
12. Graul, R.C.; Sadée, W. Evolutionary relationships among proteins probed by an iterative neighborhood cluster analysis (INCA). Alignment of bacteriorhodopsins with the yeast sequence YRO2. *Pharm. Res.* **1997**, *14*, 1533–1541. [CrossRef] [PubMed]
13. Bieszke, J.A.; Braun, E.L.; Bean, L.E.; Kang, S.C.; Natvig, D.O.; Borkovich, K.A. The nop-1 gene of *Neurospora crassa* encodes a seven transmembrane helix retinal-binding protein homologous to archaeal rhodopsins. *Proc. Natl. Acad. Sci. USA* **1999**, *96*, 8034–8039. [CrossRef] [PubMed]
14. Prado, M.M.; Prado-Cabrero, A.; Fernández-Martín, R.; Avalos, J. A gene of the opsin family in the carotenoid gene cluster of *Fusarium fujikuroi*. *Curr. Genet.* **2004**, *46*, 47–58. [CrossRef] [PubMed]
15. Brown, L.S.; Jung, K.H. Bacteriorhodopsin-like proteins of eubacteria and fungi: The extent of conservation of the haloarchaeal proton-pumping mechanism. *Photochem. Photobiol. Sci.* **2006**, *5*, 538–546. [CrossRef] [PubMed]

16. Chow, B.Y.; Han, X.; Dobry, A.S.; Qian, X.F.; Chuong, A.S.; Li, M.J.; Henninger, M.A.; Belfort, G.M.; Lin, Y.X.; Monahan, P.E.; et al. High-performance genetically targetable optical neural silencing by light-driven proton pumps. *Nature* **2010**, *463*, 98–102. [CrossRef] [PubMed]

17. Singh, R.; Sunder, S. Foot rot and bakanae of rice: An overview. *Rev. Plant Pathol.* **2012**, *5*, 565–604.

18. Avalos, J.; Cerdá-Olmedo, E.; Reyes, F.; Barrero, A.F. Gibberellins and other metabolites of *Fusarium fujikuroi* and related fungi. *Curr. Org. Chem.* **2007**, *11*, 721–737. [CrossRef]

19. Hedden, P.; Sponsel, V. A Century of gibberellin research. *J. Plant Growth Regul.* **2015**, *34*, 740–760. [CrossRef] [PubMed]

20. Gupta, A.K.; Solanki, I.S.; Bashyal, B.M.; Singh, Y.; Srivastava, K. Bakanae of rice—An emerging disease in Asia. *J. Anim. Plant Sci.* **2015**, *25*, 1499–1514.

21. Ou, S.H. Bakanae Disease and Food Rot. In *Rice Diseases*; Common Wealth Mycological Institute: Farnham Royal, UK, 1985; pp. 262–272.

22. Niehaus, E.-M.; Díaz-Sánchez, V.; von Bargen, K.W.; Kleigrewe, K.; Humpf, H.-U.; Limón, M.C.; Tudzynski, B. *Fusarins and Fusaric Acid in Fusaria*; Springer: New York, NY, USA, 2014; pp. 239–262.

23. Estrada, A.F.; Avalos, J. Regulation and targeted mutation of *opsA*, coding for the NOP-1 opsin orthologue in *Fusarium fujikuroi*. *J. Mol. Biol.* **2009**, *387*, 59–73. [CrossRef] [PubMed]

24. Brunk, M.; Sputh, S.; Doose, S.; van der Linde, S.; Terpitz, U. HyphaTracker: An ImageJ toolbox for time-resolved analysis of spore germination in filamentous fungi. *Sci. Rep.* **2018**. [CrossRef]

25. Atamna-Ismaeel, N.; Finkel, O.M.; Glaser, F.; Sharon, I.; Schneider, R.; Post, A.F.; Spudich, J.L.; von Mering, C.; Vorholt, J.A.; Iluz, D.; et al. Microbial rhodopsins on leaf surfaces of terrestrial plants. *Environ. Microbiol.* **2012**, *14*, 140–146. [CrossRef] [PubMed]

26. Cho, Y.; Ohm, R.A.; Grigoriev, I.V.; Srivastava, A. Fungal-specific transcription factor AbPf2 activates pathogenicity in *Alternaria brassicicola*. *Plant J.* **2013**, *75*, 498–514. [CrossRef] [PubMed]

27. Ghosh, A. Small heat shock proteins (HSP12, HSP20 and HSP30) play a role in *Ustilago maydis* pathogenesis. *FEMS Microbiol. Lett.* **2014**, *361*, 17–24. [CrossRef] [PubMed]

28. Caracuel, Z.; Roncero, M.I.G.; Espeso, E.A.; González-Verdejo, C.I.; García-Maceira, F.I.; Di Pietro, A. The pH signalling transcription factor PacC controls virulence in the plant pathogen *Fusarium oxysporum*. *Mol. Microbiol.* **2003**, *48*, 765–779. [CrossRef] [PubMed]

29. Alkan, N.; Espeso, E.A.; Prusky, D. Virulence regulation of phytopathogenic fungi by pH. *Antioxid. Redox Signal.* **2013**, *19*, 1012–1025. [CrossRef] [PubMed]

30. Gabriel, R.; Kesselmeier, J.; Planck, M.; Box, P.O. Apoplastic solute concentrations of organic acids and mineral nutrients in the leaves of several fagaceae. *Plant Cell Physiol.* **1999**, *40*, 604–612. [CrossRef]

31. Capurro, V.; Gianotti, A.; Caci, E.; Ravazzolo, R.; Galietta, L.J.V.; Zegarra-Moran, O. Functional analysis of acid-activated Cl⁻ channels: Properties and mechanisms of regulation. *Biochim. Biophys. Acta Biomembr.* **2015**, *1848*, 105–114. [CrossRef] [PubMed]

32. Bieszke, J.A.; Spudich, E.N.; Scott, K.L.; Borkovich, K.A.; Spudich, J.L. A eukaryotic protein, NOP-1, binds retinal to form an archaeal rhodopsin-like photochemically reactive pigment. *Biochemistry* **1999**, *38*, 14138–14145. [CrossRef] [PubMed]

33. Sakmar, T.P.; Franke, R.R.; Khorana, H.G. The role of the retinylidene Schiff base counterion in rhodopsin in determining wavelength absorbance and Schiff base pKa. *Proc. Natl. Acad. Sci. USA* **1991**, *88*, 3079–3083. [CrossRef] [PubMed]

34. Váró, G.; Lanyi, J.K. Photoreactions of bacteriorhodopsin at acid pH. *Biophys. J.* **1989**, *56*, 1143–1151. [CrossRef]

35. Stergiopoulos, I.; Kourmpetis, Y.A.I.; Slot, J.C.; Bakker, F.T.; De Wit, P.J.G.M.; Rokas, A. In silico characterization and molecular evolutionary analysis of a novel superfamily of fungal effector proteins. *Mol. Biol. Evol.* **2012**, *29*, 3371–3384. [CrossRef] [PubMed]

36. Schoch, C.L.; Sung, G.-H.; Lopez-Giraldez, F.; Townsend, J.P.; Miadlikowska, J.; Hofstetter, V.; Robbertse, B.; Matheny, P.B.; Kauff, F.; Wang, Z.; et al. The ascomycota tree of life: A phylum-wide phylogeny clarifies the origin and evolution of fundamental reproductive and ecological traits. *Syst. Biol.* **2009**, *58*, 224–239. [CrossRef] [PubMed]

37. Yu, H.-Y.; Seo, J.-A.; Kim, J.-E.; Han, K.-H.; Shim, W.-B.; Yun, S.-H.; Lee, Y.-W. Functional analyses of heterotrimeric G protein Gα and Gβ subunits in *Gibberella zeae*. *Microbiology* **2008**, *154*, 392–401. [CrossRef] [PubMed]

38. Avalos, J.; Nordzieke, S.; Parra, O.; Pardo-Medina, J.; Limón, M.C. Carotenoid Production by Filamentous Fungi and Yeasts. In *Biotechnology of Yeasts and Filamentous Fungi*; Sibirny, A.A., Ed.; Springer International Publishing: Cham, Germany, 2017; pp. 225–279. ISBN 978-3-319-58829-2.

39. Niehaus, E.-M.; Kim, H.-K.; Münsterkötter, M.; Janevska, S.; Arndt, B.; Kalinina, S.A.; Houterman, P.M.; Ahn, I.-P.; Alberti, I.; Tonti, S.; et al. Comparative genomics of geographically distant *Fusarium fujikuroi* isolates revealed two distinct pathotypes correlating with secondary metabolite profiles. *PLoS Pathog.* **2017**, *13*, e1006670. [CrossRef] [PubMed]

40. Kihara, J.; Tanaka, N.; Ueno, M.; Arase, S. Cloning and expression analysis of two opsin-like genes in the phytopathogenic fungus *Bipolaris oryzae*. *FEMS Microbiol. Lett.* **2009**, *295*, 289–294. [CrossRef] [PubMed]

41. Lyu, X.; Shen, C.; Fu, Y.; Xie, J.; Jiang, D.; Li, G.; Cheng, J. The microbial opsin homolog Sop1 is involved in *Sclerotinia sclerotiorum* development and environmental stress response. *Front. Microbiol.* **2015**, *6*, 1504. [CrossRef] [PubMed]

42. Kuo, H.C.; Hui, S.; Choi, J.; Asiegbu, F.O.; Valkonen, J.P.; Lee, Y.H. Secret lifestyles of *Neurospora crassa*. *Sci. Rep.* **2014**, *4*, 5135. [CrossRef] [PubMed]

43. Wulff, E.G.; Sørensen, J.L.; Lübeck, M.; Nielsen, K.F.; Thrane, U.; Torp, J. *Fusarium* spp. associated with rice Bakanae: Ecology, genetic diversity, pathogenicity and toxigenicity. *Environ. Microbiol.* **2010**, *12*, 649–657. [CrossRef] [PubMed]

44. Wiemann, P.; Sieber, C.M.K.; von Bargen, K.W.; Studt, L.; Niehaus, E.-M.; Espino, J.J.; Huß, K.; Michielse, C.B.; Albermann, S.; Wagner, D.; et al. Deciphering the cryptic genome: Genome-wide analyses of the rice pathogen *Fusarium fujikuroi* reveal complex regulation of secondary metabolism and novel metabolites. *PLoS Pathog.* **2013**, *9*, e1003475. [CrossRef] [PubMed]

45. Braga, G.U.L.; Rangel, D.E.N.; Fernandes, É.K.K.; Flint, S.D.; Roberts, D.W. Molecular and physiological effects of environmental UV radiation on fungal conidia. *Curr. Genet.* **2015**, *61*, 405–425. [CrossRef] [PubMed]

46. Hassan, A.E.M.; Dillon, R.J.; Charnley, A.K. Influence of accelerated germination of conidia on the pathogenicity of *Metarhizium anisopliae* for *Manduca sexta*. *J. Invertebr. Pathol.* **1989**, *54*, 277–279. [CrossRef]

47. Aver'yanov, A.A.; Lapikova, V.P.; Pasechnik, T.D.; Abramova, O.S.; Gaivoronskaya, L.M.; Kuznetsov, V.V.; Baker, C.J. Pre-illumination of rice blast conidia induces tolerance to subsequent oxidative stress. *Fungal Biol.* **2014**, *118*, 743–753. [CrossRef] [PubMed]

48. Canessa, P.; Schumacher, J.; Hevia, M.A.; Tudzynski, P.; Larrondo, L.F. Assessing the effects of light on differentiation and virulence of the plant pathogen *Botrytis cinerea*: Characterization of the white collar complex. *PLoS ONE* **2013**, *8*. [CrossRef] [PubMed]

49. Yu, S.M.; Ramkumar, G.; Lee, Y.H. Light quality influences the virulence and physiological responses of *Colletotrichum acutatum* causing anthracnose in pepper plants. *J. Appl. Microbiol.* **2013**, *115*, 509–516. [CrossRef] [PubMed]

50. Alvarez-Loayza, P.; White, J.F., Jr.; Torres, M.S.; Balslev, H.; Kristiansen, T.; Svenning, J.C.; Gil, N. Light converts endosymbiotic fungus to pathogen, influencing seedling survival and niche-space filling of a common tropical tree, *Iriartea deltoidea*. *PLoS ONE* **2011**, *6*, e16386. [CrossRef] [PubMed]

51. Borkovich, K.A.; Ebbole, D.J. *Cellular & Molecular Biology of Filamentous Fungi*; Borkovich, K.A., Ed.; American Society for Microbiology: Washington, DC, USA, 2010; ISBN 9786613034342.

52. Spudich, J.L.; Sineshchekov, O.A.; Govorunova, E.G. Mechanism divergence in microbial rhodopsins. *Biochim. Biophys. Acta* **2014**, *1837*, 546–552. [CrossRef] [PubMed]

53. Lórenz-Fonfría, V.A.; Heberle, J. Channelrhodopsin unchained: Structure and mechanism of a light-gated cation channel. *Biochim. Biophys. Acta Bioenerg.* **2014**, *1837*, 626–642. [CrossRef] [PubMed]

54. Nigovic, B.; KojicProdic, B.; Antolic, S.; Tomic, S.; Puntarec, V.; Cohen, J.D. Structural studies on monohalogenated derivatives of the phytohormone indole-3-acetic acid (auxin). *Acta Crystallogr. Sect. B* **1996**, *52*, 332–343. [CrossRef]

55. Nielsen, J. *Physiological Engineering Aspects of Penicillium Chrysogenum*; World Scientific: Singapore, 1997; ISBN 978-981-02-2765-4.

56. Bean, R.C.; Shepherd, W.C.; Chan, H. Permeability of lipid bilayer membranes to organic solutes. *J. Gen. Physiol.* **1968**, *52*, 495–508. [CrossRef] [PubMed]

57. Stewart, A.K.; Chernova, M.N.; Kunes, Y.Z.; Alper, S.L. Regulation of AE2 anion exchanger by intracellular pH: Critical regions of the NH2-terminal cytoplasmic domain. *Am. J. Physiol.* **2001**, *281*, C1344–C1354. [CrossRef] [PubMed]

58. Lörinczi, E.; Verhoefen, M.K.; Wachtveitl, J.; Woerner, A.C.; Glaubitz, C.; Engelhard, M.; Bamberg, E.; Friedrich, T. Voltage- and pH-dependent changes in vectoriality of photocurrents mediated by wild-type and mutant proteorhodopsins upon expression in *Xenopus* oocytes. *J. Mol. Biol.* **2009**, *393*, 320–341. [CrossRef] [PubMed]

59. Terpitz, U.; Raimunda, D.; Westhoff, M.; Sukhorukov, V.L.; Beaugé, L.; Bamberg, E.; Zimmermann, D.; Beauge, L.; Bamberg, E.; Zimmermann, D. Electrofused giant protoplasts of *Saccharomyces cerevisiae* as a novel system for electrophysiological studies on membrane proteins. *Biochim. Biophys. Acta. Biomembr.* **2008**, *1778*, 1493–1500. [CrossRef] [PubMed]

60. Lanyi, J.K. Halorhodopsin: A light-driven chloride ion pump. *Annu. Rev. Biophys. Biophys. Chem.* **1986**, *15*, 11–28. [CrossRef] [PubMed]

61. Tsavkelova, E.; Oeser, B.; Oren-Young, L.; Israeli, M.; Sasson, Y.; Tudzynski, B.; Sharon, A. Identification and functional characterization of indole-3-acetamide-mediated IAA biosynthesis in plant-associated *Fusarium* species. *Fungal Genet. Biol.* **2012**, *49*, 48–57. [CrossRef] [PubMed]

62. Quazi, S.A.J.; Meon, S.; Jaafar, H.; Ahmad, Z.A.B.M. The role of phytohormones in relation to bakanae disease development and symptoms expression. *Physiol. Mol. Plant Pathol.* **2015**, *90*, 27–38. [CrossRef]

63. Terakita, A. The opsins. *Genome Biol.* **2005**, *6*, 213. [CrossRef] [PubMed]

64. Palczewski, K. G Protein–coupled receptor rhodopsin. *Annu. Rev. Biochem.* **2006**, *75*, 743–767. [CrossRef] [PubMed]

65. Avalos, J.; Casadesús, J.; Cerdá-Olmedo, E. *Gibberella fujikuroi* mutants obtained with UV radiation and N-methyl-N'-nitro-N-nitrosoguanidine. *Appl. Environ. Microbiol.* **1985**, *49*, 187–191. [PubMed]

66. Schindelin, J.; Arganda-Carreras, I.; Frise, E.; Kaynig, V.; Longair, M.; Pietzsch, T.; Preibisch, S.; Rueden, C.; Saalfeld, S.; Schmid, B.; et al. Fiji: An open-source platform for biological-image analysis. *Nat. Methods* **2012**, *9*, 676–682. [CrossRef] [PubMed]

67. Schneider, C.A.; Rasband, W.S.; Eliceiri, K.W. NIH Image to ImageJ: 25 years of image analysis. *Nat. Methods* **2012**, *9*, 671–675. [CrossRef] [PubMed]

68. Kraehmer, H.; Thomas, C.; Vidotto, F. Rice Production in Europe. In *Rice Production Worldwide*; Chauhan, B.S., Jabran, K., Mahajan, G., Eds.; Springer International Publishing: Cham, Germany, 2017; pp. 93–116; ISBN 978-3-319-47516-5.

69. Rigaud, J.; Puppo, A. Indole-3-acetic acid catabolism by soybean bacteroids. *J. Gen. Microbiol.* **1975**, *88*, 223–228. [CrossRef]

70. Lichtenthaler, H.K.; Buschmann, C. Chlorophylls and carotenoids: Measurement and characterization by UV-VIS. *Curr. Protoc. Food Anal. Chem.* **2001**. [CrossRef]

71. Geissman, T.A.; Verbiscar, A.J.; Phinney, B.O.; Cragg, G. Studies on the biosynthesis of gibberellins from (−)-kaurenoic acid in cultures of *Gibberella fujikuroi*. *Phytochemistry* **1966**, *5*, 933–947. [CrossRef]

72. Candau, R.; Avalos, J.; Cerdá-Olmedo, E. Gibberellins and carotenoids in the wild type and mutants of *Gibberella fujikuroi*. *Appl. Environ. Microbiol.* **1991**, *57*, 3378–3382. [PubMed]

73. Di Tommaso, P.; Moretti, S.; Xenarios, I.; Orobitg, M.; Montanyola, A.; Chang, J.-M.; Taly, J.-F.; Notredame, C. T-Coffee: A web server for the multiple sequence alignment of protein and RNA sequences using structural information and homology extension. *Nucleic Acids Res.* **2011**, *39*, W13–W17. [CrossRef] [PubMed]

74. Guindon, S.; Dufayard, J.F.; Lefort, V.; Anisimova, M.; Hordijk, W.; Gascuel, O. New Algorithms and methods to estimate maximum-likelihood phylogenies: Assessing the performance of PhyML 3.0. *Syst. Biol.* **2010**, *59*, 307–321. [CrossRef] [PubMed]

75. Dereeper, A.; Guignon, V.; Blanc, G.; Audic, S.; Buffet, S.; Chevenet, F.; Dufayard, J.-F.; Guindon, S.; Lefort, V.; Lescot, M.; et al. Phylogeny.fr: Robust phylogenetic analysis for the non-specialist. *Nucleic Acids Res.* **2008**, *36*, W465–W469. [CrossRef] [PubMed]

76. BLAST:_ Basic Local Alignment Search Tool. Available online: https://blast.ncbi.nlm.nih.gov/Blast.cgi (accessed on 15 October 2017).

77. Thompson, J.D.; Gibson, T.J.; Plewniak, F.; Jeanmougin, F.; Higgins, D.G. The CLUSTAL_X windows interface: Flexible strategies for multiple sequence alignment aided by quality analysis tools. *Nucleic Acids Res.* **1997**, *25*, 4876–4882. [CrossRef] [PubMed]

78. NJplot. Available online: http://doua.prabi.fr/software/njplot (accessed on 15 October 2017).

International Journal of
Molecular Sciences

MDPI

Article

Core Microbiome of Medicinal Plant *Salvia miltiorrhiza* Seed: A Rich Reservoir of Beneficial Microbes for Secondary Metabolism?

Haimin Chen [1,2], Hongxia Wu [2], Bin Yan [2], Hongguang Zhao [3], Fenghua Liu [3], Haihua Zhang [1,2], Qing Sheng [2], Fang Miao [1,*] and Zongsuo Liang [1,2,*]

[1] College of Life Sciences, Northwest A&F University, Yangling 712100, China; chenhm@zstu.edu.cn (H.C.); hhzhang@zstu.edu.cn (H.Z.)
[2] College of Life Sciences, Zhejiang Sci-Tech University, Hangzhou 310018, China; wuhx@zstu.edu.cn (H.W.); yanb@zstu.edu.cn (B.Y.); csheng@zstu.edu.cn (Q.S.)
[3] Tianjin Tasly Holding Group Co., Ltd., Tianjin 300410, China; zhaohongguang@tasly.com (H.Z.); liufh@tasly.com (F.L.)
*   Correspondence: miaofangmf@nwafu.edu.cn (F.M.); Liangzs@ms.iswc.ac.cn (Z.L.)

Received: 25 January 2018; Accepted: 23 February 2018; Published: 27 February 2018

Abstract: Seed microbiome includes special endophytic or epiphytic microbial taxa associated with seeds, which affects seed germination, plant growth, and health. Here, we analyzed the core microbiome of 21 *Salvia miltiorrhiza* seeds from seven different geographic origins using 16S rDNA and ITS amplicon sequencing, followed by bioinformatics analysis. The whole bacterial microbiome was classified into 17 microbial phyla and 39 classes. Gammaproteobacteria (67.6%), Alphaproteobacteria (15.6%), Betaproteobacteria (2.6%), Sphingobacteria (5.0%), Bacilli (4.6%), and Actinobacteria (2.9%) belonged to the core bacterial microbiome. Dothideomycetes comprised 94% of core fungal microbiome in *S. miltiorrhiza* seeds, and another two dominant classes were Leotiomycetes (3.0%) and Tremellomycetes (2.0%). We found that terpenoid backbone biosynthesis, degradation of limonene, pinene, and geraniol, and prenyltransferases, were overrepresented in the core bacterial microbiome using phylogenetic examination of communities by reconstruction of unobserved states (PICRUSt) software. We also found that the bacterial genera *Pantoea*, *Pseudomonas*, and *Sphingomonas* were enriched core taxa and overlapped among *S. miltiorrhiza*, maize, bean, and rice, while a fungal genus, *Alternaria*, was shared within *S. miltiorrhiza*, bean, and Brassicaceae families. These findings highlight that seed-associated microbiomeis an important component of plant microbiomes, which may be a gene reservoir for secondary metabolism in medicinal plants.

Keywords: seed-associated microbiome; 16S rRNA and ITS2 gene amplicons; Illumina sequencing; diversity; PICRUSt; *Salvia miltiorrhiza* Bge

1. Introduction

Seed production is one of the most important stages of plant life history. Seeds harbor high diversity of microbial taxa, known as seed-associated microbiomes, which are the endophytic or epiphytic microbial communities associated with seeds. Seed microbiomes can allow vertical transmission across generations, and have profound impacts on plant ecology, health, and productivity [1–3].

The concept of core microbiome was firstly established for human microbiome, and further expanded to other host-associated microbiomes such as plants. In addition, this concept was even used to describe microbial members shared across soils, lakes, and wastewater [4–6]. The composition and function of plant core microbiomes have been achieved for several model plants, such as *Arabidopsis*, maize, rice, barley, and soybean. Several studies showed that soil types and host plant genotypes are the main factors affecting the microbial community assemblage [5,7–11]. Studies on human microbiomes showed that human association of microbial communities have a huge impact on host metabolism [12–14], but few studies have analyzed the effects of plant microbiome on host metabolism. Existing plant microbiome studies have focused on rhizosphere and phyllosphere microbial communities, and our understanding of the seed microbiome has remained limited. Some studies have shown that seed core microbiome is specific for terroir and emergence [15,16]. Seed microbiomes have diverse seed–microbe interactions, and properties, such as being fast-growing, use as bio-fertilizer, antagonistic properties, and ability to cope with environmental stress [17–23], and are predicted to be an important biological resource for sustainable agriculture [24].

Danshen (*Salvia miltiorrhiza* Bge) is an important medicinal plant, mainly used to treat coronary heart diseases and cerebrovascular diseases, and has been used in China, Japan, and other east Asian countries for hundreds of years [25]. Tanshinone (a diterpenoid quinones compound) and salvianolic acid are two important active constituents of *S. miltiorrhiza*. Several studies on the associated microbes of danshen mainly focused on in vitro activities of endophytes and mycorrhizal fungi, and some endophytes could produce the similar active constituents in host plants [26–34]. However, the composition and function of root, leaf, or seed-associated microbiome in *S. miltiorrhiza* have not been deciphered yet.

We collected different seeds from different geographic cultivation areas and characterized the seed-associated microbiome by deep-sequencing approach to decipher the seed-associated microbiome in *S. miltiorrhiza*. Sampling was performed across the main planting zones for *S. miltiorrhiza*, including Shaanxi, Shanxi, Henan, and Shandong provinces. We used IlluminaMiSeq platform to sequence 16S ribosomal RNA (rRNA) gene, and ITS2 amplicons for DNA prepared from seven diverse geographic sources of *S. miltiorrhiza* seeds. Later, we analyzed the overlap between different source seeds and their common microbial taxa. In addition, we also analyzed the overlap in microbial taxa among danshen, maize, bean, rice, and Brassicaceae. Furthermore, we also predicted the bacterial functional profiles of core microbiome in *S. miltiorrhiza* seeds using phylogenetic investigation of communities by reconstruction of unobserved states (PICRUSt) software.

2. Results

2.1. Genetic Diversity of Different S. miltiorrhiza Seeds

The genetic diversity of the seven *Salvia* cultivars used in this study was firstly assessed through 10 simple sequence repeat (SSR) markers (S3). According to the SSR data analyses, the individual number was suitable to represent the cultivar level of genetic diversity. At the population level, the number of different alleles (*Na*) and effective alleles (*Ne*) ranged between from 2.80 to 4.10 and 2.21 to 3.28, respectively (mean 5.305 and 3.113, respectively). This finding indicates allele differences among the *S. miltiorrhiza* groups, but these differences were insignificant. Among the population genetic diversity parameters, the ranges of Shannon's Information Index (*I*), Observed Heterozygosity (*Ho*) and Expected Heterozygosity (*He*) ranged from 0.86 to 1.24, from 0.63 to 0.85, and from 0.53 to 0.66, respectively. These results showed that the genetic diversity in different populations of *S. miltiorrhiza* was high. There were differences among the groups but the difference between groups was low. All of the observed heterozygosity (*Ho*) values were higher than expected heterozygosity (*He*), indicating the existence of significant excess heterozygosity in the population. Analysis of molecular variance (AMOVA) results indicated that the genetic diversity among cultivars was 96% much higher than genetic diversity within cultivar (4%) (Table S4). These results confirmed that the genetic variation of these *S. miltiorrhiza* populations were mainly due to the genetic differences within the population. SSR cluster analysis chart of these *S. miltiorrhiza* seeds from different geographic sources were displayed in Figure S2.

2.2. The Bacterial 16S rRNA and Fungal ITS Sequencing Data Set

Bacterial 16S rRNA and fungal ITS gene profiling of 21 seed samples from seven different producing area were subjected to Illumina Miseq sequencing to identify bacterial and fungal seed-associated core microbiome of cultivated *S. miltiorrhiza*. Later, bioinformatics analyses were carried out, and seed-associated core microbiome analyses were carried out among six source *S. miltiorrhiza* seeds, besides the seed of *Salvia miltiorrhiza* Bge. f. alba (a variant of *S. miltiorrhiza*) from Laiwu city in Shandong province.

The bacterial 16S rRNA sequencing resulted in 662,164 raw reads, and 662,098 of them passed the quality and length filtering. The data set comprised of 11,632–48,933 (the mean: 30,576) sequences per sample, clustered into 2548 OTUs (97%) (Table S3). The data set was rarefied, as showed in Supplementary Figure S3a. Richness estimation of seed sample complete data set revealed that Illumina 16S rDNA sequencing attained 57.0–87.4% of the estimated richness (Table S5).

The fungal ITS sequencing resulted in 1,729,121 raw reads and 1,665,477 of them passed the quality and length filtering. The data set comprised of 49,796–107,206 (the mean: 79,308) sequences per sample clustered into 222 OTUs (97%). The data set was rarefied as shown in Supplementary Figure S3b. Richness estimation of complete data set revealed that Illumina ITS sequencing attained 77.9–96.8% of estimated richness (Table S6).

2.3. Diversity of the Seed-Associated Bacterial Microbiome in S. miltiorrhiza

Alpha-diversity of the seed-associated bacterial microbiome of each sample was estimated using the observed species, community richness (Chao 1, expressed as the projected total number of OTU in each sample), Shannon diversity index, and evenness (Simpson's index). The observed species, Chao 1, and Shannon diversity indices suggested that bacterial community richness showed significant difference between *S. miltiorrhiza* seed samples from different geographic origins, and community evenness showed by Simpson's index also suggested the presence of significant differences (Tables S5 and S7, and Figure S6). DS4-LG seeds that came from Langao county, Shaanxi province had the highest community diversity within these seed samples (Shannon diversity indices: 6.63 ± 0.73).

The variation of seed-associated microbiome diversity was explained by cultivated area (sampling location) (Figure 1). The β-diversity of the seed-associated bacterial microbiome among the different sampling locations was statistically significant. Moreover, the bacterial microbiome from different sampling locations was clustered well by unweighted unifrac distance matrix cluster analysis (Figure S1).

Altogether, the bacterial microbiome was classified into 17 microbial phyla and candidate divisions, and 39 classes (Figures 2 and 3). At phylum level, the most dominant bacterial phyla were Proteobacteria (85.9%), Bacteroidetes (6.3%), Firmicutes (4.8%), and Actinobacteria (2.9%). At class level, the bacterial microbiome was dominated by Gammaproteobacteria (67.6%), Alphaproteobacteria (15.6%), Betaproteobacteria (2.6%), Sphingobacteriia (5.0%), Bacilli (4.6%), and Actinobacteria (2.9%) (Figure 2a, Figure S4).

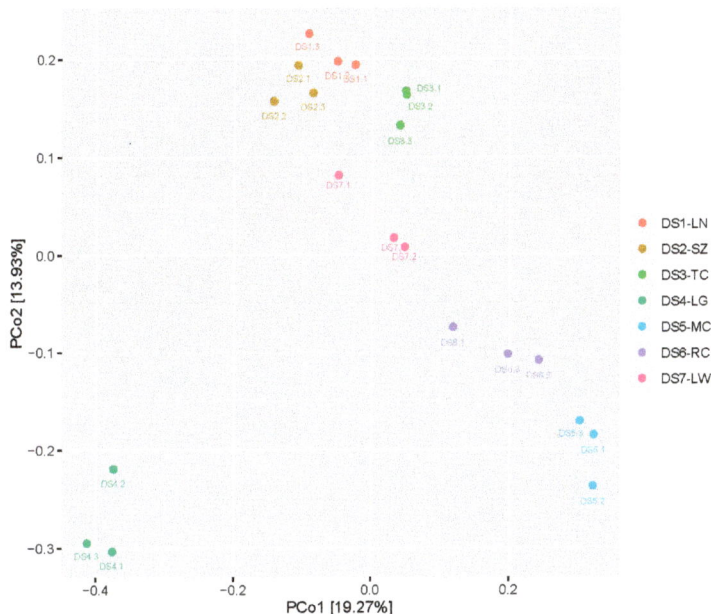

Figure 1. Comparison of seed-associated bacterial microbiome with cultivated area (sampling geographic origins) by principal coordinates analysis (PCoA). PCoA plot is based on unweighted unifrac distance matrix of the 16S rRNA gene amplicons. The color of the symbols indicates samples with their IDs: pale red (DS1-LG, Luonan, Shaanxi, China), light brown (DS2-SZ, Shangzhou, Shaanxi, China), light green (DS3-TC, Tongchuan, Shaanxi, China), Wathet (DS4-LG, Langao, Shaanxi, China), blue (DS5-MC, Mianchi, Henan, China), purple (DS6-RC, Ruicheng , Shanxi, China), and pink (DS7-LW, Laiwu, Shandong, China).

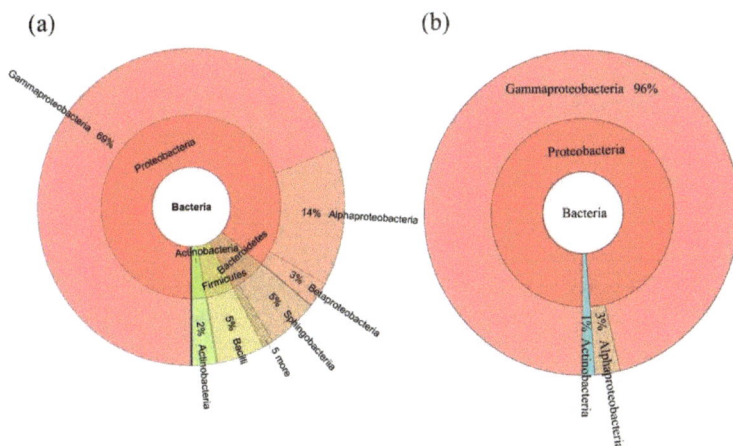

Figure 2. Taxonomic composition of the seed-associated (**a**) whole and (**b**) core bacterial microbiome of *S. miltiorrhiza* at the class level. Pie charts represent relative abundances of bacterial classes for the whole and core microbiome.

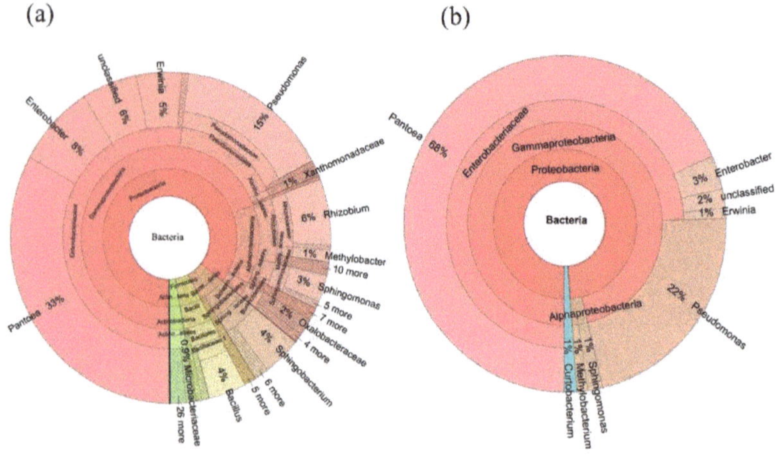

Figure 3. Taxonomic composition of the seed-associated (**a**) whole and (**b**) core bacterial microbiome of *S. miltiorrhiza* at the genus level. Pie charts represent relative abundances of bacterial genera for the whole and core microbiome.

2.4. Diversity of the Seed-Associated Fungal Microbiome in S. miltiorrhiza

The fungal community richness showed significant differences between *S. miltiorrhiza* seed samples from different geographic origins, as with the bacterial microbiome, which is indicated by the observed species, Chao 1, and Shannon diversity indices. Community evenness also showed some significant differences between different seeds (Tables S6 and S8, and Figure S7).

The variation of seed-associated fungal microbiome diversity was explained by cultivated area (sampling location). The beta-diversity of the seed-associated fungal microbiome among the different sampling locations had statistical significance. However, these seed samples were not clustered well by Euclidean distance matrix, unweighted unifrac distance matrix, and weighted unifrac distance matrix.

The fungal microbiome was mainly classified into 4 phyla and candidate divisions and 19 classes, whereas 3.9% and 9.0% of the fungal microbiome remained unassigned at phylum and class level, respectively (Figures 4 and 5). At the phylum level, the most dominant fungal phyla were Ascomycota (92.4%), Basidiomycota (3.6%), and other unclassified phylum (3.9%) (Figure 4, Figure S5). At the class level, fungal microbiome was dominated by Dothideomycetes (73.5%), Sordariomycetes (11.4%), Tremellomycetes (3.0%), and three other unidentified classes (two Ascomycota classes: 3.9 and 1.2%, respectively, and another unassigned class: 3.9%).

2.5. Determination of the Core Bacterial Microbiome (Bacteriome) of S. miltiorrhiza Seeds

We used the persistence method to identify the OTUs present across these seed samples and determine the core bacterial microbiome (bacteriome) in *S. miltiorrhiza* seed. This core bacterial microbiome contained 16 OTUs (233,225 seq.) and corresponded to 54.5% of the whole microbiome. Taxonomic composition of the core microbiomewasmore concentrated than those of the whole microbiome at both the class and genus levels (Figures 2 and 3). Gammaproteobacteria took the absolute advantage, and ran up to 96% of the core bacterial microbiome in *S. miltiorrhiza* seed at the class level. However, Alphaproteobacteria and Actinobacteria contributed 3% and 1%, respectively. At the genus level, *Pantoea* and *Pseudomonas* contained 68% and 22%, respectively. In addition, *Enterobacter* occupied 3%, whereas *Erwinia*, *Sphingomonas*, *Methylobacterium*, and *Curtobacterium* and an unclassified genus exceeded 1%. These results revealed that the seeds of *S. miltiorrhiza* shared the dominant microbiota within their microbiome (Figures 2 and 3, Table S9).

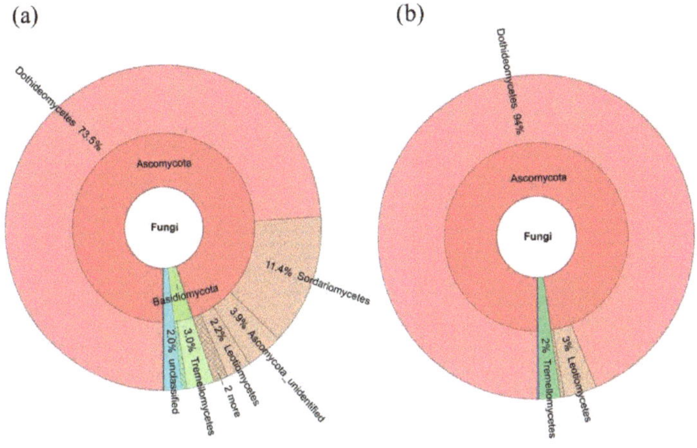

Figure 4. Taxonomic composition of the seed-associated (**a**) whole and (**b**) core fungal microbiome of *S. miltiorrhiza* at the class level. Pie charts represent relative abundances of fungal classes for the whole and core microbiome.

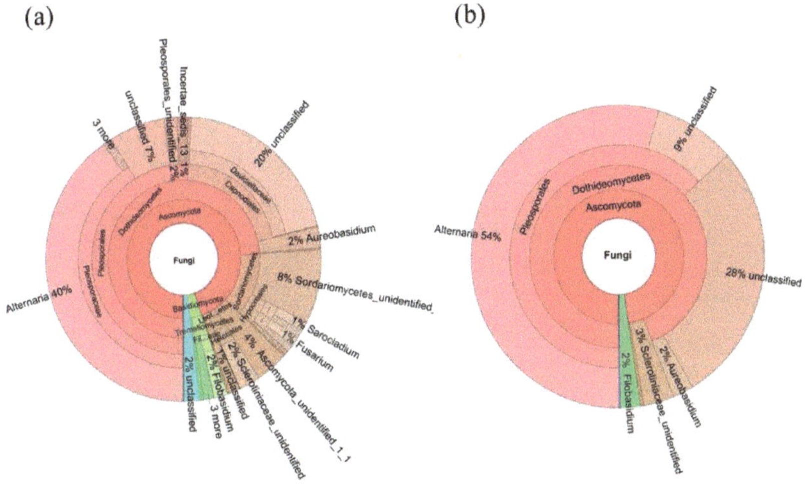

Figure 5. Taxonomic composition of the seed-associated (**a**) whole and (**b**) core fungal microbiome of *S. miltiorrhiza* at the genus level. Pie charts represent relative abundances of fungal genera for the whole and core microbiome.

2.6. Determination of the Core Fungal Microbiome (Mycobiome) of S. miltiorrhiza Seeds

We deciphered the core fungal microbiome (mycobiome) for *S. miltiorrhiza* seed using the same persistence approach. The core fungal microbiome of *S. miltiorrhiza* seed contained 3 OTUs (544,258 seq.) and contributed to 39.5% of the total fungal microbiome abundance in *S. miltiorrhiza* seed. At the class level, Dothideomycetes took up 94% of the core fungal microbiome in *S. miltiorrhiza* seed, and another two dominant classes were Leotiomycetes and Tremellomycetes. *Alternaria* (54%), another two unclassified genera of Dothideomycetes (28% and 9%), and a genus of Leotiomycetes (3%) were the dominant genera in the core fungal microbiome in *S. miltiorrhiza* seed, whereas *Aureobasidium* and *Filobasidium* also occupied 2% (Figures 4 and 5, Table S10).

2.7. Predictive Function of Core Bacterial Microbiome in S. miltiorrhiza Seeds

We predicted the functional profiles of bacterial core microbiome based on the 16S rRNA gene copy number of deciphered core bacterial taxa using PICRUSt according to the KEGG Ortholog groups (KOs). We mainly focused on predicted abundances of KOs assigned to metabolism of terpenoids and polyketides, and biosynthesis of secondary metabolites. The overrepresented group included terpenoid backbone biosynthesis, limonene, pinene, and geraniol degradation, prenyltransferases in metabolism of terpenoids and polyketides, and streptomycin biosynthesis in biosynthesis of secondary metabolites. Moreover, biosynthesis of siderophore group nonribosomal peptides, tetracycline, polyketide sugar unit, tropane, piperidine, pyridine alkaloid, novobiocin, and phenylpropanoid had also a certain abundance (Figure 6). The metabolism of functional profiles was similar between different geographic origin seeds of *S. miltiorrhiza*.

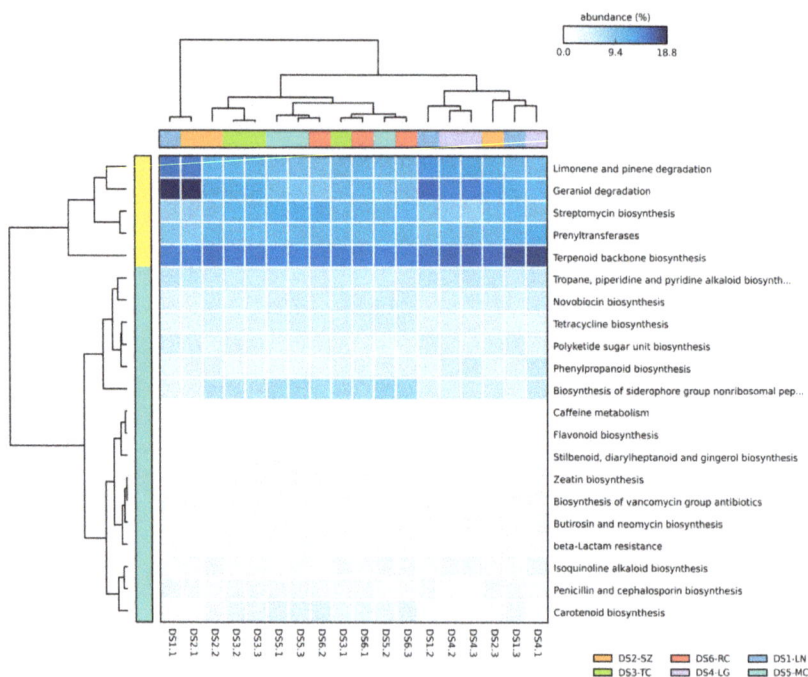

Figure 6. The heatmap of normalized relative abundance of imputed functional profiles of KOs assigned to biosynthesis of secondary metabolites and metabolism of terpenoids and polyketides within *S. miltiorrhiza* seed-associated core bacterial microbiome using PICRUSt grouped into level-3 functional categories.

3. Discussion

The seed-associatedmicrobiome may play an important role in plant growth and fitness by vertical transmission, influencing the primary assemblage of the plant microbiota [1,2]. Culture-independent methods using the next-generation sequencing platforms provide a high-resolution microbial community profiles for seed microbiome of maize (*Zea mays*) [35], bean (*Phaseolus vulgaris*) [15], rice (*Oryza sativa*) [36], and barley (*Hordeum vulgare*) [37], and family Brassicaceae [16] in recent years. In these studies, plant species and genotype, and growth environment were the main determinants of the seed-associated (endophyte and epiphyte) microbiota community structure. Both horizontally (acquired from the surrounding environment) and vertically (acquired directly from the parent) mode may contribute to the final composition of seed microbiome.

Despite significant differences between the seed microbiome of different plant species or varieties, the seed-associated microbiome consists a core set of microbial taxa. By comparing the present study of seed-associated microbiome of *S. miltiorrhiza* with maize (*Z. mays*) [35], bean (*P. vulgaris*) [15], and rice (*O. sativa*) [36], we found some enriched core taxa (genera) overlap among these plant seeds, which include the bacterial genera *Pantoea, Pseudomonas, Sphingomonas*, and a fungal genus *Alternaria* (Figures 7 and 8). A recently interesting study by Rybakova et al. confirmed that bacterial genera *Sphingomonas, Pseudomonas,* and *Bacillus* were also the most abundant taxa in the seed microbiome of *Brassica napus* [38]. These finding suggested an interesting possibility of a long association and coevolution between some seed-associated microbial taxa and their hosts.

Figure 7. Seed-associated bacterial genera shared among danshen, maize, bean, and rice. The font color of genera was color-coded by phyla. Maize bacterial genera are based on [35]. Bean bacterial genera are based on [15]. Rice bacterial genera are based on [36]. * We isolated many *Bacillus* strains from danshen seeds, but *Bacillus* was not the dominant genus according to the 16S rRNA sequencing results. This may be due to the spore formed microbes being more easily culturable, or the bias caused by 16S rDNA primer specificity.

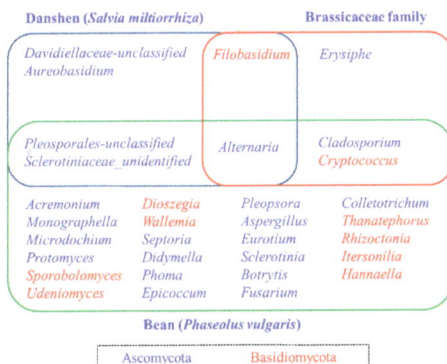

Figure 8. Seed-associated fungal genera shared among danshen, Brassicaceae, and bean. The font color of genera was color-coded by phyla. Brassicaceae fungal genera are based on [16]. Bean fungal genera are based on [15].

The predominance of *Pantoea* was especially apparent among the overlapping bacterial core taxa. In recent years, many studies have shown that some strains of *Pantoea* isolated from rice [39], maize [40], wheat [41], and *Brassica* seeds expressed a few better antagonistic activities, whereas other isolates of *Pantoea* exhibited neutral or weak pathogenic activities. Therefore, the function of *Pantoea* strains contained in seed-associated microbiome needs to be further evaluated. The other dominant genus, *Pseudomonas*, which is a kind of important plant growth-promoting bacteria (PGPR), were widely distributed in rhizosphere and endosphere of plants, and they can promote plant growth and drive root development [42,43]. *Pseudomonas* spp. represent one of the most abundant genera of the root microbiome [5,7,44,45]. Another overlapping genus, *Sphingomonas*, was also enriched taxa in some plant root systems which display plant growth promoting and bioremediation activities [46–48]. Therefore, the genera *Pseudomonas* and *Sphingomonas* in the seed microbiome are likely to be important reservoirs of rhizosphere or endosphere microbiome. Regarding the shared fungal core taxa, genus *Alternaria* can be a potential plant pathogen, but it also includes beneficial endophytes as biocontrol agents or other active compounds producing microbes [49–52]. Therefore, *Alternaria* can affect the germination of seeds and the assemblage of plant microbiome, and consequently, the growth and fitness of plants.

PICRUSt analysis of the seed core microbiome in *S. miltiorrhiza* showed high relative abundances of some secondary metabolism pathways or key enzymes which are closely related to terpenoid biosynthesis. Terpenoid backbone biosynthesis can provide many important precursors for terpenoid biosynthesis, which are common in upstream metabolic pathway. Limonene and pinene degradations are important component of monoterpene biosynthesis pathway. Geraniol degradation is also an important terpenoid metabolism pathway [53]. Prenyltransferases are key enzymes in many primary and secondary metabolism [54]. These pathways are common terpenoid metabolic pathways in microorganisms, and their overrepresentation in seed-associated microbiome indicates their potential for secondary metabolism gene repository in *S. miltiorrhiza*. Although the PICRUSt predicted results could not fully reflect the actual metabolic capacities of the microbial community, the seed microbiome might be considered as enriched species that are closely related to terpenoid metabolism, just as Del Giudice et al. found that the microbial community of Vetiver root involved its essential oil biogenesis [55], *S. miltiorrhiza* seed-associated microbiota might influence the secondary metabolism of host plant, and participate in the biological process of plant stress resistance and immunity.

4. Material and Methods

4.1. Sampling of Salvia miltiorrhiza Seeds

We collected 18 seeds from diverse geographic origins within the northwest of China in August 2015 to decipher core seed-associated microbiome of cultivated *S. miltiorrhiza*. We covered the main producing areas of danshen by choosing the seeds from Luonan county (34°05′26.19″ N, 110°02′6.09″ E), Shangzhou district (33°57′42.11″ N, 109°58′7.41″ E), Tongchuan city (34°54′15.02″ N, 108°57′5.61″ E), Langao county (32°18′50.68″ N, 108°54′31.81″ E) in Shaanxi province, Mianchi County (34°46′27.43″ N, 111°46′6.34″ E) in Henan province, and Ruicheng County (34°38′11.09″ N, 110°19′8.77″ E) in Shanxi province (Table S1, Figure S1). Moreover, we collected a seed of *Salvia miltiorrhiza* Bge. f. alba, a variant of *S. miltiorrhiza* from Laiwu city in Shandong province (36°12′49.00″ N, 117°40′14.86″ E), as a control of closely related species. Seeds from each geographic origin were collected as three independent replicates. All seeds were collected from *S. miltiorrhiza* standard planting base of Tasly Group Company (Tianjin, China). The quality of collected seeds was shown in Table S1. These seeds were stored separately in plastic bags at −20 °C before DNA extraction.

4.2. Plant DNA Exaction and SSR Genotyping

A total of 27 danshen plants were genotyped using 10 SSR molecular markers to analyze the genetic diversity existing between and within the seven *S. miltiorrhiza* cultivars (Table S2). Danshen SSR genotyping analysis method used established by Dr. Qi ZC [56,57].

Briefly, the plant seedlings were grown in a greenhouse from the seed stage and 50 mg of fresh leaf tissue were collected from each individual seedling. Genomic DNA were extracted following a modified cetyltrimethyl ammonium bromide (CTAB) protocol [58], which uses a more efficient Plant DNAzol® kit (Thermo Fisher Scientific, Waltham, MA, USA). Later, DNA quality was examined on 1% agarose gel and concentration were assessed through spectrophotometry by using NanoDrop 2000 (Thermo Fisher Scientific, Waltham, MA, USA).

The SSR genotyping was carried out using 10 SSR markers covering all the species linkage groups (Table S2). PCR amplifications were performed on a T100 Thermal Cycler (Applied Biosystem, Foster City, CA, USA) with a 10 μL reaction mixture containing the following protocol: 1 μL template genomic DNA, 5 μL 2× Master Mix (TSINGKE, Hangzhou, China), 0.2 μM of each primer. The PCR protocol used was as follows: 94 °C for 3 min; followed by 35 cycles of 94 °C for 30 s, a locus-specific Ta (Table S2) for 30 s, and 72 °C 45 s, and a final extension at 72 °C for 10 min. Amplification products were checked on 2 % agarose gel stained with Gene Green Nucleic Acid dye (TIANGEN, Beijing, China).

Afterwards, PCR products were sent to TSINGKE (Hangzhou, China) where genotyped by ABI 3730 sequencer (Thermo Fisher Scientific, Waltham, MA, USA). Genetic diversity parameters, including the number of allele (*Na*), observed and expected heterozygosity (*Ho*, *He*) and polymorphism information content (PIC), which were estimated using GenAlEx 6.502 software [59]. Deviations from Hardy-Weinberg equilibrium (HWE) were tested by GENEPOP 4.2 software [60].

4.3. Microbial DNA Extraction

DNA from different samples was extracted using PowerPlant® DNA Isolation kit (13400-50, MOBIO, Inc., Germantown, MD, USA) according to manufacturer's instructions. Sample blanks consisted unused swabs processed through DNA extraction, and they were tested to contain no 16S amplicons. The total DNA was eluted in 50 μL of elution buffer by a modification of the procedure described by the manufacturer (MOBIO), and stored at −80 °C until measurement in the PCR by LC-Bio Technology Co., Ltd., Hangzhou, China.

4.4. PCR Amplification, 16S rDNA or ITS Sequencing and Data Analysis

We amplified the V3–V4 region of the bacterial 16S rRNA gene and ITS2 region of the eukaryotic (fungal) small-subunit rRNA gene using the total DNA from 21 *S. miltiorrhiza* seed samples as a template and the primer (319F 5′-ACTCCTACGGGAGGCAGCAG-3′; 806R 5′-GGACTA CHVGGGTWTCTAAT-3′) for bacterial microbiota and primer (ITS7F 5′-ACTCCTACGGGAGGCAG CAG-3′; ITS4R 5′-GGACTACHVG GGTWTCTAAT-3′) for fungal microbiota. The 5′ ends of the primers were tagged with specific barcodes per sample and sequencing universal primers.

All reactions were carried out in 25 μL (total volume) mixtures containing approximately 25 ng of genomic DNA extract, 12.5 μL PCR Premix, 2.5 μL of each primer, and PCR-grade water to adjust the volume. PCR reactions were performed in a master cycler gradient thermocycler (Eppendorf, Hamburg, Germany) set to the following conditions: initial denaturation at 98 °C for 30 s; 35 cycles of denaturation at 98 °C for 10 s, annealing at 54/52 °C for 30 s, and extension at 72 °C for 45 s; and then a final extension at 72 °C for 10 min. The PCR products were confirmed with 2% agarose gel electrophoresis. Throughout the DNA extraction process, ultrapure water instead of a sample solution, was used to exclude the possibility of false-positive PCR results as a negative control. The PCR products were normalized by AxyPrep TM Mag PCR Normalizer (Axygen Biosciences, Union City, CA, USA), which allowed the skipping of the quantification step, regardless of the PCR volume submitted for sequencing. The amplicon pools were prepared for sequencing with AMPure XT beads (Beckman

Coulter Genomics, Danvers, MA, USA), and the size and quantity of the amplicon library were assessed on the LabChip GX (Perkin Elmer, Waltham, MA, USA) and with the Library Quantification Kit for Illumina (Kapa Biosciences, Woburn, MA, USA), respectively. PhiX Control library (V3) (Illumina) was combined with the amplicon library (expected at 30%). The library was clustered to a density of approximately 570 K/mm². The libraries were sequenced either on 300PE MiSeq runs, and one library was sequenced with both protocols using the standard Illumina sequencing primers, which eliminated the need for a third (or fourth) index read.

Samples were sequenced on an Illumina MiSeq platform according to the manufacturer's recommendations provided by LC-Bio. Paired-end reads were assigned to samples based on their unique barcode, and truncated by cutting off the barcode and primer sequence. Paired-end reads were merged using PEAR (v.0.9.6) (Heidelberg, Germany) [61]. Quality filtering on the raw tags was performed under specific filtering conditions to obtain the high-quality clean tags according to the FastQC (v.0.10.1) (New Delhi, India) [62]. Chimeric sequences were filtered using VSEARCH (v.2.3.4) (Oslo, Norway) and sequences with ≥97% similarity were assigned to the same operational taxonomic units (OTUs) using the same software [63]. Representative sequences were chosen for each OTU, and taxonomic data were then assigned to each representative sequence using the RDP (Ribosomal Database Project) classifier. To examine the differences of the dominant species in different groups, multiple sequence alignments were conducted using PyNAST (v.1.2) software (Boulder, CO, USA) [64] to study phylogenetic relationships of different OTUs. Abundance information of OTUs was normalized using a standard of sequence number corresponding to the sample with the least sequences. Alpha diversity was applied in analyzing complexity of species diversity for a sample through 4 indices, including Chao 1, Shannon, Simpson, and Observed species. All these indices in our samples were calculated with QIIME software (Boulder, CO, USA) in Python (v.1.8.0) (La Jalla, CA, USA) [65]. Beta diversity analysis was used to evaluate differences of samples in species complexity. Beta diversity was calculated by principle coordinates analysis (PCoA) and cluster analysis by QIIME [66].

4.5. Determination of Core Microbiome of S. miltiorrhiza Seed

Metagenomics Core Microbiome Exploration Tool (MetaCoMET) was applied to decipher the core microbiome across cultivated *S. miltiorrhiza* seeds according to the membership and persistence methods, especially by focusing on the latter methods [67]. In brief, we input the OUT BIOM file generated by QIIME, which described the above and metadata file in turn. We uploaded these files to the website, and selected parameters and Venn type. Later, we submitted them to the web platform to obtain results.

4.6. Predict Microbial Functional Profiles of Core Microbiome

PICRUSt software (http://picrust.github.io/picrust) was used to predict the microbial functional profiles of core microbiome in *S. miltiorrhiza* seeds. We modified the sequence data format according to the platform requirements, and then performed the functional prediction using the method provided by PICRU St. Briefly, the OUT BIOM table of seed-associated core microbiome was used as an input file for metagenome imputation of *S. miltiorrhiza* seed samples, and predicted gene class abundances were analyzed at KEGG Orthology group levels 3 [68]. Results from PICRUSt were analyzed in statistical analysis of taxonomic and functional profiles (STAMP) [69].

5. Conclusions

In conclusion, deciphering the core microbiome across different cultivated *S. miltiorrhiza* seeds indicated that seed microbiome is a distinctive genetic resource for the host plant. Although some studies had indicated seed microbiomes have significant impacts on host plant health and productivity [1,2,16–18], our study provides the first insights into the seed-associated core microbiome of a medicinal plant. Our PICRUSt prediction analysis revealed that these microbial core taxa can influence the growth and quality of *S. miltiorrhiza*. Especially, we found the seed-associated microbiome could be a reservoir and supplement of secondary metabolic capabilities, in addition to the host plant genome. Just as Aleti et al. found, the secondary metabolite genes encoded by potato rhizosphere microbiomes were diverse and vary with the different samples and vegetation stage, which influence on the growth and metabolism of the host plant [70]. Our study suggested that some core taxa of seed microbiome not only promoted seed germination and plant growth, but also regulated and participated in the secondary metabolism of host plants.

Supplementary Materials: Supplementary materials can be found at http://www.mdpi.com/1422-0067/19/3/672/s1.

Acknowledgments: The authors would like to thank Qiulei Lang, Xiaogang Zhang, Jianfang Zhu, Zhiyuan Huang, Changli Ge, and other staff members in LC-Bio for 16S rRNA and ITS sequencing and analysis. We thank Zhechen Qi, Ruizhen Liu and Chao Shen for their help us in performing SSR analysis. This research was supported by the National Natural Science Foundation of China (81773835), and the Major Program of the State Administration of Traditional Chinese Medicine of People's Republic of China (ZYBZH-C-TJ-55), and the National innovative and entrepreneurship training program for College Students in China (201710338015).

Author Contributions: Haimin Chen, Fang Miao and Zongsuo Liang designed this study. Hongguang Zhao and Fenghua Liu collected the seeds. Haimin Chen, Hongxia Wu and Bin Yan was completed plant DNA extraction and SSR analysis. Haimin Chen, Hongxia Wu, Bin Yan, and Haihua Zhang isolated total DNA, and prepared it for sequencing. Hongxia Wu, Hongguang Zhao and Qing Sheng analyzed microbial community sequences. Hongxia Wu performed the P analysis. Haimin Chen drafted the manuscript. All authors contributed to writing and finalizing the paper, and also read and approved the final manuscript.

Conflicts of Interest: The authors declare no conflict of interest.

References

1. Nelson, E.B. The seed microbiome: Origins, interactions, and impacts. *Plant Soil* **2017**, *422*, 7–34. [CrossRef]
2. Shade, A.; Jacques, M.A.; Barret, M. Ecological patterns of seed microbiome diversity, transmission, and assembly. *Curr. Opin. Microbiol.* **2017**, *37*, 15–22. [CrossRef] [PubMed]
3. Truyens, S.; Weyens, N.; Cuypers, A.; Vangronsveld, J. Bacterial seed endophytes: Genera, vertical transmission and interaction with plants. *Environ. Microbiol. Rep.* **2015**, *7*, 40–50. [CrossRef]
4. Shade, A.; Handelsman, J. Beyond the Venn diagram: The hunt for a core microbiome. *Environ. Microbiol.* **2012**, *14*, 4–12. [CrossRef] [PubMed]
5. Lundberg, D.S.; Lebeis, S.L.; Paredes, S.H.; Yourstone, S.; Gehring, J.; Malfatti, S.; Tremblay, J.; Engelbrektson, A.; Kunin, V.; del Rio, T.G.; et al. Defining the core *Arabidopsis thaliana* root microbiome. *Nature* **2012**, *488*, 86–90. [CrossRef] [PubMed]
6. Astudillo-Garcia, C.; Bell, J.J.; Webster, N.S.; Glasl, B.; Jompa, J.; Montoya, J.M.; Taylor, M.W. Evaluating the core microbiota in complex communities: A systematic investigation. *Environ. Microbiol.* **2017**, *19*, 1450–1462. [CrossRef] [PubMed]
7. Bulgarelli, D.; Rott, M.; Schlaeppi, K.; ver Loren van Themaat, E.; Ahmadinejad, N.; Assenza, F.; Rauf, P.; Huettel, B.; Reinhardt, R.; Schmelzer, E.; et al. Revealing structure and assembly cues for *Arabidopsis* root-inhabiting bacterial microbiota. *Nature* **2012**, *488*, 91–95. [CrossRef] [PubMed]
8. Peiffer, J.A.; Spor, A.; Koren, O.; Jin, Z.; Tringe, S.G.; Dangl, J.L.; Buckler, E.S.; Ley, R.E. Diversity and heritability of the maize rhizosphere microbiome under field conditions. *Proc. Natl. Acad. Sci. USA* **2013**, *110*, 6548–6553. [CrossRef] [PubMed]
9. Mendes, L.W.; Kuramae, E.E.; Navarrete, A.A.; van Veen, J.A.; Tsai, S.M. Taxonomical and functional microbial community selection in soybean rhizosphere. *ISME J.* **2014**, *8*, 1577–1587. [CrossRef] [PubMed]
10. Bulgarelli, D.; Garrido-Oter, R.; Munch, P.C.; Weiman, A.; Droge, J.; Pan, Y.; McHardy, A.C.; Schulze-Lefert, P. Structure and function of the bacterial root microbiota in wild and domesticated barley. *Cell Host Microbe* **2015**, *17*, 392–403. [CrossRef] [PubMed]
11. Edwards, J.; Johnson, C.; Santos-Medellin, C.; Lurie, E.; Podishetty, N.K.; Bhatnagar, S.; Eisen, J.A.; Sundaresan, V. Structure, variation, and assembly of the root-associated microbiomes of rice. *Proc. Natl. Acad. Sci. USA* **2015**, *112*, E911-20. [CrossRef] [PubMed]
12. Sonnenburg, J.L.; Backhed, F. Diet-microbiota interactions as moderators of human metabolism. *Nature* **2016**, *535*, 56–64. [CrossRef] [PubMed]

13. Utzschneider, K.M.; Kratz, M.; Damman, C.J.; Hullarg, M. Mechanisms linking the gut microbiome and glucose metabolism. *J. Clin. Endocrinol. Metab.* **2016**, *101*, 1445–1454. [CrossRef] [PubMed]

14. Wang, Z.; Koonen, D.; Hofker, M.; Fu, J. Gut microbiome and lipid metabolism: From associations to mechanisms. *Curr. Opin. Lipidol.* **2016**, *27*, 216–224. [CrossRef] [PubMed]

15. Klaedtke, S.; Jacques, M.A.; Raggi, L.; Preveaux, A.; Bonneau, S.; Negri, V.; Chable, V.; Barret, M. Terroir is a key driver of seed-associated microbial assemblages. *Environ. Microbiol.* **2016**, *18*, 1792–1804. [CrossRef] [PubMed]

16. Barret, M.; Briand, M.; Bonneau, S.; Preveaux, A.; Valiere, S.; Bouchez, O.; Hunault, G.; Simoneau, P.; Jacquesa, M.A. Emergence shapes the structure of the seed microbiota. *Appl. Environ. Microbiol.* **2015**, *81*, 1257–1266. [CrossRef] [PubMed]

17. Links, M.G.; Demeke, T.; Grafenhan, T.; Hill, J.E.; Hemmingsen, S.M.; Dumonceaux, T.J. Simultaneous profiling of seed-associated bacteria and fungi reveals antagonistic interactions between microorganisms within a shared epiphytic microbiome on *Triticum* and *Brassica* seeds. *New Phytol.* **2014**, *202*, 542–553. [CrossRef] [PubMed]

18. Khalaf, E.M.; Raizada, M.N. Taxonomic and functional diversity of cultured seed associated microbes of the cucurbit family. *BMC Microbiol.* **2016**, *16*, 131. [CrossRef] [PubMed]

19. Hameed, A.; Yeh, M.W.; Hsieh, Y.T.; Chung, W.C.; Lo, C.T.; Young, L.S. Diversity and functional characterization of bacterial endophytes dwelling in various rice (*Oryza sativa* L.) tissues, and their seed-borne dissemination into rhizosphere under gnotobiotic P-stress. *Plant Soil* **2015**, *394*, 177–197. [CrossRef]

20. Sorty, A.M.; Meena, K.K.; Choudhary, K.; Bitla, U.M.; Minhas, P.S.; Krishnani, K.K. Effect of plant growth promoting bacteria associated with halophytic weed (*Psoralea corylifolia* L.) on germination and seedling growth of wheat under saline conditions. *Appl. Biochem. Biotechnol.* **2016**, *180*, 872–882. [CrossRef] [PubMed]

21. Herrera, S.D.; Grossi, C.; Zawoznik, M.; Groppa, M.D. Wheat seeds harbour bacterial endophytes with potential as plant growth promoters and biocontrol agents of *Fusarium graminearum*. *Microbiol. Res.* **2016**, *186*, 37–43. [CrossRef] [PubMed]

22. Chimwamurombe, P.M.; Gronemeyer, J.L.; Reinhold-Hurek, B. Isolation and characterization of culturable seed-associated bacterial endophytes from gnotobiotically grown Marama bean seedlings. *FEMS Microbiol. Ecol.* **2016**, *92*, fiw083. [CrossRef] [PubMed]

23. Sanchez-Lopez, A.S.; Pintelon, I.; Stevens, V.; Imperato, V.; Timmermans, J.P.; Gonzalez-Chavez, C.; Carrillo-Gonzalez, R.; van Hamme, J.; Vangronsveld, J.; Thijs, S. Seed endophyte microbiome of *Crotalaria pumila* unpeeled: Identification of plant-beneficial methylobacteria. *Int. J. Mol. Sci.* **2018**, *19*, 291. [CrossRef] [PubMed]

24. Lugtenberg, B.J.J.; Caradus, J.R.; Johnson, L.J. Fungal endophytes for sustainable crop production. *FEMS Microbiol. Ecol.* **2016**, *92*, fiw194. [CrossRef] [PubMed]

25. Su, C.Y.; Ming, Q.L.; Rahman, K.; Han, T.; Qin, L.P. *Salvia miltiorrhiza*: Traditional medicinal uses, chemistry, and pharmacology. *Chin. J. Nat. Med.* **2015**, *13*, 163–182. [CrossRef]

26. Li, X.Q.; Zhai, X.; Shu, Z.H.; Dong, R.F.; Ming, Q.L.; Qin, L.P.; Zheng, C.J. *Phoma glomerata* D14: An endophytic fungus from *Salvia miltiorrhiza* that produces salvianolic acid C. *Curr. Microbiol.* **2016**, *73*, 1–7. [CrossRef] [PubMed]

27. Ming, Q.; Han, T.; Li, W.; Zhang, Q.; Zhang, H.; Zheng, C.; Huang, F.; Rahman, K.; Qin, L. Tanshinone IIA and tanshinone I production by *Trichoderma atroviride* D16, an endophytic fungus in *Salvia miltiorrhiza*. *Phytomedicine* **2012**, *19*, 330–333. [CrossRef] [PubMed]

28. Yan, Y.; Zhang, S.C.; Zhang, J.Y.; Ma, P.D.; Duan, J.L.; Liang, Z.S. Effect and mechanism of endophytic bacteria on growth and secondary metabolite synthesis in *Salvia miltiorrhiza* hairy roots. *Acta Physiol. Plant.* **2014**, *36*, 1095–1105. [CrossRef]

29. Sun, J.; Xia, F.; Cui, L.; Liang, J.; Wang, Z.; Wei, Y. Characteristics of foliar fungal endophyte assemblages and host effective components in *Salvia miltiorrhiza* Bunge. *Appl. Microbiol. Biotechnol.* **2014**, *98*, 3143–3155. [CrossRef] [PubMed]

30. Yan, X.; He, L.; Song, G.; Wang, R. Antagonistic bioactivity of endophytic strains isolated from *Salvia miltiorrhiza*. *Afr. J. Biotechnol.* **2013**, *10*, 15117–15122. [CrossRef]

31. Ming, Q.; Su, C.; Zheng, C.; Jia, M.; Zhang, Q.; Zhang, H.; Rahman, K.; Han, T.; Qin, L. Elicitors from the endophytic fungus *Trichoderma atroviride* promote *Salvia miltiorrhiza* hairy root growth and tanshinone biosynthesis. *J. Exp. Bot.* **2013**, *64*, 5687–5694. [CrossRef] [PubMed]

32. Li, X.J.; Tang, H.Y.; Duan, J.L.; Gao, J.M.; Xue, Q.H. Bioactive alkaloids produced by *Pseudomonas brassicacearum* subsp. Neoaurantiaca, an endophytic bacterium from *Salvia miltiorrhiza*. *Nat. Prod. Res.* **2013**, *27*, 496–499. [CrossRef] [PubMed]

33. Li, X.L.; Yuan, H.M.; Qi, S.S.; Feng, M.L.; Liu, H.W.; Zhang, X.P. Isolation and genetic diversity of the endophytic actinomycetes from *Salvia miltiorrhiza* Bge. and *Polygonatum sibiricum* Red. *Microbiol. China* **2010**, *37*, 1341–1346.

34. He, X.L.; Wang, L.Y.; Ma, J.; Zhao, L.L. AM fungal diversity in the rhizosphere of *Salvia miltiorrhiza* in Anguo city of Hebei province. *Biodivers. Sci.* **2010**, *18*, 187–194.

35. Liu, Y.; Zuo, S.; Zou, Y.Y.; Wang, J.H.; Song, W. Investigation on diversity and population succession dynamics of endophytic bacteria from seeds of maize (*Zea mays* L. Nongda108) at different growth stages. *Ann. Microbiol.* **2013**, *63*, 71–79. [CrossRef]

36. Midha, S.; Bansal, K.; Sharma, S.; Kumar, N.; Patil, P.P.; Chaudhry, V.; Patil, P.B. Genomic resource of rice seed associated bacteria. *Front. Microbiol.* **2016**, *6*, 1551. [CrossRef] [PubMed]

37. Yang, L.; Danzberger, J.; Scholer, A.; Schroder, P.; Schloter, M.; Radl, V. Dominant groups of potentially active bacteria shared by barley seeds become less abundant in root associated microbiome. *Front. Plant Sci.* **2017**, *8*, 1005. [CrossRef] [PubMed]

38. Rybakova, D.; Mancinelli, R.; Wikström, M.; Birch-Jensen, A.-S.; Postma, J.; Ehlers, R.-U.; Goertz, S.; Berg, G. The structure of the *Brassica napus* seed microbiome is cultivar-dependent and affects the interactions of symbionts and pathogens. *Microbiome* **2017**, *5*, 104. [CrossRef] [PubMed]

39. Wu, L.; Liu, R.; Niu, Y.; Lin, H.; Ye, W.; Guo, L.; Hu, X. Whole genome sequence of *Pantoea ananatis* R100, an antagonistic bacterium isolated from rice seed. *J. Biotechnol.* **2016**, *225*, 1–2. [CrossRef] [PubMed]

40. Sheibani-Tezerji, R.; Naveed, M.; Jehl, M.A.; Sessitsch, A.; Rattei, T.; Mitter, B. The genomes of closely related *Pantoea ananatis* maize seed endophytes having different effects on the host plant differ in secretion system genes and mobile genetic elements. *Front. Microbiol.* **2015**, *6*, 440. [CrossRef] [PubMed]

41. Town, J.; Dumonceaux, T.J. High-quality draft genome sequences of *Pantoea agglomerans* isolates exhibiting antagonistic interactions with wheat seed-associated fungi. *Genome Announc.* **2016**, *4*, e00511–e00516. [CrossRef] [PubMed]

42. Zamioudis, C.; Mastranesti, P.; Dhonukshe, P.; Blilou, I.; Pieterse, C.M. Unraveling root developmental programs initiated by beneficial *Pseudomonas* spp. bacteria. *Plant Physiol.* **2013**, *162*, 304–318. [CrossRef] [PubMed]

43. Van de Mortel, J.E.; de Vos, R.C.; Dekkers, E.; Pineda, A.; Guillod, L.; Bouwmeester, K.; van Loon, J.J.; Dicke, M.; Raaijmakers, J.M. Metabolic and transcriptomic changes induced in arabidopsis by the rhizobacterium *Pseudomonas fluorescens* SS101. *Plant Physiol.* **2012**, *160*, 2173–2188. [CrossRef] [PubMed]

44. Sessitsch, A.; Hardoim, P.; Doring, J.; Weilharter, A.; Krause, A.; Woyke, T.; Mitter, B.; Hauberg-Lotte, L.; Friedrich, F.; Rahalkar, M.; et al. Functional Characteristics of an Endophyte Community Colonizing Rice Roots as Revealed by Metagenomic Analysis. *Mol. Plant Microbe* **2012**, *25*, 28–36. [CrossRef] [PubMed]

45. Mendes, R.; Kruijt, M.; de Bruijn, I.; Dekkers, E.; van der Voort, M.; Schneider, J.H.; Piceno, Y.M.; DeSantis, T.Z.; Andersen, G.L.; Bakker, P.A.; et al. Deciphering the rhizosphere microbiome for disease-suppressive bacteria. *Science* **2011**, *332*, 1097–1100. [CrossRef] [PubMed]

46. Hartman, K.; van der Heijden, M.G.; Roussely-Provent, V.; Walser, J.C.; Schlaeppi, K. Deciphering composition and function of the root microbiome of a legume plant. *Microbiome* **2017**, *5*, 2. [CrossRef] [PubMed]

47. Pan, F.; Meng, Q.; Wang, Q.; Luo, S.; Chen, B.; Khan, K.Y.; Yang, X.; Feng, Y. Endophytic bacterium *Sphingomonas* SaMR12 promotes cadmium accumulation by increasing glutathione biosynthesis in *Sedum alfredii* Hance. *Chemosphere* **2016**, *154*, 358–366. [CrossRef] [PubMed]

48. Gao, J.L.; Sun, P.; Wang, X.M.; Cheng, S.; Lv, F.; Qiu, T.L.; Yuan, M.; Sun, J.G. *Sphingomonaszeicaulis* sp. nov. an endophytic bacterium isolated from maize root. *Int. J. Syst. Evol. Microbiol.* **2016**, *66*, 3755–3760. [PubMed]

49. Ismaiel, A.A.; Ahmed, A.S.; Hassan, I.A.; El-Sayed, E.R.; Karam El-Din, A.A. Production of paclitaxel with anticancer activity by two local fungal endophytes, *Aspergillus fumigatus* and *Alternaria tenuissima*. *Appl. Microbiol. Biotechnol.* **2017**, *101*, 5831–5846. [CrossRef] [PubMed]

50. Bian, G.; Yuan, Y.; Tao, H.; Shi, X.; Zhong, X.; Han, Y.; Fu, S.; Fang, C.; Deng, Z.; Liu, T. Production of taxadiene by engineering of mevalonate pathway in *Escherichia coli* and endophytic fungus *Alternaria alternata* TPF6. *Biotechnol. J.* **2017**, *12*, 1600697. [CrossRef] [PubMed]

Int. J. Mol. Sci. **2018**, 19, 672

51. Soltani, J.; Hosseyni Moghaddam, M.S. Antiproliferative, antifungal, and antibacterial activities of endophytic *Alternaria* species from cupressaceae. *Curr. Microbiol.* **2014**, 69, 349–356. [CrossRef] [PubMed]

52. Egan, J.M.; Kaur, A.; Raja, H.A.; Kellogg, J.J.; Oberlies, N.H.; Cech, N.B. Antimicrobial fungal endophytes from the botanical medicine goldenseal (*Hydrastis canadensis*). *Phytochem. Lett.* **2016**, 17, 219–225. [CrossRef] [PubMed]

53. Kanehisa, M.; Goto, S.; Sato, Y.; Furumichi, M.; Tanabe, M. KEGG for integration and interpretation of large-scale molecular data sets. *Nucleic Acids Res.* **2012**, 40, D109–D114. [CrossRef] [PubMed]

54. Winkelblech, J.; Fan, A.; Li, S.M. Prenyltransferases as key enzymes in primary and secondary metabolism. *Appl. Microbiol. Biotechnol.* **2015**, 99, 7379–7397. [CrossRef] [PubMed]

55. Del Giudice, L.; Massardo, D.R.; Pontieri, P.; Bertea, C.M.; Mombello, D.; Carata, E.; Tredici, S.M.; Talà, A.; Mucciarelli, M.; Groudeva, V.I.; et al. The microbial community of Vetiver root and its involvement into essential oil biogenesis. *Environ. Microbiol.* **2008**, 10, 2824–2841. [CrossRef] [PubMed]

56. Liu, R.Z.; Shen, C.; Pan, Y.Y.; Han, Y.W.; Yang, M.; Yu, W.D.; Liu, J.L.; Wu, X.; Li, M.D.; Qi, Z.C.; et al. Development of 40 novel microsatellites in wild populations of *Salvia miltiorrhiza* Burge based on a transcriptome database. *Conserv. Genet. Resour.* **2017**, accepted.

57. Qi, Z.C.; Shen, C.; Han, Y.W.; Shen, W.; Yang, M.; Liu, J.; Liang, Z.S.; Li, P.; Fu, C.X. Development of microsatellite loci in Mediterranean sarsaparilla (*Smilax aspera*; Smilacaceae) using transcriptome data. *Appl. Plant Sci.* **2017**, 5, 1700005. [CrossRef] [PubMed]

58. Narzary, D.; Verma, S.; Mahar, K.S.; Rana, T.S. A rapid and effective method for isolation of genomic DNA from small amount of silica-dried leaf tissues. *Natl. Acad. Sci. Lett.* **2015**, 38, 441–444. [CrossRef]

59. Kalinowski, S.T.; Taper, M.L.; Marshall, T.C. Revising how the computer program CERVUS accommodates genotyping error increases success in paternity assignment. *Mol. Ecol.* **2007**, 16, 1099–1106. [CrossRef] [PubMed]

60. Rousset, F. GENEPOP'007: A complete re-implementation of the GENEPOP software for Windows and Linux. *Mol. Ecol. Resour.* **2008**, 8, 103–106. [CrossRef] [PubMed]

61. Zhang, J.; Kobert, K.; Flouri, T.; Stamatakis, A. PEAR: A fast and accurate Illumina paired-end read merger. *Bioinformatics* **2014**, 30, 614–620. [CrossRef] [PubMed]

62. Patel, R.K.; Jain, M. NGS QC toolkit: A toolkit for quality control of next generation sequencing data. *PLoS ONE* **2012**, 7, e30619. [CrossRef] [PubMed]

63. Rognes, T.; Flouri, T.; Nichols, B.; Quince, C.; Mahe, F. VSEARCH: A versatile open source tool for metagenomics. *PeerJ* **2016**, 4. [CrossRef] [PubMed]

64. Caporaso, J.G.; Bittinger, K.; Bushman, F.D.; DeSantis, T.Z.; Andersen, G.L.; Knight, R. PyNAST: A flexible tool for aligning sequences to a template alignment. *Bioinformatics* **2010**, 26, 266–267. [CrossRef] [PubMed]

65. Sanner, M.F. Python: A programming language for software integration and development. *J. Mol. Graph. Model.* **1999**, 17, 57–61. [PubMed]

66. Caporaso, J.G.; Kuczynski, J.; Stombaugh, J.; Bittinger, K.; Bushman, F.D.; Costello, E.K.; Fierer, N.; Peña, A.G.; Goodrich, J.K.; et al. QIIME allows analysis of high-throughput community sequencing data. *Nat. Methods* **2010**, 7, 335. [CrossRef] [PubMed]

67. Wang, Y.; Xu, L.; Gu, Y.Q.; Coleman-Derr, D. MetaCoMET: A web platform for discovery and visualization of the core microbiome. *Bioinformatics* **2016**, 32, 3469–3470. [CrossRef] [PubMed]

68. Langille, M.G.; Zaneveld, J.; Caporaso, J.G.; McDonald, D.; Knights, D.; Reyes, J.A.; Clemente, J.C.; Burkepile, D.E.; Vega Thurber, R.L.; Knight, R.; et al. Predictive functional profiling of microbial communities using 16S rRNA marker gene sequences. *Nat. Biotechnol.* **2013**, 31, 814–821. [CrossRef] [PubMed]

69. Parks, D.H.; Tyson, G.W.; Hugenholtz, P.; Beiko, R.G. STAMP: Statistical analysis of taxonomic and functional profiles. *Bioinformatics* **2014**, 30, 3123–3124. [CrossRef] [PubMed]

70. Aleti, G.; Nikolić, B.; Brader, G.; Pandey, R.V.; Antonielli, L.; Pfeiffer, S.; Oswald, A.; Sessitsch, A. Secondary metabolite genes encoded by potato rhizosphere microbiomes in the Andean highlands are diverse and vary with sampling site and vegetation stage. *Sci. Rep.* **2017**, 7, 2330. [CrossRef] [PubMed]

International Journal of
Molecular Sciences

MDPI

Article

Seed Endophyte Microbiome of *Crotalaria pumila* Unpeeled: Identification of Plant-Beneficial Methylobacteria

Ariadna S. Sánchez-López [1,2], Isabel Pintelon [3], Vincent Stevens [1], Valeria Imperato [1], Jean-Pierre Timmermans [3], Carmen González-Chávez [2], Rogelio Carrillo-González [2], Jonathan Van Hamme [4], Jaco Vangronsveld [1] and Sofie Thijs [1,*]

[1] Centre for Environmental Sciences, Hasselt University, Agoralaan building D, 3590 Diepenbeek, Belgium; ariadnas@colpos.mx (A.S.S.-L.); vincent.stevens@uhasselt.be (V.S.); valeria.imperato@uhasselt.be (V.I.); jaco.vangronsveld@uhasselt.be (J.V.)
[2] Laboratory of Environmental Chemistry and Environmental Microbiology, Edaphology, Colegio de Postgraduados, Campus Montecillo, Carretera Mexico-Texcoco km 36.5, Montecillo 56230, Mexico; carmeng@colpos.mx (C.G.-C.); crogelio@colpos.mx (R.C.-G.)
[3] Laboratory of Cell Biology and Histology, University of Antwerp Campus Drie Eiken, Universiteitsplein 1, 2610 Wilrijk, Antwerp, Belgium; isabel.pintelon@uantwerpen.be (I.P.); jean-pierre.timmermans@uantwerpen.be (J.-P.T.)
[4] Department of Biology, Thompson Rivers University, 950 McGill Road, Kamloops, BC V2C0E1, Canada; jvanhamme@tru.ca
* Correspondence: sofie.thijs@uhasselt.be; Tel.: +32-11-268-225

Received: 15 December 2017; Accepted: 15 January 2018; Published: 19 January 2018

Abstract: Metal contaminated soils are increasing worldwide. Metal-tolerant plants growing on metalliferous soils are fascinating genetic and microbial resources. Seeds can vertically transmit endophytic microorganisms that can assist next generations to cope with environmental stresses, through yet poorly understood mechanisms. The aims of this study were to identify the core seed endophyte microbiome of the pioneer metallophyte *Crotalaria pumila* throughout three generations, and to better understand the plant colonisation of the seed endophyte *Methylobacterium* sp. Cp3. Strain Cp3 was detected in *C. pumila* seeds across three successive generations and showed the most dominant community member. When inoculated in the soil at the time of flowering, strain Cp3 migrated from soil to seeds. Using confocal microscopy, Cp3-mCherry was demonstrated to colonise the root cortex cells and xylem vessels of the stem under metal stress. Moreover, strain Cp3 showed genetic and *in planta* potential to promote seed germination and seedling development. We revealed, for the first time, that the seed microbiome of a pioneer plant growing in its natural environment, and the colonisation behaviour of an important plant growth promoting systemic seed endophyte. Future characterization of seed microbiota will lead to a better understanding of their functional contribution and the potential use for seed-fortification applications.

Keywords: metalliferous soil; trace metals; *Methylobacterium*; seed core microbiome; plant growth-promoting endophyte; xylem

1. Introduction

Metal contaminated sites are a threat to human health when left untreated and lead to significant economic costs [1]. In Europe, an estimated 137,000 km^2 or 6.24% of the agricultural soils is contaminated with trace metals [1]. In China, as much as 10.18% of farmland soil is heavily contaminated and about 13.86% of cereal production is affected [2]. Besides anthropogenically contaminated soils, natural metalliferous soils exist, including serpentine soils (enriched in Ni, Cr, Co) and calamine soils (enriched in Cd, Pb, Zn), which are interesting for mining. However, at the same time these activities destroy the soil structure and life. On natural metalliferous soils and mine tailings (waste heaps), metallophyte plants can be found which are tolerant to high concentrations of trace metals. Some of these metallophytes are able to (hyper)accumulate trace metals in their aboveground tissues in high concentrations without showing any symptoms of toxicity [3,4]. Several mechanisms are at the basis of metal detoxification, including the sequestration of trace metals in the vacuole or apoplast and/or the complexation of metals with metal-binding peptides, such as metallothioneines and phytochelatins [5,6]. In contrast to these metallophytes, most agricultural crops are very sensitive to elevated metal concentrations in soils. Therefore, it is of crucial importance to develop methods to reclaim heavily disturbed metal contaminated sites, and improve plant growth and metal tolerance.

Plants are colonized by an enormous diversity of microorganisms with a range of metabolic functions that allow for soil microorganisms to affect metal uptake, transformation, and accumulation [7,8]. In this respect, rhizosphere micro-organisms have been extensively studied for their interactions with metalliferous plants [8,9]. They were shown to contribute to metal accumulation in plants either directly or indirectly by stimulating plant growth, increasing the surface area of roots, the release of nutrients, and affecting metal uptake, or by (im)mobilizing and/or complexing metals [5,8,9]. Endophytic bacteria are also recognized as very important in respect of stress tolerance and plant growth [10]. Endophytes can promote plant growth and metal uptake directly by producing plant growth beneficial substances, phytohormones, siderophores and specific enzymes, metal mobilizing compounds, and biosurfactants; and, indirectly through controlling plant pathogens or by improving plant stress tolerance by producing 1-aminocyclopropane-1-carboxylate (ACC)-deaminase [10,11]. In contrast to the available amount of information on rhizospheric and shoot endophytic bacteria of metalliferous plants, very little is known about seed endophytes. Mastretta et al. [12] demonstrated that seeds of tobacco grown on a Cd containing growth medium carried beneficial endophytes, which improved biomass production under conditions of Cd exposure, and resulted in higher plant Cd concentrations when compared to non-inoculated plants [12]. Truyens et al. [13] have shown that the seed endophytic community of *Arabidopsis thaliana* exposed to Cd for several generations, contained different bacterial taxa and different functional properties, e.g., metal tolerance and ACC-deaminase dominated in the strains isolated from seeds grown on a Cd enriched growth substrate, while siderophore production, IAA production, and organic acids was more prevalent in endophytes from seeds of plants grown in absence of Cd [13,14]. This suggests that certain endophytes and traits can be transferred to next generations and might be of high importance for seed germination and seedling development.

Crotalaria pumila is an annual herbaceous (sub)tropical plant species with wide environmental tolerance. It is a potential accumulator and phytoextractor of Zn, growing on metalliferous soils in the semi-arid region in Zimapan, Mexico. The plant has unique adaptations to deal with metal stress and accumulates up to 300 mg Zn per kg of dry weight (DW) in the shoots, which reflects the total Zn concentrations in the soil [15]. To better understand the abilities of this metallophyte to grow and proliferate under these harsh environmental conditions, we sampled and characterized the seed microbiota over three successive years which led to the interesting discovery of a high abundance of Methylobacteria present in the samples [16]. But so far, we have yet an incomplete understanding of how these Methylobacteria can contribute to plant growth and health. Therefore, we performed a detailed investigation of the seed microbiome of *C. pumila*, in order to figure out which are the dominant re-occurring seed endophytes (seed core microbiome), which representative cultured *Methylobacterium* spp. can be characterized and which plant growth promoting properties they have. In addition, what is the origin of *Methylobacterium* in the seed, is it a systemic endophyte or a stochastic phenomenon. In addition, the effects of inoculation with *Methylobacterium* sp. strain Cp3 on seed germination were evaluated.

This paper describes for the first time the seed core microbiome of *Crotalaria pumila*, and presents the full characterized potential of its dominant colonizer, *Methylobacterium* sp. Cp3. We found that strain Cp3 is able to migrate from sand to seeds, produces a plethora of plant growth promoting (PGP) compounds and has multiple metal resistant elements in its genome. Moreover, it has the potential to perform aerobic anoxygenic photosynthesis and can consume 1C-compounds from plants. Most importantly, strain Cp3 can improve seed germination and significantly increases root radicle length. Future studies on other plant species growing on the same site, such as *Brickellia veronicifolia*, *Dalea bicolor*, *Dichondra argentea*, and *Pteridium* sp., and seed microbiome interaction studies can further develop our knowledge on the importance of seed microbiota of metalliferous plants, and may lead to the development of seeds characterized by improved germination and seedling/plant development on trace metal contaminated sites.

2. Results

2.1. Seed Microbiome of C. pumila

The seed microbiome of *C. pumila* was identified in seed pods that are collected over three consecutive years. *Methylobacterium* is the most abundant genus of the seed microbiome, constituting 48.90% in 2011, 37.62% in 2012, and 29.91% in 2013 (Figure 1A). In 2012, also Enterobacteriaceae accounted for a large fraction (Figure 1B). In addition to Methylobacteria and Enterobacteriaceae, other dominant taxa identified in the seeds were Firmicutes (Staphylococcus), and Actinobacteria with Corynebacterium. Interestingly, within the Methylobacteria group, a single OTU_4434806 was the most dominant member across all of the years (Figure 1B). A sample per sample comparison showed that this specific OTU occurred in 11 of 12 different seed pods collected. Analyses of the seed core microbiome, defined as the bacterial taxa occurring in at least 50% of the samples over three consecutive years, confirmed that *Methylobacterium* OTU_4434806 was the most abundant OTU (57%) of the seed bacterial community (Figure 1C). Because of the high prevalence of this OTU in the seeds and the potential importance for seed germination and plant growth, we queried our culturable collection for representatives matching at least 99% of the partial 16S rDNA sequence with OTU_4434806. This led to the identification of isolate *Methylobacterium* sp. Cp3, the candidate for studying the plant colonisation mechanism of Methylobacteria.

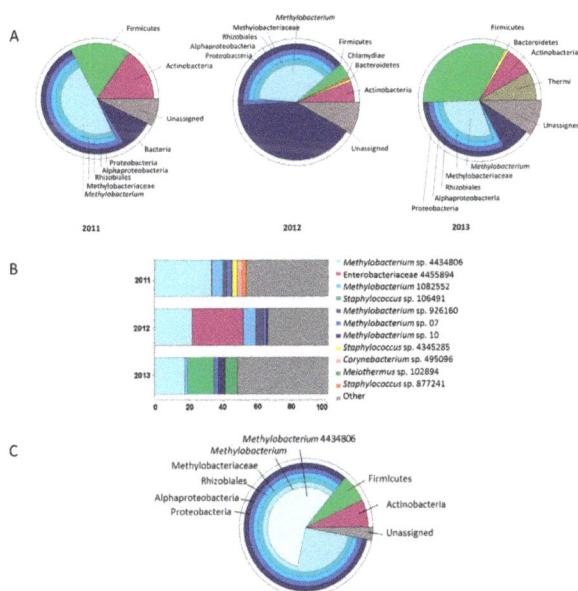

Figure 1. Composition of the core microbiome of *Crotalaria pumila* seeds throughout three consecutive generations. Average of core microbiome in each generation (**A**); and, identification of the most abundant taxon in each generation (**B**) and in the core microbiome over the three generations (**C**).

2.2. Endophytic Colonisation of Methylobacterium sp. Cp3 from Soil Substrate to Seed

To gain insights in the mode of plant colonisation, strain Cp3 was inoculated via the nutrient solution added to the soil on which *Arabidopsis thaliana* plants were growing. We chose *A. thaliana* for these experiments because of the long generation time of *Crotalaria* (six months), when compared to *A. thaliana* (10 weeks), and moreover, *Methylobacterium* has been shown to be part of the endophytic community *A. thaliana* seeds as well [13]. Control plants were watered with only the nutrient solution. Ten weeks after inoculation, strain Cp3 was found in the seeds by *Methylobacterium* specific Automated Ribosomal Intergenic Spacer Analysis (ARISA)-fingerprint, while the strain was not present in non-inoculated plants (Figure 2A, fragment of 620 bp). Strain Cp3 was also highly abundant in the inoculated soil, while native Methylobacteria species were present in the non-inoculated non-sterile soil, as shown by DNA fingerprints of different sizes (Figure 2A). *Methylobacterium* Cp3 was not detected in the shoot of mature inoculated plants, indicating that at the time of seed ripening and drying of the shoot, the seed is a more conducive and protective habitat. We performed also ARISA using general bacteria primers (Figure 2B). Although this gives a more complicated and rich fingerprint profile, the 750-bp PCR-fragment corresponding to *Methylobacterium* Cp3 could be distinguished in the seeds of inoculated plants, in the soil and also in the shoot, providing an additional confirmation of the systemic spread and the presence of the inoculated strain throughout the plant. Because the general bacteria primers target a broad range of bacteria, amplifying many more fragments not necessarily corresponding to Methylobacteria, we consider the fingerprint profiles with Methylobacteria-specific primers as more confirmative for Cp3 colonisation, while the general bacteria primers provide an indication of the total bacterial community diversity in the seeds, shoot, and sand substrate.

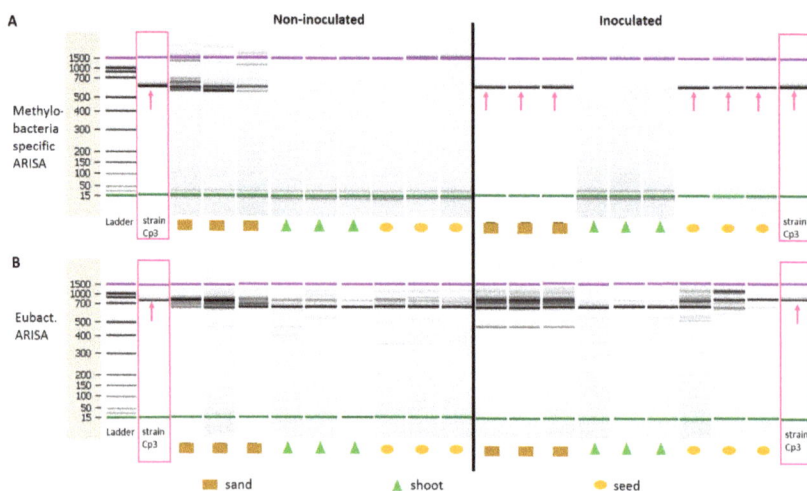

Figure 2. Methylobacteria specific ARISA fingerprints and general bacteria Automated Ribosomal Intergenic Spacer Analysis (ARISA) fingerprints of the control non-inoculated *A. thaliana* plants and plants inoculated with *Methylobacterium* sp. Cp3. Pink boxes indicate the fingerprint of the inoculated strain Cp3 and the arrows point to the specific amplicon for Cp3.

In addition to ARISA fingerprinting, the presence of strain Cp3 in seeds was assessed by counting the abundance of Methylobacteria colony forming units (CFU) on methanol impregnated medium and BOX fingerprint analyses. Inoculated plants were significantly more colonised with $8.6 \times 10^5 \pm 1.5$ pink-colored CFU g^{-1} seed, against $2.1 \times 10^4 \pm 3.5$ CFU g^{-1} for seeds of non-inoculated plants. BOX fingerprint profiles of randomly picked colonies confirmed that the inoculated *Methylobacterium* sp.

Cp3 indeed was present in surface-sterilised macerated seeds, and that it was alive and actively growing (Figure S1).

2.3. Establishment of Methylobacterium sp. Cp3 in Plant Tissue in Presence of Trace Metals

Confocal laser scanning microscopy was used to study where and how strain Cp3-mCherry is colonising *C. pumila*. As can be observed from Figure 3, radicles of inoculated seeds were intensively colonised by Cp3 10 days after inoculation when exposed to the trace metals Cd and Zn. More specifically, a biofilm of bacterial cells can be observed on the root surface and root hairs (Figure 3). Interestingly, tagged Cp3 also was detected as endobacterium inside root cortex cells in extreme dense colonisation (Figure 4A–C). In addition, xylem vessels were colonised with lower numbers of bacteria, mainly solitary cells, as demonstrated using three-dimensional (3D)-volume rendering (Figure 5). During these microscopic analyses, higher numbers of tagged cells were observed in roots in comparison to stems. This was confirmed after isolating tagged cells from surface sterilized plant tissues (Figure S2). The number of CFU of mCherry-*Methylobacterium* sp. Cp3 cells in roots was $3.2 \times 10^6 \pm 0.1$, in comparison to $0.17 \times 10^6 \pm 0.01$ in stems of *C. pumila*.

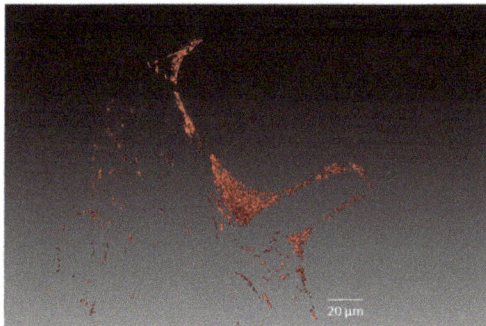

Figure 3. Confocal microscopy picture in the red-channel with mCherry-tagged *Methylobacterium* sp. Cp3 lining the root hair surface of *Crotalaria pumila* in medium supplemented with Zn and Cd.

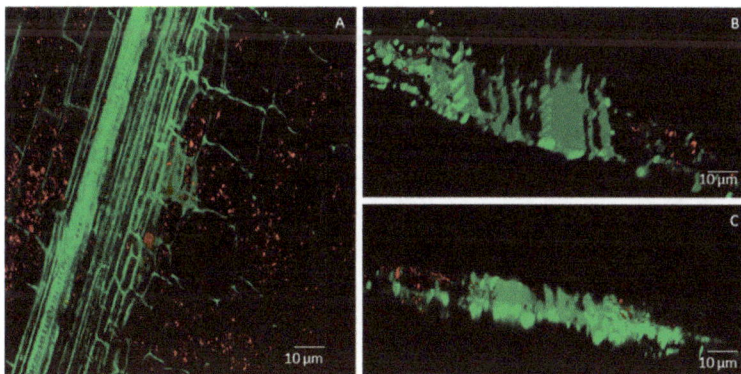

Figure 4. Confocal images of combined m-Cherry fluorescence (red) and plant autofluorescence (green) showing root colonisation by *Methylobacterium* strain Cp3. Maximum intensity projection (**A**) and volume rendering (**B,C**) where Cp3-mCherry is localized intracellularly in root cortex of *Crotalaria pumila*. Confocal stack thickness is 58 μm and was acquired with the Ultra VIEW VoX (PerkinElmer, Zaventem, Belgium) using the CFI Plan Apochromat VC objective 20.0 × 0.75. Z-step was 1 μm. Three-dimensional models were created with the software Amira 6.0.1 (FEI software, Hillsboro, OR, USA).

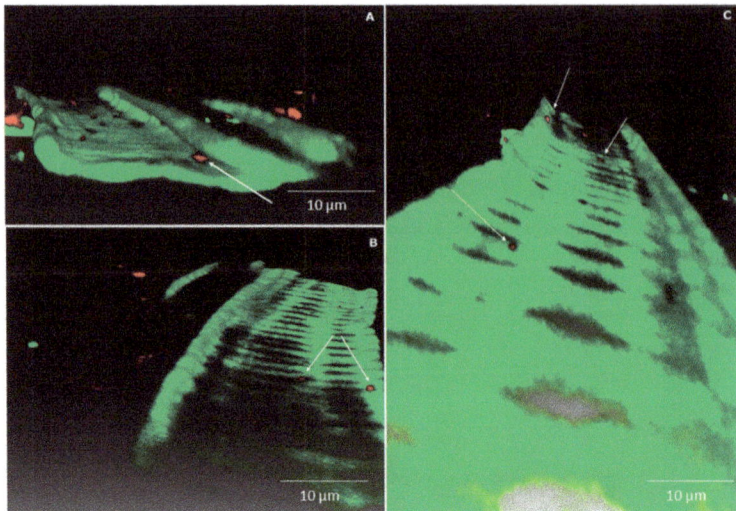

Figure 5. Confocal images with combined mCherry fluorescence (red) and plant autofluorescence (green). Volume rendering (**A–C**) of *Methylobacterium* sp. Cp3-mCherry colonising the xylem vessels in the stem of *Crotalaria pumila* growing in medium supplemented with Zn and Cd. White arrows indicate strain Cp3. Confocal stack has a thickness of 54 µm, acquired with a Ultra VIEW VoX (PerkinElmer) using the CFI Plan Fluor objective 40.0 × 0.75. Z-step was 1 µm. Three-dimensional models were created with the software Amira 6.0.1 (FEI software, Hillsboro, OR, USA).

2.4. Strain Cp3 Inoculation Improves Seed Germination and Plantlet Survival under Cadmium Stress

To assess whether strain Cp3 influences the germination of *C. pumila* seeds, surface-sterilised seeds were inoculated with 10^6 CFU mL^{-1}. This treatment resulted in a significantly higher germination rate in the inoculated seeds (90 ± 2.6%) when compared to the control (75 ± 6.3%) (*t*-test, $p < 0.05$, $n = 50$). Moreover, at five days after the start of germination, radicles of inoculated seedlings were significantly longer (1.8 ± 0.3 cm) than radicles of non-inoculated seedlings (1.2 ± 0.12 cm) (*t*-test, $p < 0.05$, Figure 6). To assess whether colonisation by Cp3 also protects the plantlets against Cd and Zn stress, inoculated and non-inoculated seeds were sown on a trace metal contaminated soil. After 10 days, the percentages of survival were determined: strain Cp3 inoculated seedlings had a significantly higher survival rate (95 ± 4.8%) than those that were not inoculated (68 ± 2.3%). At the end of the experimental period (60 days after transplantation), the inoculated plants produced higher amounts of both fresh and dry biomass (863 ± 66.3 and 94 ± 9.7 mg per plant, respectively) than plants without inoculation (503 ± 42.6 mg of fresh and 61 ± 8.9 mg of dry biomass), demonstrating that strain Cp3 indeed has protective effects for plants under trace metal exposure (*t*-test, $p < 0.05$).

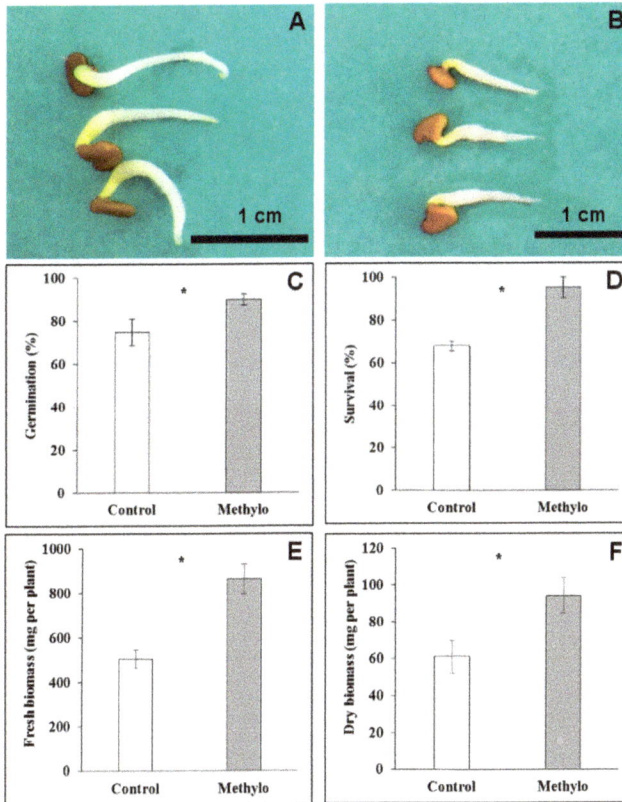

Figure 6. Control seedlings of *Crotalaria pumila* (**A**) and seedlings inoculated with *Methylobacterium* sp. Cp3 (**B**) at five days after start of germination. Percentage of germination (**C**); survival rate (**D**); and, fresh (**E**) and dry biomass (**F**) of inoculated and non-inoculated plants. * significant difference between treatments according to *t*-test ($p < 0.05$), mean ± standard deviation.

2.5. Genes Related to Cadmium Tolerance, Plant Colonisation and Plant-Growth Promotion

In the 5.72 Mb draft genome of *Methylobacterium* sp. Cp3, several genes that were related to plant-growth promotion were found including IAA production, acetoin production, and ACC-deaminase (ACCD). The presence of two homologues of indole acetamide hydrolase suggest that the biosynthesis of auxin occurs via the indole-3-acetamide pathway. A homologue of the butanediol-dehydrogenase enzyme (*adh*) was present, and in vitro tests confirmed the production of (R,R)-2,3-butanediol via (R)-acetoin, both important volatile plant hormones. One gene coding for ACCD was detected, which could improve plant growth under stress conditions. In vitro assays for these PGP traits proved that the genes are functional and can be expressed. Furthermore, the strain Cp3 genome contains three copies of the leucyl aminopeptidase (*pepA*) involved in seed germination protein turnover. Several genes with a role in carbohydrate metabolism were found (272 CDS, Figure 7), including genes involved in xylan degradation, cellulose degradation, D-mannose, galactose, glucose, glycogen, L-arabinose, trehalose, and xylose degradation; these sugars are important components in plant seeds. Also Biolog GN2 plates inoculated with strain Cp3 confirmed the utilisation of D-mannose, D-galactose, D-glucose, glycogen, L-arabinose, and D-trehalose in addition to the use of methyl pyruvate, and carboxylic acids like acetic acid, formic acid, β- and γ-hydroxybutyric acid, α-ketogluaric acid, D,L-lactic acid, malonic acid, propionic acid and succinic acid, the amide succinamic acid, and the amino acids, and L-asparagine and L-aspartic acid. Amino acid transport and metabolism account for a total of 447 CDS of the annotated ones in the genome (Figure 7). Interestingly, also several genes coding for enzymes that are involved in superoxide radical degradation are present including five catalases and two superoxide dismutases *sodB* and *sodC*. Strain Cp3 also possesses 24 genes involved in metal tolerance, such as copper resistance protein CopZ, and several metal transporters, including metal ABC permeases, heavy metal RND transporters in addition to resistance and binding proteins, arsenate reductase, and arsenite oxidase operon. Plating on minimal medium supplemented with high concentrations of trace metals showed that Cp3 is tolerant to 2 mM Cd, 5 mM Zn, 0.5 mM Pb, and 0.4 mM Cu. Furthermore, genes coding for the methanol dehydrogenase and methanol oxidation system are present, which allows the strain to use methanol, an 1-C compound, as sole source of energy. Genes that encode for proteins involved in photosynthesis were located, including bacterial light-harvesting complex (one gene), photosynthetic reaction center (two genes), as well as biosynthesis of chorophyll (four genes), bacteriochlorophyll (10 genes), and carotenoids (10 genes). To confirm the presence of the photosystems we recorded the absorbance and emission fluorescence spectra of strain Cp3 grown under light and dark regime. We observed a strong absorbance at 360 nm, and using this wavelength as excitation, we recorded several emission peaks, one at 450, at 520, and 650, and a smaller one in the near infrared 820 nm (Figure S3). These wavelengths can correspond to bacteriochlorophyll and the ability to perform aerobic anoxygenic photosynthesis.

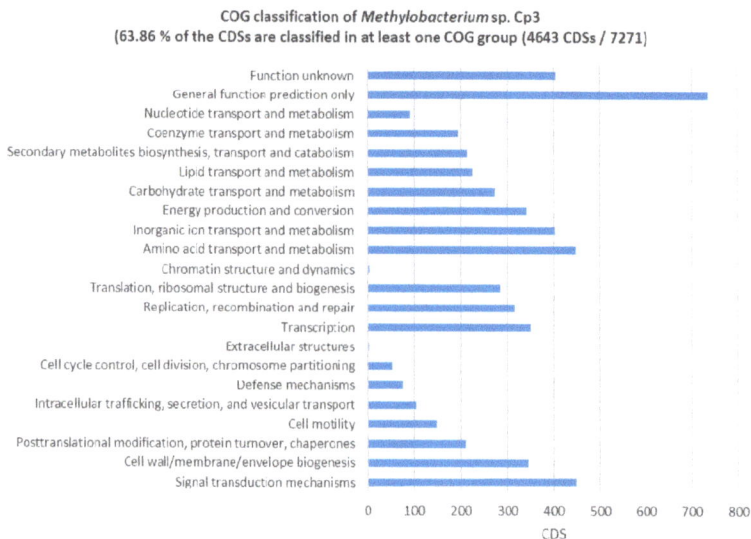

Figure 7. COG classification of the CDS of *Methylobacterium* sp. Cp3.

Methylobacterium Cp3 is a dominant member of the *C. pumila* seed endophyte community, so its genes and genomic content potentially contribute significantly to the total seed microbiome function. Hence, we used Phylogenetic Investigation of Communities by Reconstruction of Unobserved States (PICRUSt) to explore the metagenome functional content of *C. pumila* seeds. Metabolic functions of bacterial 16S rRNA genes in *Methylobacterium* were the most abundant and predicted for 47.0% of the genome (Figure 8). Within these functions, at level 2, the most dominant processes were related to carbohydrate metabolism (20.6%), amino acids (20.2%), and energy metabolism (11.5%), with the latter possessing genes coding for enzymes involved in carbon fixation, methanol and nitrogen metabolism.

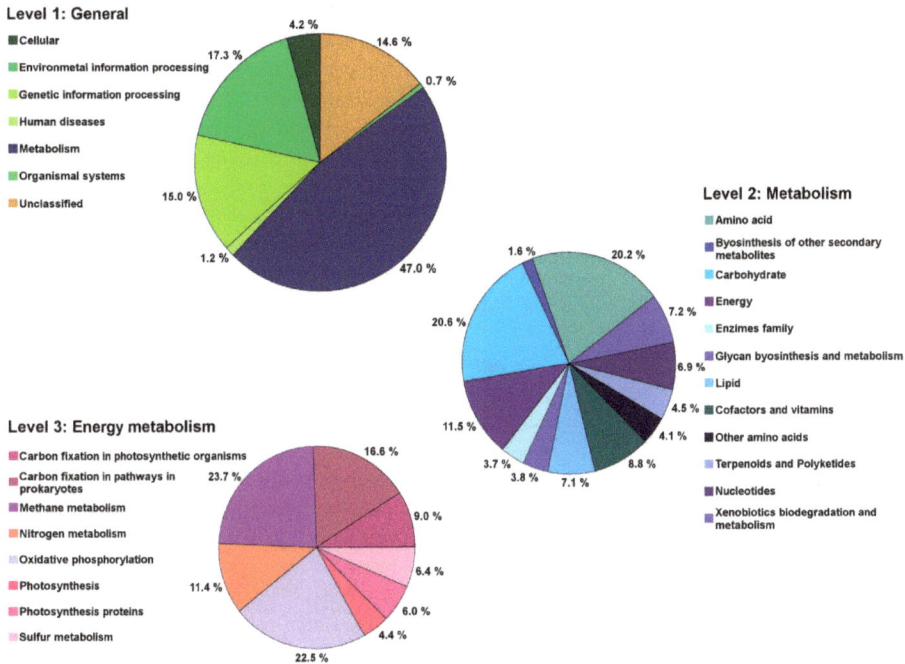

Figure 8. Predicted metabolic functions of *Methylobacterium*, the dominant member of the seed endophyte core microbiome of *Crotalaria pumila* across three generations.

3. Discussion

In this study, we examined the seed endophyte microbiome of *C. pumila* and investigated the manner of plant colonization of a representative *Methylobacterium* sp. (strain Cp3), in addition to the study of its ability to influence seed germination and seedling growth. The systemic approach followed here, from characterization of the seed microbiome in the field using 454 pyrosequencing, to in plantae colonization tests under non-sterile conditions, and in gnotobiotic conditions using confocal microscopy, seed-inoculation experiments, and genetic and phenotypytic characterization of strain Cp3, allowed for an integrated and holistic picture of the seed endophyte microbiome of *C. pumila* and the origin, colonisation, and behavior of one of its predominant seed endophytes.

The results showed the steps of the colonisation by strain Cp3: entrance from the sand substrate into plant root tissues, migration to the aboveground plant tissues through the xylem vessels, and finally establishment in the seeds. This process of plant colonisation by endophytes has been suggested previously by Compant et al. [17] and Truyens et al. [18]. However, in this work, the mentioned process was demonstrated taking into account conditions of metal exposure. Although other seed endophytes of *Crotalaria* could be studied more in detail, *Methylobacterium* attracted our attention because it was a highly abundant taxon in the seeds of *C. pumila* growing on a mining site, and was consistently present over three seed generations. Hence, we hypothesize that this microorganism plays an important role for the plant, for example in protecting young seedlings from metal stress. In other studies, *Methylobacterium* has been reported as a coloniser of plant leaf surfaces, but also as an endophyte of diverse plant species growing on metal containing substrates, which might confirm our hypothesis [19–21].

The core microbiome of a plant is considered as being a group of microorganisms shared among plants of a population under study; changes occurring over time should be taken into account to define the core microbiome [22]. In this work, we defined the seed core microbiome of *C. pumila*, taking into account both aspects, a population of plants colonising metal ore mining residues and changes across three consecutive years (Figure 1). *Methylobacterium* was found as the taxon dominating the seed core microbiome of this pioneer plant species colonising metal-contaminated mine residues (Figure 1). The structure of the core microbiome suggests that they are not random guests in the plant habitat; they seem to play essential roles, interacting with the plant host, and influencing plant physiology, as suggested by Gaiero et al. [23]. Especially the seed endophytic microbiome can be considered to be a key player during the acclimatisation to local conditions [22]. Recently, using 16S rRNA gene amplicon sequencing of seed samples, Truyens et al. [14] identified a small subset of the *A. thaliana* seed microbiome that is conserved across generations. Functional traits of the microbiome (IAA production and ACCD activity) were found to be more important than genotypes for subsequent bacterial seed community composition [14].

According to the results obtained in this work, the core microbiome of *C. pumila* seeds holds genes that are related to nitrogen fixation, photosynthesis, and methanol metabolism (Figure 8). We observed genes for nitrogen fixation in our strain, previously others have described that Methylobacteria contain nitrogen-fixation related genes which might contribute to ammonium provision to its host plant [24]; therefore, strain Cp3 can supply nitrogen to its host plant *C. pumila* when growing on metal-contaminated mine residues. The presence of methanol metabolism related genes, especially in the case of *Methylobacterium* (Figure 8), represents a potential advantage for the core microbiome. During plant colonisation, methylotrophic endophytes can take advantage of the methanol that is released by plants and use it as an additional energy source [19]. Therefore, colonisation of the host plant is more efficient in comparison to other plant-associated bacteria [25]. In turn, the core microbiome provides beneficial plant interactions; Abanda-Nkpwatt et al. [26] demonstrated that *M. extorquens* strains can sustain themselves using the methanol released by the host plant and simultaneously improve the growth of the seedlings. The prediction of functions indicated that Cp3 can perform aerobic anoxygenic photosynthesis, and this is in line with earlier reports that mentioned that *Methylobacterium* contains core genes that are related to photosynthesis, including those encoding for the light-harvesting complex [27], and those involved in the synthesis of bacteriochlorophyll and carotenoids [28]. Some authors mentioned that the carotenoid pink pigment, which is characteristic of Methylobacteria, is also associated with resistance to reactive oxygen species and UV light [29–31]. As a dominant member of the core microbiome (Figure 1) and taking into account the functional potential of *Methylobacterium* (Figure 8), it is expected that this methylotrophic seed endophyte provides its host plant multiple plant growth promotion benefits during germination and growth on metal mine residues. These findings, methylotrophic metabolism, involvement in phytohormone synthesis, metal-tolerance, and photosynthesis, may explain the relatively high abundance and transmission of *Methylobacterium* in *C. pumila* seeds and the success of both, the endophyte and its host, in harsh conditions.

Results obtained in this work demonstrated that *Methylobacterium* sp. Cp3 can be transmitted from a soil substrate to the seeds (Figure 2). The soil, besides providing nutrients, water, and imposing potential stressors to plant growth (in this case metals), is also an important reservoir for endophytic bacteria, including seed endophytes, which are recruited from this pool [32–35]. Previously, it has been reported that Methylobacteria can colonise roots of different plant species [36,37], and also stems and leaves [38,39]. As also mentioned by Araújo et al. [40], the initial step in the colonisation process seems to be the formation of biofilms on roots and on root hairs. Subsequently, the entrance into root cells occurs, as it has been demonstrated, that proliferating root hairs and side roots are important entry points for endophytic colonisation [41–43]. In this study, we present evidence that *Methylobacterium* sp. Cp3 not only colonises roots (Figure 3), and thereby protects *C. pumila* locally from stress when growing in a metal-contaminated substrate, but that the studied strain spreads systemically through the plant through the xylem (Figures 4 and 5) and also ends up in the seeds in order to protect future generations of *C. pumila* plants from metal toxicity.

As a mutualistic symbiont, *Methylobacterium* sp. Cp3 used in this work, was shown to be tolerant to metals (5 mM Zn, 2 mM Cd, 0.4 mM Cu), able to solubilize phosphate and to produce plant hormones (IAA) (which can promote plant growth), and to produce ACCD (which can decrease stress of their host) [44]. However, it remains to be investigated whether *Methylobacterium* sp. Cp3 possesses anti-fungal properties or can act as a bio-control agent in *C. pumila* seeds, besides protection against metal-stress and promotion of plant growth. Altogether, the functional analyses suggest that Methylobacteria possess traits that assist them to survive in the metal contaminated mine residues in the semi-arid area of Zimapan (Mexico), and have multiple traits to help plant growth and development. Therefore, from a plant perspective, the most tolerant bacterial phenotypes are primarily recruited from the original environment from which microbes can be selected.

It has to be noted that despite the participation of seed endophyte Cp3 in the establishment of its host plant, the mechanisms of metal tolerance of the plant itself still remain unclear. Since *C. pumila* is reported to possess high antioxidant activity [45,46], it is logical to think that such characteristics might result helpful in a metal contaminated environment as well. However, this mechanism is studied only from a medicinal point of view, the antioxidant activity of *C. pumila* in metal tolerance can be investigated in the future.

To the best of our knowledge, the present study is the first in which a seed endophyte was observed colonising the stem xylem vessels in the presence of metal stress as environmental variable (Figures 5 and 6), suggesting that the exposure of the host plant to metals promotes the migration of specific endophytes to the aboveground plant parts. In several other studies, bacterial strains isolated from the rhizosphere [47,48], and some endophytes from stems [49–51] and seeds [52–54] were shown to colonise the internal root and shoot tissues (including cortex and xylem) under control conditions. Using gfp-tools and under metal exposure conditions, Zhang et al. [55] reported that labelled root endophytes of *Sedum alfredii* (*Burkholderia* and *Variovorax*) were observed inside the root cortex, but no colonisation of plant vascular system was reported. In addition, according to the criteria of a real endophyte, as defined by Schulz and Boyle [56], we conclude that *Methylobacterium* sp. Cp3 is a real endophyte. It was originally isolated from surface sterilised seeds of *C. pumila*, inoculated to plantlets of its host plant species, it was observed colonizing inner plant tissues (Figures 4–6) and could be re-isolated (Figure S2).

Both plant species used during this study, *C. pumila* [16,44] and *A. thaliana* [13,14], have been shown to contain endogenous endophytic Methylobacteria, which were able to survive strong seed sterilisation. We found that three times inoculation of *Methylobacterium* sp. Cp3 at the roots during the flowering phase, is effective to enrich seeds with desired *Methylobacterium* strains, and importantly without causing any harm to the host plants. Our results imply that the soil is an adequate route to augment bacteria in seeds, which might have important consequences for geographic influences on seed stock production, crop growth, and exploiting this property for generating bacteria-fortified seeds. Because soils are much richer in bacteria, the importance of the stage of inoculation will have to be investigated more in depth in forthcoming studies. Migration kinetics during seed development should be investigated too. Moreover, the transfer of endophytes from seeds to soil was recently demonstrated in the case of maize [57]; thus, inoculation of a substrate by endophytes from germinating seeds can occur, and endophytes may be recruited by neighbouring plants. This process is highly important when studying colonization of neglected or contaminated soils by pioneer plant species.

To date, endophytes via flower pathway and exogenous seed coating [58,59], are in use to produce better seeds [60,61] and to reduce the incidence of pathogens and pests [62,63]. In our study, the seeds of *C. pumila* naturally enriched with *Methylobacterium* can be an important resource for endophyte-enhanced phytoremediation and the reclamation of metal contaminated soils, as well as inoculant for agricultural crops growing on metal contaminated soils. The developments in new generation sequencing technologies and declining prices have already enabled the study of transgenerational seed-endophyte association in maize [52], wheat [64], and *A. thaliana* [14]. A better knowledge about the overall natural plant associates (diversity, function, and interactions) is thus a good starting point to explore features of other seed endophytes, and intrinsic interactions of plants and their associated microbiomes.

4. Materials and Methods

4.1. Crotalaria pumila Seed Core Microbiome Analyses

The generation of the 16S rRNA gene amplicon pyrosequencing data used for the core seed microbiome analyses was described in our previous study [16]. The data are available under accession number SRP080874 (GenBank, NCBI, https://www.ncbi.nlm.nih.gov/sra/SRX1998247[accn]). Briefly, closed pods of *C. pumila* were collected from field conditions from at least 40 different individual plants, in laboratory pods were open and unripen and damaged seeds were eliminated. Then, seeds were surface sterilized by washing with phosphorus free detergent and tap water, immersion in NaClO 0.1% solution supplemented with 0.1% Tween 80 for 10 s, and finally rinsed in sterile deionized water (8 × 100 mL). Sterilization was confirmed by plating 100 µL of the last rinsing water on solid medium [16] and by running a PCR on the last rinsing water [14]. Genomic DNA used for the pyrosequencing was extracted from 150 mg of surface sterilized *C. pumila* seeds, and then subjected to PCR reactions using primers 799F and 1391R, according to the conditions reported previously [16].

The *C. pumila* seed core microbiome was determined using Quantitative Insights Into Microbial Ecology (QIIME) and was defined as the phylotypes consistently present in at least 50% of the samples (n = 12 samples, four per year) across three consecutive seed generations. Gene prediction with Phylogenetic Investigation of Communities by Reconstruction of Unobserved States (PICRUSt) was applied to predict the functional content from the 16S rRNA dataset of the core microbiome [65]. Function predictions were categorized on the Kyoto Encyclopedia of Genes and Genomes (KEGG) classification.

4.2. Methylobacterium Cp3 Isolate Sanger Sequence

Methylobacterium Cp3 was previously isolated from *C. pumila* seeds [44]. Blast search was performed using 16S rRNA gene Sanger sequences of Methylobacteria isolates and the 454 pyrosequencing data, to determine the closest culturable representative of Methylobacteria in the seed microbiome using Blast after sequences were quality trimmed and aligned in Geneious v 4.8.5. The 16S rDNA gene sequence of *Methylobacterium* sp. Cp3 was Sanger sequenced before and the sequence was deposited in Genbank with accession number KX056917.

4.3. Plant Colonisation Experiment

A pot experiment was performed using *A. thaliana* (ecotype Columbia-0) and *Methylobacterium* sp. Cp3. Seeds were sown in pots with quartz sand (particle size of 0.4–0.8 mm) in the greenhouse at 22/18 °C day/night temperature, with a photoperiod of 14 h, relative humidity of 60%, and plants were supplied regularly with 1/10 diluted Hoagland nutrient solution [66]. At the moment of appearance of the inflorescence stem, the plants were inoculated for the first time with *Methylobacterium* sp. Cp3. To avoid contact with, and thus inoculation of, the inflorescence 2 mL of an exponentially grown *Methylobacterium* sp. Cp3 culture (10^9 cells mL^{-1} of Hoagland solution) were added carefully with a pipet to the root zone of each plant. Subsequently, once a week for the next two weeks, the plants received a second and third inoculation in the same way as the first one. Aracons with plastic transparent flower sleeves were placed over single plants at the time of inoculation, so that all plant inflorescences were maintained within a sleeve to avoid cross-pollination. When flowering was complete, no more Hoagland solution was given. Seeds were harvested when the soil and plant inflorescence were dry. The seeds were shaken off the plants in a bag, and sieved to separate them from the chaff. Non-inoculated control plants were grown under the same conditions but instead of bacterial inoculum, they were watered with the same amount of 1/10 Hoagland solution. The experiment was performed with 50 plants per condition.

4.4. Seed Sample Preparation and DNA-Extraction

Arabidopsis thaliana seeds were rinsed and surface sterilised prior to DNA-extraction to remove surface contamination. Briefly, 300 mg of seeds (per sample) were washed for 1 min in sterile deionized H_2O followed by 1 min in 1% NaClO, and 5×5 min in sterile deionized H_2O. Subsequently, the surface sterilized seeds were homogenized in 2 mL sterile 10 mM $MgSO_4$ using a mortar and pestle, and frozen at $-80\,°C$ until DNA-extraction. DNA was extracted using the Invisorb Spin Plant Mini Kit (Invitek, Berlin, Germany). In the first step, two stainless steel beads were added to each sample and ground using the Retsch Mixer Mill MM400 (Retsch, Haan, Germany) for 2×1 min at maximum frequency (30/s). Subsequently, lysis buffer was added to the homogenised samples, and the standard protocol according to manufacturer's instructions was followed thereafter. For each condition, five biological replicates were prepared.

4.5. Specific Automated Ribosomal Intergenic Spacer Analyses (ARISA)

To monitor the fate of *Methylobacterium* sp. Cp3 in *A. thaliana* tissues during colonisation, *Methylobacterium*-specific ARISA was used. For the PCR assay, primer 1319fGC20 (5′ GCC CCC CGC CCC CGC CGC CCA CTC GRG TGC ATG AAG GCG G 3′) and 45r (5′ GAC GGG ATC GAA CCG ACG ACC 3′) were used [67]. The primers amplify a variable PCR product from a partial fragment of the 16S rRNA gene and the 16S–23S intergenic spacer region as the 45r reverse primer binds to *tRNAALA* gene, upstream of the 23S. The generated PCR products have different lengths for different species, between 440 and 810 bp [67], which allows for the discrimination of the inoculated *Methylobacterium* sp. Cp3 strain and resolving the overall *Methylobacterium* community composition. In addition to *Methylobacterium*-specific ARISA, the bacterial 16S–23S ITS DNA was also amplified with general bacteria primers S-D-Bact-1522-b-S-20 (5′ TGCGGCTGGATCCCCTCCTT 3′) and L-D-bact-132-a-A-18 (5′ CCGGGTTTCCCCATTCGG 3′) [68]. Briefly, each PCR reaction contained $1\times$ High Fidelity PCR-buffer (Invitrogen, Carlsbad, CA, USA), 2 mM $MgSO_4$, 0.2 mM of each dNTP, 0.2 μM of each primer, and 1 μL of DNA (1–10 ng per μL), in 25 μL total reaction volume. Cycling conditions for the general bacteria primer pair consisted of a hot start at $94\,°C$ for 3 min and 35 subsequent cycles consisting of $94\,°C$ for 1 min, annealing at $55\,°C$ for 30 s, elongation at $72\,°C$ for 1 min, and final elongation step at $72\,°C$ for 5 min. Amplification conditions for the *Methylobacterium* specific primers were similar, except that an annealing temperature of $69.5\,°C$ was used, after six cycles of touchdown from $72\,°C$ with a $0.5\,°C$ temperature decrease for each cycle.

Amplified reaction mixtures were loaded onto DNA-1000-chips that were prepared, according to the manufacturer's recommendations, and size sorting of the PCR amplicons was performed on an Agilent 2100 Bioanalyzer (Agilent Technologies, Santa Clara, CA, USA). Expert Software (Agilent Technologies) was used to digitalize the ARISA fingerprints, resulting in electropherograms in ASCII formats that were processed using the StatFingerprints package [69], in the 2.13.0 version of the R project (The R Foundation for Statistical Computing, Vienna, Austria). Peaks of the same size were grouped together. Correct assignment of the peaks was verified by visual inspection of the chromatograms. The manually corrected procedure allowed deciphering whether peaks of different samples, correctly matched the *Methylobacterium* sp. Cp3 unique fragment.

4.6. Isolation of Methylobacteria and BOX-Fingerprint Analysis

The seeds from the inoculated and non-inoculated control *A. thaliana* plants were surface sterilized during 0.5 min in 0.1% NaClO supplemented with 0.1% Tween 80 and rinsed thoroughly in sterile deionized water. Seed endophytes were isolated by crushing 50 mg seeds in 500 μL 10 mM $MgSO_4$. Dilutions of 0 to 10^{-2} were plated onto 284 medium [70], and incubated in the presence of 1% methanol in a gas-tight containers during 1 week at $25\,°C$. For each pink-pigmented colony appearing, six representative colonies were chosen for BOX-fingerprint analysis. Briefly, DNA was extracted using the DNeasy 96 Blood and Tissue Kit (Qiagen, Venlo, The Netherlands).

A BOX-PCR was used to generate length-variable PCR products using the protocol of Weyens et al. [70]. Fingerprints were visualised under UV illumination on a gel with 1.5% agarose, and gel red nucleic acid stain. Fingerprint band patterns of the isolated colonies and *Methylobacterium* sp. Cp3 were compared through visual inspection.

4.7. Confocal Microscopy Analyses

Strain Cp3 was equipped with the mCherry plasmid [71] using a triparental conjugation with *E. coli* DH5A as donor strain, and *E. coli* PRK 2013 as helper bacterium. Briefly, the three strains were mixed in equal volumes and were pipetted on a 0.4 μm Isopore™ membrane filters (Millipore, Billerica, MA, USA) on an LB medium plate. After overnight incubation, the cells were washed from the filter and plated on 284 medium with tetracycline. Pink colonies growing on the selective plate were picked up, and checked for fluorescence using a Nikon epifluorescence microscope (Nikon Eclipse 80i, Nikon Instruments Inc., Melville, NY, USA).

Crotalaria pumila seeds used for the colonisation assay were collected from metal-contaminated mine residues [15]. In order to increase and homogenize the germination rate, seeds were submerged for 30 min in concentrated H_2SO_4 [72], subsequently rinsed eight times with sterile distilled water and were surface sterilized according to Sánchez-López [44]. *Methylobacterium* mCherry-labeled inoculum was grown in 1/10 diluted 869 liquid medium; subsequently, the culture was washed with $MgSO_4$ 10 mM, centrifuged at 2000 rpm for 10 min and resuspended in sterile distilled water. 1 mL of inoculum (10^7 CFU mL^{-1}) was spread on the sterile Petri dish on which seeds were germinating. After 72 h, seedlings were sufficiently developed, and were placed to grow in a gnotobiotic system in vertical agar plates (VAP) using 50-fold dilution of Gamborg's B5 sterile medium and in the presence of 0.2 mM Cd ($CdSO_4$) and 0.4 mM Zn ($ZnSO_4$) [73]. Two days later, a second inoculation of the roots was performed (1 mL of 10^7 CFU mL^{-1} solution). Plants were kept on VAP for in total 10 days and exposed.

Sections of the root, root hairs, and stem were hand cut; subsequently, the outer layer was carefully removed to avoid smearing the bacteria from outside. Longitudinal, transversal, and leaned sections were made and placed on glass plates with coverslip. Z stacks (1 μm) of samples were collected using a spinning disk confocal laser microscope Ultra VIEW VoX, PerkinElmer (Zaventem, Belgium). An excitation wavelength of 561 nm (red) was used for mCherry, and 405 (Dapi) for plant cell wall structures. Lenses used for image acquisition included a 40× CFI Plan Fluor lens (numerical aperture of 0.75; working distance of 0.72 mm) and a 20× CFI Plan Apochromat VC lens (numerical aperture of 0.75, and working distance of 0.72 mm). Images were taken using a Hamamatsu C9100-50 camera (Hamamatsu Photonics K.K., Hamamatsu, Japan).

In order to verify the endophytic colonisation, mCherry tagged bacterial cells were isolated from surface sterilized root and stem sections. Three 1 cm long segments of both roots and stems, from seedlings grown on VAP with and without metals were surface sterilized, as follows: 30 s immersion in 70% (*v/v*) ethanol solution, then 1 min in 1% active chloride solution supplemented with Tween 80 (1 droplet per 100 mL solution), and rinsed eight times with sterile distilled water. To verify the effectiveness of the sterilization protocol, 100 μL of the last rinsing water was plated on 1/10 diluted 869 medium. Subsequently, tissues were crushed using a sterile mortar and pestle containing 3 mL of 10 mM sterile $MgSO_4$. 100 μL aliquots of the suspension and from the 1/10 and 1/100 dilutions were plated on 1/10 diluted 869 medium supplemented with 0.4 mM Cd and 0.5 mL of tetracycline per liter of medium (10 mg tetracycline mL^{-1} of methanol). The plates were incubated at 28 °C for two days, after which the numbers of CFU were determined.

4.8. Image Processing and Three-Dimensional (3D) Visualisation

The confocal pictures were analyzed using ImageJ software and Amira 3D visualisation software version 6.1.0 (FEI Visualisation Sciences Group, Hillsboro, OR, USA). The plant autofluorescence recorded at 405 nm was pseudocolored in green and the mCherry channel (561 nm) was put in red. 3D renderings were created using a volume-rendering visualisation technique (Voltex), based on the emission-absorption of light in every voxel in the stack. Different color maps were then assigned to each channel to distinguish individual fluorescent signals.

4.9. In Vitro Plant Growth Promotion Traits Detection

The production of ACCD by strain Cp3 was detected, according to a previously described method [74]. The production of IAA was examined when the strain was grown in a minimal salts medium supplemented with 0.5 mg mL^{-1} tryptophan and was detected by the Salkowski's reagent reaction [75]. The isolate was screened on National Botanical Research Institute's phosphate growth solid medium to determine phosphate solubilisation ability [76]. The ability of strain Cp3 to produce siderophores was checked according to the chrome azurol-S assay [77]. Nitrogen-fixing capacity was tested in a semi-solid malate-sucrose medium with bromothymol blue as a pH indicator [78]. The same medium supplemented with 0.12 g L^{-1} NH$_4$Cl was used as a positive control. To detect if the isolate produce acetoin the protocol proposed by Romick and Fleming [79] was used. Trace metal tolerance was evaluated by measuring the minimal inhibitory concentration of different heavy metal ions (ZnSO$_4$, CdSO$_4$, CuSO$_4$, NiCl$_2$, Pb(NO$_3$)$_2$) in 284 liquid selective medium pH 7 [70], incubated at 28 °C for five days.

4.10. Genome Sequencing, Assembly and Annotation

Strain Cp3 was inoculated into 10 mL of LB medium and grown at 30 °C for 3 days. Cells were collected by centrifugation and used for genomic DNA extraction using the Qiagen Blood and tissue kit (Qiagen) prior to digesting and ligating sequencing adaptors/barcodes using an Ion Xpress Plus Fragment Library Kit (Thermo Fisher Scientific, Waltham, MA, USA). Processed DNA was size-selected (480 bp) on a 2% E-Gel SizeSelect agarose gel and purified using Agencourt AMPure XP beads (Beckman Coulter, Brea, CA, USA). The library dilution factor was determined using an Ion Universal Library Quantitation Kit prior to amplification and enrichment with an Ion PGM Hi-Q Template OT2 400 Kit on an Ion OneTouch 2 System. The enriched Ion Sphere Particles were quantified using an Ion Sphere Quality Control Kit. Sequencing was performed on an Ion 316 Chip v2 (Ion PGM System) with an Ion PGM Hi-Q View Sequencing Kit (Thermo Fisher Scientific). Reads were assembled using SPAdes v3.8.2 (uniform coverage mode; k-mers = 21, 33, 55, 77, 99, 127). Genes were predicted using RAST [80], according to the databases GO, COB, and KEGG. Further annotation was performed using MetaCyc [81] and in the MicroScope platform, under locus tag *Methylobacterium extorquens* Cp3 (METCP3). The draft genome sequence was deposited into NCBI with the accession number MNAO00000000.

5. Conclusions

In conclusion, we found that Methylobacteria are important members of the *C. pumila* seed core microbiome and that the plant growth promoting *Methylobacterium* sp. Cp3 can be trans-generationally transmitted, indicating that seeds offer a significant potential for the discovery of vertically transmitted endophytic strains that are important for the growth of next generations of their host plants. *Methylobacterium* sp. Cp3 was able to move from the soil to seeds, when inoculated in the substratum during plant flower development. Interestingly, the studied bacterial strain colonised its host plant via xylem vessels in case of metal exposure through the growth medium and improves seedling growth. Future experiments using a similar approach to identify other systemic endophytes will be valuable tools to increase our knowledge about natural seed endophytes, and whether and how they are vertically transmitted. In addition, further research can focus more on the actively transcribed phenotypic properties of the core microbiome (seed metatranscriptome) and produced compounds (seed metabolome), time-point of entry, and characterising both soil and fungal seed-communities trans-generationally to confirm the links.

Supplementary Materials: Supplementary materials can be found at www.mdpi.com/1422-0067/19/1/291/s1.

Acknowledgments: This research was supported by a BOF-BILA grant from Hasselt University, the UHasselt Methusalem project 08M03VGRJ. We sincerely thank Roland Valcke (UHasselt) for performing the bacteriochlorophyll measurements. We thank Nele Weyens for assistance in sample preparation for the confocal microscopy imaging.

Author Contributions: The experimental performance and elaboration of this manuscript was possible only through the input of different research fields. Jaco Vangronsveld and Carmen González-Chávez thoroughly added their profound knowledge about plant-microbe interactions during phytoremediation of metals. Rogelio Carrillo-González contributed with his expertise in the behavior and fate of metals in the environment. Isabel Pintelon and Jean-Pierre Timmermans provided confocal laser microscope facilities, technical and operational guidance and performed images acquirement. Valeria Imperato was profoundly involved in the experimental part with *A. thaliana* plants. Vincent Stevens helped performing the records of fluorescence and absorbance of the studied strain, and helped in the bioinformatics analysis. Jonathan Van Hamme provided the facilities, instruments and guidance for the genome sequencing of *Methylobacterium*. Ariadna S. Sánchez-López and Sofie Thijs conceived and designed the experiments and performed the evaluation of results. All authors read and approved the final version of the manuscript to be published.

Conflicts of Interest: The authors declare no conflict of interest. The founding sponsors had no role in the design of the study; in the collection, analyses, or interpretation of data; in the writing of the manuscript, and in the decision to publish the results.

References

1. Tóth, G.; Hermann, T.; Da Silva, M.R.; Montanarella, L. Heavy metals in agricultural soils of the European Union with implications for food safety. *Environ. Int.* **2016**, *88*, 299–309. [CrossRef] [PubMed]
2. Zhang, X.; Zhong, T.; Liu, L.; Ouyang, X. Impact of soil heavy metal pollution on food safety in China. *PLoS ONE* **2015**, *10*, e0135182. [CrossRef] [PubMed]
3. Sessitsch, A.; Puschenreiter, M. Endophytes and rhizosphere bacteria of plants growing in heavy metal-containing soils. In *Microbiology of Extreme Soils Volume 13*; Dion, P., Nautiyal, C.S., Eds.; Springer: Berlin/Heidelberg, Germany, 2008.
4. Mench, M.; Lepp, N.; Bert, V.; Schwitzguébel, J.P.; Gawronski, S.W.; Schröder, P.; Vangronsveld, J. Successes and limitations of phytotechnologies at field scale: Outcomes, assessment and outlook from COST Action 859. *J. Soils Sediments* **2010**, *10*, 1039–1070. [CrossRef]
5. Sessitsch, A.; Kuffner, M.; Kidd, P.; Vangronsveld, J.; Wenzel, W.W.; Fallmann, K.; Puschenreiter, M. The role of plant-associated bacteria in the mobilization and phytoextraction of trace elements in contaminated soils. *Soil Biol. Biochem.* **2013**, *60*, 182–194. [CrossRef] [PubMed]
6. Thijs, S.; Langill, T.; Vangronsveld, J. The bacterial and fungal microbiota of hyperaccumulator plants: Small organisms, large influence. *Adv. Bot. Res.* **2017**, *83*, 43–86.
7. Visioli, G.; D'Egidio, S.; Sanangelantoni, A.M. The bacterial rhizobiome of hyperaccumulators: Future perspectives based on omics analysis and advanced microscopy. *Front. Plant. Sci.* **2015**, *5*. [CrossRef] [PubMed]
8. Kidd, P.S.; Álvarez-López, V.; Becerra-Castro, C.; Cabello-Conejo, M.; Prieto-Fernández, Á. Potential role of plant-associated bacteria in plant metal uptake and implications in phytotechnologies. *Adv. Bot. Res.* **2017**, *83*, 87–126.
9. Kidd, P.; Barceló, J.; Bernal, M.P.; Navari-Izzo, F.; Poschenrieder, C.; Shilev, S.; Clemente, R.; Monterroso, C. Trace element behaviour at the root-soil interface: Implications in phytoremediation. *Environ. Exp. Bot.* **2009**, *67*, 243–259. [CrossRef]
10. Weyens, N.; van der Lelie, D.; Taghavi, S.; Vangronsveld, J. Phytoremediation: Plant-endophyte partnerships take the challenge. *Curr. Opin. Biotechnol.* **2009**, *20*, 248–254. [CrossRef] [PubMed]
11. Montalbán, B.; Thijs, S.; Lobo, M.C.; Weyens, N.; Ameloot, M.; Vangronsveld, J.; Pérez-Sanz, A. Cultivar and metal-specific effects of endophytic bacteria in *Helianthus tuberosus* exposed to Cd and Zn. *Int. J. Mol. Sci.* **2017**, *18*, 2026. [CrossRef] [PubMed]
12. Mastretta, C.; Taghavi, S.; van der Lelie, D.; Mengoni, A.; Galardi, F.; Gonnelli, C.; Barac, T.; Boulet, J.; Weyens, N.; Vangronsveld, J. Endophytic bacteria from seeds of *Nicotiana tabacum* can reduce cadmium phytotoxicity. *Int. J. Phytoremediat.* **2010**, *11*, 37–41.
13. Truyens, S.; Weyens, N.; Cuypers, A.; Vangronsveld, J. Changes in the population of seed bacteria of transgenerationally Cd-exposed *Arabidopsis thaliana*. *Plant Biol.* **2013**, *15*, 971–981. [CrossRef] [PubMed]

14. Truyens, S.; Beckers, B.; Thijs, S.; Weyens, N.; Cuypers, A.; Vangronsveld, J. Cadmium-induced and trans-generational changes in the cultivable and total seed endophytic community of *Arabidopsis thaliana*. *Plant Biol.* **2016**, *18*, 376–381. [CrossRef] [PubMed]
15. Sánchez-López, A.S.; González-Chávez, M.D.C.A.; Carrillo-González, R.; Vangronsveld, J.; Díaz-Garduño, M. Wild flora of mine tailings: Perspectives for use in phytoremediation of potentially toxic elements in a semi-arid region in Mexico. *Int. J. Phytoremediat.* **2015**, *17*, 476–484. [CrossRef] [PubMed]
16. Sánchez-López, A.S.; Thijs, S.; Beckers, B.; González-Chávez, M.C.; Weyens, N.; Carrillo-González, R.; Vangronsveld, J. Community structure and diversity of endophytic bacteria in seeds of three consecutive generations of *Crotalaria pumila* growing on metal mine residues. *Plant Soil* **2017**. [CrossRef]
17. Compant, S.; Clément, C.; Sessitsch, A. Plant growth-promoting bacteria in the rhizo- and endosphere of plants: Their role, colonization, mechanisms involved and prospects for utilization. *Soil Biol. Biochem.* **2010**, *42*, 669–678. [CrossRef]
18. Truyens, S.; Weyens, N.; Cuypers, A.; Vangronsveld, J. Bacterial seed endophytes: Genera, vertical transmission and interaction with plants. *Environ. Microbiol. Rep.* **2015**, *7*, 40–50. [CrossRef]
19. Dourado, M.N.; Camargo Neves, A.A.; Santos, D.S.; Araújo, W.L. Biotechnological and agronomic potential of endophytic pink-pigmented methylotrophic methylobacterium spp. *BioMed Res. Int.* **2015**, *2015*. [CrossRef] [PubMed]
20. Idris, R.; Trifonova, R.; Puschenreiter, M.; Wenzel, W.W.; Sessitsch, A. Bacterial communities associated with flowering plants of the Ni hyperaccumulator *Thlaspi goesingense*. *Appl. Environ. Microbiol.* **2004**, *70*, 2667–2677. [CrossRef] [PubMed]
21. Chen, L.; Luo, S.; Chen, J.; Wan, Y.; Li, X.; Liu, C.; Liu, F. A comparative analysis of endophytic bacterial communities associated with hyperaccumulators growing in mine soils. *Environ. Sci. Pollut. Res.* **2014**, *21*, 7538–7547. [CrossRef] [PubMed]
22. Vandenkoornhuyse, P.; Quaiser, A.; Duhamel, M.; Le Van, A.; Dufresne, A. The importance of the microbiome of the plant holobiont. *New Phytol.* **2015**, *206*, 1196–1206. [CrossRef] [PubMed]
23. Gaiero, J.R.; McCall, C.A.; Thompson, K.A.; Day, N.J.; Best, A.S.; Dunfield, K.E. Inside the root microbiome: Bacterial root endophytes and plant growth promotion. *Am. J. Bot.* **2013**, *100*, 1738–1750. [CrossRef] [PubMed]
24. De Voogd, N.J.; Cleary, D.F.R.; Polónia, A.R.M.; Gomes, N.C.M. Bacterial community composition and predicted functional ecology of sponges, sediment and seawater from the thousand islands reef complex, West Java, Indonesia. *FEMS Microbiol. Ecol.* **2015**, *91*, fiv019. [CrossRef] [PubMed]
25. Sy, A.; Timmers, A.C.J.; Knief, C.; Vorholt, J. Methylotrophic metabolism is advantageous for *Methylobacterium extorquens* during colonization of *Medicago truncatula* under competitive conditions. *Appl. Environ. Microbiol.* **2005**, *71*, 7245–7252. [CrossRef] [PubMed]
26. Abanda-Nkpwatt, D.; Müsch, M.; Tschiersch, J.; Boettner, M.; Schwab, W. Molecular interaction between *Methylobacterium extorquens* and seedlings: Growth promotion, methanol consumption, and localization of the methanol emission site. *J. Exp. Bot.* **2006**, *57*, 4025–4032. [CrossRef] [PubMed]
27. Marx, C.J.; Bringel, F.; Chistoserdova, L.; Moulin, L.; Farhan Ul Haque, M.; Fleischman, D.E.; Gruffaz, C.; Jourand, P.; Knief, C.; Lee, M.C.; et al. Complete genome sequences of six strains of the genus *Methylobact*. *J. Bacteriol.* **2012**, *194*, 4746–4748. [CrossRef] [PubMed]
28. Cervantes-Martínez, J.; López-Díaz, S.; Rodríguez-Garay, B. Detection of the effects of *Methylobacterium* in *Agave tequilana* Weber var. azul by laser-induced fluorescence. *Plant Sci.* **2004**, *166*, 889–892. [CrossRef]
29. Sundin, G.W.; Jacobs, J.L. Ultraviolet radiation (UVR) sensitivity analysis and UVR survival strategies of a bacterial community from the phyllosphere of field-grown peanut (*Arachis hypogeae* L.). *Microb. Ecol.* **1999**, *38*, 27–38. [CrossRef] [PubMed]
30. Umeno, D.; Tobias, A.V.; Frances, H.; Umeno, D.; Tobias, A.V.; Arnold, F.H. Diversifying carotenoid biosynthetic pathways by directed evolution diversifying carotenoid biosynthetic pathways by directed evolution. *Microbiol. Mol. Biol. Rev.* **2005**, *69*, 51–78. [CrossRef] [PubMed]
31. Mohammadi, M.; Burbank, L.; Roper, M.C. Biological role of pigment production for the bacterial phytopathogen *Pantoea stewartii* subsp. stewartii. *Appl. Environ. Microbiol.* **2012**, *78*, 6859–6865. [CrossRef] [PubMed]

32. Truyens, S.; Beckers, B.; Thijs, S.; Weyens, N.; Cuypers, A.; Vangronsveld, J. The effects of the growth substrate on cultivable and total endophytic assemblages of *Arabidopsis thaliana*. *Plant Soil* **2016**, *405*, 325–336. [CrossRef]

33. Bulgarelli, D.; Rott, M.; Schlaeppi, K.; Ver Loren van Themaat, E.; Ahmadinejad, N.; Assenza, F.; Rauf, P.; Huettel, B.; Reinhardt, R.; Schmelzer, E.; et al. Revealing structure and assembly cues for *Arabidopsis* root-inhabiting bacterial microbiota. *Nature* **2012**, *488*, 91–95. [CrossRef] [PubMed]

34. Yergeau, E.; Sanschagrin, S.; Maynard, C.; St-Arnaud, M.; Greer, C.W. Microbial expression profiles in the rhizosphere of willows depend on soil contamination. *ISME J.* **2014**, *8*, 344–358. [CrossRef] [PubMed]

35. Croes, S.; Weyens, N.; Janssen, J.; Vercampt, H.; Colpaert, J.V.; Carleer, R.; Vangronsveld, J. Bacterial communities associated with *Brassica napus* L. grown on trace element-contaminated and non-contaminated fields: A genotypic and phenotypic comparison. *Microb. Biotechnol.* **2013**, *6*, 371–384. [CrossRef] [PubMed]

36. Omer, Z.S.; Tombolini, R.; Broberg, A.; Gerhardson, B. Indole-3-acetic acid production by pink-pigmented facultative methylotrophic bacteria. *Plant Growth Regul.* **2004**, *43*, 93–96. [CrossRef]

37. Poonguzhali, S.; Madhaiyan, M.; Yim, W.J.; Kim, K.A.; Sa, T.M. Colonization pattern of plant root and leaf surfaces visualized by use of green-fluorescent-marked strain of *Methylobacterium suomiense* and its persistence in rhizosphere. *Appl. Microbiol. Biotechnol.* **2008**, *78*, 1033–1043. [CrossRef] [PubMed]

38. Andreote, F.D.; Lacava, P.T.; Gai, C.S.; Araújo, W.L.; Maccheroni, W.J.; van Overbeek, L.S.; van Elsas, J.D.; Azevedo, J.L. Model plants for studying the interaction between *Methylobacterium mesophilicum* and *Xylella fastidiosa*. *Can. J. Microbiol.* **2006**, *52*, 419–426. [CrossRef] [PubMed]

39. Gai, C.S.; Lacava, P.T.; Quecine, M.C.; Auriac, M.C.; Lopes, J.R.S.; Araújo, W.L.; Miller, T.A.; Azevedo, J.L. Transmission of *Methylobacterium mesophilicum* by *Bucephalogonia xanthophis* for paratransgenic control strategy of Citrus variegated chlorosis. *J. Microbiol.* **2009**, *47*, 448–454. [CrossRef] [PubMed]

40. Araújo, W.L.; Santos, D.S.; Dini-Andreote, F.; Salgueiro-Londoño, J.K.; Camargo-Neves, A.A.; Andreote, F.D.; Dourado, M.N. Genes related to antioxidant metabolism are involved in *Methylobacterium mesophilicum*-soybean interaction. *Antonie Van Leeuwenhoek* **2015**, *108*, 951–963. [CrossRef] [PubMed]

41. Rouws, L.F.M.; Meneses, C.H.S.G.; Guedes, H.V.; Vidal, M.S.; Baldani, J.I.; Schwab, S. Monitoring the colonization of sugarcane and rice plants by the endophytic diazotrophic bacterium *Gluconacetobacter diazotrophicus* marked with gfp and gusA reporter genes. *Lett. Appl. Microbiol.* **2010**, *51*, 325–330. [CrossRef] [PubMed]

42. Zhang, X.; Li, E.; Xiong, X.; Shen, D.; Feng, Y. Colonization of endophyte *Pantoea agglomerans* YS19 on host rice, with formation of multicellular symplasmata. *World J. Microbiol. Biotechnol.* **2010**, *26*, 1667–1673. [CrossRef]

43. Prieto, P.; Schilirò, E.; Maldonado-González, M.M.; Valderrama, R.; Barroso-Albarracín, J.B.; Mercado-Blanco, J. Root hairs play a key role in the endophytic colonization of olive roots by *Pseudomonas* spp. with biocontrol activity. *Microb. Ecol.* **2011**, *62*, 435–445. [CrossRef] [PubMed]

44. Sánchez-López, A.S. Basis for the Remediation of Sites Polluted by Potentially Toxic Elements in Zimapan, Mexico: Interdisciplinary Approach. Ph.D. Thesis, Colegio de Postgraduados, Texcoco, Mexico, Hasselt University, Hasselt, Belgium, 2015.

45. Khan, M.; Harun, N.; Rehman, A.H.N.A.; Elhussein, S.A. In vitro antioxidant evaluation of extracts of three wild Malaysian plants. *Procedia Eng.* **2013**, *53*, 29–36. [CrossRef]

46. Villa-Ruano, N.; Zurita-Vásquez, G.G.; Pacheco-Hernández, Y.; Betancourt-Jiménez, M.G.; Cruz-Durán, R.; Duque-Bautista, H. Anti-Iipase and antioxidant properties of 30 medicinal plants used in Oaxaca, México. *Biol. Res.* **2013**, *46*, 153–160. [CrossRef] [PubMed]

47. Compant, S.; Reiter, B.; Sessitsch, A.; Clément, C.; Barka, E.A.; Nowak, J. Endophytic colonization of *Vitis vinifera* L. by plant growth-promoting bacterium *Burkholderia* sp. strain PsJN. *Appl. Environ. Microbiol.* **2005**, *71*, 1685–1693. [CrossRef] [PubMed]

48. Compant, S.; Kaplan, H.; Sessitsch, A.; Nowak, J.; Ait Barka, E.; Clément, C. Endophytic colonization of *Vitis vinifera* L. by *Burkholderia phytofirmans* strain PsJN: From the rhizosphere to inflorescence tissues. *FEMS Microbiol. Ecol.* **2008**, *63*, 84–93. [CrossRef] [PubMed]

49. Filho, A.S.F.; Quecine, M.C.; Bogas, A.C.; de Rossetto, P.B.; de Lima, A.O.S.; Lacava, P.T.; Azevedo, J.L.; Araújo, W.L. Endophytic *Methylobacterium extorquens* expresses a heterologous β-1,4-endoglucanase A (EglA) in *Catharanthus roseus* seedlings, a model host plant for *Xylella fastidiosa*. *World J. Microbiol. Biotechnol.* **2012**, *28*, 1475–1481. [CrossRef] [PubMed]

50. De Procópio, R.E.L.; Araújo, W.L.; Andreote, F.D.; Azevedo, J.L. Characterization of a small cryptic plasmid from endophytic *Pantoea agglomerans* and its use in the construction of an expression vector. *Genet. Mol. Biol.* **2011**, *34*, 103–109.

51. Anand, R.; Chanway, C.P. Detection of GFP-labeled *Paenibacillus polymyxa* in autofluorescing pine seedling tissues. *Biol. Fertil. Soils* **2013**, *49*, 111–118. [CrossRef]

52. Johnston-Monje, D.; Raizada, M.N. Conservation and diversity of seed associated endophytes in Zea across boundaries of evolution, ethnography and ecology. *PLoS ONE* **2011**, *6*. [CrossRef] [PubMed]

53. Verma, S.C.; Singh, A.; Chowdhury, S.P.; Tripathi, A.K. Endophytic colonization ability of two deep-water rice endophytes, *Pantoea* sp. and *Ochrobactrum* sp. using green fluorescent protein reporter. *Biotechnol. Lett.* **2004**, *26*, 425–429. [CrossRef] [PubMed]

54. Ferreira, A.; Quecine, M.C.; Lacava, P.T.; Oda, S.; Azevedo, J.L.; Araújo, W.L. Diversity of endophytic bacteria from Eucalyptus species seeds and colonization of seedlings by *Pantoea agglomerans*. *FEMS Microbiol. Lett.* **2008**, *287*, 8–14. [CrossRef] [PubMed]

55. Zhang, X.; Lin, L.; Zhu, Z.; Yang, X.; Wang, Y.; An, Q. Colonization and modulation of host growth and metal uptake by endophytic bacteria of *Sedum alfredii*. *Int. J. Phytoremediat.* **2013**, *15*, 51–64. [CrossRef] [PubMed]

56. Schulz, B.; Boyle, C. The endophytic continuum. *Mycol. Res.* **2005**, *109*, 661–686. [CrossRef] [PubMed]

57. Johnston-Monje, D.; Lundberg, D.S.; Lazarovits, G.; Reis, V.M.; Raizada, M.N. Bacterial populations in juvenile maize rhizospheres originate from both seed and soil. *Plant Soil* **2016**, *405*, 337–355. [CrossRef]

58. Mitter, B.; Sessitsch, A.; Naveed, M. Method for Producing Plant Seed Containing Endophytic Micro-Organisms 2013. European Patent #2,676,536 A1, 26 November 2015.

59. Bashan, Y.; de-Bashan, L.E.; Prabhu, S.R.; Hernandez, J.P. Advances in plant growth-promoting bacterial inoculant technology: Formulations and practical perspectives (1998–2013). *Plant Soil* **2014**, *378*, 1–33. [CrossRef]

60. Lodewyckx, C.; Taghavi, S.; Mergeay, M.; Vangronsveld, J.; Clijsters, H.; van der Lelie, D. The effect of recombinant heavy metal-resistant endophytic bacteria on heavy metal uptake by their host plant. *Int. J. Phytoremediat.* **2001**, *3*, 173–187. [CrossRef]

61. Berg, G.; Zachow, C.; Müller, H.; Philipps, J.; Tilcher, R. Next-generation bio-products sowing the seeds of success for sustainable agriculture. *Agronomy* **2013**, *3*, 648–656. [CrossRef]

62. Keyser, C.A.; Thorup-Kristensen, K.; Meyling, N.V. Metarhizium seed treatment mediates fungal dispersal via roots and induces infections in insects. *Fungal Ecol.* **2014**, *11*, 122–131. [CrossRef]

63. Gonzalez, F.; Tkaczuk, C.; Dinu, M.M.; Fiedler, Ż.; Vidal, S.; Zchori-Fein, E.; Messelink, G.J. New opportunities for the integration of microorganisms into biological pest control systems in greenhouse crops. *J. Pest. Sci.* **2016**, *89*, 295–311. [CrossRef] [PubMed]

64. Huang, Y.; Kuang, Z.; Wang, W.; Cao, L. Exploring potential bacterial and fungal biocontrol agents transmitted from seeds to sprouts of wheat. *Biol. Control* **2016**, *98*, 27–33. [CrossRef]

65. Langille, M.G.I.; Zaneveld, J.; Caporaso, J.G.; McDonald, D.; Knights, D.; Reyes, J.A.; Clemente, J.C.; Burkepile, D.E.; Vega Thurber, R.L.; Knight, R.; et al. Predictive functional profiling of microbial communities using 16S rRNA marker gene sequences. *Nat. Biotechnol.* **2013**, *31*, 814–821. [CrossRef] [PubMed]

66. Smeets, K.; Ruytinx, J.; Van Belleghem, F.; Semane, B.; Lin, D.; Vangronsveld, J.; Cuypers, A. Critical evaluation and statistical validation of a hydroponic culture system for *Arabidopsis thaliana*. *Plant. Physiol. Biochem.* **2008**, *46*, 212–218. [CrossRef] [PubMed]

67. Knief, C.; Frances, L.; Cantet, F.; Vorholt, J.A. Cultivation-independent characterization of *Methylobacterium* populations in the plant phyllosphere by automated ribosomal intergenic spacer analysis. *Appl. Environ. Microbiol.* **2008**, *74*, 2218–2228. [CrossRef] [PubMed]

68. Normand, P.; Ponsonnet, C.; Nesme, X.; Neyra, M.; Simonet, P. ITS analysis of prokaryotes. In *Molecular Microbial Ecology Manual*; Akkermans, D., van Elsas, J.D., de Bruijn, E.I., Eds.; Kluwer Academic Publishers: Amsterdam, The Netherlands, 1996.

69. Michelland, R.J.; Dejean, S.; Combes, S.; Fortun-Lamothe, L.; Cauquil, L. StatFingerprints: A friendly graphical interface program for processing and analysis of microbial fingerprint profiles. *Mol. Ecol. Resour.* **2009**, *9*, 1359–1363. [CrossRef] [PubMed]

70. Weyens, N.; Taghavi, S.; Barac, T.; van der Lelie, D.; Boulet, J.; Artois, T.; Carleer, R.; Vangronsveld, J. Bacteria associated with oak and ash on a TCE-contaminated site: Characterization of isolates with potential to avoid evapotranspiration of TCE. *Environ. Sci. Pollut. Res.* **2009**, *16*, 830–843. [CrossRef] [PubMed]

Int. J. Mol. Sci. **2018**, *19*, 291

71. Lagendijk, E.L.; Validov, S.; Lamers, G.E.M.; De Weert, S.; Bloemberg, G.V. Genetic tools for tagging Gram-negative bacteria with mCherry for visualization in vitro and in natural habitats, biofilm and pathogenicity studies. *FEMS Microbiol. Lett.* **2010**, *305*, 81–90. [CrossRef] [PubMed]

72. Linding-Cisneros, R.; Lara-Cabrera, S. Effect of scarification and growig media on seed germination of *Crotalaria pumila* (Ort.). *Seed Sci. Technol.* **2004**, *32*, 231–234. [CrossRef]

73. Zhang, H.; Forde, B.G. An Arabidopsis MADS box gene that controls nutrient-induced changes in root architecture. *Science* **1998**, *279*, 407–409. [CrossRef] [PubMed]

74. Belimov, A.A.; Hontzeas, N.; Safronova, V.I.; Demchinskaya, S.V.; Piluzza, G.; Bullitta, S.; Glick, B.R. Cadmium-tolerant plant growth-promoting bacteria associated with the roots of Indian mustard (*Brassica juncea* L. Czern.). *Soil Biol. Biochem.* **2005**, *37*, 241–250. [CrossRef]

75. Gordon, S.A.; Weber, R.P. Colorimetric estimation of Inodoleacetic Acid. *Plant Physiol.* **1951**, *26*, 192–195. [CrossRef] [PubMed]

76. Nautiyal, C.S. An efficient microbiological growth medium for screening phosphate solubilizing microorganisms. *FEMS Microbiol. Lett.* **1999**, *170*, 265–270. [CrossRef] [PubMed]

77. Schwyn, B.; Neilands, J. Universal chemical assay for the detection and determination of siderophores. *Anal. Biochem.* **1987**, *160*, 47–56. [CrossRef]

78. Xie, G.H.; Cui, Z.; Yu, J.; Yan, J.; Hai, W.; Steinberger, Y. Identification of nif genes in N2-fixing bacterial strains isolated from rice fields along the Yangtze River Plain. *J. Basic Microbiol.* **2006**, *46*, 56–63. [CrossRef] [PubMed]

79. Romick, T.L.; Fleming, H.P. Acetoin production as an indicator of growth and metabolic inhibition of *Listeria monocytogenes*. *J. Appl. Microbiol.* **1998**, *84*, 18–24. [CrossRef] [PubMed]

80. Aziz, R.K.; Bartels, D.; Best, A.A.; DeJongh, M.; Disz, T.; Edwards, R.A.; Formsma, K.; Gerdes, S.; Glass, E.M.; Kubal, M.; et al. The RAST server: Rapid annotations using subsystems technology. *BMC Genom.* **2008**, *9*, 75. [CrossRef] [PubMed]

81. Caspi, R.; Billington, R.; Ferrer, L.; Foerster, H.; Fulcher, C.A.; Keseler, I.M.; Kothari, A.; Krummenacker, M.; Latendresse, M.; Mueller, L.A.; et al. The MetaCyc database of metabolic pathways and enzymes and the BioCyc collection of pathway/genome databases. *Nucleic Acids Res.* **2016**, *44*, D471–D480. [CrossRef] [PubMed]

International Journal of
Molecular Sciences

MDPI

Article

Belowground Interactions Impact the Soil Bacterial Community, Soil Fertility, and Crop Yield in Maize/Peanut Intercropping Systems

Qisong Li [1,2,3], Jun Chen [2,3], Linkun Wu [2,3], Xiaomian Luo [1,2,3], Na Li [2,3], Yasir Arafat [2,3], Sheng Lin [1,2,3,*] and Wenxiong Lin [1,2,3,*]

[1] College of crop Sciences, Fujian Agriculture and Forestry University, Fuzhou 350002, China; liqisong0591@gmail.com (Q.L.); luoxiaomian01@163.com (X.L.)
[2] Fujian Provincial Key Laboratory of Agroecological Processing and Safety Monitoring, College of Life Sciences, Fujian Agriculture and Forestry University, Fuzhou 350002, China; chenjunfafu@163.com (J.C.); wulinkun619@fafu.edu.cn (L.W.); lina20101020@163.com (N.L.); stanadar2012@gmail.com (Y.A.)
[3] Key Laboratory of Crop Ecology and Molecular Physiology (Fujian Agriculture and Forestry University), Fujian Province University, Fuzhou 35002, China
* Correspondence: linsh@fafu.edu.cn (S.L.); lwx@fafu.edu.cn (W.L.); Tel.: +86-0591-8378-9301 (W.L.)

Received: 12 December 2017; Accepted: 15 February 2018; Published: 22 February 2018

Abstract: Intercropping has been widely used to control disease and improve yield in agriculture. In this study, maize and peanut were used for non-separation intercropping (NS), semi-separation intercropping (SS) using a nylon net, and complete separation intercropping (CS) using a plastic sheet. In field experiments, two-year land equivalent ratios (LERs) showed yield advantages due to belowground interactions when using NS and SS patterns as compared to monoculture. In contrast, intercropping without belowground interactions (CS) showed a yield disadvantage. Meanwhile, in pot experiments, belowground interactions (found in NS and SS) improved levels of soil-available nutrients (nitrogen (N) and phosphorus (P)) and enzymes (urease and acid phosphomonoesterase) as compared to intercropping without belowground interactions (CS). Soil bacterial community assay showed that soil bacterial communities in the NS and SS crops clustered together and were considerably different from the CS crops. The diversity of bacterial communities was significantly improved in soils with NS and SS. The abundance of beneficial bacteria, which have the functions of P-solubilization, pathogen suppression, and N-cycling, was improved in maize and peanut soils due to belowground interactions through intercropping. Among these bacteria, numbers of *Bacillus*, *Brevibacillus brevis*, and *Paenibacillus* were mainly increased in the maize rhizosphere. *Burkholderia*, *Pseudomonas*, and *Rhizobium* were mainly increased in the peanut rhizosphere. In conclusion, using maize and peanut intercropping, belowground interactions increased the numbers of beneficial bacteria in the soil and improved the diversity of the bacterial community, which was conducive to improving soil nutrient (N and P) supply capacity and soil microecosystem stability.

Keywords: soil nutrition; soil bacterial community; microbial diversity; intercropping; T-RFLP; qPCR

1. Introduction

Intercropping is becoming common in the Americas, Asia, Africa, and Europe. It plays an important role in maintaining farmland ecosystem biodiversity and stability, improving resource efficiency, and achieving high and stable yields in the agroecosystem [1,2]. Most intercropping systems (i.e., maize/peanut, wheat/maize, faba bean/wheat, etc.) show yield advantages. However, some intercropping systems show yield disadvantages due to strong interspecific competition [3]. It is very necessary to explore the underlying mechanisms of yield advantages under intercropping systems.

Intercropping involves the aboveground and belowground interaction of crops. In previous studies, the aboveground interaction in intercropping systems altered the canopy micro-ecology, resulting in improved solar light use efficiency and soil water storage capacity, and a reduction in the evaporation of soil moisture [4,5]. In recent years, more studies have been focused on soil nutrients and root exudates [6]. In legume/cereal intercropping systems in alkaline calcareous soils with low phosphorus, legume root exudates including malic acid and citric acid have been shown to acidify the rhizosphere and mobilize insoluble P while improving legume nodulation and nitrogen fixation [7]. Meanwhile, increased rhizosphere phosphorus availability was also observed in durum wheat and chickpea intercropping in neutral soil, but rhizosphere acidification was not observed; in fact, the pH value increased [8]. Moreover, in acidic soil, He et al. [9] suggested that the increase in plant P uptake was due to the changes in the microbial community composition in maize/chickpea and maize/soybean intercropping systems. Therefore, belowground interspecific interactions are complex, and further research is needed.

Soil microorganisms play crucial roles in the rhizosphere ecosystem, being involved in soil nutrient cycling, suppression of soil-borne pathogenic microorganisms, and the decomposition of organic matter, which is closely associated with the aboveground performance of plants [10,11]. In recent years, the root-associated microbial community has been the focus of many plant studies. Mortel et al. [12] found that *Pseudomonas fluorescens* was able to induce the salicylic acid signaling pathway in *Arabidopsis*, and further improved the resistance of plants to pests and diseases. Sanguin et al. [13] found that take-all decline disease in wheat was closely associated with the shift of the bacterial community in long-term monoculture. Weidner et al. [14] suggested that high soil microbial diversity was favorable for positive plant-soil feedback and nitrogen nutrient supply in soil.

Previous studies have demonstrated that plants can modify their rhizosphere microbial community through their root exudates [15,16], and that the shift of rhizosphere microbial can further affect the soil enzymes and soil fertility [17]. Therefore, the rhizosphere microbial community can be affected by different plants in intercropping, and the changes in soil microbial communities may play important roles in the benefits of intercropping. Intercropping of maize and peanut has been previously demonstrated to provide significant advantages, and has been commonly used [18,19]. However, little is known about the alternation of the rhizosphere microbial community at the species and genera level in the intercropping systems, and even less information is available about this in maize/peanut intercropping specifically.

Molecular methods can assay the microbial community structure in phylum, genera, and even species in the rhizosphere soil of plants [20]. The terminal restriction fragment length polymorphism (T-RFLP) coupled with cloning sequences quantitative PCR (qPCR) technique offers the best possibilities for analyzing the structure and diversity of soil microbial communities [21,22]. In this study, maize and peanut were intercropped in three patterns of belowground interactions with the same aboveground canopy structure. Firstly, the changes in agronomic traits, photosynthetic characteristics, soil-available nutrients, and soil enzymes between different treatments were analyzed to explore the key factor (aboveground or belowground interactions) contributing to the yield advantage. Then, the shifts of soil bacterial community structure and functional diversity were assessed using qPCR and T-RFLP. The aim of this study is to explore the relationships between the changes of soil bacterial community and their related soil enzyme activities and yield advantage in maize/peanut intercropping regimes, in order to provide a theoretical basis and practical guide for reasonable intercropping and maintaining biodiversity in agricultural ecosystems.

2. Results

2.1. Yields in Field Experiments and Plant Properties in Pot Experiments

The plant yields of two crops in field were measured in 2011 and 2012 under different treatments (Table 1). Two-year yield results showed that the maize yields using non-separation intercropping (NS) and semi-separation intercropping (SS) were significantly improved when compared with complete separation intercropping (CS) and monoculture treatment (MS). In 2011, the peanut yield was significantly higher when using NS and slightly higher when using SS and MS as compared to CS. In 2012, a similar trend was found in peanut yield between the treatments, and no significant differences were observed. Compared with monoculture, land equivalent ratio (LER) results showed yield advantages due to belowground interactions in the NS and SS patterns in the two-year experiments, while the intercropping pattern without belowground interactions (CS) showed a yield disadvantage.

Table 1. Maize and peanut yields under different intercropping treatments.

Treatments	Maize Yield (Mg·ha^{-1})	Peanut Yield (Mg·ha^{-1})	LER
	2011		
NS	8.62 ± 0.39a	3.42 ± 0.14a	1.31
SS	9.00 ± 0.34a	3.32 ± 0.14ab	1.33
CS	4.69 ± 0.29b	3.14 ± 0.06b	0.95
MS	4.70 ± 0.30b	3.38 ± 0.07ab	1
	2012		
NS	10.22 ± 0.33a	3.69 ± 0.35a	1.11
SS	9.95 ± 0.22a	3.61 ± 0.15a	1.08
CS	8.15 ± 0.30b	3.40 ± 0.17a	0.97
MS	8.06 ± 0.39b	3.61 ± 0.27a	1

NS: non-separation intercropping; SS: semi-separation intercropping; CS: complete separation intercropping; MS: monoculture treatment; LER: land equivalent ratios. Different letters show significant differences determined by the LSD (least significant difference) test ($p < 0.05$, $n = 3$).

In the pot experiment, the NS and SS patterns also showed the same trend when compared with CS (Table 2). The shoot biomass, root biomass, and net photosynthetic rate in maize with NS and SS were significantly higher than in maize with CS. Peanut root biomass, nodule number per plant, and dry weight per nodule were significantly higher in maize with NS and SS.

Table 2. Plant biomass, root nodulation, and net photosynthetic rate of both crops in three different intercropping treatments.

Treatments	Biomass (g)	Shoot Biomass (g)	Root Biomass (g)	Pn (μmol CO$_2$·m^{-2}·s^{-1})	Nodule Number per Plant	Dry Weight per Nodule (mg)
			Maize			
NS	118.40 ± 6.26a	95.13 ± 4.71a	23.27 ± 1.55a	40.38 ± 1.81a	/	/
SS	120.22 ± 5.56a	94.95 ± 4.81a	25.27 ± 0.75a	40.23 ± 3.52a	/	/
CS	94.19 ± 2.98b	76.16 ± 2.58b	18.03 ± 0.40b	34.48 ± 1.08b	/	/
			Peanut			
NS	13.95 ± 0.99a	12.36 ± 0.76a	1.59 ± 0.23a	26.60 ± 1.53a	528.33 ± 25.42a	0.51 ± 0.03ab
SS	14.60 ± 1.07a	12.83 ± 0.70a	1.77 ± 0.37a	27.60 ± 1.63a	626.00 ± 40.00a	0.57 ± 0.06a
CS	14.30 ± 0.72a	13.17 ± 0.66a	1.13 ± 0.06b	24.98 ± 1.67a	310.00 ± 50.20b	0.43 ± 0.04b

NS: non-separation intercropping; SS: semi-separation intercropping; CS: complete separation intercropping; Pn: net photosynthetic rate. Different letters show significant differences determined by the LSD test ($p < 0.05$, $n = 3$).

2.2. Soil Nutrition and Soil Enzyme Activities

Differences in soil nutrient contents and soil enzyme activities were detected between the distinct treatments in the pot experiment (Figure 1). In maize rhizosphere soils, belowground interactions (in NS and SS) significantly enhanced soil-available nutrients (N, P and K), urease, and acid phosphomonoesterase (PME) as compared to CS. In peanut rhizosphere soils, the activities of soil urease, acid PME and invertase, and available P were significantly enhanced in the NS and SS treatments as compared with CS. Available N levels were significantly higher in the NS soil and slightly higher in the SS soil as compared to the CM soil. However, NS and SS reduced the available K contents in peanut soil as compared to the CS soil.

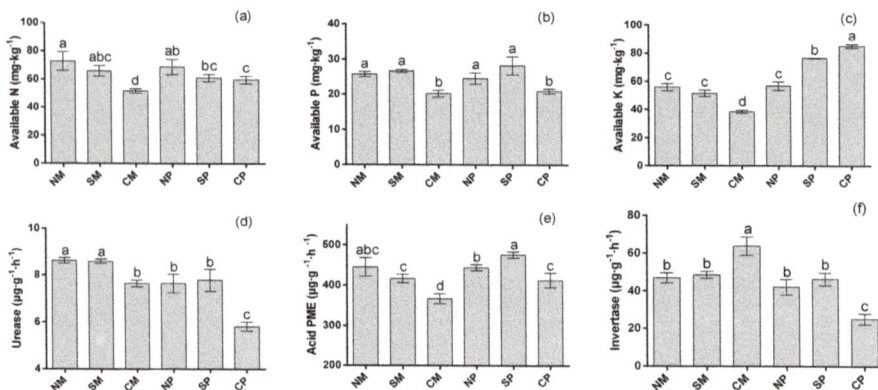

Figure 1. Soil available N (**a**); available P (**b**); available K (**c**); urease activity (**d**); acid PME (phosphomonoesterase) activity (**e**); and invertase activity (**f**) under different intercropping treatments. NM: non-separated maize; SM: semi-separated maize; CM: completely separated maize; NP: non-separated peanut; SP: semi-separated peanut; CP: completely separated peanut. Bars with different letters indicate significant differences (LSD test, $p < 0.05$, $n = 3$).

2.3. Shifts of the Soil Microbial Community

T-RFLP was used to analyze the rhizosphere bacterial communities of maize and peanut under different intercropping treatments. We obtained bacterial T-RFLP profiles using the digestion of four restriction enzymes (*Msp*I, *Hae*III, *Afa*I, and *Alu*I) (Figure S1). The terminal restriction fragments (T-RFs) of four enzymes were combined for matrix calculation. We analyzed the diversity of rhizosphere soil bacterial communities under different intercropping patterns (Table 3). Simpson and Shannon–Wiener indices were improved in plant soils under NS and SS as compared with CS. Meanwhile, the lowest value for CS was found in peanut. The result of non-metric multi-dimensional scaling (NMDS) and cluster analyses (Figure 2) showed that NS and SS crop soil bacterial communities in both plants clustered together and were considerably different from those of CS crops.

Table 3. Diversity analysis of rhizosphere soil bacterial communities under different intercropping patterns.

Treatments	Simpson Index (J)	Shannon-Wiener Index (H)
	Maize	
NS	0.988a	5.921a
SS	0.984b	5.84b
CS	0.982c	5.71c
	Peanut	
NS	0.985a	5.826a
SS	0.986a	5.862a
CS	0.982b	5.596c

NS: non-separation intercropping; SS: semi-separation intercropping; CS: complete separation intercropping. Different letters indicate significant differences (LSD test, $p < 0.05$, $n = 3$).

Based on T-RFs, we grouped the identified bacteria into 11 phyla (Figure 3) using the phylogenetic assignment tool (PAT). Proteobacteria, Firmicutes, and Actinobacteria were the most abundant bacteria in the soil samples. Proteobacteria were the dominant bacteria in maize soil with CS. The presence of Proteobacteria gradually decreased in SS and NS maize soils, while quantities of Firmicutes and Actinobacteria gradually increased in maize soils with SS and NS. Firmicutes was the dominant bacteria in peanut soil with CS. The quantities of Firmicutes gradually decreased, while those of Proteobacteria gradually increased in peanut soils with SS and NS.

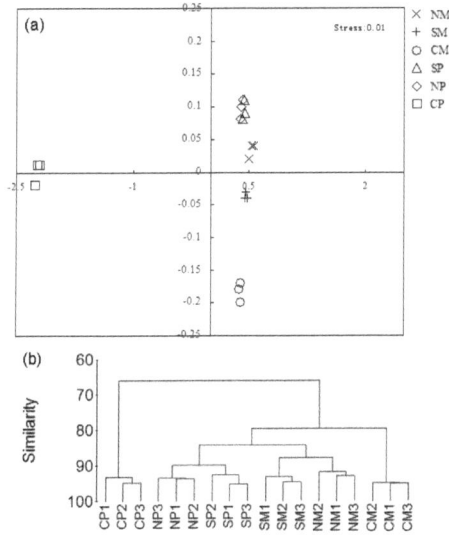

Figure 2. NMDS ordinations (**a**) and clustering analysis (**b**) of bacterial communities in soil. NM: non-separated maize; SM: semi-separated maize; CM: completely separated maize; NP: non-separated peanut; SP: semi-separated peanut; CP: completely separated peanut.

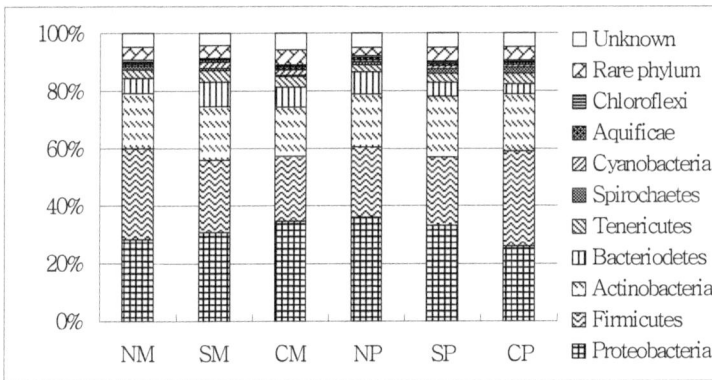

Figure 3. The groups of maize and peanut rhizosphere soil bacterial phyla under different intercropping patterns. NM: non-separated maize; SM: semi-separated maize; CM: completely separated maize; NP: non-separated peanut; SP: semi-separated peanut; CP: completely separated peanut.

Furthermore, the similarity percentages (SIMPERs) of T-RFs was determined. NS and SS were combined as the planting pattern with belowground interactions (BI) and CS was the planting pattern without interactions (WI). Both maize and peanut were analyzed under these two patterns (BI and WI). The results showed that there was a dissimilarity of 19.33% between BI and WI for the maize soil bacterial community and 33.52% for peanut. The top T-RFs with 20% cumulative contribution to the dissimilarity are shown in Table 4 and Table S1. The results showed that under the BI pattern (SS and NS), numbers of beneficial bacteria (i.e., *Bacillus*, *Burkholderia*, *Pseudomonas*, etc.)

were increased in maize and peanut soils. Among them, *Bacillus*, *Brevibacillus brevis*, and *Paenibacillus*, all of which belong to Firmicutes, mainly increased in the maize rhizosphere. *Burkholderia*, *Pseudomonas*, *Sphingomonas*, and *Rhizobium*, all of which belong to Proteobacteria, mainly increased in the peanut rhizosphere.

Table 4. Top terminal restriction fragments (T-RFs) with 20% cumulative contribution to the dissimilarity between belowground interaction (NS and SS) and complete separation (CS).

Contribution (%)	TRFLP-PAT Assignment	Relative Abundance (%)			Functions
		NS	SS	CS	
		Maize			
4.63	*Brevibacillus brevis* (D78457)	3.53	3.63	0	Improving root growth, nodulation and pathogen antagonism
3.33	*Paenibacillus* sp.	2.8	2.35	0	N-fixation and pathogen antagonism
1.87	*Bacillus* sp.	6	5.65	4.38	Improving nodulation, P-solubilization and pathogen antagonism
1.68	Clone OCS155 (AF001652)	1.25	0	3.46	Unknown
1.58	No Match	0.85	0.96	0.18	Unknown
1.28	*Polyangium* sp.	0.76	0.79	1.62	C cycle
1.21	*Bacillus subtilis* (AL009126)	4.99	4.77	3.95	Improving nodulation, P-solubilization and pathogen antagonism
1.19	Clone T33 (Z93960)	1.33	0.32	1.55	Unknown
1.12	*Pseudomonas* sp.	3.64	4.32	3.26	P-solubilizing and pathogen antagonism
1.08	*Acidosphaera* (D86512)	1.4	0.8	1.02	Unknown
1.02	*Sphingomonas* sp.	1.52	1.24	1.89	C cycle
		Peanut			
2.23	*Bacillus* sp.	8.33	8.89	5.81	Improving nodulation, P-solubilization and pathogen antagonism
1.98	*Burkholderia* sp.	5.78	5.62	3.49	Legume N-fixing symbiont. Pathogen antagonism and plant growth promotion
1.81	*Clostridium* sp.	1.53	1.32	3.98	C cycle
1.6	*Pseudomonas* sp.	1.25	1.44	0.62	P-solubilization and pathogen antagonism
1.59	str. AS2988.(AF060671)	1.4	0.94	0	Unknown
1.5	*Nocardia crassostrae* (U92800)	0.99	1.13	3.44	Unknown
1.31	clone Sva0556.	0.49	0	1.8	Unknown
1.29	*Brevibacillus brevis* (D78457)	3.45	3.08	2.65	Improving root growth, nodulation and pathogen antagonism
1.2	No match	0	0	0.87	Unknown
1.02	*Xylophilus ampelinus* (AF078758)	0.77	1.38	2.98	Plant pathology
0.99	*Cytophaga lytica* (M62796)	1.24	1.24	0	C cycle
0.95	*Mesorhizobium loti* (D14514)	2.71	3.3	1.81	Legume nodulation and N fixation
0.9	*Rhizobium hainanense* (U71078)	1.53	1.7	0.89	Legume nodulation and N fixation
0.86	*Afipia clevelandensis* (M69186)	1.9	2.3	2.5	Nitrification
0.86	*Sphingomonas* sp. (U52146)	1.35	1.59	0.66	C cycle

NS: non-separation intercropping; SS: semi-separation intercropping; CS: complete separation intercropping; TRFLP-PAT: terminal restriction fragment length polymorphism-phylogenetic assignment tool. The references of bacterial functions are shown in Table S1.

2.4. qPCR of Specific Bacterial Groups

Effective use of nitrogen is considered the most important factor with respect to the advantages of intercropping. Based on previous results, soil-available N, urease activity, and the soil bacterial community was improved in the belowground interaction planting pattern. Therefore, seven genes encoding key enzymes in N cycling (*nifH*, *amoA* (AOA), *amoA* (AOB), *narG*, *nirK*, *nirS* and *nosZ*) were analyzed (Figure 4 and Figure S2). As for peanut, for *nirK*, there were no significant differences among the treatments, while abundances of the other six genes were significantly improved ($p < 0.05$) in conditions of belowground interaction through intercropping (NS and SS). In the maize rhizosphere, the results showed that the abundances of the nitrogen fixation-related gene (*nifH*), the ammonia oxidation-related gene (*amoA* (AOA), *amoA* (AOB)) and the nitrate reductase gene (*narG*) were significantly improved ($p < 0.05$) under the planting patterns with belowground interactions, while the other three genes (*nirK*, *nirS*, and *nosZ*) showed no significant differences.

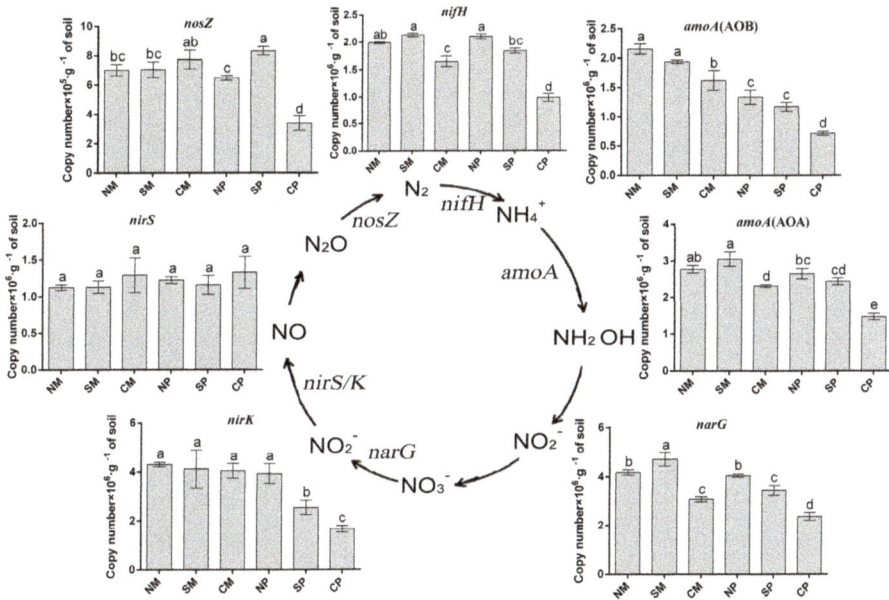

Figure 4. Quantification of genes involved in N-cycling from rhizosphere soil under different intercropping patterns. NM: non-separated maize; SM: semi-separated maize; CM: completely separated maize; NP: non-separated peanut; SP: semi-separated peanut; CP: completely separated peanut. Bars with different letters indicate significant differences (LSD's test, $p < 0.05$, $n = 4$).

Quantitative PCR further confirmed the changes in the abundance of the main bacterial phyla (α-Proteobacteria, β-Proteobacteria, Firmicutes, and Actinobacteria) and beneficial bacteria (*Bacillus*, *Burkholderia*, *Pseudomonas*, and *Rhizobium*) in all six soil samples (Figure 5 and Figure S3). The numbers of *Bacillus*, *Burkholderia*, *Pseudomonas*, *Rhizobium*, α-Proteobacteria, and β-Proteobacteria were significantly higher in peanut soil under the planting pattern with belowground interactions. Meanwhile, significantly higher abundances of *Bacillus*, *Rhizobium*, Firmicutes, and Actinobacteria were observed in belowground interactions in maize soil. The results were consistent with the T-RFLP analysis (Figure 3 and Table 4).

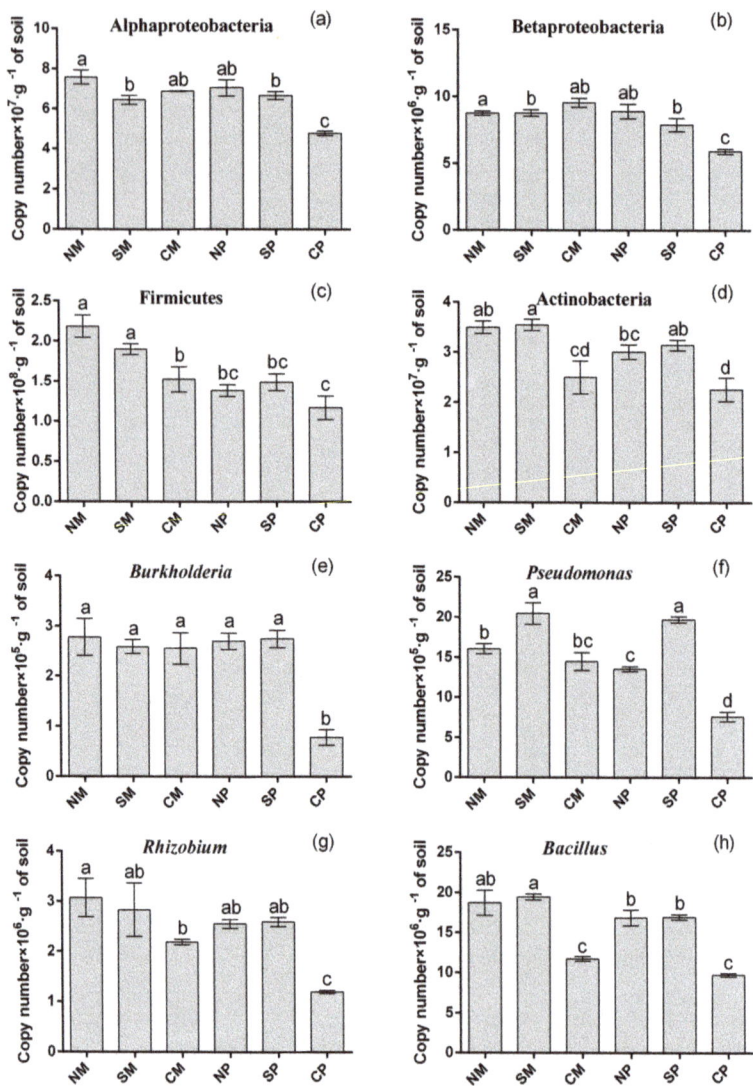

Figure 5. Quantification of genes involved in major microbial communities under different intercropping patterns. (**a**): Alphaproteobacteria; (**b**): Betaproteobacteria; (**c**): Firmicutes; (**d**): Actinobacteria; (**e**): *Burkholderia*; (**f**): *Pseudomonas*; (**g**): *Rhizobium*; (**h**): *Bacillus*. NM: non-separated maize; SM: semi-separated maize; CM: completely separated maize; NP: non-separated peanut; SP: semi-separated peanut; CP: completely separated peanut. Bars with different letters indicate significant differences (LSD's test, $p < 0.05$, $n = 4$).

3. Discussion

Legume and cereal intercropping has been widely applied in agriculture, and the mechanism of beneficial effects extensively studied [23]. According to previous studies, most intercropping systems display a yield advantage [6]. However, there is still no consensus on the underlying mechanisms of beneficial effects in intercropping regimes. In some works, the beneficial effects were attributed to the improvement of soil-available P by the acidification of root exudation (i.e., organic acids) in alkaline soil [7]. Meanwhile, other studies found that the improvement of soil-available P is not related to the pH value in acidic and neutral soils [8,9] because the organic acids secreted by the root are quickly fixed by the soil, and the concentrations of organic acids are too low (1~50) μM to activate insoluble P [24,25]. Hence, the advantages of maize/peanut intercropping need to be confirmed before studying their underlying mechanisms in a new soil environment. In a recent study, two-year LER in field experiments showed yield advantages in the presence of belowground interactions (NS and SS), while the without belowground interactions (CS) pattern showed yield disadvantages as compared to the monoculture. The results indicated that the advantages of maize/peanut intercropping were due to the belowground interactions, and the advantages may be derived from the enhancement of soil nutrient, enzymes activity and microbial community composition.

Soil enzymes are important bioactive proteins in soil that are mainly derived from microorganisms [17]. They directly participate in soil nutrient cycling and are closely related to soil fertility and soil environmental quality [26]. Urease participates in soil nitrogen cycling and indicates soil quality and fertility [27]. Phosphomonoesterase plays an important role in converting organic phosphate into inorganic phosphate, which can improve soil available P [28]. Invertase activitiy is correlated with soil carbon cycling, phosphorus content, microbial biomass, and soil respiration intensity [29]. Previous studies have demonstrated that interspecific interactions in intercropping systems could effectively improve the mobilization and uptake of nitrogen and phosphorus as compared to monocropping [8,30,31]. Higher soil enzyme activities can improve available N and P supply capacity in the plant soil. Our study found that urease, acid PME activities, and available N and P contents in both crops were significantly higher in the presence of belowground interactions (NS and SS) as compared to intercropping without belowground interactions (CS). Similar changes were observed in other intercropping systems [32–34]. Invertase were significantly enhanced in the NS and SS treatments as compared with CS in peanut soil, while decreased in maize soil. These results are supported by the finding of Dai et al. (2013), that the activity of invertase was significantly increased in peanut when intercropped with *Atractylodes lancea* [35]. The results indicated that intercropping of maize and peanut improved soil fertility through belowground interactions and the improvements were closely related to with the shifts in the soil microbial community, also indicating soil microbial community and enzymatic activity is sensitive to different plants' interspecific interactions.

The microbial community is a key component of the rhizosphere soil ecosystem and can be used for evaluating soil quality [36]. Changes in soil microbial communities have been observed in mulberry/soybean, maize/chickpea, maize/soybean, and sorghum/peanut intercropping [9,37,38]. Our previous study demonstrated that Gram-positive bacteria (G$^+$) were significantly improved in maize soils when intercropped with peanut, while Gram-negative (G$^-$) bacteria were significantly improved in peanut soil [39]. Thus, the soil bacterial community in intercropping was studied by T-RFLP analysis and similar results were observed in the present study. The results showed that Proteobacteria (G$^-$), Firmicutes (G$^+$), and Actinobacteria (G$^+$) were the most abundant in the different soil samples. Numbers of Firmicutes (G$^+$) and Actinobacteria (G$^+$) were increased in intercropped maize soils with belowground interactions (NS and SS), while Proteobacteria (G$^-$) was increased in intercropped peanut. These trends were further confirmed by qPCR (Figure 5). Meanwhile, multi-dimensional scaling (NMDS) and cluster analyses clearly demonstrated that NS and SS soil bacterial communities in both plants clustered together and were considerably different from CS bacterial communities.

Moreover, plant growth-promoting rhizobacteria have an important role in the agroecological system. They can suppress plant soil-borne pathogens, improve the concentration of available nutrients, and promote the growth of plants [40]. The SIMPER analysis of T-RFs delivered an insight into key variation bacterial between planting pattern with belowground interactions and without interactions. The result (Table 4 and Table S1) showed that higher abundances of beneficial bacteria (*Bacillus*, *Burkholderia*, *Pseudomonas* and *Rhizobium* etc.) were observed in belowground interaction intercropping treatments (NS and SS). *Bacillus*, *Brevibacillus brevis*, and *Paenibacillus* were found to have functions of phosphate solubilization, nitrogen fixation, suppression of pathogenic microorganisms, and improvement of legume nodulation. *Burkholderia* and *Pseudomonas* exist widely in healthy and disease-free soil, and they can effectively control the occurrence of soil diseases and improve the solubilization of fixed soil phosphorus (Table S1); *Sphingomonas* is involved in decomposition of aromatic compounds; *Rhizobium* is involved in nitrogen fixation and is closely related to peanut nodulation. We found numbers of *Bacillus*, *Brevibacillus brevis*, and *Paenibacillus*, belonging to Firmicutes (G$^+$), mainly to be increased in the maize rhizosphere as a result of belowground interaction. Meanwhile, *Paenibacillus*, *Burkholderia*, *Pseudomonas*, *Sphingomonas* and *Rhizobium*, belonging to Proteobacteria (G$^-$), were mainly increased in the peanut rhizosphere. In addition, *Bacillus*, *Burkholderia*, *Pseudomonas*, and *Rhizobium* were considered to be the key plant growth-promoting rhizobacteria [41] and were further analyzed by qPCR in six soil samples. The qPCR results were consistent with the T-RFLP analysis (Figure 5). The results revealed that belowground interaction in intercropping can induce changes in soil bacterial community structure and attract more beneficial bacteria, which participate in nutrient cycling, legume nodulation, and suppression of plant soil-borne pathogens. These findings explained the reason for higher soil nutrient (N and P) supply capacity and crop disease resistance in intercropping.

The level of soil bacterial community diversity plays an important role in maintaining agricultural ecosystem stability, and improving crop resistance, crop growth and yield formation [42]. Plant diversity ensures soil bacterial community diversity and reduces the number of pathogenic microorganisms [43]. Qiao et al. [44] found that the intercropping of oats and vetch could improve the diversity of soil bacterial communities in the system. Conversely, consecutive monoculture of plants (cucumber and *Pseudostellaria heterophylla*) could decrease the diversity of the rhizosphere bacterial community [45,46]. Moreover, increased soil microbial diversity not only suppresses soil-borne diseases, but also improves soil nitrogen nutrient supply capacity [14]. Pankaj et al. [47] also found that significantly higher levels of nitrogen metabolic cycles could be observed in healthy soil as compared to soil with soil-borne diseases. It has been proven that legumes and non-legumes can efficiently utilize N sources in intercropping systems [48]. In this study, higher soil bacterial diversity indices were observed in the rhizosphere soil of both maize and peanut under belowground interaction intercropping (NS and SS). Moreover, genes related to soil N-cycling were improved by belowground interaction intercropping (NS and SS). These results demonstrated that the belowground interaction in intercropping systems improved soil bacterial community diversity and N-cycling bacteria, which were conducive to soil N nutrient supply capacity and soil health.

Based upon the discussion above, NS and SS show the same trends when compared with CS. Under NS treatment (no separation), the interactions between each crop include soil nutrition, microorganisms, root exudates exchanges, etc. The nylon net in SS treatment allowed root exudates, soil microorganisms, and soluble matter to flow across the net. Previous studies have demonstrated that plants release enormous amounts of chemicals through their roots, which affect soil microorganisms [49]. Because of the differences in the composition of root exudates, different crops and even genotypes can determine their unique soil microbial communities [15,50–52]. As such, we supposed that shifts in the soil microbial community in NS and SS were mainly due to root exudate exchanges in the intercropping.

4. Materials and Methods

4.1. Experimental Design

The experiment was carried out in the experimental field and greenhouse of the College of Crop Sciences, Fujian Agricultural and Forestry University, Fuzhou, China (26°08′ N, 119°23′ E) in 2011 and 2012. In addition to monoculture (MS), maize (*Zea mays* L.) and peanut (*Arachis hypogaea* L.) were grown under three different intercropping treatments with the same canopy structure. Maize and peanut were planted with no separation intercropping (NS), semi-separation intercropping (SS) with a nylon net (50 μm), and complete separation intercropping (CS) with a plastic sheet (Figure S4). The experimental soil was sandy loam soil (pH value 5.5) containing total nitrogen 1.12 g·kg^{-1}, total phosphorus 0.47 g·kg^{-1}, total K 6.59 g·kg^{-1}, available nitrogen 36.15 mg·kg^{-1}, available phosphorus 8.23 mg·kg^{-1}, and available K 30.46 mg·kg^{-1}.

The yields of two crops were measured in 2011 and 2012 (Figure S4) in field experiments. The area for each pattern type (NS, SS, CS, and MS) was 16 m^2 (4 × 4 m), with three replicates. The same field management and fertilization processes (including 180 kg·hm^{-2} CO(NH$_2$)$_2$, 52 kg·hm^{-2} Ca(H$_2$PO$_4$)$_2$ and 75 kg·hm^{-2} KCl) were implemented in treatments during the whole experimental period. The pot experiment was carried out in 2012. Three peanut and one maize were grown in each pot with 13 kg of soil under three different intercropping treatments (NS, SS and CS), each with six replicates (Figure S5). Basal fertilizers were mixed in soil before planting, including (mg·kg^{-1} soil): N 100 (Ca (NO$_3$)$_2$·4H$_2$O), P 150 (KH$_2$PO$_4$), K 150 (KCl), Mg 50 (MgSO$_4$·7H$_2$O), Cu 5(CuSO$_4$·5H$_2$O), and Zn 5(ZnSO$_4$·7H$_2$O).

4.2. Field Yields, Experimental Plant Property Determination and Soil Sampling

At maturity, five plants in each row were randomly selected in all treatments, and then the grain yields in each treatment were detected. The land equivalent ratio (LER) is the total land area of sole crops required to achieve the same yields as the intercrops.

$$LER = Y_{im}/Y_m + Y_{ip}/Y_p \qquad (1)$$

where Y_{im} and Y_m are the yields of intercropped and sole maize crops, and Y_{ip} and Y_p are the yields of intercropped and sole peanut crops, respectively. Intercropping systems exhibit yield advantages when LER > 1, while LER < 1 indicates yield disadvantages [53].

At the maize flowering stage (60 days after sowing) in the pot experiments, four maize and four peanut plants were randomly selected. The net photosynthetic rate of maize and peanut function leaves (the first fully expanded leaf from the top of the maize plant and last two or three leaves on the peanut main stem) were measured with Li-6400 (LI-COR, Lincoln, NE, USA) from 9:00 a.m. to 11:00 a.m. Peanut and maize were cut near the soil surface. The roots were carefully uprooted from the soil and gently shaken to remove loosely attached soil. The rhizosphere soils, at depths of 5~15 cm and closely attached to the roots, were carefully brushed down and collected, and then the rhizospheric soils were sieved by 0.9-mm mesh. Soil samples were separated into two parts: one part stored at −80 °C for soil DNA extraction, and other part stored at 4 °C for soil enzyme and soil nutrient analysis. The plant roots and peanut nodules were collected and immediately flushed with water. The flesh peanut nodules were detached from the roots and counted. These plants (shoots and roots) and nodules were first dried (105 °C for 30 min, 60 °C for 48 h), and then the biomasses were measured.

4.3. Measurement of Soil Nutrient and Enzymatic Activities

The measurement methods of soil nutrients (available nitrogen, phosphorus, and potassium) were proposed by Jackson [54]. Available nitrogen was measured by the alkaline hydrolysable method. Available phosphorus was measured by molybdenum blue method. Available potassium was extracted by incubating 5 g soil with 50 mL of 1 mol·L^{-1} CH$_3$COONH$_4$ (pH 7) for 30 min. The supernatant was filtered through the filter paper and then measured by flame photometry. The measurement methods of soil enzymes (urease, invertase, and acid PME) activities were proposed by Guan [29]. Urease (EC 3.5.1.5) activity was determined by incubating 5 g soil with 10 mL of 10% urea solution and 20 mL of citrate buffer (dissolved 184 g citric acid and 147.5 g KOH in 300 mL ddH$_2$O respectively, combined the two solutions and then adjusted the pH to 6.7 with 1 mol·L^{-1} NaOH, and then diluted with ddH$_2$O to 1 L) at 37 °C for 24 h. The formation of ammonium was measured by spectrophotometer at 578 nm and expressed as μg·g^{-1} soil·h^{-1}. Soil invertase (EC 3.2.1.26) activity was determined by mixing 5 g soil with 15 mL of 8% sucrose and 5 mL of 66.7 mmol·L^{-1} phosphate buffer (pH 5.5) at 37 °C for 24 h. The supernatant was rapidly filtered through the filter paper and measured by 3,5-dinitrosalicylic acid colorimetric method (DNS method). Invertase activity expressed as µg glucose·g^{-1} soil·h^{-1}. Acid PME (EC: 3.1.3.2) activity was determined by incubating 1 g soil with 4 mL of acidic buffer (20 mmol·L^{-1} Tris(hydroxymethyl)methyl aminomethane, 20 mmol·L^{-1} butenedioic acid, 14.6 mmol·L^{-1} citric acid, 20 mmol·L^{-1} boronic acid, pH 6.0), 0.2 mL of methylbenzene, and 1 mL of 0.05 mol·L^{-1} *para*-nitrophenyl phosphate (*p*NPP) at 37 °C for 24 h. The formation of *para*-nitrophenol was measured by spectrophotometer at 410 nm and expressed as µg·g^{-1} soil·h^{-1}.

4.4. DNA Extraction and Terminal Restriction Fragment Length Polymorphism (T-RFLP) Analysis

The Soil Genomic DNA Extraction Kit (Bioer Technology Co., Ltd., Hangzhou, China) was used to extract soil whole genome DNA. DNA concentration was measured by Nanodrop and stored in −80 °C for T-RFLP [55] and quantification PCR analysis. Primers 27F-FAM (5′-AGAGTTTGATCCTGGCTCAG-3′) and 1492R (5′-GGTTACCTTGTTACGACTT-3′) with 6-carboxyfurescein-label were used to amplify the bacterial 16S rRNA gene. The PCR reaction mixture (50 µL in final volume) included 25 µL Taq PCR Mix (2×) (TianGen Biotech Co., Ltd., Beijing, China), 1 µL of each primer (10 pmol·mL^{-1}), 1 µL 0.1% BSA (Bovine Serum Albumin), 20 ng DNA template, and ddH$_2$O. The PCR program was as follows: 5 min at 95 °C, followed by 35 cycles of 95 °C for 1 min, 60 °C for 90 s, and 72 °C for 90 s, and then 1 cycle of 72 °C for 10 min. To purify PCR products, 1.2% agarose gel electrophoresis and the Gel Extraction Kit (OMEGA Bio-Tek, Norcross, GA, USA) were used. The purified PCR products were digested separately with four enzymes (*Msp*I, *Hae*III, *Afa*I and *Alu*I) [55]. The digestion mixture (15 µL in final volume) included 4 µL of enzyme buffer, 1.5 µL of restriction endonuclease, 8 µL of PCR products and ddH$_2$O. The endonuclease digestion products were determined by the ABI 3730xl DNA sequencer (Applied Biosystems, Foster City, CA, USA). PCR amplification and enzyme digestion took place in dark conditions to avoid fluorescence decay. Gene Marker software Version 1.2 (SoftGenetics LLC, State College, PA, USA) was used to analyze T-RFLP profiles. Lengths of terminal restriction fragments (T-RFs) between 30 and 600 bp were selected for further analysis. The T-RFs measuring ±1 bp were combined and considered as the same operational taxonomic units [56]. Based on the T-RFs, bacteria were identified by the phylogenetic assignment tool (PAT) [57]. Only three or four restriction enzyme T-RF lengths matching phylogenetic assignments were used in this study.

4.5. Quantification PCR of Bacterial Communities

Quantification PCR (qPCR) was used to analyze the bacterial genes (*nifH*, *amoA* (AOA), *amoA* (AOB), *narG*, *nirK*, *nirS*, and *nosZ*), which encode the key enzymes in nitrogen cycling (Table S2). *AmoA* (AOB) and *amoA* (AOA) encode ammonia monooxygenase; *narG* encodes membrane-bound nitrate reductase; *nirK* and *nirS* encode nitrite reductase; *nosZ* encodes nitrous oxide reductase; and *nifH* encode the nitrogenase. Meanwhile, the qPCR of the main bacterial phylum (α-Proteobacteria, β-Proteobacteria, Firmicutes and Actinobacteria) and beneficial bacteria (*Bacillus*, *Burkholderia*, *Pseudomonas*, *Rhizobium*) measured by the methods are described in Table S2. qPCR reaction mixture (15 µL in final volume) included 0.5 µL of each primer (10 µM), 7.5 µL (2×) SYBR green I SuperReal Premix (TIANGEN, Beijing, China), and template DNA (20 ng of total soil DNA or plasmid DNA for standard curves).

4.6. Statistical Analyses

SPSS V11.5 software (IBM Corporation, Chicago, IL, USA) was used for statistical analysis, and ANOVA (analysis of variance) was used to determine the significance of difference with the LSD's test ($p < 0.05$). PRIMER V5 software (PRIMER-E Ltd., Plymouth, UK) was used for non-metric multidimensional scaling (NMDS) and similarity percentage analysis (SIMPER) of T-RFLP data. NMDS is a superior analysis method for investigating microbial community data [58], and is considered an accurate and reliable method of analysis when Kruskal's stress value <0.1. The similarities of microbial communities were evaluated by NMDS in this study. The contribution (%) of each T-RF to the dissimilarity between samples was carried out by means of similarity percentage analysis [58].

5. Conclusions

The advantages of maize/peanut intercropping can be attributed to belowground interactions. Intercropping of maize and peanut improved the soil nutrient supply capacity and soil health. This improvement was driven by the shifts of the soil microbial community, including the improvement in the abundance of beneficial bacteria (i.e., *Bacillus*, *Burkholderia*, *Pseudomonas*, and *Rhizobium* etc.), bacterial diversity, and N-cycling bacteria. Besides plant nutrition, our study suggested that the soil microbial community could be a key factor for reasonable intercropping, and that the interaction of root exudates may play a key role in modifying the soil microbial community. This study also provided a clue to solving problems stemming from excessive monoculture in current agricultural production. Additional work is needed to explore the interactions of root exudates in intercropping, and the mechanisms of how root exudates shape the soil microbial community.

Supplementary Materials: Supplementary materials can be found at www.mdpi.com/1422-0067/19/2/622/s1.

Acknowledgments: This work was supported by grants from the National Natural Science Foundation of China (No. 81573530, 31401950) and the Fujian-Taiwan Joint Innovative Centre for Germplasm Resources and Cultivation of Crops (grant No. 2015-75. Fujian 2011 Program, China).

Author Contributions: Wenxiong Lin, Sheng Lin and Qisong Li conceived and designed the experiments; Qisong Li, Xiaomian Luo, Na Li, and Jun Chen performed field and lab experiments; Qisong Li, Linkun Wu and Yasir Arafat analyzed the data; and Qisong Li and Wenxiong Lin wrote the paper. All authors discussed and commented on the results in the manuscript.

Conflicts of Interest: The authors declare no conflict of interest.

Abbreviations

NS	Non-separation intercropping
SS	Semi-separation intercropping
CS	Complete separation intercropping
MS	Monoculture
NM	Non-separation intercropping maize
SM	Semi-separation intercropping maize
CM	Complete separation intercropping maize
NP	Non-separation intercropping peanut
SP	Semi-separation intercropping peanut
CP	Complete separation intercropping peanut
LER	Land equivalent ratio

References

1. Hauggaard-Nielsen, H.; Jørnsgaard, B.; Kinane, J.; Jensen, E.S. Grain legume–cereal intercropping: The practical application of diversity, competition and facilitation in arable and organic cropping systems. *Renew. Agric. Food Syst.* **2008**, *23*, 3–12. [CrossRef]
2. Zhang, X.; Huang, G.; Bian, X.; Jiang, X.; Zhao, Q. Review of researches on advantages of intercropping. *Asian Agric. Res.* **2012**, *114*, 126–132.
3. Li, L.; Sun, J.; Zhang, F.; Li, X.; Rengel, Z.; Yang, S. Wheat/maize or wheat/soybean strip intercropping: Ii. Recovery or compensation of maize and soybean after wheat harvesting. *Field Crop. Res.* **2001**, *71*, 173–181. [CrossRef]
4. Wang, Z.; Zhao, X.; Wu, P.; He, J.; Chen, X.; Gao, Y.; Cao, X. Radiation interception and utilization by wheat/maize strip intercropping systems. *Agric. For. Meteorol.* **2015**, *204*, 58–66. [CrossRef]
5. Morris, R.; Garrity, D.P. Resource capture and utilization in intercropping: Water. *Field Crop. Res.* **1993**, *34*, 303–317. [CrossRef]

6. Li, L.; Tilman, D.; Lambers, H.; Zhang, F.S. Plant diversity and overyielding: Insights from belowground facilitation of intercropping in agriculture. *New Phytol.* **2014**, *203*, 63–69. [CrossRef] [PubMed]

7. Li, L.; Li, S.M.; Sun, J.H.; Zhou, L.L.; Bao, X.G.; Zhang, H.G.; Zhang, F.S. Diversity enhances agricultural productivity via rhizosphere phosphorus facilitation on phosphorus-deficient soils. *Proc. Nat. Acad. Sci. USA* **2007**, *104*, 11192–11196. [CrossRef] [PubMed]

8. Betencourt, E.; Duputel, M.; Colomb, B.; Desclaux, D.; Hinsinger, P. Intercropping promotes the ability of durum wheat and chickpea to increase rhizosphere phosphorus availability in a low p soil. *Soil Biol. Biochem.* **2012**, *46*, 181–190. [CrossRef]

9. He, Y.; Ding, N.; Shi, J.; Wu, M.; Liao, H.; Xu, J. Profiling of microbial plfas: Implications for interspecific interactions due to intercropping which increase phosphorus uptake in phosphorus limited acidic soils. *Soil Biol. Biochem.* **2013**, *57*, 625–634. [CrossRef]

10. Berg, G.; Smalla, K. Plant species and soil type cooperatively shape the structure and function of microbial communities in the rhizosphere. *FEMS Microbiol. Ecol.* **2009**, *68*, 1–13. [CrossRef] [PubMed]

11. Marschner, P. Plant-microbe interactions in the rhizosphere and nutrient cycling. In *Nutrient Cycling in Terrestrial Ecosystems*; Springer: Berlin/Heidelberg, Germany, 2007; pp. 159–182.

12. Van de Mortel, J.E.; de Vos, R.C.; Dekkers, E.; Pineda, A.; Guillod, L.; Bouwmeester, K.; van Loon, J.J.; Dicke, M.; Raaijmakers, J.M. Metabolic and transcriptomic changes induced in arabidopsis by the rhizobacterium pseudomonas fluorescens ss101. *Plant Physiol.* **2012**, *160*, 2173–2188. [CrossRef] [PubMed]

13. Sanguin, H.; Sarniguer, A.; Gazengel, K.; Moënne-Loccoz, Y.; Grundmann, G.L. Rhizosphere bacterial communities associated with disease suppressiveness stages of take-all decline in wheat monoculture. *New Phytol.* **2009**, *184*, 694–707. [CrossRef] [PubMed]

14. Weidner, S.; Koller, R.; Latz, E.; Kowalchuk, G.; Bonkowski, M.; Scheu, S.; Jousset, A. Bacterial diversity amplifies nutrient-based plant–soil feedbacks. *Funct. Ecol.* **2015**, *29*, 1341–1349. [CrossRef]

15. Kourtev, P.; Ehrenfeld, J.; Häggblom, M. Experimental analysis of the effect of exotic and native plant species on the structure and function of soil microbial communities. *Soil Biol. Biochem.* **2003**, *35*, 895–905. [CrossRef]

16. Baudoin, E.; Benizri, E.; Guckert, A. Impact of artificial root exudates on the bacterial community structure in bulk soil and maize rhizosphere. *Soil Biol. Biochem.* **2003**, *35*, 1183–1192. [CrossRef]

17. Sinsabaugh, R. Enzymic analysis of microbial pattern and process. *Biol. Fertil. Soils* **1994**, *17*, 69–74. [CrossRef]

18. Awal, M.; Koshi, H.; Ikeda, T. Radiation interception and use by maize/peanut intercrop canopy. *Agric. For. Meteorol.* **2006**, *139*, 74–83. [CrossRef]

19. Zuo, Y.; Zhan, F. Iron and zinc biofortification strategies in dicot plants by intercropping with gramineous species. A review. *Agron. Sustain. Dev.* **2009**, *29*, 63–71. [CrossRef]

20. Nannipieri, P.; Ascher, J.; Ceccherini, M.T.; Landi, L.; Pietramellara, G.; Renella, G. Microbial diversity and soil functions. *Eur. J. Soil Sci.* **2010**, *54*, 655–670. [CrossRef]

21. Wang, M.; Ahrné, S.; Antonsson, M.; Molin, G. T-rflp combined with principal component analysis and 16s rrna gene sequencing: An effective strategy for comparison of fecal microbiota in infants of different ages. *J. Microbiol. Methods* **2004**, *59*, 53. [CrossRef] [PubMed]

22. Courtney, K.C.; Bainard, L.D.; Sikes, B.A.; Koch, A.M.; Maherali, H.; Klironomos, J.N.; Hart, M.M. Determining a minimum detection threshold in terminal restriction fragment length polymorphism analysis. *J. Microbiol. Methods* **2012**, *88*, 14–18. [CrossRef] [PubMed]

23. Xue, Y.; Xia, H.; Christie, P.; Zhang, Z.; Li, L.; Tang, C. Crop acquisition of phosphorus, iron and zinc from soil in cereal/legume intercropping systems: A critical review. *Ann. Bot.* **2016**, *117*, 363. [CrossRef] [PubMed]

24. Strobel, B.W. Influence of vegetation on low-molecular-weight carboxylic acids in soil solution—A review. *Geoderma* **2001**, *99*, 169–198. [CrossRef]

25. Ström, L.; Owen, A.; Godbold, D.; Jones, D. Organic acid behaviour in a calcareous soil: Sorption reactions and biodegradation rates. *Soil Biol. Biochem.* **2001**, *33*, 2125–2133. [CrossRef]

26. Dick, R.P. Soil enzyme activities as indicators of soil quality. *Soil Sci. Soc. Am. J.* **1994**, *58*, 107–124.

27. Zhang, F.; Shen, J.; Li, L.; Liu, X. An overview of rhizosphere processes related with plant nutrition in major cropping systems in china. *Plant Soil* **2004**, *260*, 89–99. [CrossRef]

28. Pang, P.; Kolenko, H. Phosphomonoesterase activity in forest soils. *Soil Biol. Biochem.* **1986**, *18*, 35–39. [CrossRef]

29. Guan, S.; Zhang, D.; Zhang, Z. *Soil Enzyme and Its Research Methods*; China Agriculture Press: Beijing, China, 1986.

30. Gao, Y.; Duan, A.; Sun, J.; Li, F.; Liu, Z.; Liu, H.; Liu, Z. Crop coefficient and water-use efficiency of winter wheat/spring maize strip intercropping. *Field Crop. Res.* **2009**, *111*, 65–73. [CrossRef]

31. Li, L.; Zhang, F.; Li, X.; Christie, P.; Sun, J.; Yang, S.; Tang, C. Interspecific facilitation of nutrient uptake by intercropped maize and faba bean. *Nutr. Cycl. Agroecosyst.* **2003**, *65*, 61–71. [CrossRef]

32. Xiao, X.; Cheng, Z.; Meng, H.; Liu, L.; Li, H.; Dong, Y. Intercropping of green garlic (*allium sativum* L.) induces nutrient concentration changes in the soil and plants in continuously cropped cucumber (*cucumis sativus* L.) in a plastic tunnel. *PLoS ONE* **2013**, *8*, e62173. [CrossRef] [PubMed]

33. Xiao, X.; Cheng, Z.; Meng, H.; Khan, M.A.; Li, H. Intercropping with garlic alleviated continuous cropping obstacle of cucumber in plastic tunnel. *Acta Agric. Scand. Sect. B Soil Plant Sci.* **2012**, *62*, 696–705. [CrossRef]

34. Khan, M.A.; Zhihui, C.; Khan, A.R.; Rana, S.J.; Ghazanfar, B. Pepper-garlic intercropping system improves soil biology and nutrient status in plastic tunnel. *Int. J. Agric. Biol.* **2015**, *17*, 869–880. [CrossRef]

35. Dai, C.C.; Chen, Y.; Wang, X.X.; Li, P.D. Effects of intercropping of peanut with the medicinal plant atractylodes lancea on soil microecology and peanut yield in subtropical china. *Agrofor. Syst.* **2013**, *87*, 417–426. [CrossRef]

36. Sharma, S.K.; Ramesh, A.; Sharma, M.P.; Joshi, O.P.; Govaerts, B.; Steenwerth, K.L.; Karlen, D.L. Microbial community structure and diversity as indicators for evaluating soil quality. In *Biodiversity, Biofuels, Agroforestry and Conservation Agriculture*; Springer: Berlin/Heidelberg, Germany, 2010; pp. 317–358.

37. Yang, Z.; Yang, W.; Li, S.; Hao, J.; Su, Z.; Sun, M.; Gao, Z.; Zhang, C. Variation of bacterial community diversity in rhizosphere soil of sole-cropped versus intercropped wheat field after harvest. *PLoS ONE* **2016**, *11*, e0150618. [CrossRef] [PubMed]

38. Li, X.; Sun, M.; Zhang, H.; Xu, N.; Sun, G. Use of mulberry–soybean intercropping in salt–alkali soil impacts the diversity of the soil bacterial community. *Microb. Biotechnol.* **2016**, *9*, 293–304. [CrossRef] [PubMed]

39. Li, Q.S.; Wu, L.K.; Chen, J.; Khan, M.A.; Luo, X.M.; Lin, W.X. Biochemical and microbial properties of rhizospheres under maize/ peanut intercropping. *J. Integr. Agric.* **2016**, *15*, 101–110. [CrossRef]

40. Bhattacharyya, P.N.; Jha, D.K. Plant growth-promoting rhizobacteria (pgpr): Emergence in agriculture. *World J. Microbiol. Biotechnol.* **2012**, *28*, 1327–1350. [CrossRef] [PubMed]

41. Chauhan, H.; Bagyaraj, D.J.; Selvakumar, G.; Sundaram, S.P. Novel plant growth promoting rhizobacteria—prospects and potential. *Appl. Soil Ecol.* **2015**, *95*, 38–53. [CrossRef]

42. Bossio, D.; Scow, K.; Gunapala, N.; Graham, K. Determinants of soil microbial communities: Effects of agricultural management, season, and soil type on phospholipid fatty acid profiles. *Microb. Ecol.* **1998**, *36*, 1–12. [CrossRef] [PubMed]

43. Bennett, A.E.; Daniell, T.J.; White, P.J. Benefits of breeding crops for yield response to soil organisms. In *Molecular Microbial Ecology of the Rhizosphere: Volume 1 & 2*; Wiley: Hoboken, NJ, USA, 2013; pp. 17–27.

44. Qiao, Y.J.; Li, Z.Z.; Wang, X.; Zhu, B.; Hu, Y.G.; Zeng, Z.H. Effect of legume-cereal mixtures on the diversity of bacterial communities in the rhizosphere. *Plant Soil Environ.* **2012**, *58*, 174–180. [CrossRef]

45. Yao, H.; Jiao, X.; Wu, F. Effects of continuous cucumber cropping and alternative rotations under protected cultivation on soil microbial community diversity. *Plant Soil* **2006**, *284*, 195–203. [CrossRef]

46. Zhao, Y.P.; Sheng, L.; Chu, L.; Gao, J.T.; Azeem, S.; Lin, W. Insight into structure dynamics of soil microbiota mediated by the richness of replanted pseudostellaria heterophylla. *Sci. Rep.* **2016**, *6*, 26175. [CrossRef] [PubMed]

47. Trivedi, P.; He, Z.; Van Nostrand, J.D.; Albrigo, G.; Zhou, J.; Wang, N. Huanglongbing alters the structure and functional diversity of microbial communities associated with citrus rhizosphere. *ISME J.* **2012**, *6*, 363–383. [CrossRef] [PubMed]

48. Jensen, E.S. Grain yield, symbiotic n_2 fixation and interspecific competition for inorganic n in pea-barley intercrops. *Plant Soil* **1996**, *182*, 25–38. [CrossRef]

49. Badri, D.V.; Weir, T.L.; Lelie, D.V.D.; Vivanco, J.M. Rhizosphere chemical dialogues: Plant–microbe interactions. *Curr. Opin. Biotechnol.* **2009**, *20*, 642–650. [CrossRef] [PubMed]

50. Bouffaud, M.L.; Kyselková, M.; Gouesnard, B.; Grundmann, G.; Muller, D.; Moënne-Loccoz, Y. Is diversification history of maize influencing selection of soil bacteria by roots? *Mol. Ecol.* **2012**, *21*, 195–206. [CrossRef] [PubMed]

51. Houlden, A.; Timms-Wilson, T.M.; Day, M.J.; Bailey, M.J. Influence of plant developmental stage on microbial community structure and activity in the rhizosphere of three field crops. *FEMS Microbiol. Ecol.* **2008**, *65*, 193–201. [CrossRef] [PubMed]

52. Gschwendtner, S.; Esperschütz, J.; Buegger, F.; Reichmann, M.; Müller, M.; Munch, J.C.; Schloter, M. Effects of genetically modified starch metabolism in potato plants on photosynthate fluxes into the rhizosphere and on microbial degraders of root exudates. *FEMS Microbiol. Ecol.* **2011**, *76*, 564–575. [CrossRef] [PubMed]

53. Mead, R.; Willey, R. The concept of a 'land equivalent ratio' and advantages in yields from intercropping. *Exp. Agric.* **1980**, *16*, 217–228. [CrossRef]

54. Jackson, M.L. *Soil Chemical Analysis*; Prentice-Hall: Upper Saddle River, NJ, USA, 1958.

55. Wu, L.; Wang, J.; Huang, W.; Wu, H.; Chen, J.; Yang, Y.; Zhang, Z.; Lin, W. Corrigendum: Plant-microbe rhizosphere interactions mediated by rehmannia glutinosa root exudates under consecutive monoculture. *Sci. Rep.* **2016**, *6*, 19101. [CrossRef] [PubMed]

56. Egert, M.; Marhan, S.; Wagner, B.; Scheu, S.; Friedrich, M.W. Molecular profiling of 16s rrna genes reveals diet-related differences of microbial communities in soil, gut, and casts of lumbricus terrestris l.(oligochaeta: Lumbricidae). *FEMS Microbiol. Ecol.* **2004**, *48*, 187–197. [CrossRef] [PubMed]

57. Kent, A.D.; Smith, D.J.; Benson, B.J.; Triplett, E.W. Web-based phylogenetic assignment tool for analysis of terminal restriction fragment length polymorphism profiles of microbial communities. *Appl. Environ. Microbiol.* **2003**, *69*, 6768–6776. [CrossRef] [PubMed]

58. Rees, G.N.; Baldwin, D.S.; Watson, G.O.; Perryman, S.; Nielsen, D.L. Ordination and significance testing of microbial community composition derived from terminal restriction fragment length polymorphisms: Application of multivariate statistics. *Antonie Van Leeuwenhoek* **2004**, *86*, 339–347. [CrossRef] [PubMed]

International Journal of
Molecular Sciences

MDPI

Article

Spatial Distribution Patterns of Root-Associated Bacterial Communities Mediated by Root Exudates in Different Aged Ratooning Tea Monoculture Systems

Yasir Arafat [1,2,3,4], Xiaoya Wei [1,2,3], Yuhang Jiang [1,2,3,4], Ting Chen [1,2,3,4],
Hafiz Sohaib Ahmed Saqib [4], Sheng Lin [1,2,3,4,]* and Wenxiong Lin [1,2,3,4,]*

[1] Key Laboratory of Fujian Province for Agroecological Process and Safety Monitoring,
Fujian Agriculture and Forestry University, Fuzhou 35002, China; arafat_pep@yahoo.com (Y.A.);
xlwei112930@163.com (X.W.); janmiky@163.com (Y.J.); iamchenting@126.com (T.C.)
[2] Key Laboratory of Ministry of Education for Crop Genetics/Breeding and Integrative Utilization,
Fujian Agriculture and Forestry University, Fuzhou 35002, China
[3] College of Life Science, Fujian Agriculture and Forestry University, Fuzhou 35002, China
[4] Institute of Agroecological Ecology, Fujian Agriculture and Forestry University, Fuzhou 35002, China;
sohaibsaqib@gmail.com
* Correspondence: linsh@fafu.edu.cn (S.L.); lwx@fafu.edu.cn (W.L.); Tel.: +86-0591-837-789301 (S.L. & W.L.)

Received: 16 June 2017; Accepted: 4 August 2017; Published: 8 August 2017

Abstract: Positive plant–soil feedback depends on beneficial interactions between roots and microbes for nutrient acquisition; growth promotion; and disease suppression. Recent pyrosequencing approaches have provided insight into the rhizosphere bacterial communities in various cropping systems. However; there is a scarcity of information about the influence of root exudates on the composition of root-associated bacterial communities in ratooning tea monocropping systems of different ages. In Southeastern China; tea cropping systems provide the unique natural experimental environment to compare the distribution of bacterial communities in different rhizo-compartments. High performance liquid chromatography–electrospray ionization–mass spectrometry (HPLC–ESI–MS) was performed to identify and quantify the allelochemicals in root exudates. A high-throughput sequence was used to determine the structural dynamics of the root-associated bacterial communities. Although soil physiochemical properties showed no significant differences in nutrients; long-term tea cultivation resulted in the accumulation of catechin-containing compounds in the rhizosphere and a lowering of pH. Moreover; distinct distribution patterns of bacterial taxa were observed in all three rhizo-compartments of two-year and 30-year monoculture tea; mediated strongly by soil pH and catechin-containing compounds. These results will help to explore the reasons why soil quality and fertility are disturbed in continuous ratooning tea monocropping systems; and to clarify the associated problems.

Keywords: monoculture; allelochemicals; microbiomes; rhizo-compartments; high-throughput sequence; redundancy analysis (RDA); high performance liquid chromatography-electrospray ionization-mass spectrometry (HPLC–ESI–MS)

1. Introduction

Plant–soil feedback is a two-step process in which the presence of a plant alters the structure and composition of the rhizosphere microorganism community, and then that change in the microorganism community alters the growth rate and development of the plant [1–6]. In rhizosphere soil there exists a positive soil feedback mechanism, in which microorganisms mediate plant efficiency by producing plant growth hormones, promoting plant nutrient uptake, competitively suppressing plant pathogens, and fixing nitrogen [7]. On the other hand, a negative plant–soil feedback also exists, caused by the accumulation of phytotoxic compounds in soil. This feedback inhibits beneficial microbes but promotes parasites and pathogen outbreaks, and in turn results in autotoxicity or soil sickness, and the hindrance of plant growth and development, which reduces plant yield and quality [8–10]. Autotoxicity and soil sickness are the typical results of negative plant–soil interactions, mainly driven by agricultural landscape simplification, such as continuous monoculture which is the cropping of the same plant in the same field for many consecutive years [11–14]. Production of autotoxins, soil nutrient imbalance, and the alteration of soil-associated microbial community composition are considered as fundamental issues related to autotoxicity and soil sickness [15,16].

Autotoxicity-related problems are observed when practicing consecutive monoculture of agronomic crops, forages, fruits, medicinal, and horticultural plants [6,13,14,16–22]. Several groups of chemicals have been identified for instigating autotoxicity, such as steroids, flavonoids, cyanogenic glycosides, alkaloids, terpenoids, and phenolic acids. Among these autotoxins, various flavonoids are known for their antimicrobial activity [23–25]. Catechin, one of the major flavonoids, is widely reported in wine, traditional Chinese medicine, green tea, natural fruits [26–29], and also in invasive plant species such as *Centuria muculusa* [30,31] and *Rhododendron formosanum* [32]. Studies have also shown that catechins play a vital role in the successful invasion and establishment of invasive species due to its allelopathic action [32,33], but the influential role of catechins in causing problems, by altering the distribution of microbes associated with modern monocropping systems, especially in consecutive ratooning tea gardens, has not yet been studied.

Camellia sinensis (L.) O. Kuntze is one of the most economically valuable mountainous crops in Southern China and is also widely grown in Africa and Asia. It has been commercially grown for almost 1500 years in China. However, the problems related to tea cultivation, such as reduction of tea crop productivity and quality, have increased overtime after the establishment of tea plants in the same soil for several years [34,35]. Researchers found that tea leaves and roots produce various biologically active catechins, which can act as allelochemicals [25,36,37]. Studies also proposed that catechins appear to degrade quickly into other chemical compounds after introduction into the soil, which can affect the population dynamics and growth of several bacterial taxa [38,39]. Most studies addressing the problems related to continuous tea monocropping systems have simply investigated the numerical responses of microbial communities mediated by the soil physiochemical properties [35,36]. Moreover, the influence of root exudates on the distribution of rhizosphere bacterial communities in a continuous monocultured tea garden has not been studied. This study was designed to explore the distribution of bacterial taxa present in the rhizosphere of new and continuous monocropping tea systems and an adjacent uncultivated field. We also wanted to investigate the influence of root exudates and soil physiochemical properties on the abundance of those bacterial taxa. Specifically, we hypothesized that different lengths of time of ratooning tea monoculture fields share bacteria to varying degrees and potentially this would be mediated over time by the variations in soil physiochemical properties and root exudates.

2. Results

2.1. The Problems of Camellia sinensis (L.) Plantation under Continuous Monoculture

When compared to a new tea field planted two years ago, the tea field which was continuously monocultured for 30 years showed poor growth, chlorosis, wilting, and ratooning problems (Figure S1a,b). Moreover, the quality of tea leaves was significantly lower in the continuously monocultured tea fields than in the newly planted tea garden (Table 1).

Table 1. Quality parameters of tea leaves from tea plantations of different ages. TNN, TPY, TPP, TAA, and WC represent the anine, theophylline, total polyphenols, total free amino acids, and water content, respectively. 2YTL and 30YTL represent the two-year-old tea field and the 30-year-old tea field leaves, respectively. Different letters ([a] and [b]) in columns show a significant difference determined by Tukey's test ($p \leq 0.05$, $n = 2$).

Treatments	TNN (g/kg)	TPY (mg/kg)	TPP (g/kg)	TAA (g/kg)	WC (%)
2YTL	1.70 [a]	0.12 [a]	106.40 [a]	12.44 [a]	0.76 [a]
30YTL	0.47 [b]	0.02 [b]	30.66 [b]	8.87 [b]	0.56 [a]

2.2. Soil Physio-Chemical Properties and Root Exudates

Soil physio-chemical properties and root exudates were analyzed to determine the main impacts of tea monoculture in a field. Most of the soil nutrients, including nitrogen (N), phosphorus (P), and potassium (K), did not show a significant difference in bulk soil between the newly planted tea field and the 30-year continuously monocultured tea field, but soil pH was significantly lower in the 30-year-old tea plantation compared to the bulk soil in the two-year tea plantation (Table S1). Meanwhile, soil moisture in the 30-year-old tea field was higher in the bulk soil than in the two-year-old tea plantation (Table S1). Seven compounds including protocatechuic acid (PCA), epigallocatechin (EGC), epigallocatechin gallate (EGCG), epicatechin (EC), (+)-catechin (C), epicatechingallate (ECG), and taxifolin (TF) were identified in the root exudates in both the newly planted tea field and the 30-year continuously monocultured tea field (Figure 1). Moreover, in the young plantation, the concentrations of PCA, EGC, EGCG, EC, C, ECG, and TF were 2.04, 1.73, 6.95, 11.98, 21.31, 1.18, and 1.61 mg/kg, respectively. Meanwhile in the 30-year continuous monoculture field, these concentrations were 14.58, 3.20, 3.27, 21.38, 3.70, 2.62, and 3.65 mg/kg, respectively. The recovery percentage of PCA, EGC, EGCG, EC, C, ECG, and TF were in an acceptable range of between 60% and 120% (Table S2).

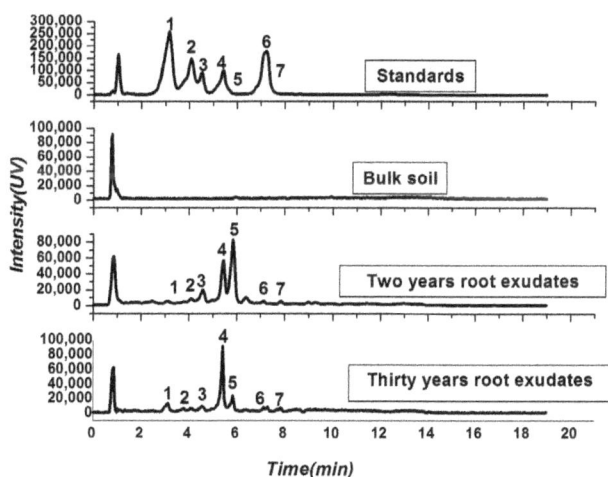

Figure 1. High performance liquid chromatography–electrospray ionization–mass spectrometry (HPLC–ESI–MS) spectra of catechins in root exudates collected from different plantations of different ages tea fields and bulk soil; "1"represents protocatechuic acid (PCA) with a retention time of 3.12 min; "2"represents epigallocatechin (EGC) with a retention time of 4.07 min; "3" represents epigallocatechin-3-gallate (EGCG) with a retention time of 4.53 min; "4" represents epicatechin (EC) with a retention time of 5.43 min; "5" represents (+)-catechin (C) with a retention time of 5.72 min; "6" represents epicatechin-3-gallate (ECG) with a retention time of 7.13 min; and "7" represents taxifolin (TF) with a retention time of 7.32 min.

2.3. 16S rDNA-Based Meta-Genomic Analysis of Tea Root-Associated Bacteria

According to rarefaction analysis, the number of Operational Taxonomic Units (OTUs) for 16S rRNA plateaued after 45,000 sequences at 97% similarity (Figure 2). This implied that the sequence depth was sufficient to relatively and accurately capture the diversity and richness of these samples. A total of 1,452,893 (average: 69,185) reads were analyzed by pyrosequencing. In total, 1,442,103 (average: 68,672) reads were paired successfully with 16S rRNA. After removing short and low-quality reads, singletons, replicates, and chimeras, 1,363,660 high quality reads (average: 64,936) were obtained from 21 samples. In total 1,210,023 tags (average: 67,223.5) were classified into 1,166,269 (average: 64,792.72) taxa tags. Based on ≥97% similarity, taxa tags were clustered into 54,719 (average: 2606) OTUs ranging from 1695 to 3684 per sample. In these taxa tags, 99.1% were classified as various bacteria, which primarily consist of 45 phyla, whereas the remaining 0.9% were classified as *Archaea* (Figure S2, Supplemental Dataset 1–5). Heat map analysis (Figure 3a) showed that the dominant bacterial phyla in bulk soil (CK) were *Proteobacteria (Phenylobacterium, Haliangium, Halomonas, Delftia,* and *Sorangium)* and *Acidobacteria (Bryobacter* and *Candidatus_solibacter).* In rhizosphere of the 30-year-old tea plantation (RS30) *Mizugakiibacter, Rodanobacter, Acidobacterium, Granulicella, Telmatobacter,* and *Verrucomicobia,* while in rhizosphere of the two-year-old tea plantation (RS2) *Halomonas, Acidobacter, Mizugakiibacter, Acidobacterium, Granulicella, Telmatobacter,* and *Acidothermus* genera were the most dominant. In rhizoplane of the two-year-old tea plantation (RP2) *Cyanomonas* and *Arthobacter,* while in rhizoplane of the 30-year-old tea plantation (RP30) *Arthobacter* and *Sphingomonas* were the most dominant genera. In endosphere of the two-year-old tea plantation (ES2) *Burkholderia, Bradyrhizobium, Dyella, Sphingomonas, Rhizobium, Cyanobacteria,* and *Actinospica* were dominant, while in endosphere of the 30-year-old tea plantation (ES30) *Novosphingobium, Pseudomonas, Shinella, Rhizobium, Delfia, Sphingobacterium, Dyadobacter, Flavanobacterium, Chryseobacterium,* and *Oerskovia* were the most dominant genera. The most dominant phyla across all the samples were *Proteobacteria, Acidobacteria, Cyanobacteria, Chloroflexi, Actinobacteria, Bacteroides, Nitrospirae, Verrucomicobia* (WD272), *Gemmatimonadetes,* and *Firmicutes* accounting for 90–96% of the bacterial sequences (Figure 3b).

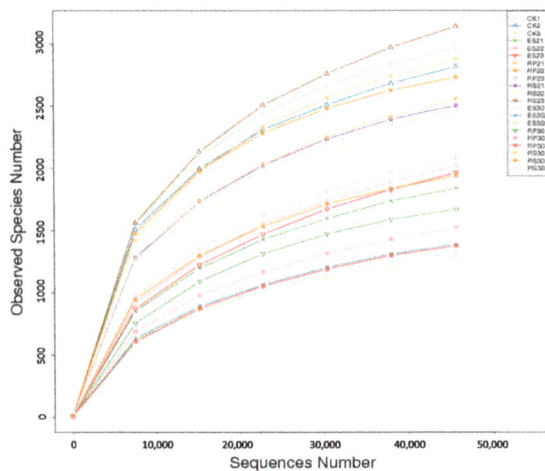

Figure 2. The rarefaction curve based on 97% similarity. CK1, CK2, CK3 refer to the bulk soil, RS21, RS22, RS23 represent rhizosphere of the two-year-old tea plantation, RP21, RP22, RP23 represent rhizoplane of the two-year-old tea plantation, and ES21, ES22, ES23 represent endosphere of the two-year-old tea plantation respectively. While RS301, RS302, RS303 represent rhizosphere of the thirty-year-old tea plantation, RP301, RP302, RP303 represent rhizoplane of the thirty-year-old tea plantation, and ES301, ES302, ES303 represent endosphere of the thirty-year-old tea plantation respectively.

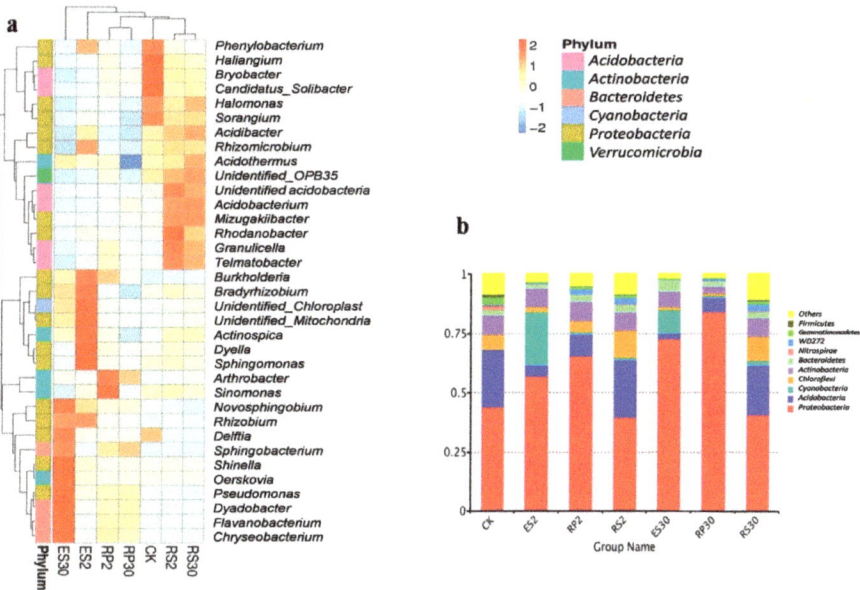

Figure 3. (**a**) Heat map showing the distribution of the 35 most abundant genera. (**b**) Distribution of the top 10 most abundant phyla in bulk soil (CK), rhizosphere of the two-year-old tea plantation (RS2), rhizoplane of the two-year-old tea plantation (RP2), endosphere of the two-year-old tea plantation (ES2), rhizosphere of the thirty-year-old tea plantation (RS30), rhizoplane of the thirth-year-old tea plantation (RP30) and endosphere of the thirty-year-old tea plantation (ES30).

2.4. Distinct and Overlapping Bacterial Communities of Root-Associated Bacterial Communities across Different Tea Plantations of Different Ages

According to Chao1 and ACE estimators (Figure 4a,b), the rhizosphere, rhizoplane, and endosphere of the newly planted tea garden showed higher bacterial community richness than the continuous (30-year) monocropping tea field. Although the Shannon and Simpson diversity indices decreased significantly in the rhizoplane and endosphere of the continuous monoculture plants (30-year) when compared to the fresh tea plantation and the uncultivated field (Figure 4c,d), no significant differences were observed for the rhizosphere. Moreover, in the rhizo-compartments of the same type of tea garden, bacterial community richness showed a decreasing gradient from the rhizosphere toward the endosphere (Figure 4), but the alpha diversity in the same type of tea garden showed varying gradients. These results suggest that continuous monoculture could impact the variance in diversity and abundance of the bacterial communities in root-associated microbiomes within the samples.

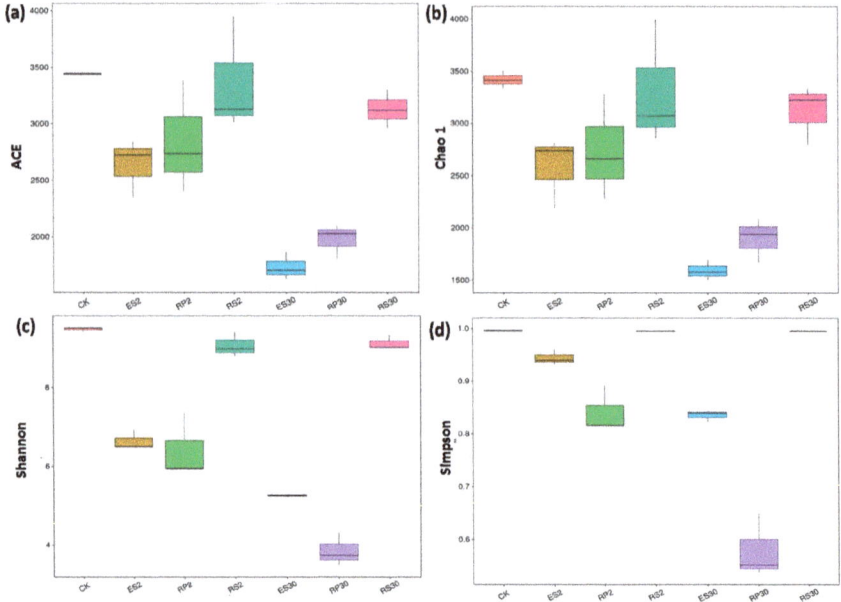

Figure 4. α-diversity indices including calculations of (**a**) abundance-based coverage estimators (ACE), (**b**) Chao1, (**c**) Shannon, and (**d**) Simpson in bulk soil (CK) and in various rhizo-compartments of different tea plantations of different ages: RS2, RP2, ES2, RS30, EP30 and ES30.

β-diversity was estimated by using both weighted (based on the abundance of taxa) and unweighted (sensitive to rare taxa) UniFrac distance matrices between samples. The weighted UniFrac principal coordinate analysis (PCoA) depicted that the bacterial communities in bulk soil and the rhizosphere were separated from the rhizoplane and endosphere and clustered along the principal coordinate axis-1, while the RP and ES were clustered along axis-2 (Figure 5a). Moreover, unweighted UniFrac PCoA analysis showed that the bacterial communities in the bulk soil and rhizosphere were different from the RP and ES and clustered along principle coordinate axis-1, and the RP and ES clustered along principle component axis-2 (Figure 5b).The unweighted pair group method with arithmetic mean (UPGMA) clustering analysis clearly indicated the differences in bacterial community structure among different samples, and similar results were observed for the same sample when analyzed in triplicate (Figure 5c). The weighted and unweighted UniFrac distances between CK and RS2 were 0.184 and 0.478, respectively and were 0.223 and 0.479 between CK and RS30, respectively. In comparison with CK and newly planted tea (RS2, RP2, and ES2) the weighted and unweighted UniFrac distances increased in the 30-year-old tea plantation (RS30, RP30, and ES30) (Figure 5d).

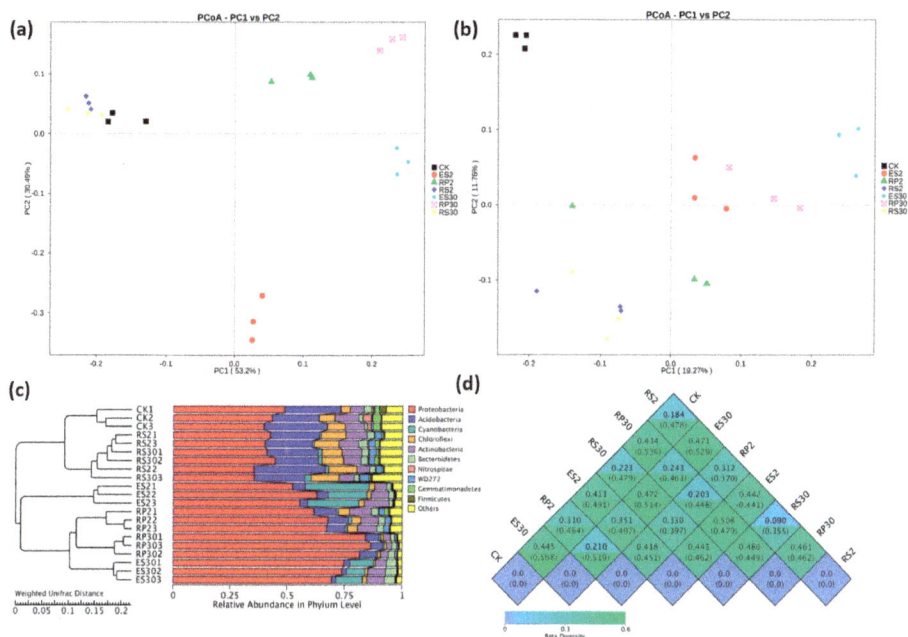

Figure 5. Biplot ordination of (**a**) weighted UniFrac principal coordinate analysis (PCoA) and (**b**) Unweighted UniFrac (UUF) principal coordinate analysis (PCoA) among various samples. CK represents a nearby uncultivated field, while RS2, RP2, and ES2 represent the rhizosphere, rhizoplane, and endosphere of the two-year tea plantation, and RS30, RP30, and ES30 represent the rhizosphere, rhizoplane, and endosphere of the 30-year-old tea plantation. (**c**) UPGMA/hierarchical clustering analysis based on weighted UniFrac distances showing the relative abundance of the most abundant bacterial phylum in various rhizo-compartments of different tea plantations of different ages. (**d**) β-diversity heat map based on weighted (WUF) and unweighted (UUF) UniFrac distances. Values in the upper and lower corners represented the WUF and UUF distances.

2.5. Interactions of Bacterial Abundance with Soil Physio-Chemical Properties and Root Exudates

Redundancy analysis was performed to study the relationships between root exudates, soil physical and chemical properties, and the abundance of bacterial phylum. Strong associations were found among available nitrogen (AN), total nitrogen (TN), available potassium (AK), and available phosphorus (AP) with the abundance of *WD272*, *Cyanobacteria*, and *Actinobacteria*, clustering along axis-1 in the two-year-old tea plantation (Figure 6). Although most of the bacterial phylum depicted a strong negative correlation with the soil pH, the *Acidobacteria*, *Gemmatimonadetes*, and *Nitrospirae* had a strong positive association with soil pH, clustering along axis-2 in both the bulk soil and the 30-year-old tea plantation (Figure 6). The abundance of *Bacteroidetes* and *Proteobacteria* were found to be highly associated with soil moisture (MOS) in the 30-year-old tea plantation compared to the two-year-old tea plantation (Figure 6). However, *Chloroflexi* and *Actinobacteria* showed a strong positive association with the total phosphorus (TP), but were negatively associated with the soil moisture.

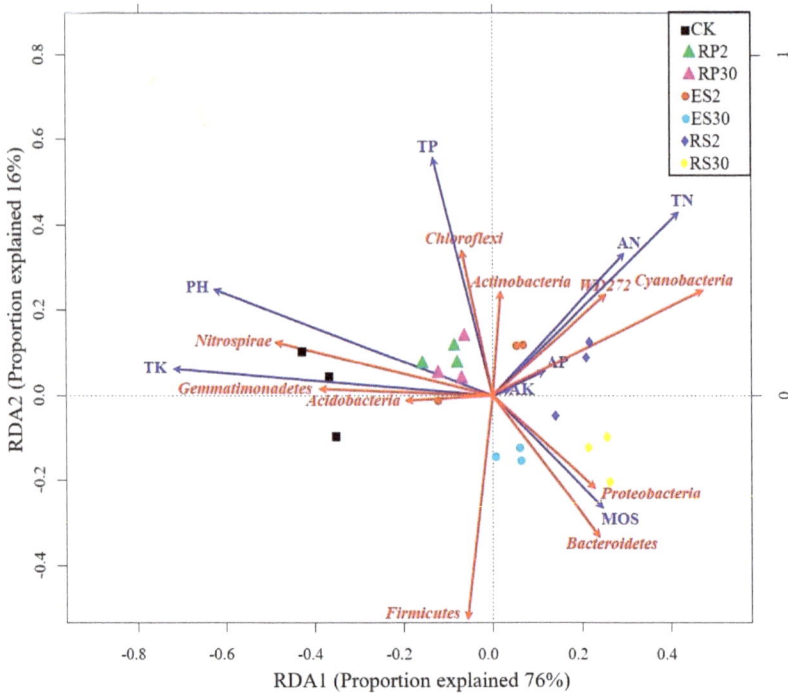

Figure 6. Redundancy analysis (RDA) Triplot (RDA on a covariance matrix) of the correlation between the most abundant phylum of bacteria and soil physiochemical properties, such as total phosphorus (TP), total nitrogen (TN), available nitrogen (AN), available potassium (AK), available phosphorus (AP), and pH across the various rhizo-compartments of the different tea plantations of different ages, where rhizosphere RS2 and RS30 are the rhizosphere, RP2 and RP30 are the rhizoplane, and endosphere ES2 and ES30 are the endosphere in two-year and 30-year-old tea gardens, respectively. The arrow length and direction correspond to the variance that can be explained by the environmental and response variables. The direction of an arrow indicates the extent to which the given factor is influenced by each RDA variable. The perpendicular distance between the abundance of bacterial phyla and environmental variable axes in the plot reflects their correlations. The smaller the distance, the stronger the correlation.

Redundancy analysis (RDA) ordination between root exudates, allelechemicals, and bacterial phylum abundance depicted that the abundance of most bacterial phylum, except *Proteobacteria*, *Bacteroidetes* and *Firmicutes*, had a strong negative association with PCA, TF, EC, and EGC cluster along the first axis (RDA1) in the 30-year-old tea plantation (Figure 7). On the other hand, C was clustered along RDA axis-2 was strongly associated with most of the bacterial phyla (Figure 7). These results clearly indicate that the higher concentration of allelochemicals (PCA, TF, EC, and EGC) play a significant role in the declining bacterial communities in the 30-year-old tea plantation.

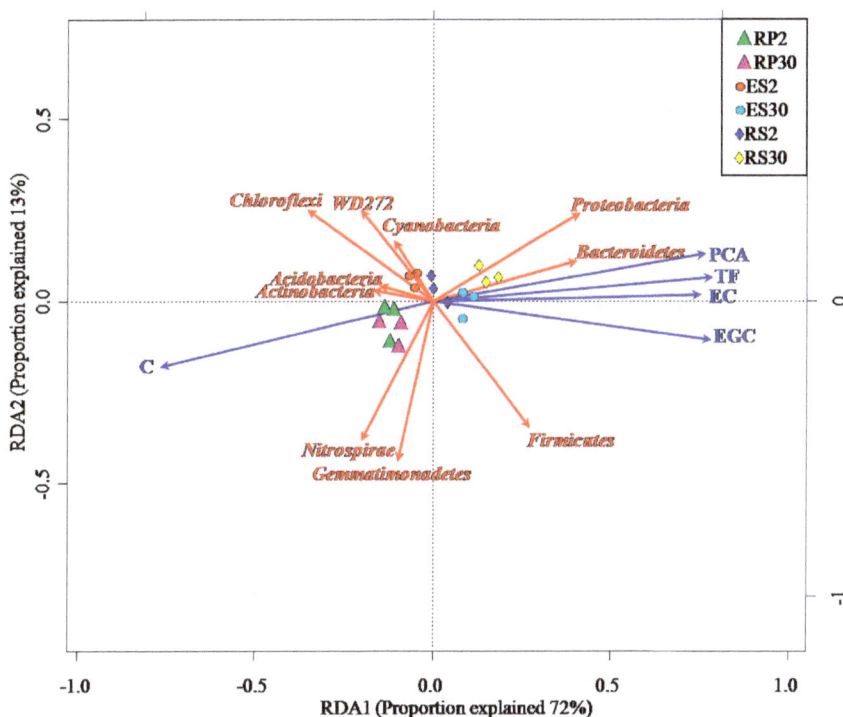

Figure 7. RDA Triplot (RDA on a covariance matrix) of the correlation between the most abundant phylum of bacteria and root exudates, such as C, EGC, EC, TF, and PCA, across various rhizo-compartments of different tea plantations of different ages, where the rhizosphere is represented by RS2 and RS30, the rhizoplane by RP2 and RP30, and the endosphere by ES2 and ES30, in two-year and 30-year-old tea gardens, respectively. The arrow length and direction correspond to the variance that can be explained by the environmental and response variables. The direction of the arrow indicates the extent to which the given factor is influenced by each RDA variable. The perpendicular distance between abundance of the bacterial phyla and the environmental variable axes in the plot reflects their correlations. The smaller the distance, the stronger the correlation.

3. Discussion

The results shown here provide an insight for addressing the problems related to continuously growing a ratooning tea monoculture through the characterization of the microbiome, combined with the finer population structural details. Our design allowed us to successfully elaborate on the influence of a continuously monocultured tea plantation on its microbial community associated with all three rhizo-compartments. Moreover, this process also permitted us to reveal the extent to which plant-microbial interactions were mediated by the soil physio-chemicals and root exudates. Specifically, we characterized the microbial compositions of three distinct rhizo-compartments: the rhizosphere, rhizoplane, and endosphere, and showed the influence of external factors on each of these rhizo-compartments in bulk soil, on a fresh tea plantation, and an old tea plantation. Although these impacts were not significant, we found that the α- and β-diversity of microbes differ significantly across plantations of different ages. Depiction of RDA ordinations illustrated the association of bacterial phylum over time with the changing concentration of soil physiochemical properties and plant root exudates.

Recent studies on microbiomes have illustrated the power of deeper sequencing in describing the structural composition of plant microbiomes covering a geographical range of cultivation [40]. By adapting a deeper sequencing approach [35,41] most studies showed only the microbial abundance and impact of soil physiochemical properties on the microbiota present in the rhizosphere. However, the diversity and abundance of microbiota present in all three rhizo-compartments, and the influence of both soil physio-chemicals and root exudates, have not been previously studied together in for tea crops in plantations of different ages. To address this, we successfully adapted previously-described protocols for the removal of microbes from all three rhizo-compartments of new and old tea plantations. We found that the microbial communities, associated with rhizo-compartments under three field conditions, were chiefly clustered by the difference in the age of the tea plantations, which is in accordance with the previous work showing the effects of continuous plantation on soil-associated microbes [41,42]. However, the microbiota associated with the rhizosphere, rhizoplane, and endosphere in the same field did not show a significant difference. Peiffer et al. [42] also showed that field conditions did not affect the patterns of variations among the rhizo-compartments. Furthermore, we observed a greater α- and β-diversity of microbiota across plantations of different ages. These results suggest that the increasing age of the tea plants affects the microbiota in all three rhizo-compartments. Heat map analysis showed that most of the beneficial microbial genera, such as *Halomonas*, *Cyanomonas*, *Burkholderia*, *Bradyrhizobium*, *Dyella*, *Rhizobium*, *Cyanobacteria*, and *Actinospica* that are involved in the carbon, sulfur, and nitrogen cycles as well asprobiotics, decreased with the increasing age of the ratooning tea monoculture tea plantation. Zhao et al. [35] also showed that beneficial bacteria decreased and harmful bacteria increased with increasing age of *Pseudostellaria heterophylla* plantations. As the richness and β-diversity of microbiota clearly indicate differences across different-aged tea plantations, we hypothesized that these variations are chiefly governed-over time by the imbalance of soil physiochemical and/or root exudates in tea fields.

Our results showed no significant difference in the soil nutrients (N, P, and K).Only soil pH significantly decreased in the 30-year-old tea plantation when compared with the bulk soil and the two-year-old tea field, which resulted in lowering the abundance of rhizospheric microbiota, such as *Actinobacteria*, *Cyanobacteria*, *Chloroflexi*, and *WD272* [35], and showed direct association with *Nitrospirae*, *Gemmatimonadetes*, and *Acidobacteria*. On the other hand, *Proteobacteria* and *Bacteroidetes* were the most dominant bacterial group present in the 30-year-old tea plantation [43,44], which may be because these bacterial taxa prefer to live in a root vicinity with nutrient deficiency, high soil moisture (MOS), and lower pH. Previous studies on bacterial communities have clearly shown that nutrient deficiency did not significantly impact the soil microbiota, however soil pH was proven to be the most influential factor in shaping the soil microbiota [45–47]. The actual cause of a reduction in pH in older tea plantations was not studied until now, so we hypothesized that soil pH in the 30-year-old soil decreases with the increasing level of root exudates (allelochemicals).

Our results depicted higher levels of PCA, TF, EC, and EGC in ratooning 30-year-old tea monocropping fields, which may cause the lowering of soil pH over time [48]. In previous studies [32], C and EC were the carbon sources of some protobacteria, such as *Pseudomonas*, which either directly affected growth of the plant or indirectly affected it by biotransformation into more toxic compounds like TF and PCA. In our results, we showed that accumulation of these organic acids, such as PCA, could change the pH of the 30-year-old tea plantation over time. In the RDA study, the *Proteobacteria*, *Bacteroidetes*, and *Firmicutes* showed strong association with increasing levels of EC, EGC, TF, and PCA. Whereas C, which is secreted by the roots of young tea plants in significant amounts, could act as a raw source for catechin's degradation of bacteria. For example, these results provide evidence that the rhizospheric microbiota in different aged tea plantations is influenced by pH and is indirectly affected by the carbon source-to-sink relationships within the plants [42]. In addition, these results suggested that the excessive accumulation of catechins can be prevented by removing the carbon sink sources, such as removal of fallen tea leaves and other plant debris from the soil.

In conclusion, the dynamic shift in microbiomes across a continuously monocultured tea plantation were related to root exudates. Moreover, soil pH changed with the changing level of root exudates (PCA), ultimately causing a decline in the population of pH-sensitive microbiota with continuous tea monocropping. Our findings suggest that removal of plant debris and pruned tea leaves from the soil can prevent the accumulation of PCA, TF, and EC, which may delay the reduction in soil pH. However, future research is needed to further explore the underlying mechanisms of rhizoshperic microbial interactions mediated by root exudates and to identify the contributing factors to soil sickness in continuous tea monocropping.

4. Materials and Methods

4.1. The Collection of Plant and Rhizosphere Soil Samples

Plant leaves, roots, and soil samples of tea were collected from freshly planted (2 years old) and old (30 years old) tea fields, and bulk soil was obtained from nearby uncultivated fields, located in the observation station in Anxi county (117°36′–118°17′ E, 24°50′–25°26′ N) on the south coast of Fujian province, China. These tea fields and nearby uncultivated tea fields shared the same environmental conditions and agronomic management. The sampling depth was about 30 cm in a range of about 25 cm from the root of the plant. Tea roots were carefully uprooted from the soil with a forked spade and slightly shaken to remove loosely attached soil. The rhizosphere soil tightly attached to the roots was brushed off and collected. The distance between the two plants was 0.5–0.7 m. The tea plants were 1.2 m in height and 0.9–1.2 m in width. At the time of sampling, the samples were taken at random, and each of the treatments included 15 rhizosphere soil samples. An ice box was used to bring the collected soil samples back to the laboratory. In order to reduce the error caused by spatial heterogeneity, five random sampling points within the 15 sampling sites were mixed into one soil sample, and three soil samples were obtained for each treatment [22]. To determine the microbial flora, we followed the method proposed by Edwards et al. [49] with little modification. Briefly, bacterial communities were derived from three different soil and tea plant compartments; the RS comprised of the soil tightly adhering to the root surface, the RP consisted of the suite of microbes present on the root surface by sonication, and the ES covered the interior of the same plant roots after sonication. A five-point sampling method was used [13] to make one sample with three replicates. All the samples were stored at −80 °C.

4.2. Quality Parameters and Soil Properties Determination

Quality parameters such as the anine (TNN), theophylline (TPY), total polyphenols (TPP), and total free amino acids (TAA) of tea plantations of different ages were determined by the method of Peng et al. [50]. Soil pH was determined using a glass electrode pH meter (1:2.5 soil to water suspensions). Soil moisture (MOS) and water content (WC) of tea leaves were measured by the following Oven Drying Method. Total Potassium (TK), TN and TP were measured by the NaOH melt flamer method, dichromate oxidation, and Kjeldahl digestion respectively [51]. AP was extracted by using ammonium fluoride and hydrochloric acid and measured by the following Molybdenum Blue Method. AN was calculated by the alkaline hydrolysable method. AK was extracted using ammonium acetate and determined by flame photometry [52].

4.3. Identification and Quantification of Allelochemicals from Tea Root Exudates

To identify and quantify allelochemical concentration in tea roots exudate samples, ≥98% pure C, ECG, EGC, EGCG, EC, PCA, and TF were purchased from Cayman Chemical (1180 E. Ellsworth Road, Ann Arbor, MI, USA) as external standards. HPLC–MS grade methanol and formic acid were purchased from Sigma-Aldrich (St. Louis, MO, USA). Firstly, the stock solutions were prepared by dissolving 1–2 mg of each standard in 1–2 mL of solution containing 99.9% methanol and 0.1% acetic acid by volume. To obtain the calibration curves, serial dilutions were performed using the same solvent as used to prepare the stock solutions at different concentrations ranging between 1.25–10 µg/mL. A 10-µL aliquot of all solutions were injected for HPLC–ESI–MS analysis. Good linearity over the calibration range was achieved with all coefficients of correlation >0.998. The concentration of catechins from roots and soil was determined by following the method of Wang et al. [32] with slight modifications. Ten grams of soil, and 5 grams of roots and leaves were vortexed in a 15-mL solution containing 0.1% acetic acid and 50% methanol by volume at 150 rpm for 24 h. The sample was then centrifuged for 10 min at 4500 rpm and supernatant fluid was moved to a sample collection vial to perform liquid chromatography (LC). Following this, HPLC–ESI–MS was carried out using T3 RP-18 column (100 × 2.1 mm; 5 µm; Waters, Milford, MA, USA) eluted with buffer A (Sigma-Aldrich Co., St. Louis, MO, USA) (0.1% acetic acid) and buffer B (EMD Milliopre Corporation, Billerica, MA, USA) (100% methanol) at a flow rate of 300 µL/min at 25 °C. Initially, the column was eluted with 95% buffer B, followed by a linear increase in buffer A to 35% from 0–10 min, and further maintained in 90% buffer A for 5 min. Then, a linear increase in buffer B to 95% was maintained. Finally, the column was maintained in 95% buffer B for 8 min. The total time for running one sample was 19 min. The negative ionization mode was selected to perform mass spectrometry at a temperature of 100 °C, and ion scans were carried out at low-energy collision (20 eV) using nitrogen as the collision gas. All the data from HPLC–ESI–MS were processed to determine the mean concentrations of the selected catechin compounds in each sample, by using Bruker Daltonics Data analysis software version 4.0 (Thermo Fisher Scientific, Waltham, MA, USA).

4.4. DNA Extraction and Purification

Whole genome DNA was extracted from bulk soil, rhizosphere, rhizoplane, and endospheric samples by using Fast DNATM Spin Kit, specialized for extracting DNA from soil, following the manufacturer's instructions manual (MP Biomedical, Santa Ana, CA, USA). Then all the DNA samples were subjected to gel electrophoresis and further purification, using Universal DNA Purification Kits according to the manufacturer's instructions (Tiangen Biotech Co., Ltd., Beijing, China). DNA was quantified by using Nanodrop (Thermo Fisher Scientific, Waltham, MA, USA) before being stored at −20 °C for further molecular analysis.

4.5. The Metagenomic Analysis of the Root-Associated Rhizo-Compartment Bacteria

Purified DNA samples were sent to Novogene Bioinformatics Technology Co., Ltd. (Beijing, China) to determine the bacterial community structure. Prior to high throughput sequencing, 16S V4, a distinct gene region of 16S rRNA, was amplified using specific primer 515F-806R with barcodes [53]. All PCR reactions were conducted using 30 µL total reaction volume with 15 µL of Phusion® High-Fidelity PCR Master Mix (New England Biolabs (Beijing) Ltd., Beijing, China) containing ~10 ng template DNA and 0.2 µM of each primer pair. The PCR condition set was denaturation at 98 °C for 1 min, followed by 30 cycles of denaturation at 98 °C for 10 s, annealing at 50 °C for 30 s, and elongation at 72 °C for 60 s, with a final extension at 72 °C for 5 min. Then electrophoresis using 2% agarose gel solution was performed to verify the successful DNA amplification mixing PCR products with the same amount of 1× loading buffer containing SYB green (TIANGEN, Beijing, China). Samples showing main strip brightness ranging between 400–450 bp was selected for further sequencing. PCR products were purified by using Gene JET Gel Extraction Kit (Qiagen, Hilden, Germany) prior to sequencing. Sequencing libraries were created in Illumina using specialized NEB Next® Ultra™ DNA Library Prep Kit (New England Biolabs (Beijing) Ltd., Beijing, China) according to the manufacturer's instructions and index codes were added. The quality of the developed sequencing library was checked in both the Agilent Bioanalyzer 2100 system (Agilent, Santa Clara CA, USA) and the Qubit® 2.0 Fluorimeter (Thermo Fisher Scientific, Waltham, MA, USA). Finally, 250/300bp paired-end reads were generated on an Illumina MiSeq platform (Illumina, San Diego, CA, USA).

4.6. Statistical and Bioinformatics Analysis

Raw sequences were classified according to the specific barcode assigned to each sample, using Quantitative Insights Into Microbial Ecology (QIIME) (CO, USA) [53]. Paired-end reads were merged from the original DNA segments using FLASH (Baltimore, MD, USA) [54]. Paired-end reads were assigned to each sample according to the unique barcodes attached with DNA fragments. Sequences were analyzed using UPARSE-OTU and UPARSE-OUT reference algorithms with UPARSE software package (CA, USA). Alpha (within samples) and beta diversity (among samples) were calculated using QIIME (CO, USA) [53]. The same Operational Taxonomic Units (OTUs) were assigned to the sequences with ≥97% in each sample. One representative sequence was selected for each OTU to annotate the taxonomic information of each representative sequence by using the RDP classifier.

To measure the Alpha diversity within the sample, we rarified the OUT table and then four diversity matrices were calculated: Chao1 estimates the species abundance, and the Observed Species, Simpson, and Shannon indices were used to determine the community diversity. Moreover, rarefaction curves were developed for each of these four indices. Abundance of each bacterial taxa, from phylum to species, was shown graphically using a Krona Chart. Beta diversity (among samples) was measured for both weighted and unweighted UniFrac distances using QIIME (Version 1.7.0) (CO, USA). Principal Component Analysis (PCA) and Principal Coordinate Analysis (PCoA) were performed and visualized using R (Version 2.15.3) packages; stat, WGCNA and ggplot2 (Elegants graphics for data analysis, New York, NY, USA). We identified the association of the most abundant bacterial phylum with selected soil physiochemical properties and root exudates using partial-RDA [55], and triplots were generated using vegan and ggplot2 packages in R software (R version 3.3.1) (Foundation for Statistical Computing, Vienna, Austria).

Supplementary Materials: Supplementary materials can be found at www.mdpi.com/1422-0067/18/8/1727/s1.

Acknowledgments: This work was supported by 948 program from Ministry of Agriculture (2014-Z36) China, National Key Research and Development (R & D), China plan 2016YFD0200900, and Major agricultural extension services (KNJ-153015), Fujian Province, China.

Author Contributions: Wenxiong Lin, Sheng Lin and Yasir Arafat conceived the study; Yasir Arafat and Wenxiong Lin wrote the paper; Yasir Arafat and Yuhang Jiang performed field sampling and lab experiments; Yasir Arafat and Hafiz Sohaib Ahmed Saqib performed the statistical analyses. All authors discussed the results and commented on the manuscript.

Conflicts of Interest: The authors declare no conflict of interest.

References

1. Bever, J.D.; Westover, K.M.; Antonovics, J. Incorporating the soil community into plant population dynamics: The utility of the feedback approach. *J. Ecol.* **1997**, *85*, 561–573. [CrossRef]
2. Bever, J.D. Soil community feedback and the coexistence of competitors: Conceptual frameworks and empirical tests. *New Phytol.* **2003**, *157*, 465–473. [CrossRef]
3. Kulmatiski, A.; Beard, K.H.; Stevens, J.R.; Cobbold, S.M. Plantntitors: Conceptual fra-analytical review. *Ecol. Lett.* **2008**, *11*, 980–992. [CrossRef] [PubMed]
4. Putten, W.H.; Bardgett, R.D.; Bever, J.D.; Bezemer, T.M.; Casper, B.B.; Fukami, T.; Kardol, P.; Klironomos, J.N.; Kulmatiski, A.; Schweitzer, J.A. Plant–soil feedbacks: The past, the present and future challenges. *J. Ecol.* **2013**, *101*, 265–276. [CrossRef]
5. Baxendale, C.; Orwin, K.H.; Poly, F.; Pommier, T.; Bardgett, R.D. Are plant–soil feedback responses explained by plant traits? *New Phytol.* **2014**, *204*, 408–423. [CrossRef] [PubMed]
6. Huang, L.F.; Song, L.X.; Xia, X.J.; Mao, W.H.; Shi, K.; Zhou, Y.H.; Yu, J.Q. Plant-soil feedbacks and soil sickness: From mechanisms to application in agriculture. *J. Chem. Ecol.* **2013**, *39*, 232–242. [CrossRef] [PubMed]
7. Klironomos, J.N. Feedback with soil biota contributes to plant rarity and invasiveness in communities. *Nature* **2002**, *417*, 67–70. [CrossRef] [PubMed]

8. Van-der-Putten, W.H. Plant defense belowground and spatiotemporal processes in natural vegetation. *Ecology* 2003, *84*, 2269–2280. [CrossRef]
9. Wardle, D.A.; Bardgett, R.D.; Klironomos, J.N.; Setälä, H.; van-der-Putten, W.H.; Wall, D.H. Ecological linkages between aboveground and belowground biota. *Science* 2004, *304*, 1629–1633. [CrossRef] [PubMed]
10. Harrison, K.A.; Bardgett, R.D. Influence of plant species and soil conditions on plant–soil feedback in mixed grassland communities. *J. Ecol.* 2010, *98*, 384–395. [CrossRef]
11. Muller, C.H. The role of chemical inhibition (allelopathy) in vegetational composition. *Bull. Torrey Bot. Club* 1966, *93*, 332–351. [CrossRef]
12. Zhao, Y.P.; Lin, S.; Chu, L.; Gao, J.T.; Azeem, S.; Lin, W. Insight into structure dynamics of soil microbiota mediated by the richness of replanted *Pseudostellaria heterophylla*. *Sci. Rep.* 2016, *6*, 26175. [CrossRef] [PubMed]
13. Zhao, Y.; Wu, L.; Chu, L.; Yang, Y.; Li, Z.; Azeem, S.; Zhang, Z.; Fang, C.; Lin, W. Interaction of *Pseudostellaria heterophylla* with *Fusarium oxysporum* f. Sp. *Heterophylla* mediated by its root exudates in a consecutive monoculture system. *Sci. Rep.* 2015, *5*, 8197. [CrossRef]
14. Wu, L.; Chen, J.; Wu, H.; Wang, J.; Wu, Y.; Lin, S.; Khan, M.U.; Zhang, Z.; Lin, W. Effects of consecutive monoculture of *Pseudostellaria heterophylla* on soil fungal community as determined by pyrosequencing. *Sci. Rep.* 2016, *6*, 26601. [CrossRef] [PubMed]
15. Bennett, A.J.; Bending, G.D.; Chandler, D.; Hilton, S.; Mills, P. Meeting the demand for crop production: The challenge of yield decline in crops grown in short rotations. *Biol. Rev.* 2012, *87*, 52–71. [CrossRef] [PubMed]
16. Wu, L.; Wang, H.; Zhang, Z.; Lin, R.; Zhang, Z.; Lin, W. Comparative metaproteomic analysis on consecutively *Rehmannia glutinosa*-monocultured rhizosphere soil. *PLoS ONE* 2011, *6*, e20611. [CrossRef] [PubMed]
17. Santhanam, R.; Weinhold, A.; Goldberg, J.; Oh, Y.; Baldwin, I.T. Native root-associated bacteria rescue a plant from a sudden-wilt disease that emerged during continuous cropping. *Proc. Natl. Acad. Sci. USA* 2015, *112*, E5013–E5020. [CrossRef] [PubMed]
18. Qu, X.H.; Wang, J.G. Effect of amendments with different phenolic acids on soil microbial biomass, activity, and community diversity. *Appl. Soil Ecol.* 2008, *39*, 172–179. [CrossRef]
19. Utkhede, R.S. Soil sickness, replant problem or replant disease and its integrated control. *Allelopathy J.* 2006, *18*, 23–38.
20. Li, C.; Li, X.; Kong, W.; Wu, Y.; Wang, J. Effect of monoculture soybean on soil microbial community in the northeast China. *Plant Soil* 2010, *330*, 423–433. [CrossRef]
21. Liua, X.; Herbert, S.J. Fifteen years of research examining cultivation of continuous soybean in northeast china: A review. *Field Crops Res.* 2002, *79*, 1–7. [CrossRef]
22. Li, Y.; Li, Z.; Arafat, Y.; Lin, W.; Jiang, Y.; Weng, B.; Lin, W. Characterizing rhizosphere microbial communities in long-term monoculture tea orchards by fatty acid profiles and substrate utilization. *Eur. J. Soil Biol.* 2017, *81*, 48–54. [CrossRef]
23. Ferrazzano, G.F.; Roberto, L.; Amato, I.; Cantile, T.; Sangianantoni, G.; Ingenito, A. Antimicrobial properties of green tea extract against cariogenic microflora: An in vivo study. *J. Med. Food* 2011, *14*, 907–911. [CrossRef] [PubMed]
24. Sakanaka, S.; Juneja, L.R.; Taniguchi, M. Antimicrobial effects of green tea polyphenols on thermophilic spore-forming bacteria. *J. Biosci. Bioeng.* 2000, *90*, 81–85. [CrossRef]
25. Taylor, P.W.; Hamilton-Miller, J.M.T.; Stapleton, P.D. Antimicrobial properties of green tea catechins. *Food Sci. Technol. Bull.* 2005, *2*, 71. [CrossRef] [PubMed]
26. Veluri, R.; Weir, T.L.; Bais, H.P.; Stermitz, F.R.; Vivanco, J.M. Phytotoxic and antimicrobial activities of catechin derivatives. *J. Agric. Food Chem.* 2004, *52*, 1077–1082. [CrossRef] [PubMed]
27. Henning, S.M.; Fajardo-Lira, C.; Lee, H.W.; Youssefian, A.A.; Go, V.L.W.; Heber, D. Catechin content of 18 teas and a green tea extract supplement correlates with the antioxidant capacity. *Nutr. Cancer* 2003, *45*, 226–235. [CrossRef] [PubMed]
28. Carando, S.; Teissedre, P.L. Catechin and procyanidin levels in french wines: Contribution to dietary intake. *Basic Life Sci.* 1999, *66*, 725–737. [PubMed]
29. Chunmei, D.; Jiabo, W.; Weijun, K.; Cheng, P.; Xiaohe, X. Investigation of anti-microbial activity of catechin on *Escherichia coli* growth by microcalorimetry. *Environ. Toxicol. Pharmacol.* 2010, *30*, 284–288. [CrossRef] [PubMed]

30. Bais, H.P.; Walker, T.S.; Kennan, A.J.; Stermitz, F.R.; Vivanco, J.M. Structure-dependent phytotoxicity of catechins and other flavonoids: Flavonoid conversions by cell-free protein extracts of *Centaurea maculosa* (spotted knapweed) roots. *J. Agric. Food Chem.* **2003**, *51*, 897–901. [CrossRef] [PubMed]

31. Callaway, R.M.; Thelen, G.C.; Rodriguez, A.; Holben, W.E. Soil biota and exotic plant invasion. *Nature* **2004**, *427*, 731–733. [CrossRef] [PubMed]

32. Wang, C.M.; Li, T.C.; Jhan, Y.L.; Weng, J.H.; Chou, C.H. The impact of microbial biotransformation of catechin in enhancing the allelopathic effects of *Rhododendron formosanum*. *PLoS ONE* **2013**, *8*, e85162. [CrossRef] [PubMed]

33. Perry, L.G.; Thelen, G.C.; Ridenour, W.M.; Callaway, R.M.; Paschke, M.W.; Vivanco, J.M. Concentrations of the allelochemical (±)-catechin in *Centaurea maculosa* soils. *J. Chem. Ecol.* **2007**, *33*, 2337–2344. [CrossRef] [PubMed]

34. Han, W.; Kemmitt, S.J.; Brookes, P.C. Soil microbial biomass and activity in Chinese tea gardens of varying stand age and productivity. *Soil Biol. Biochem.* **2007**, *39*, 1468–1478. [CrossRef]

35. Li, Y.C.; Li, Z.; Li, Z.W.; Jiang, Y.H.; Weng, B.Q.; Lin, W.X. Variations of rhizosphere bacterial communities in tea (*Camellia sinensis* L.) continuous cropping soil by high-throughput pyrosequencing approach. *J. Appl. Microbiol.* **2016**, *121*, 787–799. [CrossRef] [PubMed]

36. Cao, P.; Liu, C.; Li, D. Effects of different autotoxins on antioxidant enzymes and chemical compounds in tea (*Camellia sinensis* L.) kuntze. *Afr. J. Biotechnol.* **2011**, *10*, 7480–7486.

37. Zhang, X.L.; Pan, Z.G.; Zhou, X.F.; Ni, W.Z. Autotoxicity and continuous cropping obstacles: A review. *Chin. J. Soil Sci.* **2007**, *4*, 33.

38. Pollock, J.L.; Callaway, R.M.; Holben, W. Phytotoxic effects of (±)-catechin in vitro, in soil, and in the field. *PLoS ONE* **2008**, *3*, e2536.

39. Van-der-Putten, W.H. Impacts of soil microbial communities on exotic plant invasions. *Trends Ecol. Evol.* **2010**, *25*, 512–519.

40. Fierer, N.; Jackson, R.B. The diversity and biogeography of soil bacterial communities. *Proc. Natl. Acad. Sci. USA* **2006**, *103*, 626–631. [CrossRef] [PubMed]

41. Buckley, D.H.; Schmidt, T.M. The structure of microbial communities in soil and the lasting impact of cultivation. *Microb. Ecol.* **2001**, *42*, 11–21. [PubMed]

42. Peiffer, J.A.; Spor, A.; Koren, O.; Jin, Z.; Tringe, S.G.; Dangl, J.L.; Buckler, E.S.; Ley, R.E. Diversity and heritability of the maize rhizosphere microbiome under field conditions. *Proc. Natl. Acad. Sci. USA* **2013**, *110*, 6548–6553. [CrossRef] [PubMed]

43. Zhou, Y.J.; Li, J.H.; Friedman, C.R.; Wang, H.F. Variation of soil bacterial communities in a chronosequence of rubber tree (*Hevea brasiliensis*) plantations. *Front. Plant Sci.* **2017**, *8*, 849. [CrossRef] [PubMed]

44. Lynn, T.M.; Liu, Q.; Hu, Y.; Yuan, H.; Wu, X.; Khai, A.A.; Wu, J.; Ge, T. Influence of land use on bacterial and archaeal diversity and community structures in three natural ecosystems and one agricultural soil. *Arch. Microbiol.* **2017**, *199*, 711–721. [CrossRef] [PubMed]

45. Lauber, C.L.; Hamady, M.; Knight, R.; Fierer, N. Pyrosequencing-based assessment of soil ph as a predictor of soil bacterial community structure at the continental scale. *Appl. Environ. Microbiol.* **2009**, *75*, 5111–5120. [CrossRef] [PubMed]

46. Berg, G.; Smalla, K. Plant species and soil type cooperatively shape the structure and function of microbial communities in the rhizosphere. *FEMS Microbiol. Ecol.* **2009**, *68*, 1–13. [CrossRef] [PubMed]

47. Khan, A.G. Role of soil microbes in the rhizospheres of plants growing on trace metal contaminated soils in phytoremediation. *J. Trace Elem. Med. Biol.* **2005**, *18*, 355–364. [CrossRef] [PubMed]

48. Jones, D.L. Organic acids in the rhizosphere–a critical review. *Plant Soil* **1998**, *205*, 25–44. [CrossRef]

49. Edwards, J.; Johnson, C.; Santos-Medellín, C.; Lurie, E.; Podishetty, N.K.; Bhatnagar, S.; Eisen, J.A.; Sundaresan, V. Structure, variation, and assembly of the root-associated microbiomes of rice. *Proc. Natl. Acad. Sci. USA* **2015**, *112*, E911–E920. [CrossRef] [PubMed]

50. Peng, L.; Song, X.; Shi, X.; Li, J.; Ye, C. An improved hplc method for simultaneous determination of phenolic compounds, purine alkaloids and theanine in *Camellia* species. *J. Food Compos. Anal.* **2008**, *21*, 559–563. [CrossRef]

51. Bao, S.D. *Soil and Agricultural Chemistry Analysis*; China Agriculture Press: Beijing, China, 2000.

52. Pansu, M.; Gautheyrou, J. *Handbook of Soil Analysis: Mineralogical, Organic and Inorganic Methods*; Springer Science & Business Media: Berlin, Germany, 2007.

53. Caporaso, J.G.; Kuczynski, J.; Stombaugh, J.; Bittinger, K.; Bushman, F.D.; Costello, E.K.; Fierer, N.; Peña, A.G.; Goodrich, J.K.; Gordon, J.I. Qiime allows analysis of high-throughput community sequencing data. *Nat. Methods* **2010**, *7*, 335–336. [CrossRef] [PubMed]
54. Magoč, T.; Salzberg, S.L. Flash: Fast length adjustment of short reads to improve genome assemblies. *Bioinformatics* **2011**, *27*, 2957–2963. [CrossRef] [PubMed]
55. Saqib, H.S.A.; You, M.; Gurr, G.M. Multivariate ordination identifies vegetation types associated with spider conservation in brassica crops. *PeerJ Prepr.* **2017**, *5*, e3003v3001.

MDPI

St. Alban-Anlage 66

4052 Basel

Switzerland

Tel. +41 61 683 77 34

Fax +41 61 302 89 18

www.mdpi.com

International Journal of Molecular Sciences Editorial Office

E-mail: ijms@mdpi.com

www.mdpi.com/journal/ijms

www.ingramcontent.com/pod-product-compliance
Lightning Source LLC
Chambersburg PA
CBHW051725210326
41597CB00032B/5610